Springer Series in Optical Sciences

Founding Editor

H. K. V. Lotsch

Volume 242

Springer Series in Optical Sciences is led by Editor-in-Chief William T. Rhodes, Florida Atlantic University, USA, and provides an expanding selection of research monographs in all major areas of optics:

- lasers and quantum optics
- ultrafast phenomena
- optical spectroscopy techniques
- optoelectronics
- information optics
- applied laser technology
- industrial applications and
- other topics of contemporary interest.

With this broad coverage of topics the series is useful to research scientists and engineers who need up-to-date reference books.

Zinoviy Nazarchuk · Leonid Muravsky ·
Dozyslav Kuryliak

Optical Metrology and Optoacoustics in Nondestructive Evaluation of Materials

Springer

Zinoviy Nazarchuk
Department of the Theory of Wave Processes and Optical Systems of Diagnostics
Karpenko Physico-Mechanical Institute
National Academy of Sciences of Ukraine
Lviv, Ukraine

Leonid Muravsky
Department of the Theory of Wave Processes and Optical Systems of Diagnostics
Karpenko Physico-Mechanical Institute
National Academy of Sciences of Ukraine
Lviv, Ukraine

Dozyslav Kuryliak
Department of the Theory of Wave Processes and Optical Systems of Diagnostics
Karpenko Physico-Mechanical Institute
National Academy of Sciences of Ukraine
Lviv, Ukraine

ISSN 0342-4111 ISSN 1556-1534 (electronic)
Springer Series in Optical Sciences
ISBN 978-981-99-1225-4 ISBN 978-981-99-1226-1 (eBook)
https://doi.org/10.1007/978-981-99-1226-1

This Springer imprint is published by the registered company Springer Nature Singapore Pte Ltd.
The registered company address is: 152 Beach Road, #21-01/04 Gateway East, Singapore 189721, Singapore

Preface

Optical Metrology and Optoacoustic techniques are widely used in nondestructive evaluation of structural materials and technical diagnostics of constructional elements during their static or dynamic loading under laboratory and natural conditions. These techniques allow determining the real state of the studied object by analyzing its surface parameters, deforming state or by detecting various defects. The obtained result makes it possible to give some recommendations on the object danger for the safe life of a construction.

This book includes description, modeling and realization of new Optical Metrology methods implementing to nondestructive evaluation of materials. Special attention is paid to multistep phase shifting interferometry with arbitrary phase shifts between interferograms, two-step phase shifting digital speckle pattern interferometry, digital and optical-digital image correlation, dynamic speckle patterns analysis, as well as to mathematical modeling of elastic waves interaction with interface crack-type defects. All these methods were developed in Karpenko Physico-Mechanical Institute (PMI) of the NAS of Ukraine.

Methods of two-step and three-step single- and dual-wavelength phase shifting interferometry with arbitrary phase shifts between interferograms are based on the correlation approach consisting in definition the unknown phase shift between two recorded interferograms of the same object by using the Pearson correlation coefficient. Compared to conventional temporal phase shifting techniques, they are simpler due to lack of devices and algorithms for phase shift calibration. These methods were implemented for the reconstruction of 3D surface reliefs of materials and structural elements. Procedures for separation of surface roughness and waviness from the total surface relief retrieved by these methods were developed. A similar two-step digital speckle pattern interferometry method with blind phase shift of a reference wave was developed for retrieval of the rough surface displacements and deformations. This method is used for solving different tasks of nondestructive evaluation and experimental fracture mechanics, including detection of subsurface defects and blind holes in metal and composite beam specimens.

A new optical-digital speckle correlation technique that uses main principles of digital image correlation was developed in PMI to increase the reliability of a surface displacement field reconstruction. A modified technique for adaptive segmentation of specimen surface speckle patterns was proposed for recording surface displacement fields in close proximity to uneven crack faces, uneven fractures and irregular edges of specimens. Optical-digital speckle correlators implementing the developed techniques have been created. They are used for scientific research and different practical applications.

Optoacoustic technique can be treated as a separate branch of the Optical Metrology, which can solve many problems of technical diagnostics including detection and localization of subsurface defects in the laminated composite materials. The information content of this technique is significantly increased due to the implementation of acoustic wave's excitation of the object at certain frequencies. That is why an effective theoretical approach to the modeling of an elastic waves interaction with an interface defect and to the defect visualization using dynamic speckle patterns is also included to this book. The experimental proof of the proposed approaches was fulfilled using of a created hybrid optical-digital system for detection of different subsurface defects.

The book is written on the basis of joint studies of the authors and other researchers from PMI, including T. I. Voronyak, Y. L. Ivanytskyy, O. P. Ostash, O. P. Maksymenko, A. B. Kmet', O. G. Kuts, I. V. Stasyshyn, Y. P. Kulynych, M. V. Voitko, O. V. Lychak, I. S. Holynskyy, V. V. Vira and G. I. Gaskevych. The authors are also grateful to R. R. Kokot and O. D. Suriadova for assistance in preparing this book for publication.

This book is intended for engineers, researchers and Ph.D. students engaged in the field of nondestructive evaluation of materials and technical diagnostics of structural elements, phase shifting interferometry, speckle metrology and optoacoustic imaging techniques.

Lviv, Ukraine

Zinoviy Nazarchuk
Leonid Muravsky
Dozyslav Kuryliak

Contents

Abbreviations

3D DSI	3D digital speckle interferometer
3F TWIM	Three-frame two-wavelength interferometry method
3SI UPS method	Method of three-step interferometry with unknown phase shifts
ACF	Autocorrelation function
Algorithm 1	Sjödahl–Benckert algorithm
Algorithm 2	Algorithm for determining the center of gravity
CCD	Charge-coupled device
CFRP	Carbon fiber reinforced polymer
CMOS	Complementary metal–oxide–semi-conductor
CT-specimen	Compact tension specimen
DAA	Difference averaging algorithm
DC	Digital camera
DDV algorithm	Diamond diagonal vectors algorithm
DFT	Digital Fourier transform
DH	Digital holography
DIC	Digital image correlation
DLSA	Dynamic laser speckle analysis
DM	Defect map
DNN algorithm	Deep neural network algorithm
DPIV	Digital particle image velocimetry
DSDM method	Digital speckle-displacement measurement technique
DSP	Difference speckle pattern
DSPI	Digital speckle pattern interferometry
DW	Dual-wavelength
EASLM	Electrically addressed spatial light modulator
EFEMD algorithm	Enhanced fast empirical mode decomposition algorithm
ESPI	Electonic speckle pattern interferometry
EVI method (algorithm)	Extreme value of interference method (algorithm)
FAF	Fringe-adjusted filter
FPZ	Fatigue process zone

FT	Fourier transform
GB	Gauge block
GFB algorithm	Gabor filter bank algorithm
GPSI algorithm	Two-step generalized phase shifting interferometry with blind phase shift extraction algorithm
GSO method (algorithm)	Gram–Schmidt orthonormalization method (algorithm)
HHT algorithm	Hilbert–Huang transform algorithm
HODS	Hybrid optical-digital system
HODSC	Hybrid optical-digital speckle correlator
HVT algorithm	Hilbert vortex transform algorithm
IRE algorithm	Iterative robust estimator algorithm
ISP	Initial speckle pattern
ITSI UPS method	Method of iterative two-step interferometry with unknown phase shift of a reference beam
JPS	Joint power spectrum
JTC	Joint transform correlator
JTC1	Conventional JTC
JTC2	JTC with median thresholding
JTC3	JTC with subset median thresholding
JTC4	JTC with fringe-adjusted filter
JTC5	JTC with adaptive median thresholding
JTC6	JTC with ring median thresholding
KR algorithm	Kreis algorithm
LED	Light-emitting diode
LP	Low-pass
LPGF	Low-pass Gaussian filter
LSCI	Laser speckle contrast imaging
LSP	Local speckle pattern
MAE	Mean absolute error
MDSDM method	Modified digital speckle-displacement measurement method
MID algorithm	Mutual information detrending algorithm
MPD	Movement and positioning device
MTF	Modulation transfer function
NDT	Nondestructive testing
ODSC method	Optical-digital speckle correlation method
ODSI	Optical-digital speckle interferometer
Pattern 1	Real or synthesized speckle pattern r of initial surface area S_1
Pattern 2	Real or synthesized speckle pattern g of deformed or in-plane shifted surface area S_2
Pattern 3	Synthesized speckle pattern g_d obtained by adding a new synthesized speckle pattern g_u not correlated with Patterns 1 and 2 to the displaced shifted speckle pattern g containing decorrelation parameter Δ_d

PINR	Peak-to-input noise ratio
PMI	Karpenko Physico-Mechanical Institute of the National Academy of Sciences of Ukraine
PNR	Peak-to-noise ratio
PONR	Peak-to-output noise ratio
PPCA	Population Pearson correlation algorithm
PPCC	Population Pearson correlation coefficient
PS	Phase shifting
PS DSPI	Phase shifting digital speckle pattern interferometry
PS ESPI	Phase shifting electronic speckle pattern interferometry
PSA	Phase shifting algorithm
PSC algorithm	Phase step calibration algorithm
PSD	Power spectral density
PSDA	Pairwise sum of differences algorithm
PSE	Phase shifting element
PSI	Phase shifting interferomentry
PZT	Piezoelectric transducer
QPP algorithm	Quadratic phase parameter estimator algorithm
RBM	Rigid body motion
RMS	Root mean square
ROF method (algorithm)	Regularized optical flow method (algorithm)
ROI	Region of interest
RP algorithm	Random points estimation algorithm
SCAF	Sine/cosine average filter
SFCD technique	Stress field components determination technique
SI	Speckle interferogram
SIF	Stress intensity factor
SLEF algorithm	Simplified Lissajous ellipse figure algorithm
SLEF-RE algorithm	Simplified Lissajous ellipse figure by robust estimator algorithm
SPCC	Sample Pearson correlation coefficient
SRHM	Surface roughness height map
SSCA	Spatial speckle contrast algorithm
ST method (algorithm)	Self-tuning phase shifting interferometry method (algorithm)
TD	Technical diagnostics
Technique 1:	The technique based on execution of that part of the TSI UPS method, which retrieves only unwrapped phase map φ_m (k, l) of the macrorelief surface area
Technique 2:	The technique that uses the ITSI UPS method and the criterion to determine the FPZ size
Technique 3:	The technique for determining the site and time of the fatigue macrocrack initiation along given profiles

Technique 4:	The technique for determining the site and time of a fatigue macrocrack initiation in every pixel of the retrieved surface roughness height map using the roughness increments
TF DSPI method	Method of three-frame digital speckle pattern interferometry with unknown phase shifts of the reference wave
TF FPI method	Method of three-frame fringe projection interferometry with unknown phase shifts of the reference wave
TS DSPI method	Method of two-step digital speckle pattern interferometry with arbitrary phase shift of the reference beam
TS DWIM	Two-step dual-wavelength interferometry method
TSE algorithm	Tilt-shift error estimation algorithm
TSI UPS method	Method of two-step interferometry with unknown phase shift of a reference beam
US	Ultrasonic
WLSI	White light scanning interferometry
YCF	Yamaguchi correlation factor

Chapter 1
Optical Metrology and Optoacoustics Techniques for Nondestructive Evaluation of Materials

Abstract In this chapter, the authors consider the Optical Metrology and Optoacoustics techniques as a powerful and versatile tool for solving problems of nondestructive evaluation of materials and structural elements. The chapter contains a summary of all subsequent six chapters. Methods for single- and dual-wavelength two- and three-step temporal phase shifting interferometry with unknown phase shifts are described in Chap. 2. In Chap. 3, it is shown that these methods can be successfully applied to determine the surface roughness parameters of materials and to study the development of the fatigue process zone in thin metal CT-specimens under cyclic loads and the fatigue macrocrack initiation. Chapter 4 is devoted to new methods of two-step phase shifting digital speckle pattern interferometry (DSPI) and correlation DSPI, as well as to application of these techniques for studying surface deformation fields in materials and detection of surface defects in metal and composite specimens. New methods of Speckle Metrology are proposed in Chap. 5. In particular, the method for extraction the unknown phase shift between speckle interferograms using sample Pearson correlation coefficient and the method of three-frame digital speckle pattern interferometry that is used in fringe projection interferometry are described in this chapter. Chapter 6 is devoted to methods for analyzing the speckle patterns of materials surfaces. In particular, the optical-digital speckle correlation and digital image correlation methods for assessment of the stress–strain state of materials, as well as the method for detection of subsurface defects excited by ultrasonic waves are considered and analyzed. The mathematical modeling of the physical phenomena of elastic wave interaction between the crack and interfaces is considered in Chap. 7. The modeling is based on the Wiener–Hopf method, which makes it possible to find the necessary characteristics of the field with a given accuracy and provide a simple physical interpretation of their properties. Thus, combining Optical Metrology and Optoacoustics with mathematical modeling of elastic waves interacting with interface defects become the powerful tools for nondestructive testing and technical diagnostics of materials and products.

In the last decades, Optical Metrology and Optoacoustics techniques for nondestructive testing (NDT) and Technical Diagnostics (TD) of materials and structural

Z. Nazarchuk et al., *Optical Metrology and Optoacoustics in Nondestructive Evaluation of Materials*, Springer Series in Optical Sciences 242,
https://doi.org/10.1007/978-981-99-1226-1_1

elements developed the accelerated rates. Recent technical advances in creation of new high effective hardware and software, high-speed personal computers, charge-coupled devices (CCDs) and complementary metal–oxide–semiconductor (CMOS) image sensors, multifunctional video cards, as well as image/display processors became the powerful basis for creation of new directions in Optical Metrology and Optoacoustics and improvement of well-known techniques realized by using photographic, analog TV, visual and semi-manual approaches.

Essential importance for study of materials and structural elements mechanical behavior and evaluation of their rupture life have gained Optical Metrology techniques allowing retrieving the surface relief structure and its spatiotemporal changes under mechanical loading or other types of external influence (thermal, acoustic, etc.). Phase shifting interferometry (PSI), white light scanning interferometry (WLSI), digital speckle pattern interferometry (DSPI), shearography, digital image correlation (DIC), digital holography (DH) and digital holographic interferometry techniques are considered as most efficient for these purposes.

Equally important is the development of new research directions devoted to problems of the elastic wave diffraction from the finite crack-type defects in materials, which are based on the common use of optical and ultrasonic methods. These new scientific directions make it possible to analyze the displacement and stress fields and the waveguide modes in the layers, as well as to detect subsurface defects in composite structures and other dielectric structural elements.

In the Karpenko Physico-Mechanical Institute (PMI) of the National Academy of Sciences of Ukraine such Optical Metrology and Optoacoustics techniques and methods as temporal PSI [61, 81, 90, 94, 95, 99, 147], DSPI [61, 82, 90], DH [51, 74, 115], DIC [61, 65, 66, 125], optical-digital speckle correlation (ODSC) [61, 77, 87, 88, 93], dynamic speckle pattern analysis using ultrasonic excitation of materials [91, 103, 104] and acoustic wave diffraction phenomena in materials [57, 102, 148, 149] are used to solve the urgent problems of NDT, TD and experimental solid mechanics [61, 78, 90, 101]. As a rule, these problems are associated with determination of the material deformation and fracture criteria, as well as with the prediction of its fatigue life. At the same time, the combination of Optical Metrology and, in particular, Speckle Metrology techniques with various types of excitations of the material or structural element under study makes it possible to improve the efficiency of various optical-digital systems that implement the aforementioned techniques and expand their use in various problems of TD. Vivid examples can be optoacoustic imaging techniques, combining optical-digital methods and systems of the investigated object with its acoustic excitation.

This book is devoted to review of well-known PSI, Speckle Metrology and Optoacoustics techniques intended for TD and NDT of materials and to some new approaches in these directions developed in PMI.

1.1 Phase Shifting Interferometry

Recent advances in PSI are inseparably linked with development of phase shifting algorithms (PSAs) and improvement of digital cameras. Firstly, PSAs were developed for temporal single-wavelength PSI [127]; further, they were applied for other optical interferometry techniques including dynamic PSI [127], multiwavelength PSI [126], DSPI [120] and DH [52]. At present, optical coherent and noncoherent interferometry techniques, namely WLSI [126], dynamic (spatial) single- and multi-wavelength PSI [8, 70, 126] and wavelength-tuning interferometry (WTI) [70, 126] are widely used for reconstruction of surface relief of optical elements, nanostructures, microelectromechanical systems, constructional materials, manifold devices, machine components, vehicles, engines, etc.

Temporal single-wavelength PSI has some drawbacks in comparison with WLSI, two- or multiwavelength PSI and WTI. In particular, only smooth and nanorough surfaces can be exactly retrieved, if a laser light wavelength belongs to a visible or near infrared range of electromagnetic waves. Simultaneously, the temporal single-wavelength PSI has some advantages over the ones above. First of all, this technique makes it possible to retrieve the surface relief of materials with high accuracy. Besides, it can perform the full field recording of a large surface area and can be implemented by simpler and cheaper hardware.

In temporal PSI, retrieval of the test object phase map is implemented with the help of a given phase shifting algorithm and several interferograms of the same test object, which are obtained by introducing several time-sequential phase shifts $\alpha_j (j = 1, \ldots, J)$ between a reference and object wavefronts in a two-beam interferometer and recording each interferogram with a given phase shift α_j by a digital camera. The resulting phase map can be easily converted into an optical path difference map, that is a height map of the retrieved surface topography.

There are a lot of optical arrangements used in the temporal PSI. A Twyman–Green [69, 127], Mach–Zehnder and Fizeau [71, 127] two-beam interferometers' arrangements are most distributed. Phase shift in these interferometers can be produced by different methods and devices [127]. But the most popular in PSI is a piezoelectric transducer based on a piezoceramic element with a bonded movable mirror for phase shifting.

Temporal PSI is widely used in TD and NDT of materials, structural and machine elements, equipment, etc. Temporal single-wavelength techniques are applied for testing the optical surfaces of lenses, objectives and mirrors such as flats, spheres, and aspheres. Rougher surfaces can be tested by two-wavelength PSI techniques, in which phases are measured at two different wavelengths λ_1 and λ_2 to remove 2π ambiguities, and a synthesized phase map of a retrieved rough surface is equivalent to that of a longer beat wavelength $\Lambda_{12} = \lambda_1\lambda_2/|\lambda_1 - \lambda_2|$ [8]. Single-wavelength PSI is very useful for measurement of spherical surfaces [38]. Two-wavelength PSI [8] and WTI, in which phase shifts are implemented by a laser wavelength tuning [34], make it possible to measure surfaces possessing larger roughness and multiple reflective surfaces. Meanwhile, these techniques retrieve surfaces containing steps and discontinuities [126].

In the last years, a considerable quantity of commercial phase shifting interferometric devices and systems has been created. They are dedicated to control the relief of optically smooth and nanorough surfaces of optical details, crystals, devices, micro- and optoelectronic materials, microelectromechanical systems, etc. Commercial dynamic phase shifting interferometers are not sensitive to vibrations and other dynamic phenomena. ZYGO®, 4D Technology, Graham Optical Systems, and Engineering Synthesis Design, Inc. are the leading corporations and companies developing and producing phase shifting interferometric systems.

1.1.1 Two- and Three-Step PSI with Unknown Phase Shift Between Interferograms

Conventional temporal PSI techniques are gradually being suppressed, for example, by faster and more efficient generalized PSI [52, 127], dynamic or spatial PSI [127], WTI [34, 126, 127], as well as two-step PSI with an unknown (blind) phase shift [15, 21, 63, 72, 73, 156, 157]. Two-step PSI techniques with blind phase shift offer simpler technical realization because only two interferograms are used for reconstruction of a studied object, and no phase shift calibration is needed. Optical experiments have verified their effectiveness to reconstruct optical amplitude-phase transparencies [73, 157] and an optical flat [156] using standard phase shifting interferometers.

A new method of two-step interferometry with unknown phase shift (TSI UPS) of a reference wave was developed in [90, 95]. It was based on the normalization of recorded interferograms, extraction of unknown phase shift and reconstruction of the studied surface nanorelief phase map. In this method, it was first proposed to normalize the interferogram for removing background and amplitude variations [21, 95]. To extract an unknown phase shift, the correlation approach based on the calculation of the interframe correlation between two interferograms [81, 90, 95] was used. Previously, the similar approach was proposed by Bobkov for measurement of phase shift in correlation selective phase meters [4]. This approach was applied later to optical interferometry and named as the phase step calibration (PSC) algorithm [145]. Flores et al. [21] showed that this algorithm provides sufficiently small values of the systematic error in determining the unknown phase shift and can successfully compete with new similar algorithms proposed in recent years [15, 53, 63, 123, 124].

The developed TSI UPS method makes it possible to extract from the resulting total relief both a 3D surface nanorelief, that is, a 3D surface roughness, and a 3D surface macrorelief, including 3D waviness and form. However, high error level and edge effects caused by using the digital Fourier transform (DFT) filtering to extract the macrorelief phase map induce the development of an iterative method of two-step interferometry with unknown phase shift (ITSI UPS) for retrieving the nanorough surface [99]. Two iterative steps performed in the second stage of this method make it possible to suppress the random noise and edge effects using spatial low-pass filters and remove spatial singularities from the coarse phase map, as well as obtain fine phase maps of the surface nanorelief, microrelief and total relief.

Although the TSI UPS and ITSI UPS methods have a number of advantages over traditional ones, where the same and fixed phase shifts are used, it is difficult to use them when studying the kinetics of surface texture changes. In particular, they require additional registration of the reference and object waves, which complicates the optical scheme of the interferometer and significantly slows down its operation. To speed up the interferograms recording, a method of three-step interferometry with unknown phase shifts (3SI UPS) has been proposed [94]. In this method, Muravsky et al. have proved that if three interferograms with different arbitrary phase shifts between the object and reference waves are recorded, it is enough to produce two difference phase maps, in which the background is excluded. In this case, the separate registration of the object and reference wave intensity distributions is removed. The 3SI UPS method implements the preliminary extraction of unknown phase shifts from three recorded interferograms with the help of the interframe correlation between interferograms and further retrieval of the surface area phase map. Unknown phase shifts between recorded interferograms can be extracted at the range from 0 to π with a low level of systematic error, as it has been shown in [81, 90, 95]. Except the total surface relief phase map retrieval, the 3SI UPS method also contains a procedure of the nanorelief (high-frequency) and macrorelief (low-frequency) phase maps separate extraction. So, if the studied test object is a smooth or a low-roughness engineering surface, this method can be used not only to retrieve the total relief of a studied surface area, but also its areal roughness and waviness. The fulfilled computer simulations and experimental results have confirmed the reliability of the proposed method to retrieve real objects if they represent smooth or low-roughness surfaces. The 3SI UPS method can be used to record the kinetics of surface changes, and the speed of the method will depend mainly on the capabilities of the phase shifting devices performing the smooth phase shift between the reference and object beams and control units for the fringe patterns registration.

1.1.2 Dual-Wavelength Phase Shifting Interferometry

Dual-wavelength (DW) PSI is successfully used for enlargement of a measurement rate of surface relief heights [8, 13]. In single-wavelength PSI techniques, the 2π ambiguity problem does not allow us to reconstruct the surface relief if the phase jump $\delta\varphi_{\text{adj}}$ between adjacent pixels is larger than π. If a two-beam interferometer with the normal incidence of light on an object plane is used, such a jump corresponds to the height jump δh_{adj} larger than $\lambda/4$, since $\delta h_{\text{adj}} = \lambda(\delta\varphi_{\text{adj}})/4\pi$. DW PSI allows expanding a range of surface height measurement due to using a synthetic wavelength $\Lambda_{12} = \lambda_1\lambda_2/|\lambda_1 - \lambda_2|$, and the retrieved height jumps can be increased theoretically in Λ/λ_1 or Λ/λ_2 times [8].

Despite the fact that DW PSI techniques have been developed for three decades, the problem of eliminating noise that increases by about Λ/λ_1 or Λ/λ_2 times in synthesized phase maps remains relevant. There are some approaches to suppress these noises. Parshall and Kim [113] proposed to produce a coarse phase map, which is free of discontinuities and possesses a longer range (by a factor of Λ/λ_1 or Λ/λ_2), to suppress an amplified noise, and, finally, to synthesize the fine map. Tay et al. [142] proposed a similar ideology using an algorithm of phase error correction combined with windowed Fourier filtering. Another problem is caused by the necessity to produce twice as many interferograms in DW PSI as in single-wavelength PSI. For example, Kumar et al. [55] have used a five-phase step algorithm for both wavelengths to reconstruct 3D surface profiles with the help of a DW sequential illumination technique.

Kmet' et al. [50] proposed a new two-step dual-wavelength interferometry method (TS DWIM), which uses only two pairs of interferograms for the surface relief retrieval. Each pair is produced for one of two wavelengths λ_1 or λ_2, and two interferograms in each pair are differed only by an arbitrary phase shift of a reference wave. Because the method does not need any phase shift calibration, it is enough to use only one phase shift to record both pairs of interferograms. Therefore, this method can ensure fast registration of all four interferograms, suppressing simultaneously the noise in a synthesized phase map. The method makes it possible to form a synthetic beating wave Λ_{12}, thanks to which it is possible to restore the surface relief in a wide range of surface roughness up to $S_a \leq \Lambda_{12}/8$ or up to $R_a \leq \Lambda_{12}/8$. The results of computer simulations and experiments on the restoration of a test cast iron specimen corresponding to the 6th grade of roughness according to GOST 25142-82 [31] using a dual-wavelength Twyman–Green interferometer with radiation wavelengths $\lambda_1 = 0.633\,\mu\text{m}$, $\lambda_2 = 0.650\,\mu\text{m}$, and with a beat wavelength $\Lambda_{12} = 24.2\,\mu\text{m}$ showed high accuracy of phase maps retrieval.

In order to increase the speed of the interferograms recording, the three-frame two-wavelength interferometry method (3F TWIM) with unknown phase shifts between interferograms was developed. In this method, only three interferograms recorded at wavelength λ_1 and three interferograms recorded at wavelength λ_2 are used to retrieve a phase map of the studied surface area. This method, like the TS DWIM, is mainly intended for the reconstruction of 3D surface relief areas in a much wider range of surface roughness compared to single-wavelength TSI UPS and ITSI UPS methods. However, in contrast to the TS DWIM, the 3F TWIM does not require additionally to record the intensities spatial distributions of the object and reference laser beams at wavelengths λ_1 and λ_2, which significantly increases its performance. The proposed method allows using the integrating-bucket technique [33, 127, 154] to simultaneously record three interferograms for each wavelength λ_1 and λ_2 when a sawtooth voltage is applied to perform only one smooth phase shift of the reference wave in the interferometer.

TSI UPS, ITSI UPS and 3SI UPS methods of single-wavelength temporal PSI, as well as TS DWIM and 3F TWIM for DW PSI are written in Chap. 2 (Phase Shifting Interferometry Techniques for Surface Parameters Measurement) of this book.

1.2 Application of Phase Shifting Interferometry Methods for Materials Surface Diagnostics

The developed methods of single-wavelength and dual-wavelength PSI with unknown phase shifts between interferograms were used for NDT and TD of structural materials and constructional elements. Special attention was paid to the application of PSI methods to study the fatigue failure phenomena and fatigue life of compact tension specimens (CT-specimens) made of metals or alloys with round and U-shaped notches. The performed researches include defining the fatigue process zone (FPZ) size, evaluating the cyclic and monotonic plasticity zones geometric parameters, as well as determining the site and time of fatigue crack initiation.

Two techniques based on the TSI UPS and ITSI UPS methods were developed for definition the size d^* of the FPZ [81, 90, 95, 99], which is the structural mechanical parameter of materials and depends on the mechanical properties of the material and its microstructure [106–108]. First technique (Technique 1) uses the TSI UPS method, which makes it possible to retrieve two surface macrorelief areas near the notch edge before and after a certain number of cyclic loadings of the studied CT-specimen. After superimposing and subtracting these macroreliefs from each other, a field of surface deformations is obtained. The parameter d^* is determined using the resulting strain field as the distance from the notch edge to the point of maximum strain in the direction perpendicular to the direction of load application to the specimen [95, 109, 110]. This point corresponds to the point of the largest thinning of the specimen. To implement this technique, an optical-digital experimental setup based on the Twyman–Green interferometer was developed. Experiments with CT-specimens made of low-carbon steel 08kp, similar to US steels 1008, 1010 and A619, indicated that the specimen largest thinning is 5.0 μm or 0.25% after 40,000 loading cycles and $d^* = 262.5 \pm 12.5\ \mu$m at the coefficient of load cycle asymmetry value $R = 0.1$. Except the FPZ size determination, the geometric parameters of the cyclic and monotonic plastic zones also can be defined using Technique 1. However, Technique 1 did not allow one to determine the FPZ size in CT-specimens made of aluminum alloys D16T similar to alloy 2024-T4 and V95T similar to alloy 7075, due to their lower plasticity compared to steel 08kp CT-specimens.

Studying the surface nanorelief phase maps obtained by the TSI UPS method indicated the different evolution nature of the fatigue process zone, as well as the cyclic and monotonic plastic zones for CT-specimens made of aluminum alloys D16T and V95T [61]. In particular, it has been shown that the nanorelief surface roughness of both aluminum alloys specimens increases monotonically with an increase in the number of load cycles, and the largest increase in roughness was observed within the FPZ. Note that this effect was also observed in specimens of materials with high plasticity. In this regard, the surface roughness was chosen as the criterion for estimating the FPZ size in CT-specimens of various metals and alloys.

In order to accurate determine the dimensions of the FPZ and essentially the parameter d^* for any metals and alloys, regardless of the level of their plasticity, the criterion of achieving the maximum surface roughness at the FPZ boundary was proposed [99, 100]. Experimental verification of the proposed criterion was performed using the optical-digital experimental setup containing the Twyman–Green interferometer and the Technique 2 based on the ITSI UPS method. Results of manifold experiments with metal and alloy specimens possessing high, moderate and low plasticity have shown that their surface roughness reaches its maximum values at the FPZ boundary. Calculation of the arithmetic mean surface roughness R_a spatial distribution in every pixel of a surface height map confirmed the authors' assumption that the line or narrow strip containing pixels with maximum R_a values corresponds to the FPZ boundary [99]. Such operations also make it possible to calculate the FPZ dimensions and the size d^* with pinpoint accuracy.

Thus, the Technique 1 based on the TSI UPS method makes it possible to define the FPZ geometric parameters and determine the FPZ size d^*, as well as the geometric parameters of the static and monotonic plastic zones for materials with high plasticity. To define the FPZ dimensions in materials with high, moderate and low plasticity, the Technique 2 based on the ITSI UPS method can be used. This technique is more effective, since the FPZ sizes are determined using the surface roughness spatial distribution on the extracted surface nanorelief height map for a wide range of metal and alloy specimens.

One of the urgent directions of the fatigue durability evaluation of structural materials is the prediction of the site and moment of a fatigue macrocrack initiation in metal structural elements with rounded notches. The process of the macrocrack initiation is closely connected with the formation and development of the FPZ during fatigue, and the FPZ size d^* can be considered as the delimitation criterion of two main fatigue damage periods, namely the period N_i of origin of the minimum fatigue macrocrack in the material, and the period N_p of its growth. In this interpretation, the key criterion for assessing the fatigue life of structural elements with round and U-shaped notches is the moment of the fatigue macrocrack initiation at the end of the period N_i and at the beginning of the period N_p.

As shown by Muravsky et al. [90], in CT-specimens made of steel 08kp and aluminum alloy D16T, a macrocrack initiates and arises most likely in the region of interest corresponding to the FPZ boundary area where the largest increase of the nanorelief roughness occurs during cyclic loading. This result was obtained for radial surface profiles crossing the FPZ boundary. Further studies indicated that the region of interest on the surface of any CT-specimen containing a round notch is the zone D^*, which covers the FPZ and narrow band outside it. The new technique for determination of the site and time of a fatigue macrocrack initiation in every pixel of the retrieved surface roughness height map using the roughness increments of the surface relief area near the notch tip was proposed. This technique uses the 3F TWIM for retrieving phase maps of the surface nanorelief, macrorelief and total relief. To implement this technique, the experimental setup containing single-wavelength and two-wavelength Twyman–Green interferometers was used. Necessity to use such the setup was caused both by a monotonic increase in surface roughness near the notch tip in the surface zone D^* of CT-specimens during their cyclic loading within the period N_i of fatigue failure and by initiation of fatigue macrocracks for a number of materials at arithmetic mean roughness values $Sa > \lambda/8$ exceeding the allowable surface roughness, retrieved by single-wavelength PSI methods. The evolution of the surface roughness relief near the notch tip in metal and alloy CT-specimens during their fatigue was estimated by increasing the arithmetic mean roughness value *Sa* at each pixel of the surface roughness height maps retrieved from interferograms recorded after each predetermined number of load cycles, until the appearance of a fatigue macrocrack on the specimen surface. Obtained experimental results confirmed the hypothesis that the fatigue macrocracks initiation is most likely occurred within the zone D^*.

Compared with the technique to predict the fatigue crack initiation location in pure nickel developed by Sola et al. [132], the proposed technique for prediction the fatigue macrocrack initiation location in CT-specimens made of metals and aluminum alloys has some advantages. Main of these advantages are the determining the spatial distributions of the arithmetical mean height of the scale-limited surface value *Sa* and its increments ΔSa in each pixel of the 3D roughness distribution of the retrieved surface roughness height map, as well as the selection of the zone D^* as the most probable surface region of the fatigue macrocrack initiation.

The practical application of developed PSI techniques for diagnostics of materials fatigue fracture is described in Chap. 3 (Application of Phase Shifting Interferometry Methods for Diagnostics of Materials Surface).

1.3 Digital Speckle Pattern Interferometry

DSPI techniques are based on interference of an object coherent wave producing a speckle pattern of a given test object with a reference wave. In TD of materials and structural elements, the test object as a rule represents a surface area of a specimen. A speckle interferogram (SI) or a speckle fringe pattern is produced in an observation plane as a result of the object and reference wave's interference. This SI is recorded by a digital camera and introduced into a computer for further image processing. The studied surface area displacements are coded as contrast variations in the SI; moreover, any change of the surface spatial location leads to simultaneous change of its speckle intensities and locations. Two types of displacements can be fixed by using DSPI techniques, namely in-plane and out-of-plane ones, which are tangential and normal to the observation direction, respectively.

Correlation and phase shifting DSPI techniques are mostly used for NDT and TD of constructional materials and structural elements [16, 45, 61, 90, 128]. Correlation speckle interferometry methods usually are implemented as the sum or difference of two sequentially recorded SIs. These methods can fix surface displacements with accuracy comparable with half-wavelength of a coherent laser source. In this connection, they are efficient for qualitative estimations of static and quasistatic surface displacements, periodic vibrations and different static problems of NDT. On the other hand, temporal and dynamic phase shifting DSPI techniques are used for precise quantitative analysis of microdisplacements.

In temporal DSPI, several SIs (as a rule, 3–7 SIs) are recorded for initial state and each subsequent state of a studied surface area. Spatial difference between these states is identical to displacement field, which is retrieved from recorded SI series with high sensitivity reached about $\lambda/100$ [128]. SIs recording the same surface area state are differed only by equal phase shifts between adjacent SIs for the initial and deformed surface area. Intensity distributions of SIs are recorded for initial and next deformed state of the studied surface area.

The common peculiarity of all phase shifting DSPI techniques consists in the necessity of precise calibration of phase shifting devices. Besides, several fixed phase-shifted SIs of initial and deformed states of the surface area should be processed to retrieve the displacement phase map. Simplification of this procedure can be achieved by reduction of SIs quantity to two for each state of a studied surface area and fulfillment of the arbitrary phase shift for recording of the second SI.

1.3.1 Method of Two-Step Digital Speckle Pattern Interferometry with Arbitrary Phase Shift of Reference Beam

The problem of retrieval of a rough surface area displacement phase maps with the help of two SI pairs was solved by using a two-step digital speckle pattern interferometry (TS DSPI) method with an unknown phase shift of a reference wave [61, 82, 90]. The method implements the registration of two SIs before deformation of the studied surface, namely the initial SI $i_{11}(x, y)$, and the SI $i_{12}(x, y)$ with the reference wave shifted on an arbitrary unknown angle α_{21}, and two SIs after the surface deformation, namely the SI $i_{21}(x, y)$ without phase shift of a reference wave and the SI $i_{22}(x, y)$ with phase shift of the reference wave on the arbitrary unknown angle $\tilde{\alpha}_{21}$. This method theoretically assumes that $\alpha_{21} = \tilde{\alpha}_{21}$. After calculation of a coarse displacement phase map $\delta\varphi_c(x, y)$, the further its processing to obtain the fine displacement phase map $\delta\varphi(x, y)$ is fulfilled by reconstruction of only $i_{s1} = \sin(\delta\varphi_c)$ and $i_{c1} = \cos(\delta\varphi_c)$ harmonic terms using a low-pass filter. Standard software was used for fulfillment of the unwrapping procedure. The correlation approach was implemented to determine the unknown phase shift [90, 95] that was calculated with the help of the population Pearson's correlation coefficient (PPCC), as in the TSI UPS, ITSI UPS and 3SI UPS methods. It was shown [82] that the correlation approach provides small errors of the unknown phase shift, which introduce the small contribution into a total error of a retrieved displacement phase map.

To evaluate the reliability of the surface displacement phase field retrieval by the TS DSPI method, computer models of SIs generated by optically rough initial and deformed surfaces and a random flat reference beam in the Twyman–Green interferometer optical scheme were synthesized.

Results of simulation of a test displacement phase map retrieval by using this approach have shown that the systematic error of the retrieved test displacement phase map is low, and the mean value of this error is $(-3.28) \cdot 10^{-3}$ deg. On the other hand, the total phase map retrieval error does not exceed admissible critical values, if the differences between phase shifts of two interferograms at the initial and deformed states reach up to 10°. Experimental verification of this method also confirms its reliability to restore in-plane and out-of-plane displacements of real surfaces.

The developed TS DSPI method with unknown phase shift between SIs was used for reconstruction of 3D surface displacement fields. To this end, a 3D digital speckle interferometer experimental setup was developed, in which a Mach–Zehnder interferometer for measuring out-of-plane displacements and two Leendertz interferometers for measuring in-plane displacements in two mutually perpendicular directions were used. The construction of this 3D interferometer was designed in such a way as to achieve the uniformity of sensitivity to out-of-plane and in-plane surface displacements over the entire field [147]. The developed 3D digital speckle interferometer setup makes it possible to generate separate displacement fields along x, y and z directions with the same sensitivity per fringe across the displacement field, which is provided by parallel reference and object beams in all three speckle interferometers. The results of experiments to determine the 3D displacement fields of surfaces of the metal and laminated composite beam specimens using the TS DSPI method indicated its effectiveness for generating 3D displacement phase maps. They showed that in order to obtain complete data about the displacement fields along and across the direction of observation, one can limit oneself to only twelve SIs.

1.3.2 Application of Digital Speckle Pattern Interferometry for Detection of Surface and Hidden Defects in Metal and Alloy Specimens

Possibilities of correlation and phase shifting DSPI techniques to generate and analyze the spatial fields of surface displacements and deformations favorably distinguish them from the well-known methods for determining surface strains. There are different directions of DSPI techniques applications in experimental mechanics, NDT and TD [16, 20, 42, 61, 160]. In this book, the results of surface and hidden defects in metal and alloy specimens using correlation DSPI and TS DSPI methods are demonstrated.

Beam and sheet metal and alloy specimens containing certain defects were used for experimental studies in the developed experimental setups of optical-digital speckle interferometers based on architectures of the Twyman–Green, Mach–Zehnder and Leendertz interferometers. Defects were identified by local inhomogeneities in the displacement field of the specimen surface under the action of mechanical load, as well as thermal or ultrasonic (US) excitation. Both artificial (blind cylindrical holes, cuts, etc.) and real defects (surface cracks, delamination of protective and restorative coatings, internal defects in welds) were considered. Among the hidden defects, artificial and real ones were also distinguished. Artificial hidden defects were created by drilling blind holes or cutting oblique narrow cuts on one side of the specimen surface to register displacement fields on the other side. As a result, hidden defects were formed at depths of 1–8 mm from the surface under study. To analyze real subsurface defects, specimens with protective coatings, as well as specimens with weld joints, were selected.

During the experiments, the specimens were loaded using the scheme of four-point bending or heated. Since it was established that out-of-plane displacements are more effective than in-plane ones for detecting defects, a Twyman–Green interferometer was used to achieve the maximum possible sensitivity to out-of-plane displacements. SIs were recorded by using both the correlation DSPI and TS DSPI methods. During performing the experiments, it was found that in most cases, it is difficult to detect defects by correlation or phase shifting DSPI techniques at low specimen loads. It turned out that first it is necessary to apply a preload to the test specimen and then observe its surface, gradually increasing the load. The level of preload depends both on the material of the specimen, the nature of the defect (size, shape, volume, voidness, etc.), the occurrence depth and on the load circuit. Experiments have shown that preloads of the order of 100–3000 N and additional loads of the order of 100–200 N can be considered optimal for detecting artificial and some real defects in metal and composite beam specimens which thickness does not exceed 10 mm. Results of hidden defects detection in steel 20 beam specimen testify the effectiveness of the subtractive DSPI and TS DSPI methods to extract the cuts and blind holes. Detecting these defects using the subtractive DSPI method makes it possible to assert about their presence in specimens. On the other hand, the TS DSPI method is more preferable, since it allows not only to detect defects but also to determine their size and even their depth of occurrence relative to the monitored surface. It was also shown that the subtractive DSPI method can be successfully used for TD of protective and restorative coatings and identifying the coating delamination places.

1.3.3 Subtractive Synchronized Digital Speckle Pattern Interferometry Method for Detection of Subsurface Defects in Laminated Composites

The subtractive synchronized DSPI method [61, 90, 97, 103] is designed to detect subsurface defects, such as delaminations, cracks, disbonds, voids, inclusions, matrix cracking, etc., in laminated composite panels using the US excitation of the studied specimens. If the frequency of the US wave $f_{\mathrm{US}} = 1/T_{\mathrm{US}}$ is coincided with one of the resonance frequencies of the defect, it becomes oscillating, and the surface area located directly above the defect, i.e. the region of interest (ROI), is induced by this defect and becomes oscillating with the same frequency. Extremes of the US wave oscillations coincide with the maximum deviations of the ROI from the plane of the panel in both opposite directions. In the experimental setup of an optical-digital speckle interferometer (ODSI), which implements this method, two sequences of speckle interferograms are accumulated. Each of the sequences registers the maximum deviation of the ROI in one of two opposite directions, after which subtraction specklegrams are generated using the developed software. After digital processing of these specklegrams, light responses generated by vibrating ROIs and corresponding to the dislocation sites of the subsurface defects are distinguished on the defect map. Obtained experimental results demonstrate the ability of the ODSI to detect and localize subsurface delaminations, debonds and bearing strains in carbon fiber reinforced polymer and fiberglass reinforced laminated composite panels. Delaminations were also detected in the repaired laminated composite structural element of the AN aircraft wing flap.

DSPI and its practical applications for TD of materials are presented in Chap. 4 (Digital Speckle Pattern Interferometry for Investigations Surface Deformation and Fracture of Materials) of the book.

1.4 Studying Rough Surfaces of Materials by Speckle Metrology Methods

Among methods and means dedicated to studies of the optically rough surface of materials, the Speckle Metrology techniques are very important and relevant. High resolution, the ability to form multidimensional data arrays and enter them into digital devices for further processing and generating the 3D surface height maps, as well as 3D displacement and deformation fields, make it possible to successfully compete these techniques with the other ones. In particular, phase shifting DSPI techniques can be successfully used to estimate the parameters of the topography of the optically rough surface. To this end, interference fringes are projected onto the rough surface of the test specimen and, by performing calibrated phase shifts of the interference fringes, a series of fringe patterns is recorded using a coherent optical system and a digital camera [17]. To simplify the procedures for retrieving the surface nano- and microdisplacements, a new method of three-frame digital speckle pattern interferometry (TF DSPI) with unknown phase shifts of a reference beam was proposed [79]. This method can be used in projection interferometry to restore the surface texture. The PSC algorithm can be used for calculation of unknown phase shift between SIs using the PPCC. On the other hand, a new algorithm was proposed to calculate the unknown phase shifts based on a sample Pearson correlation coefficient (SPCC) [83, 84]. The SPCC use in phase shifting DSPI is a priority compared to the PPCC use, since the time required calculating the phase shift angle is significantly reduced, and the procedures for direct and inverse DFTs of SIs recorded in the form of 2D data arrays are avoided. In addition, systematic errors of the unknown phase shift extraction decrease with increase of the surface roughness. Therefore, this algorithm provides a low level of systematic errors and a higher performance compared to the PSC algorithm. Results of calculating the systematic errors of the extracted unknown phase shifts between two synthesized SIs differing only in this phase shift using the SPCC indicate a fairly low level of these errors in the entire range from 0 to π radians, which increase relatively slightly with phase shifts near 0 and π.

1.4.1 Three-Frame Digital Speckle Pattern Interferometry Method with Unknown Phase Shifts of Reference Beam

The TS DSPI method, despite being simpler in implementation compared to the multistep phase shifting DSPI methods with given phase shifts, is time consuming, since it is necessary to record the intensity distributions i_o and i_r of the object and reference wavefronts, respectively, and to delay of at least several tens of seconds after arbitrary phase shifts in order to stop the phase shifting element (PSE) oscillations. To eliminate these shortcomings, the TF DSPI method was developed [79]. Due to this method, the registration of i_o and i_r is absent, and the speed of the SIs registration increases significantly. In this method, three SIs i_{11}, i_{12} and i_{13} of a surface area in the initial state and three SIs i_{21}, i_{22} and i_{23} of the same area in the loaded state are recorded. Phase shift between all these SIs is arbitrary in the range from 0 to π. The unknown phase shifts between three SIs recorded at the first unloaded state and the second loaded one can be calculated using both the PPCC and SPCC. To find the surface displacement phase map $\delta\varphi(x, y)$, the phase surface $\varphi(x, y)$ before loading is extracted from three fundamental equations describing SIs i_{11}, i_{12} and i_{13} recorded in the unloaded state, and the phase surface $[\varphi(x, y) + \delta\varphi(x, y)]$ after loading is extracted from three fundamental equations describing SIs i_{21}, i_{22} and i_{23} in the loaded state. The searched surface displacement wrapped phase map $\delta\varphi(k, l)$ is calculated as a difference between $[\varphi(x, y) + \delta\varphi(x, y)]$ and $\varphi(x, y)$. The procedure of the phase map $\delta\varphi(k, l)$ unwrapping is similar to the unwrapping procedure performed by the TS DSPI method. All SIs were recorded with the help of the integrating-bucket technique [33, 127, 154], due to which the speed of SIs recording before and after applying the loads raises. Therefore, this method is more suitable for the fast reconstruction of in-plane and out-of-plane surface displacements and strains compared to the TS DSPI method.

1.4.2 Fringe Projection Interferometry for Surface Nondestructive Testing

Nowadays, fringe projection techniques have become very popular due to application of relatively simple equipment and the ability to study the surfaces of large sizes with significant changes in shape and form [36, 137]. At the same time, the accuracy of these techniques depends on the available equipment, as well as the conditions and methods of the surface relief on registration and restoration. The fringe projection techniques use, as a rule, parallel linear fringes with sinusoidal or structured profiles across the fringes, which are projected on the investigated surface. The main direction of these techniques' development uses commercial multimedia projectors to generate spatial frequency variable linear periodic sinusoidal fringes or structured ones possessing other types of profiles [44] and project them onto the studied surface [162]. Several other fringe projection methods are also used to produce fringe patterns on the studied surface [30]. Among them, the fringe projection method performed with the help of laser interferometric systems is very perspective, since it makes it possible to restore surfaces with big relief depth [17, 136]. Such types of interferometers generate a parallel wavefront and project patterns of interference fringes on the studied surface. Therefore, they can be used to evaluate the surface structure, because they provide the projection of interference patterns on the studied surface with a high fringe density. This makes it possible to retrieve the waviness and form of the surface and even assess its roughness.

The fringe projection interferometry is based on the principle of triangulation [36]. If parallel linear equidistant sinusoidal fringes fall on the studied diffuse 3D surface, then a subjective fringe pattern reflected from this surface is formed in the plane of the digital camera matrix photodetector. This pattern can be characterized as an interferogram or SI of the surface [17]. In the interferogram or in the SI, the linear fringes falling on the surface are distorted due to its irregularities and surface roughness. The interferogram and SI intensity distributions are presented in the form of the fundamental equation for PSI, which differs from it only by the factor $R(k, l)$ describing the reflectivity of the studied surface. Depending on how the collimated light beam is incident and reflected, one can determine the equivalent or effective wavelength Λ_e [36] for the interferometric fringe projection system. This wavelength, as a rule, exceeds the laser source wavelength λ several times. Therefore, one can possible with high reliability to retrieve the surface areas whose arithmetic mean roughness value Sa is much larger than λ.

If the studied surface is optically rough, the developed TF DSPI method can be used for its retrieval. To this end, one can use only the first part of this method, that is the first series of digital SIs recorded at the initial state of the studied surface [79]. Unknown phase shifts between SIs are calculated by using the PPCC or SPCC. After obtaining the retrieved rough surface wrapped phase map, its unwrapping is implemented by one of the well-known algorithms [22, 26, 155].

To retrieve rough surface areas using fringe projection interferometry and a developed TF DSPI method, an optical-digital experimental setup containing the Twyman–Green interferometer was developed. In this setup, parallel linear equidistant interference fringes generated by the interferometer falls on the studied surface and three SIs of this surface differing only by arbitrary phase shifts are recorded by a digital camera during the time of applying one voltage saw to the PSE. Processing of the recorded SIs and retrieving of the surface height maps is performed using the developed software. During the experiment several real surface areas were restored, for example, the retrieved 3D surface shape of the diamond drill ⌀8 mm, including its form and microrelief.

The fringe projection interferometry technique for retrieval of smooth and rough surfaces using the TF DSPI method with unknown phase shifts of reference beam is described in Chap. 5 (New Speckle Metrology Methods for Studying Rough Surfaces of Materials) of this book.

1.5 Methods for Processing and Analyzing Speckle Patterns of Materials Surface

In recent years, many techniques for processing and analyzing static and dynamic speckle patterns have been rapidly and intensively developed. In contrast to PSI, DSPI and DH techniques, they cannot capture the phase data of optical wavefronts. However, digital speckle patterns generated by rough surfaces of structural materials, structural elements and biological tissues, as well as random media, including air and liquid flows, also contain very valuable data on their characteristics [5, 27, 29, 119, 130, 139, 153, 158, 159]. Among these techniques, DIC has gained particular popularity in solving many practical problems of experimental mechanics, as well as in studying the stress–strain state of surfaces and subsurface layers of materials, structural elements and various products [10, 112, 114, 139, 140].

Dynamic laser speckle analysis (DLSA) is the technique that makes it possible to study the dynamics of speckle patterns generated by moving surfaces and various other nonstationary phenomena. The temporal activity of studied objects is performed by means of statistical processing in order to determine correlation dependencies [48, 105, 118], local speckle contrast [5, 49, 143] or different estimators [14, 118] characterizing dynamic phenomena recorded by sets of speckle patterns. Methods for detecting subsurface defects in composite structures using laser speckle contrast imaging (LSCI) and speckle decorrelation [91, 103, 104], as well as assessing fruit shelf life [161] and cellular changes in muscle tissue [67] can be considered as perspective directions of DLSA.

DIC is considered as a general scientific direction for measuring the displacement and deformation fields of various objects or their surfaces [112]. At least two speckle patterns are used to measure the surface displacement and strain fields in all speckle correlation methods. To generate the vector displacement and deformation fields, two speckle patterns of the initial and deformed surface area are recorded and divided in a computer into $M \times N$ square subsets. The subset sizes contain an odd or even number of pixels in two orthogonal directions. Due to correlation comparison of each subset pair, the displacement vector $\vec{d}_{mn}$ evaluating the shift between the corresponding reference r_{mn} and deformed g_{mn} subsets is defined. All these $M \times N$ vectors create the displacement field in the form of a matrix of local displacements.

Among various DIC methods, two approaches of determining the surface displacement and deformation fields can be distinguished. The first approach involves measuring only the displacements of the studied surface points with the subsequent determination of the strain components as derivatives of its displacements

[7, 46, 76, 77, 87, 130, 153]. The second approach consists in the simultaneous full-field determination of the surface displacements and strains [9, 112, 139]. Very often, the methods in which the first approach is implemented also refer to DIC, although some of them have their own names, for example, digital particle image velocimetry [46, 153], electronic speckle photography [130, 131], digital speckle-displacement measurement (DSDM) technique [7], as well as the modified digital speckle-displacement measurement (MDSDM) [68, 85, 87] and optical-digital speckle correlation (ODSC) [77, 85, 87, 88, 93, 96] methods developed in PMI. On the other hand, the second approach is directly related to DIC because the methods developed by Sutton's group had just such a name [112, 139].

1.5.1 Optical-Digital Speckle Correlation Method

DIC techniques are widely used to solve a variety of problems of experimental mechanics, materials science, diagnostics, and nondestructive testing of materials and structural elements. However, for many problems of fracture mechanics, most of the known DIC techniques are not able to determine satisfactory these fields in the immediate vicinity of stress concentrators, in particular, near crack tips and edges. Decrease in the DIC efficiency in these sites is mainly due to the significant speckle decorrelation near stress concentrators and the relatively large width of the correlation peaks. The developed ODSC method makes it possible to narrow the peak width and increase the signal-to-noise ratio, thereby providing a more reliable restoration of displacement and deformation fields near the crack tips and edges [77, 85, 87, 88, 93]. The ODSC method is based on the use of a cross-correlation operation between the corresponding subsets of two speckle patterns that are compared with each other to form in-plane displacement and deformation vector fields. To implement this operation, it is proposed to use the architecture of the joint transform correlator (JTC) [28, 121, 152]. In the ODSC method, linear and nonlinear spatial filters can be synthesized for transformation of a joint power spectrum (JPS) from two corresponding subsets. These filters, namely binary filters with median and subset median thresholds [43, 150], adaptive binary filters [35] and linear fringe-adjusted filters (FAFs) [1–3] are used in optical pattern recognition to improve the signal characteristics of correlation peaks and reduce noise and interference. The global and subset median thresholding methods, as well as the method for JPS filtering using FAFs, can improve the parameters of correlation peaks. However, these methods do not take into account the specific spatial structure of the Fourier spectra and the power spectral densities of speckle patterns. In order to further improve the parameters of the correlation peaks, a new method of the JPS binarization by a ring median threshold was proposed [92, 93].

To check the effectiveness of the ODSC method using the above filters and nonlinear JPS binarization procedures, it is better to choose the digital version of the JTC, since thanks to this version it is possible to work with both synthesized and real speckle patterns and generate linear and nonlinear spatial filters. Meanwhile, because the JTC has a two-stage configuration, it is possible to use its hybrid optical-digital version, in which the first stage is digital and the second one is optical. In this case, an electrically addressed spatial light modulator (EASLM) should be used at the input of the second stage. Six types of JTCs computer models implementing the ODSC method, namely conventional JTC (JTC1); JTC with median thresholding (JTC2); JTC with subset median thresholding (JTC3); JTC with FAFs (JTC4); JTC with adaptive median thresholding (JTC5); JTC with ring median thresholding (JTC6), were synthesized. The developed digital models allow obtaining all the data necessary for a detailed study of the information characteristics of the correlation signals.

1.5.2 Comparative Evaluation of Two Methods for Speckle Patterns Correlation

To compare the efficiency and reliability of developed in PMI methods for speckle patterns correlations, namely the ODSC and MDSDM methods, the digital models implementing these methods were developed. These models were used to compare the information parameters of output correlation signals generated by subsets of synthesized speckle patterns. To calculate subsets displacements with subpixel accuracy, the Sjödahl–Benckert algorithm (Algorithm 1) [129, 131] and the algorithm for determining the center of gravity (Algorithm 2) [85, 87] were used. The results of calculating the mean bias error and the random error of speckle patterns displacements during in-plane rigid body motion of the synthesized rough flat surface, taking into account the decorrelation parameter Δ_d, showed that they are approximately equal for the digital implementation of the MDSDM method, as well as JTC1, JTC2 and JTC3 digital models. On the other hand, these errors are larger for JTC4 and smaller for JTC5 and JTC6 compared to other digital models. Besides, Algorithm 1 provides greater accuracy than Algorithm 2 when calculating displacements. There was also a tendency for mean bias error to increase with increasing bias and increasing decorrelation parameter Δ_d.

A comparative analysis of the ODSC and MDSDM methods effectiveness was performed using the metrics proposed by Horner [41] and Kumar and Hassebrook [54] for optical correlators. The objective measures of the correlator efficiency is the peak-to-correlation energy metric [41], or the peak-to-output noise ratio (PONR) μ_{mn} for each pair of subsets r_{mn} and g_{mn}, and the PONR mean value $\langle\mu\rangle$ for all $M \times N$ pairs of subsets or for all correlation peaks. The objective measures of variability of the correlation peak in the output correlation field $c_{mn}\left(k', l'\right)$ are the peak-to-input noise ratio η_{mn} (PINR) [54] for each pair of subsets r_{mn} and g_{mn}, and the PINR mean value $\langle\eta\rangle$ for all correlation peaks.

According to the results of calculation, it is possible to trace the tendency to decrease $\langle\mu\rangle$ and $\langle\eta\rangle$ both for JTC1–JTC6, and for the digital implementation of the MDSDM method in those cases when the displacements by an integer number of pixels increases and when the decorrelation parameter Δ_d increases. The PONR mean values $\langle\mu\rangle$ and PINR mean values $\langle\eta\rangle$ decrease much faster for those cases when the deformed speckle pattern is shifted by a non-integer number of pixels. If $\Delta_d = 0$, the largest values of $\langle\mu\rangle$ are obtained for digital models JTC5 and JTC6, and the largest values of $\langle\eta\rangle$ are obtained for models JTC2, JTC3, JTC5 and JTC6. Similar trends can also be observed at $\Delta_d = 0.25$ and $\Delta_d = 0.5$. The values of $\langle\mu\rangle$ are approximately the same for the MDSDM method, as well as for JTC2 and JTC3, more for JTC5 and JTC6 for small displacements and much more for large ones. Therefore, it can be argued about better robustness of JTC5 and JTC6 to output noise. A similar behavior is observed for values of $\langle\eta\rangle$. Thus, JTC5 and JTC6 exhibit the highest robustness to output noise. JTC4 has the lowest robustness for large displacements. The digital implementations of the MDSDM method and all digital JTC models, with the exception of JTC4, have approximately the same robustness to input noise [85, 87].

The obtained results indicate a high coincidence of the positions of the correlation peaks with the positions of the corresponding subsets in all digital models of JTCs. All the models are characterized by approximately the same values of random and mean absolute errors. Based on the obtained data, it can be asserted that random overshoots and false signals are absent at the outputs of all JTC1–JTC6 correlators. This testifies to the high reliability of the proposed methods of the JPS linear and nonlinear filtering, as they prevent the formation of false signals in the output correlation plane. The measurement results show that the proposed JPS filtering and binarization methods are correct, and the implementation of additional filtering procedures and nonlinear spectrum transformations does not lead to an increase in systematic and random errors in determining the displacements of the speckle patterns subsets.

Experimental implementation of the MDSDM and ODSC methods was performed using the developed digital models of correlators that simulated these methods. Real speckle patterns of specimen surfaces were entered into the computer with the help of an experimental setup developed for experimental studies of the specimen surface in-plane displacements without its loading in order to simulate the rigid body motion. The results of the experimental evaluation of random errors of the studied steel specimen in-plane displacements using the digital implementation of the MDSDM method and JTC1–JTC6 digital models show that the lowest level of random errors was obtained in the JTC5 and JTC6, that is as for the synthesized speckle patterns. The highest level of errors is in the JTC1 and in the digital implementation of the MDSDM method. Experiments with the real specimen show that the highest PONR mean values are achieved for the JTC5. For the JTC6, these values are approximately two times smaller, and for the JTC2 and JTC3, they are approximately four times smaller. Higher PONR mean values were obtained for the JTC4 than for the JTC2 and JTC3. The smallest values of $\langle\mu\rangle$ were obtained in the JTC1 [77].

The conducted experimental verification of the ODSC method showed the high reliability of such digital correlator models as the JTC5 and JTC6, and their resistance to noise and interference. In comparison with the digital model of the MDSDM method, the JTC5 and JTC6 models have significantly better information parameters. The digital models JTC2 and JTC3 in terms of such indicators as the level of random errors of displacements also have an advantage over the MDSDM method.

1.5.3 Studying Surface Displacements of Structural Materials Specimens Using Optical-Digital Speckle Correlation Method

Sheet specimens made of steel 08kp were used to study their surface displacements under mechanical loads [75, 76, 80, 92, 93, 96]. The specimens were loaded using the EU-40 universal tension-testing machine. The displacement fields at the specified load values corresponded to the points on the stress–strain diagram were obtained during the experiment. To generate the displacement fields of the 08kp steel specimens' surface areas under uniaxial loads, the developed digital models of JTC1, JTC2, JTC5 and JTC6 were used. For comparison the efficiency of these JTC digital models, the PINR mean values $\langle\mu\rangle$ were calculated for each output grating of correlation peaks obtained at different tensile loads. The obtained results indicated that the largest values of $\langle\mu\rangle$ are obtained in the JTC5 and JTC6 digital models. In the JTC5 model, values of $\langle\mu\rangle$ are approximately 1.5 times larger than these values in the JTC6 model and about 12 times larger than those in the digital model of the MDSDM method. The smallest values of $\langle\mu\rangle$ were obtained in the JTC1 model.

The obtained research results indicate the high efficiency of the ODSC method in determining the surface displacement fields in real specimens of structural materials under mechanical loads. The main information parameters of the JTC2, JTC3, JTC5 and JTC6 digital models implementing this method exceed the corresponding information parameters of the correlation system implementing the MDSDM method by several times.

1.5.4 Using Digital Image Correlation Technique to Assess Stress–Strain State of Material Near Crack

DIC technique [112, 139] has some advantages in comparison with MDSDM [68, 77, 85–87] and ODSC [75, 77, 85, 87–89, 93, 96] methods. Using the DIC, it is quite accurate to determine the surface strains and displacements in a wide measurement range. The resolution limit of DIC is very high and depends on the dimensions of the subsets, which is used in the iterative procedure. DIC provides smaller errors in determining the surface displacements and strains, which can significantly improve the accuracy of diagnosing the stress–strain state of materials and structural elements containing cracks and other mechanical damages [37, 122, 141]. On the other hand, the technical implementation of surface displacements in 2D DIC is more complicated, since it requires exclusively in-plane movements and strains, and hence the absence of the normal strains.

DIC technique allows solving different problems of fracture mechanics. Among them, the studies of materials cracking and crack resistance are of particular interest. Standard measurement techniques of the materials fracture toughness [32] provide for the determination of the stress intensity factors (SIFs) critical values K_{IC}, K_{IIC}

and K_{IIIC}, which are the force criteria for crack resistance, as well as the critical values of crack opening (deformation criterion δ_c) and J-integral (energy criterion J_c). These techniques are destructive and not universal enough; therefore, they are not appropriate for the diagnosis of specific constructions, as it involves significant restrictions on a specimen size and shape. In addition, they are unsuitable for mixed types of loads, which are most often used in practice.

In contrast, DIC methods are nondestructive, universal and provide the low level of errors in the formation of displacement and deformation fields of the studied surface area. For example, Lopez–Crespo et al. [62] determined the position of the crack tip to calculate the SIF under a mixed load using the DIC. Du et al. [18] assessed the speed of crack growth at mixed load and calculated the corresponding SIF values. In these and other similar papers, the crack tip coordinates were determined using the Sobel algorithm. However, for precise determination of the stress state parameters in a material with a crack, the coordinates of its tip, obtained from the displacement fields with the help of standard algorithms for digital image processing are usually used only as a certain starting approximation. This circumstance is due both to the errors of this algorithm, and to the fact that real cracks have a complex curvilinear front in the volume of the material. More accurate and reliable data are obtained by restoring the displacement and stress fields around the crack tip with the help of appropriate optimization algorithms based on image adaptive segmentation approach [125].

To increase the accuracy of diagnosing the stress–strain state of quasi-brittle materials near the crack tip, the stress field components determination technique using the DIC was developed [61, 64–66]. This technique makes it possible to calculate the stress fields in the vicinity of the crack tip from surface strain maps of a studied material obtained by DIC and to introduce the stress field components decomposition parameters into Williams' series. A new method for evaluating random error components of Williams' series parameters was also developed. The proposed approach made it possible to reliably determine the eight first coefficients of Williams' series and coordinates of the crack tip for type I and mixed type I + II loading. Also such approach allowed to develop new methods for compensation of biases in stress field components errors for improving accuracy of SIFs K_I and K_{II} derivation. New NDT methods for determining the stress intensity factors, J-integral, crack propagation direction, boundaries of plasticity zone during a specimen loading were also developed by using the DIC. In particular, a new method was proposed for determining the direction of a crack initiation and propagation in metal and alloy structural materials by the calculated stress tensor determinant critical value [40, 61]. The method was implemented when testing loaded metal structural elements in gantry crane and other pieces of construction equipment.

1.5.5 Detecting Subsurface Defects in Composite Structures Using Speckle Decorrelation

Development of new effective and high-speed methods for NDT of composite structures to detect subsurface defects is an important scientific and technical problem. Systems implementing these methods compared to experimental setups fulfilling the synchronized reference updating electronic speckle pattern interferometry (ESPI) technique [117], additive–subtractive ESPI/shearography technique [23, 24, 116, 151] and subtractive synchronized DSPI method [91, 98, 103] should have relatively simple equipment and be able to function under natural conditions. One of the possible ways to solve it consists in the study of dynamic speckle patterns for their analysis and processing. Such speckle patterns are formed in the optical system when coherent or quasi-coherent light is reflected from the optically rough surface of the laminated composite, which is excited by harmonic US radiation, and are registered with the help of a digital camera in the form of a sequence of speckle images, after additional processing of which defects are highlighted.

Studying speckle fluctuations on the surfaces and in the subsurface layers of various objects is the urgent direction of Speckle Metrology. Due to acoustic excitation of the laminated thin composite panel, acoustic flexural waves propagating in thin layers activate bottom defects at their resonant frequencies [11, 12, 133–135], and the region of interest (ROI) placed directly above the defect begins to vibrate at one of its own resonant frequencies in the direction transverse to the plane of the surface. The ROI can be considered as a thin edge-clamped membrane placed directly above the defect, and the excitation of the defect under resonant conditions is inextricably linked to the vibrations of the ROI due to flexural waves propagating in the thin layer above the defect, i.e. in the ROI [104].

If the optically rough surface area of the laminated composite, within which there is a subsurface defect, is illuminated by an expanded laser beam and the US harmonic excitation has a frequency that coincides with one of the resonance frequencies of the defect, a series of speckle images of the excited surface is formed and registered during the vibrations of the ROI with the help of an optical system. If the ROI vibrates at one of the resonant frequencies, this frequency is not resonant for the area of the surface outside the ROI, and this area does not vibrate in contrast to the ROI. The spatial response to the vibrated ROI is highlighted as a local part of the recorded speckle pattern generated by the ROI. This local part, i.e. the local speckle pattern (LSP), changes its structure and contrast in sync with the ROI vibrations. At the same time, the rest of the speckle pattern outside the LSP, which is generated by the area of the surface around the ROI, remains unchanged. This difference between the LSP and the rest of the speckle pattern makes it possible to highlight the LSP or its spatial elements and thereby detect a subsurface defect [104].

Two main factors affect the structure and contrast of speckles. The first factor (Factor 1) is caused by the spatial frequency shift in the lens aperture plane of the optical scheme forming a speckle pattern, which leads to the transformation of this pattern spatial structure [6, 25, 158, 159]. The second factor (Factor 2) is caused by the out-of-plane movement of the diffuse surface, which leads to a decrease in the speckle contrast and the speckle blurring [111].

Influence of Factor 1 on transformation of dynamic speckle patterns structure. The transformation of the LSP generated by the ROI during its oscillations occurs due to the tilt of the ROI spatial elements, which leads to a shift of the spatial frequency spectrum in the plane of the lens aperture, which forms speckle patterns in the image plane of a coherent or quasi-coherent optical imaging system [25, 158, 159]. Decorrelation between the transformed LSP and the initial one leads to a decrease in the local speckle contrast in this pattern and can be given by the Yamaguchi correlation factor, which depends on the spatial frequency shift and, respectively, the ROI tilt. Since the ROI can be considered as a thin edge-clamped membrane, its maximum tilts at resonant frequencies occur at its nodes located along the direction transverse to the direction of the US wave propagation. Correlative comparison of two LSPs registering in two opposite phases of the vibrating ROI by their subtraction leads to changes in the contrast of these parts. As a result, these parts are highlighted with bright spots and the subsurface defect is detected in the resulting defect map. The accumulation of sequences of such subtractive LSPs makes it possible to increase the intensity of the spatial response from the desired defect relative to its surrounding background.

Influence of Factor 2 on transformation of dynamic speckle patterns structure. Factor 2 influencing change in the LSP is associated with the ROI out-of-plane displacements, which directly leads to a decrease in the speckle contrast and speckle blurring. If the surface does not move, the spatial coherence between the optical waves reflected from the surface does not change and has a constant level. If the surface begins to vibrate, the optical waves acquire random phase differences during the vibration cycle, caused by phase distortions introduced by the moving rough surface [19, 111]. As a result, the spatial coherence during the formation of a speckle pattern of a moving ROI decreases, and the LSP generated by the region of interest loses its contrast. In addition, the registration of speckle patterns generated by a rough surface containing the vibrating ROI always has some finite exposure time; and temporal averaging of dynamic speckles during their registration entails additional blurring of speckles in the obtained defect map.

Technical implementation of the method for subsurface defects detection. To check the reliability of subsurface defects detection using the developed method, an experimental setup of a hybrid optical-digital system (HODS) with US harmonic excitation of laminated composite panels was created [91, 98, 103, 104]. This setup is similar to the ODSI experimental setup in which the electronic and optoelectronic units and hardware are the same. However, the optical system producing speckle patterns is considerably simpler, since it does not contain a reference beam. Due to the absence of the reference beam, the compact version of the HODS setup can be used not only in the laboratory, but also in natural conditions, as it is resistant to vibrations and other external factors.

The developed software was used for recording the initial and accumulated total speckle patterns and for their processing to increase the signal-to-noise ratio in defect maps. The difference averaging algorithm (DAA) and pairwise sum of differences algorithm (PSDA) [91, 98, 138] designed in PMI, as well as the spatial speckle contrast algorithm (SSCA) [5, 39, 47], were selected for processing dynamic speckle patterns.

Experimental results. Experimental verification of the developed method for detecting artificial subsurface defects in thin-layered composite structures was carried out using the developed HODS experimental setup with US excitation of studied composite panels in the US frequency sweep mode. The artificial defects as round and square bottom holes were fabricated in two fiberglass composite panels. The results of artificial subsurface defects detection in Panel 1 and Panel 2 showed that LSPs stand out against the surrounding background on defect maps obtained using the aforementioned algorithms for processing total speckle patterns. In this case, the LSP generated by the ROI is approximately located within the limits determined by the ROI contour. A number of obtained LSPs can be unambiguously interpreted as vibrating nodes of the edge-clamped membrane, which, due to their tilt, form bright spots in the LSP. Real subsurface defects were also analyzed in composite and metal–composite elements of aircraft structures using the HODS experimental setup with a frequency swept US excitation. In particular, honeycomb composite panels attached to the lower part of the aircraft fuselage and metal–composite joints, which are used in the constructions of modern aircrafts, were studied. Internal defects in metal–composite joints were also detected. Using the HODS setup, a light response from an operational internal defect was detected in a 4-mm-thick carbon fiber composite panel, which is part of the metal–composite joint element [104]. The defect appeared in the composite panel after applying a static load of 4,500 Kg to the metal–composite joint, which led to the shift of the inner layers.

Thus, the new method for detection of subsurface defects in laminated composite panels has been developed using speckle decorrelation and speckle blurring phenomena. Thanks to Factor 1 and Factor 2, the LSP in the form of light spots indicates the presence of a subsurface defect within the studied composite surface area. The light responses are generated in the coherent or quasi-coherent optical system due to vibrations of the ROI placed directly above the defect. Previous studies have shown that the method can provide detection of defects with sizes from 5 to 35 mm. However, it has potentially greater capabilities and is able to detect defects of much larger sizes, since with an increase in the amplitude of ROI vibrations, the level of speckle decorrelation increases, due to which it will be possible to detect large defects. The size of defects that can be detected using this method will be limited only by the level of laser radiation intensity.

The MDSDM and ODSC methods, the optimization algorithms based on image adaptive segmentation approach, as well as the method for detection of subsurface defects in laminated composite panels and the technical implementations of the developed methods are described in Chap. 6 (Methods for Processing and Analyzing Speckle Patterns of Materials Surface) of this book.

1.6 Mathematical Modeling of Elastic Waves Interaction with Interface Crack-Type Defects

The effects of interaction of the elastic waves with the defects of materials are one of the main sources of information about their damages. The necessary information is obtained from the analysis of the influence of the defect parameters on the displacement field. Since the modern optical methods make it possible to visualize the displacement fields on the surfaces of the testing sample with high accuracy, the problem of establishing a relationship between defect parameters and displacement fields arises in order to develop new methods of recognizing the defects. Therefore, to develop new methods of material diagnostics, it is important to study the physical properties of scattering of the elastic waves from the defects under different conditions of their sounding, visualizing the field displacement and analyzing its features. The novel theoretical methods of the field scattering analysis from the defects are based on the use of the most general numerical methods with minimal restrictions on their shapes and properties. However, to use these methods in the form of commercial software packages does not always meet practical needs. It is known that finding the diffraction fields characteristics with the help of such tools requires significant computer time. Obviously, to study the physical properties of the elastic fields interaction with the defects, it is important to apply the simple models that allow for mathematically correct analysis of the problem for any parameters and which can be used to develop the approximate techniques for the solution of more complicated problems.

In this regard, we consider the diffraction of elastic waves on the interface crack-like defects that often appear in junctions of the construction elements. In order to form the theoretical bases of the interface defects detection, we suppose that such defects are thin, and as their models, we use the flat strips or half-planes at the plane interface. The main advantage of such models is that one of the most powerful mathematical techniques of theoretical analysis, namely the Wiener–Hopf method, can be applied for this purpose. The obtained solutions allow finding the necessary characteristics of the field with a given accuracy and providing the simple physical interpretation of their properties. For this purpose, we consider two key problems: the SH-wave diffraction from the finite crack at the plane interface of two homogeneous and isotropic semi-spaces [56–59, 146] and the interface crack at the plane interface of the semi-space and layer [60, 102, 144]. We regard this as an important prototype problem for developing solution methods for more complex problems.

Issues of mathematical modeling the physical phenomena of elastic wave interaction between the crack and interfaces as well as displacement and stress field analysis are considered in Chap. 7 (Mathematical Modeling of Elastic Waves Interaction with Interface Crack-type Defects) of this book.

1.7 Conclusions

Recently, Optical Metrology and Optoacoustic techniques in combination with mathematical modeling of elastic waves interaction with interface defects become the powerful tools for nondestructive evaluation and TD of materials and products. Due to high reliability and the possibility of restoration and analysis of surface relief, surface displacement and deformation fields of studied samples and structural elements, as well as detection and identification of internal defects in materials, they are used in various industrial applications and scientific studies. The widespread use of these techniques in such prominent branches such as the aerospace, engineering, automotive and construction industries testifies their importance, prospectivity and competitiveness.

References

1. Alam MS (2000) Phase-encoded fringe-adjusted joint transform correlation. Opt Eng 39(5):1169–1176
2. Alam MS, Ochilov S (2010) Spectral fringe-adjusted joint transform correlation. Appl Opt 49(10):B18–B25
3. Alam MS, Chen XW, Karim MA (1997) Distortion-invariant fringe-adjusted joint transform correlation. Appl Opt 36(29):7422–7427
4. Bobkov YuN (1984) Korreliatsionnyye mietody i ustroistva izmierieniya sostavliayushchih komponientov signalov (Correlation methods and devices for measuring signal components). Vyshcha Shkola, Lviv
5. Briers D, Duncan DD, Hirst ER, Kirkpatrick SJ, Larsson M, Steenbergen W, Stromberg T, Thompson OB (2013) Laser speckle contrast imaging: theoretical and practical limitations. J Biomed Opt 18(6):066018
6. Bruno L, Pagnotta L, Poggiani A (2000) Laser speckle decorrelation in NDT. Opt Lasers Eng 34(1):55–65
7. Chen DJ, Chiang FP, Tan FP, Don HS (1993) Digital speckle-displacement measurement using a complex spectrum method. Appl Opt 32(11):1839–1849
8. Cheng Y, Wyant JC (1984) Two-wavelength phase shifting interferometry. Appl Opt 23(24):4539–4543
9. Cheng P, Sutton MA, Schreier HW, McNeill SR (2002) Full-field speckle pattern image correlation with B-spline deformation function. Exp Mech 42(3):344–352
10. Chu TC, Ranson WF, Sutton MA (1985) Applications of digital-image-correlation techniques to experimental mechanics. Exp Mech 25(3):232–244
11. Ciampa F, Scarselli G, Meo M (2017) On the generation of nonlinear damage resonance intermodulation for elastic wave spectroscopy. J Acoust Soc Am 141(4):2364–2374
12. Ciampa F, Pickering SG, Scarselli G, Meo M (2017) Nonlinear imaging of damage in composite structures using sparse ultrasonic sensor arrays. Struct Control Health Monit 24(5):e1911
13. Cochran ER, Creath K (1988) Combining multiple-subaperture and two-wavelength techniques to extend the measurement limits of an optical surface profiler. Appl Opt 27(10):1960–1966
14. Dai Pra AL, Meschino GJ, Guzmán MN, Scandurra AG, González MA, Weber C, Trivi M, Rabal H, Passoni LI (2016) Dynamic speckle image segmentation using self-organizing maps. J Opt 18(8):085606

15. Dalmau O, Rivera M, Gonzalez A (2016) Phase shift estimation in interferograms with unknown phase step. Opt Commun 372:37–43
16. De la Torre IM, Montes MDSH, Flores-Moreno JM, Santoyo FM (2016) Laser speckle based digital optical methods in structural mechanics: a review. Opt Lasers Eng 87:32–58
17. Dhanasekar B, Ramamoorthy B (2008) Digital speckle interferometry for assessment of surface roughness. Opt Lasers Eng 46(3):272–280
18. Du Y, Diaz-Garrido FA, Burguete RL, Patterson EA (2009) Stress intensity factor evaluation using DIC on an aerospace component. In: Proceedings of the 2009 SEM annual conference and exposition on experimental and applied mechanics. Society for Experimental Mechanics
19. Eliasson B, Mottier FM (1971) Determination of the granular radiance distribution of a diffuser and its use for vibration analysis. J Opt Soc Am 61(5):559–565
20. Ettemeyer A, Wang Z, Walz T (1997, September) Applications of 3D speckle interferometry to material and component testing. In: Gorecki C (ed)Optical inspection and micromeasurements II, vol 3098. SPIE, Bellingham, WA, pp 188–194
21. Flores VH, Reyes-Figueroa A, Carrillo-Delgado C, Rivera M (2020) Two-step phase shifting algorithms: where are we? Opt Laser Technol 126:106105
22. Flynn TJ (1997) Two-dimensional phase unwrapping with minimum weighted discontinuity. J Opt Soc Am A 14(10):2692–2701
23. Fomitchov PA, Krishnaswamy S (1997) A compact dual-purpose camera for shearography and electronic speckle-pattern interferometry. Meas Sci Technol 8(5):581–583
24. Fomitchov P, Wang L-S, Krishnaswamy S (1997) Advanced image-processing techniques for automatic nondestructive evaluation of adhesively-bonded structures using speckle interferometry. J Nondestr Eval 16(4):215–227
25. Fricke-Begemann T (2003) Three-dimensional deformation field measurement with digital speckle correlation. Appl Opt 42(34):6783–6796
26. Goldstein RM, Zebker HA, Werner CL (1988) Satellite radar interferometry: two-dimensional phase unwrapping. Radio Sci 23(4):713–720
27. Goodman JW (2000) Statistical optics. Wiley, NewYork
28. Goodman JW (2005) Introduction to Fourier optics, 3rd edn. Roberts and Company Publishers, Greenwood Village, CO
29. Goodman JW (2007) Speckle phenomena in optics: theory and applications. Roberts and Company Publishers, Greenwood Village, CO
30. Gorthi SS, Rastogi P (2010) Fringe projection techniques: whither we are? Opt Lasers Eng 48(Article):133–140
31. GOST 25.142-82 (1982) Sherohovatost' poverhnosti. Terminy i opredeleniya (Surface roughness. Terms and definitions). Izd Standartov, Moscow
32. GOST 25.506-85 (1985) Raschety i ispytaniya na prochnost'. Metody mehanicheskih ispytaniy metallov. Opredeleniye kharakteristik treshchinostoykosti (vyazkosti razrusheniya) pri staticheskom nagruzhenii (Strength analyses and tests. Methods for mechanical testing of metals. Determination of the crack resistance characteristics (fracture toughness) under static loading). Izd. Standartov, Moscow
33. Greivenkamp JE (1984) Generalized data reduction for heterodyne interferometry. Opt Eng 23(4):350–352
34. Groot PD (2002) Method and system for profiling objects having multiple reflective surfaces using wavelength-tuning phase-shifting interferometry. US Patent No 6,359,692
35. Hahn WB Jr, Flannery DL (1992) Design elements of binary joint transform correlation and selected optimization techniques. Opt Eng 31(5):896–905
36. Halioua M, Liu HC (1989) Optical three-dimensional sensing by phase measuring profilometry. Opt Lasers Eng 11(3):185–215
37. Han G, Sutton MA, Chao YJ (1994) A study of stationary crack-tip deformation fields in thin sheets by computer vision. Exp Mech 34(2):125–140
38. Hariharan P (2000) Phase-shifting interferometry: minimization of systematic errors. Opt Eng 39(4):967–969

39. He XW, Briers JD (1998, July) Laser speckle contrast analysis (LASCA): a real-time solution for monitoring capillary blood flow and velocity. In: Hoffman EA (ed)Medical imaging 1998: physiology and function from multidimensional images, vol 3337. SPIE, Bellingham, WA, pp 98–107
40. Holynskyy IS, Muravsky LI, Polovynko TI (2010) Otsinyuvannia dynamiky zminy poliv deformatsij poverhni sharuvatogo kompozytu metodom prostorovo-chasovoji spekl-koreliatsiji (Assessment of change of deformation fields dynamics of the laminated composite surface by the spatial-temporal correlation method). Elektromagnitni ta akustychni metody neruynivnogo kontrolyu materialiv ta vyrobiv LEOTEST–2010 (Electromagnetic and acoustical methods for nondestructive testing of materials and products LEOTEST–2010) 15:91–92
41. Horner JL (1992) Metrics for assessing pattern-recognition performance. Appl Opt 31(2):165–166
42. Jacquot P (2008) Speckle interferometry: a review of the principal methods in use for experimental mechanics applications. Strain 44(1):57–69
43. Javidi B, Wang J (1991) Binary nonlinear joint transform correlation with median and subset median thresholding. Appl Opt 30(8):967–976
44. Jia P, Kofman J, English C (2007) Multiple-step triangular-pattern phase shifting and the influence of number of steps and pitch on measurement accuracy. Appl Opt 46(16):3253–3262
45. Jones R, Wykes C (1989) Holographic and speckle interferometry. Cambridge University Press, Cambridge
46. Keane RD, Adrian RJ (1992) Theory of cross-correlation analysis of PIV images. Appl Sci Res 49(3):191–215
47. Keene L, Chiang F-P (2009) Real-time anti-node visualization of vibrating distributed systems in noisy environments using defocused laser speckle contrast analysis. J Sound Vib 320(3):472–481
48. Kelly DP, Ward JE, Gopinathan U, Hennelly BM, O'Neill FT, Sheridan JT (2006) Generalized Yamaguchi correlation factor for coherent quadratic phase speckle metrology systems with an aperture. Opt Lett 31(23):3444–3446
49. Khaksari K, Kirkpatrick SJ (2016) Combined effects of scattering and absorption on laser speckle contrast imaging. J Biomed Opt 21(7):076002
50. Kmet' AB, Muravsky LI, Voronyak TI, Stasyshyn IV (2019) Two-step dual-wavelength interferometry for surface relief retrieval. Opt Appl XLIX(2):331–343
51. Kotsiuba Y, Muravsky L, Petrovska H, Fitio V (2019, July) Retrievieng the surface relief of the smooth objects by the digital holography. In: 2019 IEEE 2nd Ukraine conference on electrical and computer engineering (UKRCON). IEEE, Lviv, pp 775–778
52. Kreis T (2005) Handbook of holographic interferometry: optical and digital methods. WILEY–VCH Verlag GmbH & Co, Weinheim
53. Kulkarni R, Rastogi P (2018) Two-step phase demodulation algorithm based on quadratic phase parameter estimation using state space analysis. Opt Lasers Eng 110:41–46
54. Kumar BV, Hassebrook L (1990) Performance measures for correlation filters. Appl Opt 29(20):2997–3006
55. Kumar UP, Bhaduri B, Kothiyal MP, Mohan NK (2009) Two wavelength interferometry for 3-D surface profiling. Opt Lasers Eng 47(2):223–229
56. Kuryliak DB, Voytko MV (2005) Wiener-Hopf analysis of the elastic wave diffraction by the finite crack located at the plane interface between the elastic isotropic half–space medium. In: Direct and inverse problems of electromagnetic and acoustic wave theory (DIPED), Lviv–Tbilisi, pp 143–147
57. Kurylyak DB, Nazarchuk ZT, Voitko MV (2006) Analysis of the field of a plane SH-wave scattered by a finite crack on the interface of materials. Mater Sci 42(6):711–724
58. Kurylyak DB, Nazarchuk ZT, Voitko MV (2007) Stress field in a crack on the interface of materials irradiated with SH-waves. Mater Sci 43(4):464–478
59. Kurylyak DB, Nazarchuk ZT, Voitko MV (2009) On the relationship between the stress intensity factors and the far field of SH-waves diffraction on interface cracks. Mater Sci 45(3):329–338

60. Kuryliak DB, Nazarchuk ZT, Voytko MV, YaP K (2021) Dyfraktsiya SH-hvyli na mizhfaznomu defekti u z'yednanni pruzhnogo sharu z pivprostorom (Diffraction of SH-wave at the interface defect in the joint of the elastic layer with a half-space). Fizyko-Himichna Mehanika Materialiv (Physicochemical Mechanics of Materials) 57(5):13–24
61. Lobanov LM, Muravsky LI, Pivtorak VA, Voronyak TI (2017) Monitoryng napruzhenogo stanu elementiv konstruktsiy z vykorystanniam elektromagnitnyh hvyl' optychnogo diapazonu (Monitoring of structural elements stress state with the use of electromagnetic waves in the optical range), vol 3. In: Nazarchuk ZT (ed) Tehnichna diagnostyka materialiv i konstruktsiy (Technical diagnostics of materials and structures), Reference manual in 8 volumes. Publishing House "Prostir-M", Lviv
62. Lopez-Crespo P, Burguete RL, Patterson EA, Shterenlikht A, Withers PJ, Yates JR (2009) Study of a crack at a fastener hole by digital image correlation. Exp Mech 49(4):551–559
63. Luo C, Zhong L, Sun P, Wang H, Tian J, Lu X (2015) Two-step demodulation algorithm based on the orthogonality of diamond diagonal vectors. Appl Phys B 119(2):387–391
64. Lychak OV, Holyns'kyi IS (2014) Evaluation of stress field reconstruction errors near the crack tip of body under plain strain conditions. Vidbir ta Obrobka Informatsiyi (Information Extraction and Processing) 41:63–69
65. Lychak OV, Holyns'kiy IS (2016) Evaluation of random errors in Williams' series coefficients obtained with digital image correlation. Meas Sci Technol 27(3):035203
66. Lychak OV, Holyns'kiy IS (2016) Improving the accuracy of derivation of the Williams' series parameters under mixed (I+II) mode loading by compensation of measurement bias in the stress field components data. Meas Sci Technol 27(12):125203
67. Maksymenko OP, Muravsky LI, Berezyuk MI (2015) Application of biospeckles for assessment of structural and cellular changes in muscle tissue. J Biomed Opt 20(9):095006
68. Maksymenko OP, Muravsky LI, Lytvyn MO (2003) Vybir parametra inversnogo filtra dlia tsyfrovogo metodu vymiryuvannia zmishchen' spekliv (Choice of inverse filter parameter for digital method of measuring speckle displacements). Fizychni metody ta zasoby kontrolyu seredovyshch, materialiv ta vyrobiv (Physical methods and means for control of media, materials and products) 8:151–156
69. Malacara D (2007) Twyman-Green interferometer. In: Malacara D (ed) Optical shop testing, 3rd edn. Wiley, Hoboken, NJ, pp 46–96
70. Malacara D, Servín M, Malacara Z (2005) Interferogram analysis for optical testing, 2nd edn. Taylor & Francis, Boca Raton, FL
71. Mantravadi MV, Malacara D (2007) Newton, Fizeau, and Haidinger interferometers. In: Malacara D (ed) Optical shop testing, 3rd edn. Wiley, Hoboken, NJ, pp 1–45
72. Meng XF, Peng X, Cai LZ, Li AM, Guo JP, Wang YR (2009) Wavefront reconstruction and three-dimensional shape measurement by two-step dc-term-suppressed phase-shifted intensities. Opt Lett 34(8):1210–1212
73. Meng XF, Cai LZ, Wang YR, Yang XL, Xu XF, Dong GY, Shen XX, Cheng XC (2008) Wavefront reconstruction by two-step generalized phase-shifting interferometry. Opt Commun 281(23):5701–5705
74. Montresor S, Picart P, Sakharuk O, Muravsky L (2017) Error analysis for noise reduction in 3D deformation measurement with digital color holography. J Opt Soc Am B 34(5):B9–B15
75. Muravskii LI, Fit'o NV (2005) Estimating the displacements of the surfaces of deformable objects and of solid bodies by an optical speckle-correlation technique. J Opt Technol 72(5):410–414
76. Muravskii LI, Sakharuk AN, Maksimenko AP (2004) Using optical speckle correlation to study breakdown processes in structural materials. J Opt Technol 71(7):461–465
77. Muravsky LI (2010) Metody spekl-koreliatsiyi dlia doslidzhennia mehanichnyh vlastyvostey konstruktsiynych materialiv (Methods of speckle correlation for study of mechanical properties of structural materials). Naukova Dumka, Kyiv
78. Muravsky LI (2011) Rozroblennia metodiv ta zasobiv spekl-metrologiyi ta fazozsuvnoyi interferometriyi dlia doslidzhennia napruzheno-deformovanogo stanu konstruktsiynyh materialiv (Development of speckle metrology and phase-shifting interferometry methods and

means for studying the stress-strain state of structural materials). In: Panasyuk VV (ed) Fizyko-Mehanichnyy Instytut (Do 60-richchia z chasu zasnuvannia): monografiya (Physico-mechanical institute (to the 60th anniversary since the foundation): monograph). SPOLOM, Lviv, pp 213–231
79. Muravsky LI (2019) Three-step electronic speckle pattern interferometry method with arbitrary phase shifts of reference wave. Vidbir ta Obrobka Informatsiyi (Information Extraction and Processing) 47:54–58
80. Muravsky L, Fityo N, Sakharuk OM (2005, May) Estimation of surface displacement field by optical speckle correlation technique. In: Proceedings of III international conference on materials testing, Nurnberg. AMA Service GmBH, Wunstdorf, Germany, pp 163–168
81. Muravsky LI, Kmet' AB, Voronyak TI (2012) Retrieving the relief of a low-roughness surface using a two-step interferometric method with blind phase shift of a reference wave. Opt Lasers Eng 50(11):1508–1516
82. Muravsky L, Kmet' A, Voronyak T (2013) Two approaches to the blind phase shift extraction for two-step electronic speckle pattern interferometry. Opt Eng 52(10):101909
83. Muravsky L, Kotsiuba Yu, Kulynych Ya (2020, September) Assessment of unknown phase shift for speckle interferometry using sample Pearson correlation coefficient. In: Conference on computer science and information technologies. Springer, Cham, pp 671–681
84. Muravsky L, Kotsiuba Yu, Kulynych Ya (2020, September) Estimation of unknown phase shift between synthesized speckle interferograms using Pearson correlation coefficient. In: Proceedings of 2020 IEEE 15th international conference on computer sciences and information technologies (CSIT), vol 2. IEEE, Lviv, pp 58–61
85. Muravsky LI, Maksymenko OP, Sakharuk OM (2003, May) Comparative analysis of digital and optical speckle displacement techniques for study of in-plane surface strains. In: Kastinger K, Løkberg OJ, Winther S (eds)Speckle metrology 2003, vol 4933. SPIE, Bellingham, WA, pp 149–154
86. Muravsky LI, Maksymenko OP, Sakharuk OM (2003) Otsinka poperechnyh zsuviv poverhni material metodamy spekl-koreliatsiji (Estimation of in-plane displacements of the material surface by speckle correlation methods). Vidbir ta Obrobka Informatsiyi (Information Extraction and Processing) 18(94):95–99
87. Muravsky LI, Maksymenko OP, Sakharuk OM (2004) Use of a joint transform correlator architecture for study of speckle displacements. Opt Commun 240(4–6):275–291
88. Muravsky LI, Sakharuk OM, Byczkowska-Lipińska L (2009) Accurate definition of correlation peak position by optical speckle-displacement correlation technique. Opt Lasers Eng 47(6):712–720
89. Muravsky LI, Sakharuk OM, Yezhov PV (2009, June) Evaluation of aliasing influence on the performance of hybrid optical-digital speckle correlator. In: Lehmann PH (ed) Optical measurement systems for industrial inspection VI, vol 7389. SPIE, Bellingham, WA, pp 785–796
90. Muravsky LI, Voronyak TI, Kmet' AB (2014) Lazerna interferometriya poverhni dlia potreb tehnichnoyi diagnostyky (Laser interferometry of surface for needs of technical diagnostics). Spolom, Lviv
91. Muravsky L, Kuts O, Gaskevych G, Suriadova O (2019, September) Detection of subsurface defects in composite panels using dynamic speckle patterns. In: 2019 XIth international scientific and practical conference on electronics and information technologies (ELIT). IEEE, Lviv, pp 7–10
92. Muravsky LI, Sakharuk AN, Fitio NV, Pavlov AV (2006) Ring median binarization of joint power spectrum at optical correlation of speckle patterns. Bull Russ Acad Sci Phys 70(1):155–160
93. Muravsky LI, Sakharuk OM, Fityo NV, Yezhov PV (2007) Increase of reliability of surface displacement field recovery by optical speckle-displacement correlation technique. Opt Lasers Eng 45(10):993–1000
94. Muravsky LI, Kmet' AB, Stasyshyn IV, Voronyak TI, Bobitski YV (2018) Three-step interferometric method with blind phase shifts by use of interframe correlation between interferograms. Opt Lasers Eng 105:27–34

95. Muravsky LI, Ostash OP, Kmet' AB, Voronyak TI, Andreiko IM (2011) Two-frame phase shifting interferometry for retrieval of smooth surface and its displacements. Opt Lasers Eng 49(3):305–312
96. Muravsky LI, Sakharuk OM, Kostyukevych SO, Maksymenko OP, Kostyukevych KV (2007, June) Performance of optical speckle displacement technique near stress concentrators. In: Osten W, Gorecki C, Novak EL (eds) Optical measurement systems for industrial inspection V, vol 6616. SPIE, Bellingham, WA, pp 980–988
97. Muravsky LI, Voronyak TI, Ivanytsky YaL, Dzhala VR, Kuts OG (2017) Pryntsypy pobudovy aparatno-programnogo kompleksu dlia nerujnivnogo kontrolyu kompozytnyh elementiv konstruktsiy aviakosmichnoyi tehniky (Principles for development of hardware-software complex for nondestructive testing of aircraft composite structural components). Tehnichna Diagnostyka ta Neruynivnyy Kontrol (Technical Diagnostics and Non-Destructive Testing) 1:37–42
98. Muravsky LI, Kuts OG, Voronyak TI, Gaskevych GI, Kotsiuba YM, Stasyshyn IV (2019) Sposib vyyavlennia vnutrishnih defektiv u zrazkah konstruktsiynyh materialiv ta elementiv konstruktsiy (Method for detecting internal defects in specimens of structural materials and structural elements) (Ukrainian patent for a utility model #136987). Ministry of Economy of Ukraine
99. Muravsky LI, Picart P, Kmet' AB, Voronyak TI, Ostash OP, Stasyshyn IV (2016) Evaluation of fatigue process zone dimensions in notched specimens by two-step phase shifting interferometry technique. Opt Eng 55(10):104108
100. Muravsky LI, Picart P, Kmet' AB, Voronyak TI, Ostash OP, Stasyshyn IV (2016) Two-step phase shifting interferometry technique for evaluation of fatigue process zone parameters in notched specimens. In: Creath K, Burke J, Gonsalves AA (eds) Interferometry XVIII, vol 9960. SPIE, Bellingham, WA, pp 996011
101. Nazarchuk ZT, Muravsky LI (2018) Novel digital interferometry and image correlation techniques for nondestructive testing application In: Lobanov LM (ed) Nauka pro materialy: dosiagnennia ta perspektyvy (Materials science: achievements and prospects), vol 2. Akademperiodyka, Kyiv, pp 374–392
102. Nazarchuk ZT, Kuryliak DB, Voytko MV, YaP K (2013) On the interaction of an elastic SH-wave with an interface crack in the perfectly rigid joint of a plate with a half-space. J Math Sci 192(6):609–623
103. Nazarchuk Z, Muravsky L, Kuryliak D (2019) To the problem of the subsurface defects detection: theory and experiment. Procedia Struct Integrity 16:11–18
104. Nazarchuk ZT, Muravsky LI, Kuts OG (2022) Nondestructive testing of thin composite structures for subsurface defects detection using dynamic laser speckles. Res Nondest Eval 33(2):59–77
105. Nemati M, Presura CN, Urbach HP, Bhattacharya N (2014) Dynamic light scattering from pulsatile flow in the presence of induced motion artifacts. Biomed Opt Exp 5(7):2145–2156
106. Ostash OP (2006) New approaches in fatigue fracture mechanics. Mater Sci 42(1):5–19
107. Ostash OP, Panasyuk VV (2001) Fatigue process zone at notches. Int J Fatigue 23(7):627–636
108. Ostash OP, Panasyuk VV, Kostyk EM (1999) A phenomenological model of fatigue macrocrack initiation near stress concentrators. Fatigue Fract Eng Mater Struct 22(2):161–172
109. Ostash OP, Muravs'kyi LI, Voronyak TI, Kmet' AB, Andreiko IM, Vira VV (2011) Determination of the size of the fatigue prefracture zone by the method of phase-shifting interferometry. Mater Sci 46(6):781–788
110. Ostash OP, Muravsky LI, Kuts OG, Andreyko IM, Voronyak TI, Kmet AB, Vira VV (2009) Sposib vyznachennia rozmiriv vtomnoyi zony peredruynuvannia (Method for determination of the fracture process zone) (Ukrainian patent #42549). Ministry of Economy of Ukraine
111. Owner-Petersen M (1991) Decorrelation and fringe visibility: on the limiting behavior of various electronic speckle-pattern correlation interferometers. J Opt Soc Am A 8(7):1082–1089
112. Pan B, Qian K, Xie H, Asundi A (2009) Two-dimensional digital image correlation for in-plane displacement and strain measurement: a review. Meas Sci Technol 20(6):062001

113. Parshall D, Kim MK (2006) Digital holographic microscopy with dual-wavelength phase unwrapping. Appl Opt 45(3):451–459
114. Peters WH, Ranson WF (1982) Digital imaging techniques in experimental stress analysis. Opt Eng 21(3):427–431
115. Picart P, Montresor S, Sakharuk O, Muravsky L (2017) Refocus criterion based on maximization of the coherence factor in digital three-wavelength holographic interferometry. Opt Lett 42(2):275–278
116. Pouet BF, Krishnaswamy S (1994) Additive-subtractive phase-modulated electronic speckle interferometry: analysis of fringe visibility. Appl Opt 33(28):6609–6616
117. Pouet B, Chatters T, Krishnaswamy S (1993) Synchronized reference updating technique for electronic speckle interferometry. J Nondestr Eval 12(2):133–138
118. Pra ALD, Passoni IL, Sendra GH, Trivi M, Rabal H (2014) Biospeckle signal descriptors: a performance comparison. arXiv preprint arXiv:1411.7522
119. Rabal HJ, Braga RA (eds) (2018). CRC Press, Boca Raton, FL
120. Rastogi PK (ed) (2000). Wiley, Hoboken, NJ
121. Refregier P, Laude V, Javidi B (1994) Nonlinear joint-transform correlation: an optimal solution for adaptive image discrimination and input noise robustness. Opt Lett 19(6):405–407
122. Riddell WT, Piascik RS, Sutton MA, Zhao W, McNeill SR, Helm JD (1999) Determining fatigue crack opening loads from near-crack tip displacement measurements. ASTM Spec Tech Publ 1343:157–174
123. Rivera M, Dalmau O, Gonzalez A, Hernandez-Lopez F (2016) Two-step fringe pattern analysis with a Gabor filter bank. Opt Lasers Eng 85:29–37
124. Saide D, Trusiak M, Patorski K (2017) Evaluation of adaptively enhanced twoshot fringe pattern phase and amplitude demodulation methods. Appl Opt 56(19):5489–5500
125. Sakharuk OM, Muravs'kyi LI, Holyns'kyi IS, Lychak OV (2014) Determination of the field of local displacements by the digital speckle correlation method with adaptive segmentation of the images. Mater Sci 49(5):660–666
126. Schmit J, Creath K, Wyant JC (2007) Surface profilers, multiple wavelength, and white light intereferometry. In: Malacara D (ed) Optical shop testing, 3rd edn. Wiley, Hoboken, NJ, pp 667–755
127. Schreiber H, Brunning JH (2007) Phase shifting interferometry. In: Malacara D (ed) Optical shop testing, 3rd edn. Wiley, Hoboken, NJ, pp 547–666
128. Sirohi SR (2009) Optical methods of measurement: wholefield techniques, 2nd edn. Taylor & Francis, Boca Raton, FL
129. Sjödahl M (1994) Electronic speckle photography: increased accuracy by nonintegral pixel shifting. Appl Opt 33(28):6667–6673
130. Sjödahl M (1998) Some recent advances in electronic speckle photography. Opt Lasers Eng 29(2):125–144
131. Sjödahl M, Benckert LR (1993) Electronic speckle photography: analysis of an algorithm giving the displacement with subpixel accuracy. Appl Opt 32(13):2278–2284
132. Sola JF, Kelton R, Meletis EI, Huang H (2019) A surface roughness based damage index for predicting future propagation path of microstructure-sensitive crack in pure nickel. Int J Fatigue 122:164–172
133. Solodov I, Bai J, Busse G (2013) Resonant ultrasound spectroscopy of defects: case study of flat-bottomed holes. J Appl Phys 113:223512
134. Solodov I, Rahammer M, Kreutzbruck M (2018) Analytical evaluation of resonance frequencies for planar defects: effect of a defect shape. NDTE Int 102:274–280
135. Solodov I, Bai J, Bekgulyan S, Busse G (2011) A local defect resonance to enhance acoustic wave-defect interaction in ultrasonic nondestructive evaluation. Appl Phys Lett 99(21):211911
136. Stoykova E, Harizanova J, Sainov V (2008) Pattern projection profilometry for 3D coordinates measurement of dynamic scenes. In: Ozaktas HM, Onural L (eds) Three-dimensional television: capture, transmission, display. Springer, Berlin, Heidelberg, pp 85–164
137. Su X, Zhang Q (2010) Dynamic 3-D shape measurement method: a review. Opt Lasers Eng 48(2):191–204

138. Suriadova OD (2019) Analiz dynamiky lazernyh spekliv dlia vyyavlennia pidpoverhnevyh defektiv u kompozytnyh paneliah (Analysis of laser speckle dynamics for detection of subsurface defects in composite panels). In: Problems of materials science and surface engineering: conference abstract of MSSE2019, Lviv, pp 192–195
139. Sutton MA, Orteu JJ, Schreier H (2009) Image correlation for shape, motion and deformation measurements: basic concepts, theory and applications. Springer Science & Business Media, New York
140. Sutton MA, Wolters WJ, Peters WH, Ranson WF, McNeill SR (1983) Determination of displacements using an improved digital correlation method. Image Vis Comput 1(3):133–139
141. Sutton MA, Zhao W, McNeill SR, Helm JD, Piascik RS, Riddell WT (1999) Local crack closure measurements: development of a measurement system using computer vision and a far-field microscope. ASTM Spec Tech Publ 1343:145–156
142. Tay CJ, Quan C, Niu H, Bhaduri B (2011) Phase retrieval in two-wavelength DSSI using a combined filtering method. Optik 122(23):2114–2118
143. Thompson O, Andrews M, Hirst E (2011) Correction for spatial averaging in laser speckle contrast analysis. Biomed Opt Exp 2(4):1021–1029
144. Voitko MV (2014) Determination of the resonance frequencies in the course of sounding of impedance defects on the boundary of the rigid joint of a plate and a half space by elastic SH-waves. Mater Sci 49(6):849–856
145. Van Brug H (1999) Phase-step calibration for phase-stepped interferometry. Appl Opt 38(16):3549–3555
146. Voitko MV, Nazarchuk ZT, Kurylyak DB (2008) Interaction of plane SH-waves with the tips of interface cracks. Mater Sci 44(2):260–267
147. Voronyak TI, Kmet AB, Muravs'kyi LI (2009) Determination of the 3D fields of displacements by the method of phase-shifting speckle interferometry. Mater Sci 45(3):372–377
148. Voytko MV, Kulynych YaP, Kuryliak DB (2016) Resonant scattering of the SH-wave by the interface impedance defect in an elastic layer. In: Proceedings of 16th international conference on mathematical methods in electromagnetic theory (MMET-2016). Lviv, pp 264–267
149. Voytko MV, Kutlyk MM, Kuryliak DB (2015) The resonant scattering of SH-waves by a finite crack in an elastic layer. Bulletin Taras Shevchenko University of Kyiv, Special issue, Series of Physical and Mathematical Sciences, pp 51–54
150. Wang J, Javidi B (1994) Multiobject detection using the binary joint transform correlator with different types of thresholding methods. Opt Eng 33(6):1793–1805
151. Wang L-S, Krishnaswamy S (1996) Additive-subtractive speckle interferometry: extraction of phase data in noisy environments. Opt Eng 35(3):794–801
152. Weaver CS, Goodman JW (1966) A technique for optically convolving two functions. Appl Opt 5(7):1248–1249
153. Willert CE, Gharib M (1991) Digital particle image velocimetry. Exp Fluids 10(4):181–193
154. Wyant JC (1975) Use of an ac heterodyne lateral shear interferometer with real–time wavefront correction systems. Appl Opt 14(11):2622–2626
155. Xia H, Montresor S, Guo R, Li J, Yan F, Cheng H, Picart P (2016) Phase calibration unwrapping algorithm for phase data corrupted by strong decorrelation speckle noise. Opt Exp 24(25):28713–28730
156. Xu J, Li Y, Wang H, Chai L, Xu Q (2010) Phase-shift extraction for phase-shifting interferometry by histogram of phase difference. Opt Exp 18(23):24368–24378
157. Xu XF, Cai LZ, Wang YR, Yang XL, Meng XF, Dong GY, Shen XX, Zhang H (2007) Generalized phase-shifting interferometry with arbitrary unknown phase shifts: direct wavefront reconstruction by blind phase shift extraction and its experimental verification. Appl Phys Lett 90(12):121124
158. Yamaguchi I (1981) Speckle displacement and decorrelation in the diffraction and image fields for small object deformation. Opt Acta Int J Opt 28(10):1359–1376
159. Yamaguchi I (1993) Theory and applications of speckle displacement and decorrelation. In: Sirohi RS (ed) Speckle metrology. CRC Press, Boca Raton, FL, pp 1–40

160. Yang L, Ettemeyer A (2003) Strain measurement by three-dimensional electronic speckle pattern interferometry: potentials, limitations, and applications. Opt Eng 42(5):1257–1266
161. Zdunek A, Muravsky LI, Frankevych L, Konstankiewicz K (2007) New nondestructive method based on spatial-temporal speckle correlation technique for evaluation of apples quality during shelf-life. Int Agrophys 21(3):305–310
162. Zhang ZH (2012) Review of single-shot 3D shape measurement by phase calculation-based fringe projection techniques. Opt Lasers Eng 50(8):1097–1106

Chapter 2
Phase Shifting Interferometry Techniques for Surface Parameters Measurement

Abstract This chapter discusses the temporal phase shifting interferometry methods developed at the Karpenko Physico-Mechanical Institute of the NAS of Ukraine. The two-step interferometry method with an unknown phase shift of the reference wave is analyzed. In this method, the extraction of the unknown phase shift is fulfilled using the population Pearson correlation coefficient. This method allows extraction of both 3D surface nanorelief and 3D surface macrorelief, including 3D waviness and form. However, the high error level in the retrieved phase maps induced the creation of an iterative method of two-step interferometry with the unknown phase shift. To speed up the interferograms recording and remove the reference and object waves registration, a method of three-step interferometry with unknown phase shifts is proposed. In order to expand the range of surface roughness reconstruction, the two-step dual-wavelength interferometry method (TS DWIM) was elaborated. In this method, only two pairs of interferograms are generated and each pair is differed only in an arbitrary phase shift of the reference wave. Since the method does not require phase shift calibration, it is enough to use only one arbitrary phase shift to record both pairs of interferograms. The three-frame two-wavelength interferometry method with unknown phase shifts between interferograms is similar to the TS DWIM. However, it does not require additional recording of the object and reference waves, which significantly increases its performance.

Recent advances in PSI are inseparably linked with phase shifting algorithms (PSAs) development and digital cameras improvement. Firstly, PSAs were developed for temporal single-wavelength PSI [104]; further, they were applied for other optical interferometry techniques including DSPI [96] and DH [53]. Nowadays, optical coherent and noncoherent interferometry techniques, namely white light scanning interferometry (WLSI) [103], temporal and dynamic (spatial) single-wavelength PSI [66, 104], temporal two- and multiwavelength coherent and noncoherent PSI [13, 66, 103, 104], and wavelength-tuning interferometry (WTI) [66, 103] are widely used for surface relief reconstruction of optical elements, nanostructures, microelectromechanical systems, constructional materials, manifold devices, machine components, vehicles, engines, etc.

Z. Nazarchuk et al., *Optical Metrology and Optoacoustics in Nondestructive Evaluation of Materials*, Springer Series in Optical Sciences 242,
https://doi.org/10.1007/978-981-99-1226-1_2

2.1 Basics of Temporal Single-Wavelength Phase Shifting Interferometry

Temporal single-wavelength PSI has some drawbacks in comparison with WLSI, two- or multiwavelength PSI and WTI. In particular, only smooth and nanorough surfaces can be retrieved, if a laser light wavelength belongs to a visible or near-infrared range of electromagnetic waves. The temporal single-wavelength PSI make it possible to restore a phase map of a studied surface area using at least three interferograms of this area recorded by a digital camera's photodetector array. However, the reliable digital phase map retrieval is possible, if a phase difference between adjacent pixels of the map does not exceed the half-period π. This phase difference is one-quarter of the laser radiation wavelength, i.e. $\lambda/4$, for the interferometers with normal incidence of the object and reference beams on the test surface. In addition, the temporal single-wavelength PSI is sensitive to various inhomogeneities and turbulences of air flows on the path of optical beams in interferometers and to the dust carried by these flows. The retrieval quality of the surface area phase map is also affected by defects in the optical elements of the interferometer.

Simultaneously, the temporal single-wavelength PSI has some advantages over the techniques mentioned above. First of all, it can be implemented by simpler and cheaper hardware. Besides, this technique provides practically instant recording of a much larger surface area by using a photodetector array.

In temporal PSI, retrieval of the test object phase map is implemented with the help of a given phase shifting algorithm and several interferograms of the same test object. Such interferograms are obtained by introducing several consecutive in time phase shifts or phase steps $\alpha_j (j = 1, 2, \ldots, J)$ between the reference and object wavefronts in a two-beam interferometer and recording each interferogram with a given phase shift α_j by a digital camera. As a result, a time-varying signal is formed at each point of the recorded digital interferograms. It encodes all the phase differences between the reference and object wavefronts, which were produced by performing J phase shifts of one of these wavefronts (usually, the reference). The produced phase map can be easily transformed into an optical path difference map or a height map using the well-known dependence between the phase change $\Delta\varphi(x, y)$ and the surface relief height change $\Delta h(x, y)$

$$\Delta h(x, y) = \frac{\lambda[\Delta\varphi(x, y)]}{2\pi(\cos\theta + \cos\theta')}, \tag{2.1}$$

where θ and θ' are the incidence and observation angles relative to the normal to the test surface, λ is the wavelength of the laser radiation.

In some interferometers (e.g. Twyman–Green and Fizeau interferometers) $\theta = \theta' = 0$ rad, and Eq. (2.1) can be transformed to

$$\Delta\varphi(x, y) = \frac{4\pi[\Delta h(x, y)]}{\lambda}. \tag{2.2}$$

As a rule, two-beam interferometers are used to produce interferograms in temporal PSI. Complex amplitudes of the object and reference wavefronts are given by

$$e_o(x, y) = |e_o(x, y)| \exp[i\varphi_o(x, y)] \tag{2.3}$$

and

$$e_r(x, y) = |e_r(x, y)| \exp\left[i\varphi_r(x, y) - \alpha_j\right], \tag{2.4}$$

where $\varphi_o(x, y)$ is the phase distribution in the object wavefront corresponding to the object phase field, $\varphi_r(x, y)$ is the phase distribution of the reference wavefront.

A photodetector array of a digital camera records the interferogram intensity distribution at the plane (x, y), which in linear approximation is expressed as

$$\begin{aligned} I(x, y) &= |e_o(x, y) + e_r(x, y)|^2 \\ &= I_o(x, y) + I_r(x, y) + 2\sqrt{I_o(x, y) I_r(x, y)} \cos\left[\varphi_o(x, y) - \varphi_r(x, y) + \alpha_j\right] \\ &= I_B(x, y) + I_M(x, y) \cos\left[\varphi(x, y) + \alpha_j\right] \\ &= I_B(x, y)\left\{1 + V(x, y) \cos\left[\varphi(x, y) + \alpha_j\right]\right\}. \end{aligned} \tag{2.5}$$

Here $\varphi(x, y) = \varphi_o(x, y) - \varphi_r(x, y)$ is the sought wavefront phase equal to the difference between $\varphi_o(x, y)$ and $\varphi_r(x, y)$, $I_B(x, y) = I_o(x, y) + I_r(x, y)$ is the background intensity (also called average intensity or low-frequency intensity), $I_o(x, y)$ and $I_r(x, y)$ are the intensities of the object and reference wavefronts produced in the two-beam interferometer, $I_M(x, y) = 2\sqrt{I_o(x, y) I_r(x, y)}$ is the fringe or intensity modulation (visibility function), $V(x, y) = \frac{I_M(x, y)}{I_B(x, y)}$ is the fringe visibility (also called modulation or contrast).

Equation (2.5) is called as fundamental equation for PSI. This equation contains three unknowns: the background intensity $I_B(x, y)$, fringe modulation $I_M(x, y)$ or fringe visibility $V(x, y)$, and sought phase $\varphi(x, y)$. Therefore, to calculate each point or pixel of the surface phase map, it is necessary to have three Eq. (2.5). For this purpose, at least three interferograms with three different phase shifts of the reference wave should be recorded.

PSI allows considerable increasing the accuracy of optical measurements. A test object illumination changes, a matrix sensor sensitivity changes in linear range, and a fringe pattern noise has a little effect on a retrieved phase map. Besides, the problem of ambiguity of surface concave and convex parts is not proper for PSI. However, considerable errors of the phase measurement can be introduced by deviations from a given phase shift α_j, which is implemented by a calibrated phase shifting device.

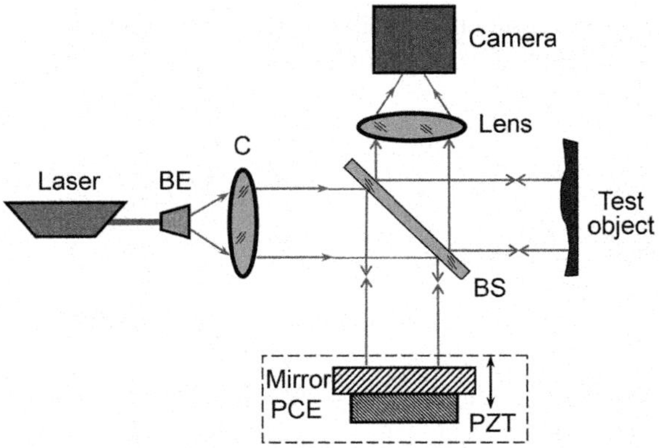

Fig. 2.1 Twyman–Green interferometer arrangement: beam expander (BE), collimator (C), beam splitter (BS), piezoceramic element (PCE), piezoelectric transducer (PZT)

2.1.1 Optical Arrangement

There are a lot of optical arrangements used in the temporal PSI. A Twyman–Green [65, 104], Mach–Zehnder and Fizeau [67, 104] two-beam interferometers' arrangements are most distributed. Phase shift in these interferometers can be produced by different methods and devices [104]. The most popular in PSI is a piezoelectric transducer (PZT) based on a piezoceramic element with a bonded movable mirror for phase shifting [65]. The Twyman–Green interferometer arrangement is shown in Fig. 2.1. In this arrangement, the reference beam's phase shift is implemented by the mirror displacement along the direction of this beam propagation. Therefore, dependence of $\Delta\varphi(x, y)$ as a function of $\Delta h(x, y)$ in this interferometer is expressed by Eq. (2.2). So, for example, the phase shift of the PZT, equal to π, corresponds to its displacement, equal to $\lambda/4$.

2.1.2 Methods for Phase Shift Implementation

Phase shifts in interferometers are implemented using phase steps or a smooth phase shift, usually of a reference wavefront. Phase steps are performed by optical or opto-electronic elements or by different laser wavelengths. Phase steps using a PZT or other optical tools occur due to a jump-like phase shift in time, and each interferogram intensity distributions can be described by Eq. (2.5). Sharp changes in phase shifts α_j lead to mechanical oscillations of a phase shifting element (PSE), which decay over time. As a result, the magnitude of the phase shift changes in time and randomly deviates from the specified value during the oscillation process. This circumstance slows down significantly the procedure for registering a series of speckle interferograms with phase shifts α_j.

In the integrating-bucket method proposed in [130], the smooth phase shift $\alpha(t)$ of the beam continuously and linearly increases in time due to the constantly increasing voltage applied to the PZT. Each interferogram is recorded for a period of the reference phase change Δ_t during exposure time. For this type of registration of interferograms, Eq. (2.5) is rewritten as follows [37]:

$$I(x, y, t) = \frac{1}{\Delta_t} \int\limits_{\alpha-0.5\Delta_t}^{\alpha+0.5\Delta_t} \{I_B(x, y) + I_M(x, y)\cos[\varphi(x, y) + \alpha(t)]\}d\alpha$$
$$= I_B(x, y) + I_M(x, y)\gamma\cos[\varphi(x, y) + \alpha(t)], \quad (2.6)$$

where $\gamma = \sin c\left(\frac{\Delta_t}{2\pi}\right)$ is the modulation transfer function (MTF) during registration of interferograms, $\sin c(u) = \frac{\sin(\pi u)}{\pi u}$. For phase step methods $\gamma = 1$, Eq. (2.6) reduces to Eq. (2.5). At small values of Δ_t, the influence of the MTF on the output signal is almost absent. In particular, if $\Delta_t = \pi/2$, we have $\gamma = 0.9$, and if $\Delta_t = \pi/4$, we have $\gamma = 0.97$ [104]. If one sets the phase shifting device so that the time of smooth voltage rise moving the PSE from 0 to π is $T = 3$ s, and the exposure time to record the j-th interferogram is $\tau_j = 0.18$ s, then in this case $\gamma = 0.9985$; i.e. γ is indistinguishable from the MTF for stepwise phase shifts. Therefore, for the ratio $\mu = \frac{\tau_j}{T} \leq 0.06$ there is no difference between MTFs for smooth and stepwise phase shifts.

Along with the linear integrating-bucket method, the sinusoidal phase shift method is also used [22, 23, 101]. The advantage of the sinusoidal phase shift method over the linear one is that, due to the continuity of the sinusoidal function, it does not require an interrupt to reverse and restart the shift, as in the case of a linear saw-tooth phase shift.

2.1.3 *Phase Shifting Algorithms*

A number of PSAs have been developed in last decades [104]. After all the variety, they have some similar properties, namely

(i) Series of interferograms are recorded, if the reference phase is varied. As a rule, phase shifts α_j with equal angles $\alpha_{j+1,j} = \alpha = 2\pi j/J$ (j is the interferogram number, J is the number of interferograms) are used in majority of PSAs.

(ii) The wavefront phase $\varphi(x, y)$, modulo π, or the wrapped phase map is calculated at each measurement point as the arctangent of a function of the interferogram intensities measured at that same point. For example, the wavefront phase $\varphi(x, y)$ of the conventional four-step PSA, which uses discrete phase values $\alpha_j = 0, \pi/2, \pi, 3\pi/2$, is calculated in every point of a retrieved phase map by using the next equation [38, 104]:

$$\varphi(x, y) = \arctan\frac{I_4 - I_2}{I_1 - I_3}, \quad (2.7)$$

where $I_1 = I_B(1 + V\cos\varphi)$, $I_2 = I_B(1 - V\sin\varphi)$, $I_3 = I_B(1 - V\cos\varphi)$, $I_4 = I_B(1 + V\sin\varphi)$.

In Eq. (2.7), x, y, t are omitted for brevity. We will use this simplification for most of the subsequent equations.

In contrast to conventional three-step, four-step, and other multistep algorithms [38, 104], list-squares multistep PSAs [9] and averaging algorithms [102, 105] possess better performance and accuracy. However, the robustness of all these algorithms to systematic phase shift errors is not high, because even insignificant deviations from rating values of angles α_j lead to the large errors in the retrieved phase map. In contrast to them, the solution of the Hariharan algorithm [42] with linear phase shifts $\alpha_j = -2\alpha, -\alpha, 0, \alpha, 2\alpha$ between frames is robust against systematic errors caused by deviations of angles α_j from their nominal values. For example, if the phase shift between two next measurements is equal to $2\pi + \varepsilon$ and $\varepsilon = 2°$, the measurement error of the phase φ is equal to about 0.02°. The same phase shift deviation for the conventional three-step algorithm leads to the error equal to 1°.

General approaches to design of PSAs, which sensitivity to systematic errors of phase shifts decreases with increasing number of phase shifts, were initiated and developed by Larkin and Oreb [59], Surrel [110], and Hibino et al. [44]. These approaches are based on calculation of sought phase $\varphi(x, y)$ as function of two polynomials ratio

$$\varphi = \arctan\left[\frac{\text{Num}}{\text{Denum}}\right], \tag{2.8}$$

in which the polynomial numerator "Num" is the sum proportional to a sine of the phase shift (imaginary part) and the polynomial denominator "Denum" is the sum proportional to the phase shift cosine (real part), i.e.

$$\text{Num} = 2kI'V\sin(\alpha_j) = \sum_j n_j I_j, \tag{2.9}$$

$$\text{Denum} = 2kI'V\cos(\alpha_j) = \sum_j d_j I_j. \tag{2.10}$$

Here constants k depend on coefficient values, which are depicted, for example, in [17].

Therefore, the resulting wavefront phase is

$$\varphi = \arctan\left[\frac{\sum_j n_j I_j}{\sum_j d_j I_j}\right], \tag{2.11}$$

where the coefficients n_j and d_j define the phase shift algorithms, and the fringe visibility is

$$V = \frac{\sqrt{(\text{Num})^2 + (\text{Denum})^2}}{2kI'}. \tag{2.12}$$

The vectors of coefficients for the numerator and denominator are window functions. For example, in the four-step algorithm, all sample weights corresponding to step numbers (phase shifts) are the same and equal to $\{1, 1, 1, 1\}$. For other algorithms with a larger number of steps, the sample weights are larger for the middle frames of the series of α_j-shifted interferograms than for the extreme ones. In a five-step algorithm, we have $\{1, 2, 2, 2, 1\}$.

2.1.4 Phase Unwrapping

The wrapped digital phase map $\varphi(k, l)$ of a test object surface is produced as a result of calculation the phase φ at each k, lth pixel ($k = 1, 2, \ldots, K; l = 1, 2, \ldots, L$) of a 2D digital array using obtained equations in the phase shifting algorithms. The phase change at any area of this map can be defined only at the range from $-\pi/2$ to $\pi/2$. However, the wrapped map has phase change data over the entire range. In addition, the phase of the test object surface can change both continuously and discretely; however, its discrete jumps are related to the features of the surface itself, and not to the limited range of the function *arctan* definition.

The first step in correcting the wrapped phase map is to increase the phase change range from $-\pi/2 < \varphi(k, l) < \pi/2$ to $-\pi < \varphi_{2\pi}(k, l) < \pi$ or to $0 < \varphi_{2\pi}(k, l) < 2\pi$ using simple trigonometric transformations. This is easy to fulfill, because the positive and negative signs of the *sine* and *cosine* functions in the numerator and denominator of all equations for calculating the phase φ can be determined and thus the correct phase value within the range 2π. Some examples of such transformations are presented, for example, in [16, 66]. To simplify next equations, we will omit the index 2π for the phase $\varphi_{2\pi}$.

The next and most important step in the correction of the wrapped phase map modulo 2π is its phase unwrapping. As a result of this procedure, the wrapped phase map $\varphi(k, l)$ is transformed into the unwrapped one $\varphi_u(k, l)$. In general, the relationship between φ and φ_u is written as follows:

$$\varphi_u(k, l) = \varphi(k, l) + 2\pi m(k, l), \tag{2.13}$$

where $m(k, l)$ is an integer called a field number.

The uncertainty of the phase modulo 2π is eliminated by comparing the phase difference between adjacent pixels $(\delta\varphi)_{\text{adj}}$ with the half-period π. If this difference is greater than π, the $2\pi m$ value must be added or subtracted to obtain a phase difference, less than π. This problem is relatively easy to solve for interference patterns with high-contrast frames, for which the following requirements are met [66]:

1. The interference pattern does not contain noise and disturbances.

2. The requirements of the sampling theorem on the number of pixels for the period 2π are not violated; i.e. the phase difference $(\delta\varphi)_{\mathrm{adj}}$ registered by two adjacent pixels should not exceed π. This requirement can be expressed mathematically by [66]:

$$\frac{\partial\varphi(x, y)}{\partial x} < \frac{\lambda}{2(\Delta x)}, \tag{2.14}$$

where Δx is the distance between two consecutive pixels. This equation sets the maximum allowable value of the wavefront slope.

The phase unwrapping procedure is complicated when the phase difference between adjacent pixels is larger than π at all points except those that characterize the discontinuities of the *arctan* function. These singularity points in the interference pattern arise due to speckle noise, boundary and internal discontinuities of the tested surface, low contrast of interference fringes, and speckle decorrelation. In potential phase map, each singular point is the center of a local vortex field. At these points, phase unwrapping errors occur and transmit to the next pixels along the selected unwrapping route. Based on different strategies, a number of algorithms have been developed to identify singular points and form an unwrapped phase map [10, 12, 14, 31–34, 45, 93, 109]. Among them, the most common are path-following and minimum-norm strategies [139].

Many new unwrapping algorithms were developed recently. Their effectiveness and performance were analyzed in several reviews [27, 141–143]. However, since our initial research on the development of new PSI and DSPI methods dates back to the beginning of the twenty-first century [75, 76, 80, 86, 123], the research team of the Karpenko Physico-Mechanical Institute of the National Academy of Sciences of Ukraine (PMI) decided to use well-known phase unwrapping algorithms discussed by Ghiglia and Pritt [32]. Verification of such nine algorithms, which software is given in [32], was performed on a phase map test model containing two slope planes tilted in opposite directions and separated by a gap equal to 20 rad [78]. In Fig. 2.2, the test models of the wrapped and unwrapped phase maps are depicted. The first line (see Fig. 2.2a) shows: (i) the unwrapped phase map of the test model with a size of 258 × 258 pixels in the range of 256 brightness gradations (first column), (ii) the same phase map in 3D axonometry (second column); (iii) the phase map cut along the A-B profile. The second line shows the wrapped phase map of the test model, and the third line shows the same wrapped phase map with the noise value of 10% relative to the signal value.

Results of the test phase maps retrieval using Flynn's method of minimum weighted sum of discontinuities [31] are shown in Fig. 2.3. A software realizing this method (software Flynn [32]) completely implemented the unwrapping procedure of both parts of the test phase map. For a phase map with noise (see Fig. 2.3b), the program generated a phase map on which traces of a phase discontinuities can be found.

Analysis of the test model unwrapped phase maps retrieved by nine algorithms has shown that the method of residues and branch points (software Gold) [32, 34],

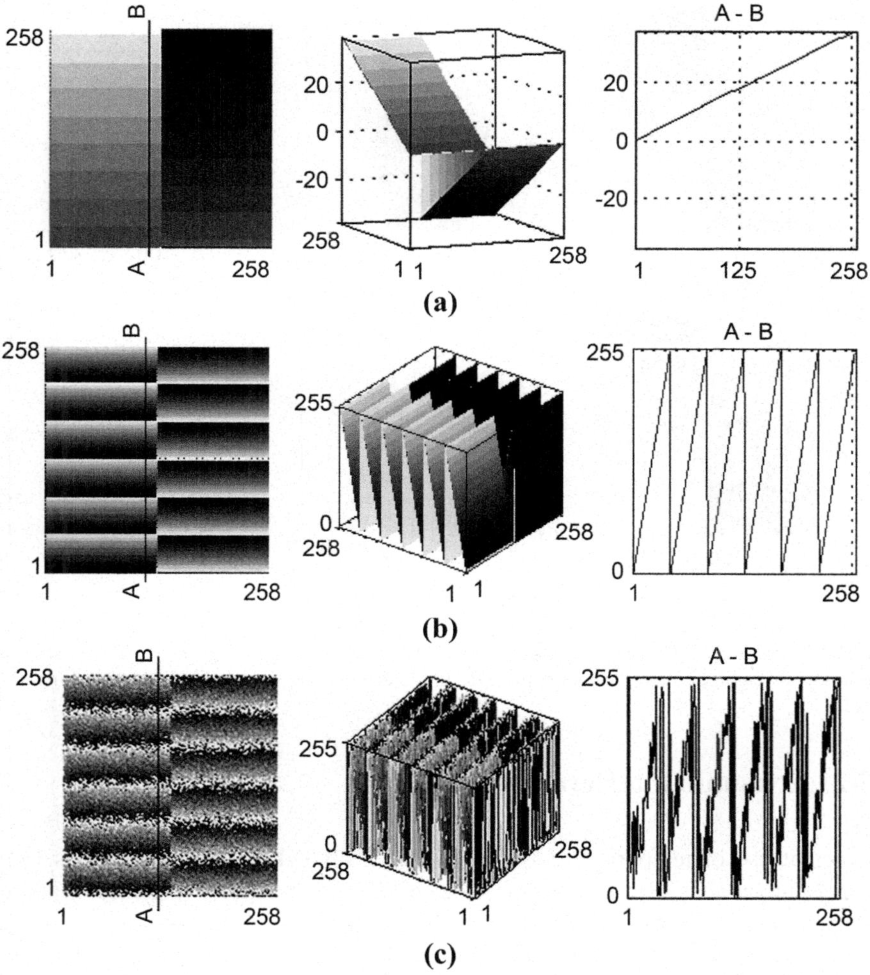

Fig. 2.2 Test models of phase maps: (**a**) unwrapped phase map; (**b**) wrapped phase map; (**c**) wrapped phase map with 10% noise

Flynn's method of minimum weighted sum of discontinuities (software Flynn) [31, 32], and the minimum *Lp*-norm method (software Lpno) [32, 33] possess the most effectiveness and best performance.

These methods reduce the level of random noise, automatically recognize isolated fragments of the phase map and perform fragment-by-fragment unwrapping with the assignment of each fragment of its base level. It has also been shown that the method of residues and branch points [34] should be used for the 10% noise level, as it is the fastest and allows tracing of phase breaks. Two other methods should be used for phase maps if their noise level is up to 40% relative to the signal level.

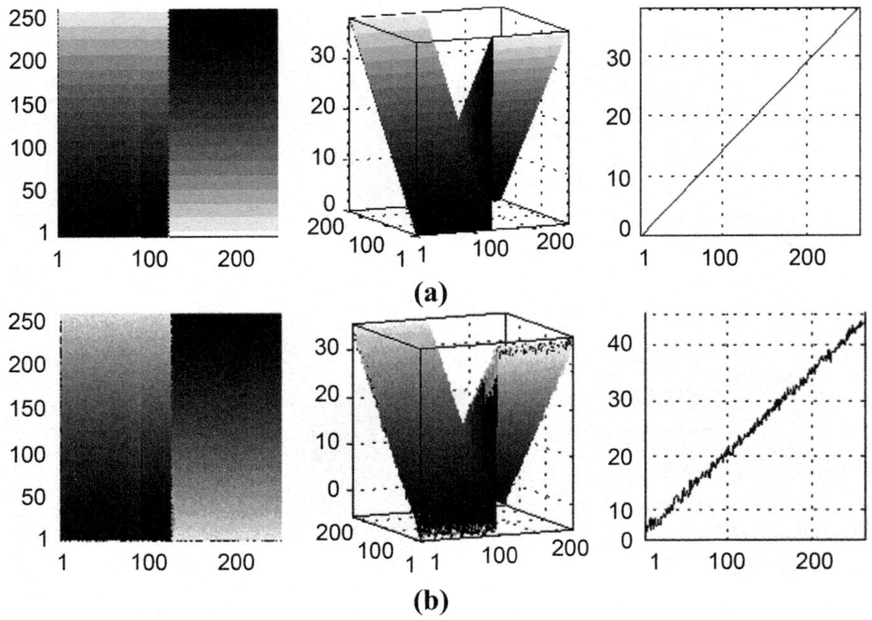

Fig. 2.3 Producing the unwrapped phase map of the test phase field using Flynn's method of minimum weighted sum of discontinuities: (**a**) phase map without noise; (**b**) phase map containing noise

2.2 Dynamic Interferometry

Efficient use of temporal PSI is possible, if a studied object or surface is fixed concerning a digital camera or other device recording temporal sequence of phase-shifted interferograms. Temporal single-wavelength PSI is used to study a surface relief that is stationary relative to the interferometric system recording the time sequence of phase-shifted interferograms. Unlike temporal PSI, in dynamic interferometry, either several interferograms with a phase shift are recorded simultaneously; either one basic interferogram is recorded with subsequent synthesis of several interferograms that differ only in phase shift, or with subsequent Fourier analysis and bandpass spatial filtering. The main advantage of dynamic interferometry over temporal PSI is the absence of strict requirements for the temporal stability of the interference pattern [8]. It is enough that this pattern does not change significantly during the exposure time, which in modern digital cameras can reach femtoseconds.

Among the known methods of dynamic PSI, there are single-out the spatial interferometry with carrier frequency [55, 90, 106], spatial synchronous detection [116, 128], spatial multichannel polarization interferometry methods [43, 73, 104, 108], and Fourier-transform method [7, 64, 112].

In the spatial interferometry with carrier frequency u_0, the basic interferogram of the surface area contains simultaneously several interferograms shifted relative to each other by a phase $\alpha_{j+1,j} = \alpha = 2\pi j/J$. This frequency is chosen so that it is inversely proportional to the integer number K of matrix pixel pitches p, that is $u_0 = (Kp)^{-1}$, and usually, $J = K$. Spatial synchronous detection is performed by multiplying the interferogram with carrier frequency u_0 and the reference plane wave incident on the recording photodetector matrix plane at an angle producing the same carrier frequency u_0. The methods of spatial multichannel polarization used separation of the basic interferogram on three or four phase-shifted interferograms with the help of optical polarizing elements.

Dynamic interferometry also includes methods for retrieval phase maps using the Fourier transform of recorded interferograms [7, 64, 107, 112]. These methods allow retrieving the surface relief phase map using only one interferogram. However, if only one registered interferogram is used for these methods, then uncertainty arises in terms of the "+" or "−" signs for the phase φ; that is, it is not known whether the surface is concave or convex. To remove the sign ambiguity, the additional interferogram shifted on the phase $\pi/3 < \alpha < 2\pi/3$ concerning the initial interferogram is used.

Because the performance of digital cameras increases, the quality and dimensionality of interferograms recorded by dynamic interferometry techniques are growing. However, their spatial resolution is below about one order than the resolution of temporal PSI techniques. Nevertheless, dynamic interferometers based on these techniques are widely used in NDT and TD of fast processes, as well as vibrating structural elements, machines and space-based optical systems [46, 72].

2.3 Practical Application of Temporal and Dynamic Phase Shifting Interferometry

Temporal PSI is widely used in NDT of materials and TD of structural and machine elements, equipment, etc. Temporal single-wavelength techniques are applied to control the roughness of optical surfaces of lenses, objectives and mirrors such as flats, spheres, and aspheres [3, 24, 41, 104, 131]. In experimental mechanics, materials science, and profilometry, these techniques are very attractive for surface relief precision retrieval [18, 46, 66]. Rougher surfaces can be tested by two- and three-wavelength PSI techniques. Two-wavelength PSI [13, 57] and wavelength-tuning PSI, in which phase shifts are implemented by a laser wavelength tuning [21], allow us to measure surfaces possessing larger roughness and multiple reflective surfaces. Meanwhile, these techniques make it possible to test and retrieve surfaces containing steps and discontinuities [57, 103].

Several PSI methods with tilt wavefront have been developed for testing optical surfaces with large numerical apertures. Liu et al. [61, 62] proposed a three-step iterative method for determining the wavefront phase and tilt phase shifts using three linear least-squares fitting steps. A vibration-insensitive iterative PSI method using iterative linear regression for fitting series of sequentially recorded interferograms to a physical model of the Fizeau interferometer and other spherical optical elements was developed by Deck [24]. This method uses robust iterative convergence to find the parameters of an optical spherical resonator. Usually, up to ten iterations are sufficient to provide a sub-nanometer error in determining the surface profile. The method is used to design all types of interferometers. Based on this method, antivibration QPSI™ technology has been developed [92], which is successfully used in ZYGO® Corporation [144]. A universal method for testing spherical surfaces using PBI with unknown phase displacements of an inclined wavefront was proposed by Zeng et al. [140]. This method is based on an improved physical model of the interferometer cavity, considering the rigid cavity motions and image distortions. The method can be used in the control of optical surfaces in the presence of low-frequency and high-amplitude vibrations.

Nowadays, a large number of commercial phase shifting interferometric devices and systems have been created. They are designed mainly to control the reliefs of optically smooth and nanorough surfaces of optical elements, crystals, devices, materials of micro- and optoelectronics, as well as microelectromechanical systems. Commercial interferometers using spatial phase shift technologies, i.e. dynamic interferometers, are insensitive to vibrations and other dynamic changes of test surfaces. Corporations and companies ZYGO®, 4D Technology, Graham Optical Systems, Engineering Synthesis Design, Inc., are among the leaders in the production of phase shifting interferometric systems. ZYGO® Corporation manufactured the VeriFireTM (XPZ, QPZ, ATZ) temporal phase shifting interferometers, as well as the DynaFizTM dynamic interferometer, operating in both temporal and spatial PSI modes [111, 144]. 4D Technology Corporation specializes in the production of temporal and spatial phase shifting interferometers for testing large optical elements [1]. In particular, dynamic laser interferometers of the PhaseCam series (4020, 4020HP, MW, NIR, IR, 5030, 6000) were manufactured using technologies for simultaneous four-channel registration of interferograms in a polarization interferometer. Graham Optical Systems Corporation has developed several series of Fizeau temporal phase shifting laser interferometers operating in a wide range of optical wavelengths (from ultraviolet to mid-infrared); in particular, two series of interferometers VPS (4 VPS and 6VPS) and HPS have been created in the visible range for $\lambda = 635$ nm [36]. The ESDI Corporation has created two series of temporal phase shifting and dynamic Fizeau interferometers Dimetior (VB, AS, PS) and Intellum (Z40, Z100) [26].

2.4 Two-Step Phase Shifting Interferometry with Unknown Phase Shift Between Interferograms

Conventional temporal PSI techniques are gradually being suppressed, for example, by faster and more efficient generalized PSI [9, 37, 58] and WTI [66, 103]. Phase interferometry with unknown phase shift between interferograms can be considered as a logical extension of the generalized PSI methods, in which the unknown phase shifts between interferograms are the same and can be extracted by least squares [9, 37]. However, two-step PSI techniques with blind phase shift are faster and have a simpler technical realization because only two interferograms are used for reconstruction of a studied object and no phase shift calibration is needed. Along with conventional retrieval algorithms, two-step algorithms that include extraction of an unknown phase shift between two recorded interferograms, have become also popular due to temporary reduction and simplification of fringe pattern phase and amplitude demodulation. In last decades, several two-step phase shifting techniques with unknown phase shift between two interferograms were proposed [20, 25, 29, 56, 63, 71, 76, 80, 98, 120–122, 127, 134–136]. Nevertheless, it is necessary to mention that some techniques were developed earlier, for example Kreis (KR) algorithm [54], simplified Lissajous ellipse figure (SLEF) algorithm [28], and phase-step calibration (PSC) algorithm [119] based on calculation of the population Pearson correlation coefficient (PPCC) [89].

Kreis and Juptner proposed the KR algorithm [54] based on the use of the Fourier transform to demodulate two interferograms with applying a quadrature width filter for low frequencies elimination, and also the sign ambiguity correction in closed interference fringes. Farrell and Player [28] developed a SLEF algorithm, which uses the Lissajous ellipse for pixelvise presentation of two interferograms intensities with continuous amplitudes and background. The robust version of this algorithm named the SLEF by robust estimator (SLEF-RE) was proposed by Flores and Rivera [29] in order to calculate phase shift using only two terms of the simplified conical equation obtained after interferograms normalization. The phase shift is determined using the terms of conic equation corresponding to the ellipse eccentricity. Van Brug [119] proposed the PSC algorithm to determine the unknown phase shift based on calculation of the PPCC $\rho[I_1(x, y), I_2(x, y)]$ between two interferograms $I_1(x, y)$ and $I_2(x, y)$, which differ only by phase shift. In this case, the phase shift is defined as the cosine of the correlation coefficient ρ. Based on mathematical modeling, it is also established that this algorithm can be used only for interferograms formed by plane waves and waves, the front of which is described by Cernike polynomials of the 5th and 13th orders. However, these conclusions are partial, refer only to the three simplest types of interferograms, and do not reflect the much broader possibilities of the correlation approach to the determining the unknown phase shift based on the Pearson correlation coefficient for both the population and the sample.

An iterative algorithm for determining an arbitrary unknown phase shift using two interferograms was proposed by Xu et al. [135]. In this least squares algorithm, the phase shift is determined with 1% random error for smooth test surfaces and 2% random error for low-rough test surfaces (standard deviation is $\lambda/30$) over 10–20 iterations. It was shown that the image of an object can be reconstructed with satisfactory accuracy in the range of phase shifts from 0.4 to 2.5 rad, and the best result was obtained for angles close to 1 rad. A similar algorithm that does not require an iterative procedure has been proposed in [136]. It is simpler than the previous one and consists in averaging the intensity distributions of both interferograms and the object and reference rays. Software verification of the test object proved the effectiveness of this algorithm. In a real experiment, the authors used only the USAF flat test measure to determine the resolution, and the recovered images were further improved by the zero-leveling method. The two-step generalized phase shifting interferometry (GPSI) with blind phase shift extraction algorithm was developed by Meng et al. [71]. To retrieve the blind phase shift by this algorithm, it is necessary to record two interferograms, as well as the object and reference wavefronts. This algorithm provides a higher resolution of restored flat test objects compared to [135, 136]. However, it is also applicable only to noiseless interferograms.

Three two-step PSI methods with unknown phase shift, namely self-tuning (ST) phase shifting interferometry, regularized optical flow (ROF) and Gram–Schmidt orthonormalization (GSO), were developed by Vargas et al. [120–122]. The authors conducted a comparative analysis with similar phase shifting methods and showed the advantages of the proposed approaches. However, the recommended general approach to filtering the diffraction center terms in the interferograms spectra seems to be insufficiently reliable in cases where only two interferograms are used. In the ST method, an algorithm for self-tuning of two interferograms restores the phase shift between interferograms by determining the minimum of the merit function, that is, a function that measures the consistency between the data and the fitting model for a certain choice of parameters [120]. In the ROF method, a regularized optical flow algorithm allows to calculate the fringe direction [121]. This algorithm is unable to calculate the phase shift equal to π. The GSO method [122] uses an algorithm that implements the Gram–Schmidt orthonormalization process of two vectors for demodulation. This algorithm also cannot calculate the phase shift equal to π. Unfortunately, these three methods are not valid for some classes of microreliefs. Standard deviations, calculated for test objects without considering the procedure of wrapped phase maps unwrapping, equal to 0.21 rad using the GSO algorithm and 0.32 rad using ST algorithm.

Another two-step PSI method with an arbitrary phase shift α was proposed by Deng et al. [25]. In this extreme value of interference (EVI) method, two sets ($k = 1, 2$) of pixels $p_1, p_2, \ldots, p_i, \ldots, p_P$ and $v_1, v_2, \ldots, v_j, \ldots, v_V$ were selected with maximum I_{k,p_i} and minimum I_{k,v_j} intensity values from two recorded interferograms that differed from each other only in phase shift α. These pixels were searched for the angle using the next equation:

$$\alpha = \frac{\sum_{i=1}^{P} \arccos\left(\frac{I_{2,p_i}}{I_{1,p_i}}\right) + \sum_{j=1}^{V} \arccos\left(\frac{I_{2,v_j}}{I_{1,v_j}}\right)}{P + V}, \tag{2.15}$$

where $P = \sum_{i=1}^{P} p_i$, $V = \sum_{j=1}^{V} v_j$.

To select the appropriate pixels $p_1, p_2, \ldots, p_i, \ldots, p_P$ and $v_1, v_2, \ldots, v_j, \ldots, v_V$ from two interferograms, it is necessary to perform additional operations of their selection from the entire set of pixels using additional software. This method gives small errors in determining the angle α only in the range from 0.5 to 2 rad.

Wielgus et al. developed the tilt-shift error estimation (TSE) algorithm [127], which can perform phase demodulation of two interferograms differing from each other by an unknown linearly inhomogeneous phase shift. This algorithm considers two interferograms in which the background is removed, but the tilt-shift error is present. First, this algorithm extracts the tilt-shift component using the intensity measure. Further, the global-error functional minimization procedure is used to extract the unknown phase shift between interferograms. Finally, an additional minimization of the tilt-shift error is performed.

The diamond diagonal vectors (DDVs) algorithm, proposed by Luo et al. [63], assumes that the norms of the two filtered interferograms are almost the same; that is, they are adjacent sides of the rhomb. Its perpendicular diagonals are considered as the sum and difference of the vectors that determine the phase shift.

Dalmau et al. [20, 98] developed a random points (RP) estimation algorithm and an iterative robust estimator (IRE) algorithm. The RP algorithm chooses pairs of randomly selected pixels on a pair of fringe patterns and calculates the phase shift from the pixels intensities [20]. The IRE algorithm is used as part of a Gabor filter bank (GFB) algorithm [98], which extracts the phase shift between interferograms. At the same time, the problem of sign ambiguity is solved by the IRE algorithm.

Saide et al. [100] and Flores et al. [30] analyzed most of the works discussed above and showed that in order to calculate the unknown phase shift between interferograms (fringe patterns); first it is desirable to perform the fringe normalization. Such the procedure allows to suppress the errors induced by spatiotemporal noise and intensity background.

If we consider two interferograms I_1 and I_2 that differ not only in the arbitrary phase shift $\alpha_{21} = \alpha$ of one of the two beams (reference or object), but also in different values of their intensities, then the intensity distributions of these interferograms are written as two fundamental interferometry equations [see Eq. (2.5)]

$$\left.\begin{array}{l} I_1(x, y) = I_{B1}(x, y) + I_{M1}(x, y)\cos[\varphi(x, y)] + n_1(x, y) \\ I_2(x, y) = I_{B2}(x, y) + I_{M2}(x, y)\cos[\varphi(x, y) + \alpha] + n_2(x, y) \end{array}\right\}, \tag{2.16}$$

where $I_{B1,2}(x, y) = I_{o1,2}(x, y)+I_{r1,2}(x, y)$, $I_{M1,2}(x, y) = 2\sqrt{I_{o1,2}(x, y)I_{r1,2}(x, y)}$, φ is the phase to be recovered, α is the unknown (arbitrary) phase shift, $n_{1,2}$ is the noise.

If time gap between two interferograms recording is short and there are no high-frequency vibrations or other dynamic external influences affecting a test object, one can assume that the intensity distributions of the object and reference beams do not change, that is, $I_{B1} = I_{B1} = I_B$ and $I_{M1} = I_{M1} = I_M$. In this case, Eq. (2.16) is rewritten as

$$\left.\begin{array}{l} I_1(x, y) = I_B(x, y) + I_M(x, y)\cos[\varphi(x, y)] \\ I_2(x, y) = I_B(x, y) + I_M(x, y)\cos[\varphi(x, y) + \alpha] \end{array}\right\}. \tag{2.17}$$

Fringe normalization process consists in eliminating the background intensity $I_B(x, y)$, normalizing contrast variations (i.e. normalizing intensity modulation $I_M(x, y)$), and, if necessary, suppressing spatial noise [29, 100]. After normalization, Eq. (2.17) can ideally be represented as

$$\left.\begin{aligned}\hat{I}_1(x, y) &= \cos[\varphi(x, y)] \\ \hat{I}_2(x, y) &= \cos[\varphi(x, y) + \alpha]\end{aligned}\right\}. \tag{2.18}$$

We assume that normalization of modulation suppresses and distorts high frequencies, which ultimately leads to an increase in errors in the restored surface microrelief of real test objects. Therefore, normalization was used only to remove the background intensity. After such normalization, Eq. (2.17) is given by

$$\left.\begin{aligned}\tilde{I}_1(x, y) &= I_M(x, y)\cos[\varphi(x, y)] \\ \tilde{I}_2(x, y) &= I_M(x, y)\cos[\varphi(x, y) + \alpha]\end{aligned}\right\}, \tag{2.19}$$

where $\tilde{I}_1 = I_1 - I_B$, $\tilde{I}_2 = I_2 - I_B$.

After fringe normalization and filtering, the unknown phase shift is calculated using the two-step algorithms discussed above. The sought unwrapped phase map φ can be calculated with the help of the relation proposed by Muravsky et al. [80] (see also [30])

$$\varphi(x, y) = \arctan\left[\frac{\hat{I}_1(x, y)\cos(\alpha) - \hat{I}_2(x, y)}{\hat{I}_1(x, y)\sin(\alpha)}\right], \tag{2.20}$$

if relationships (2.18) are used, and with the help of Eq.

$$\varphi(x, y) = \arctan\left[\frac{\tilde{I}_1(x, y)\cos(\alpha) - \tilde{I}_2(x, y)}{\tilde{I}_1(x, y)\sin(\alpha)}\right], \tag{2.21}$$

if expressions (2.19) are used.

Saide et al. [100] have analyzed the performance of six two-step algorithms, namely EVI [25], KR [54], ST [120], ROF [121], GSO [122] and TSE [127]. To reduce phase and amplitude demodulation errors caused by noise, uneven spatial amplitude modulation, and background intensity, as well as to improve the quality of phase map retrieval, the following algorithms for filtering and normalizing interference patterns were used:

- Gaussian filtering algorithm;
- Enhanced fast empirical mode decomposition (EFEMD) algorithm [118];
- EFEMD algorithm and mutual information detrending (MID) algorithm [117];
- Automated noise filtering method based on the 2D empirical mode decomposition algorithm [138] (Auto EFEMD algorithm [100]);
- Fringe pattern normalization by Hilbert vortex transform (HVT) algorithm [60];

- Quiroga algorithm for fringe pattern normalization [94].

Saide et al. [100] have studied the effect of the phase step between a pair of simulated fringe patterns on the phase demodulation results and showed that the root mean square (RMS) value of retrieved phase maps is stable in the phase shift range $[0, 2\pi]$ for most of the six selected algorithms, while the EVI, TSE and GSO have the lowest RMS value if the random noise is absent on fringe patterns. Moreover, for these algorithms the RMS remains almost unchanged when changing the phase shift within these limits. The results of phase demodulation of the simulated fringe patterns corrupted by additive random noise that is suppressed by the Auto EFEMD algorithm have shown the smallest errors for the RMS noise in the range from 0 to 2 rad are achieved using the same EVI, TSE and GSO algorithms. Testing of background and modulation influence on simulated fringe patterns proved the TSE preference. The best accuracy for simulated data without preprocessing for amplitude demodulation also was achieved by the TSE algorithm. However, the best results for filtered data were obtained by the ROF algorithm. The experimental data results show the most precise phase demodulation is provided by the TSE algorithm, which, however, is the most time-consuming. The best amplitude demodulation of experimental data provided the EVI method with EFEMD/Gaussian filtering.

Flores et al. [30] performed a detailed comparative analysis of algorithms for extraction of the unknown phase shift between two fringe patterns (i.e. interferograms) and Hilbert–Huang transform (HHT) [118], GFB [98], and deep neural network (DNN) [97] algorithms for their normalization. Except considered above six two-step algorithms (EVI, KR, ST, ROF, GSO and TSE), they analyzed also PSC [119], RP [20], SLEF–RE [29], DDV [63], GPSI [71] and IRE [98] algorithms, as well as quadratic phase parameter (QPP) estimator algorithm [56], in which the phase shift is calculated using a quadratic phase approximation in a small window of prefiltered interferograms.

Results of a comparative analysis of these 12 algorithms for extraction of the phase shift angle between two fringe patterns (interferograms) with different levels of the introduced Gaussian noise $\sigma = 0–1.0$ rad, normalized using the GFB algorithm, showed good stability of six phase shift algorithms (EVI, KR, DDV, PSC, GPSI and GSO) by the level of mean absolute error (MAE) of phase shift estimation. The use of DNN normalization also showed that the lowest MAE for the same noise levels is achieved using the TSE, KR, DDV and PSC algorithms. When using HHT normalization, the lowest MAE for $\sigma = 0–1.0$ rad was achieved by TSE, KR, DDV, PSC and GPSI algorithms for noise levels $\sigma > 0.5–1.0$ rad.

As we can see, the PSC algorithm is one of the most efficient for extracting the unknown phase shift. This algorithm is of particular interest, since the two- and three-step PSI methods with unknown phase shifts between interferograms considered below use its variants and modifications. As mentioned above, the algorithm is based on calculation the PPCC between two non-normalized interferograms, that is [89]

$$\rho(I_1, I_2) = \frac{\langle (I_1 - \langle I_1 \rangle)(I_2 - \langle I_2 \rangle) \rangle}{\sigma_{I1}\sigma_{I2}}, \tag{2.22}$$

and the desired arbitrary phase shift α is defined as

$$\alpha = \arccos[\rho(I_1, I_2)] = \arccos \frac{\langle (I_1 - \langle I_1 \rangle)(I_2 - \langle I_2 \rangle) \rangle}{\sigma_{I1}\sigma_{I2}}, \tag{2.23}$$

where $\rho(I_1, I_2)$ is the PPCC; $\langle I_1 \rangle$, $\langle I_2 \rangle$ are the average values of interferogram intensities spatial distributions; σ_{I1}, σ_{I2} are the standard deviations of I_1 and I_2.

For normalized fringe patterns represented by Eq. (2.19), the arbitrary phase shift is given by

$$\tilde{\alpha} = \arccos\left[\rho\left(\tilde{I}_1, \tilde{I}_2\right)\right] = \arccos \frac{\left\langle \left(\tilde{I}_1 - \left\langle \tilde{I}_1 \right\rangle\right)\left(\tilde{I}_2 - \left\langle \tilde{I}_2 \right\rangle\right) \right\rangle}{\tilde{\sigma}_{\tilde{I}1}\tilde{\sigma}_{\tilde{I}2}}, \tag{2.24}$$

where $\tilde{\sigma}_{\tilde{I}1}$, $\tilde{\sigma}_{\tilde{I}2}$ are the standard deviations of $\tilde{I}_1$ and $\tilde{I}_2$.

The PPCC in formulas (2.23) and (2.24) can be interpreted as a normalized dot product of two centered multidimensional vectors [4], which is treated as a cosine of phase shift between these vectors [99], i.e. between fringe patterns [76, 77]. Thus, this correlation approach to calculating the unknown phase shift is based on interframe correlation between two interferograms using the PPCC.

Note that the similar correlation approach, most likely, was first proposed by Bobkov and Molodov for extraction of the phase shift between input temporal signals in correlation rotational coordinate selective phase meters [5, 6]. Van Brug later proposed the PSC algorithm to extract the unknown phase shift between two straight equidistant fringe interferograms with no spatial noise [119]. He proved that the more fringes in two interferograms, the smaller the error of the phase shift extraction, and the largest error is when only one fringe or half of the fringe is present. For nonequidistant fringes, this error also increases sharply. However, Muravsky et al. showed that such an algorithm can extract the phase shift between any two interferograms with sufficient accuracy for engineering calculations [76, 78, 80], which was eventually shown by Flores et al. [30]. Moreover, it was shown that this algorithm is suitable for determining the unknown phase shift between two speckle fringe patterns or between two fringe patterns in which additive phase Gaussian noise with RMS error $\sigma_\varphi \leq 2\pi$ is introduced [77]. Without denying the priority of the Bobkov–Molodov algorithm [5, 6] and the PSC algorithm [119] and taking into account its modifications [76, 80], we will call the procedure for calculating the unknown phase shift between interferograms the population Pearson correlation algorithm (PPCA). The suitability of the PPCC as well as the sample Pearson correlation coefficient (SPCC) for extracting the phase shift between speckle fringe patterns (speckle interferograms) will be shown in Chaps. 4 and 5.

2.5 Retrieval of Smooth and Nanorough Surfaces Using Two-Step Phase Shifting Interferometry Methods

Modern methods of engineering surfaces analysis are based on the selection of their components using spatial frequency characteristics. There are high-frequency components of profiles and areals (three-dimensional surface areas) of the surface texture, called roughness, low-frequency components, called form, and mid-frequency components, called waviness [48, 51]. Each component of the profile or areal is characterized by standardized ISO parameters. The particular importance for the analysis of nanorough surfaces profiles height parameters has the arithmetic mean deviation of the roughness profile [48]

$$Ra = \frac{1}{l}\int_0^l |z(x)|\mathrm{d}x, \tag{2.25}$$

and RMS deviation of the assessed profile

$$Rq = \sqrt{\frac{1}{l}\int_0^l z^2(x)\mathrm{d}x}, \tag{2.26}$$

where $z(x)$ is the surface height values within a sampling length l.

The areal height parameters Sa (arithmetical mean height of the scale-limited surface) and Sq (root mean square height of the scale-limited surface), which are similar to Ra and Rq, are given as [51]

$$Sa = \frac{1}{A}\iint_A |z(x, y)|\mathrm{d}x\mathrm{d}y, \tag{2.27}$$

$$Sq = \sqrt{\frac{1}{A}\iint_A z^2(x, y)\mathrm{d}x\mathrm{d}y}, \tag{2.28}$$

where A is the definition area.

Considering the relief of a nanorough quasi-flat surface, we will distinguish only two surface texture components of spatial frequency features: high-frequency, which will be called "roughness" or "nanorelief", and low-frequency, covering the form and waviness, which will be called "macrorelief". The name "nanorelief" is justified by the fact that the roughness height parameters Ra and Rq, as well as Sa and Sq of such surfaces, which are illuminated by coherent light in the visible wavelength range, do not exceed several tens of nanometers. To the macrorelief, one can add the slope of the quasi-flat surface at a small angle ω relative to the plane that is

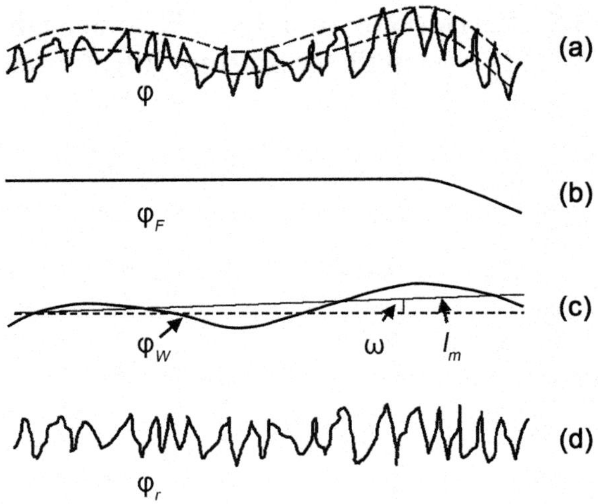

Fig. 2.4 Phase profiles of the nanorough surface relief φ (**a**) and its components, namely form φ_F (**b**), waviness φ_W, plane slop (**c**) and roughness (**d**); l_m is the middle line of the waviness profile, ω is the angle of the plane slope

normal to the direction of observation in the interferometer. This slope may occur when recording fringe patterns in the interferometer due to imperfect alignment of the test surface perpendicular to the direction of observation. If necessary, the slope of the surface can be removed from the macrorelief, highlighting the waviness and form. Schematically, the phase profile of the nanorough surface and its components, namely the phase profiles of macrorelief and nanorelief (roughness) are shown in Fig. 2.4.

As already mentioned in Sect. 2.3, conventional methods of single-wavelength PSI with known equal phase shifts between interferograms are successfully used to control the surface roughness of optical elements [3, 24, 41, 104, 131] and to restore the nanorough surface relief for problems of experimental mechanics and materials science [18, 46, 66]. Two-wavelength, multiwavelength and wavelength-tuning PSI methods make it possible to restore the surface relief with much greater roughness, as well as sharp differences and jumps in some surface areas [13, 21, 57, 103]. Single-wavelength two-step PSI methods with arbitrary phase shift are also used to control nanorough surfaces of materials. First, let's note the works of Muravsky et al. [75, 76, 78–81] and Ostash et al. [86], in which the relief of the nanorough surface of metal and alloy compact tension specimens (CT-specimens) with round notches was retrieved and the spatial distributions of roughness and macrorelief were estimated. Tian and Liu proposed a method for demodulation of two-shot fringe patterns with random phase shifts [114] based on the Gram–Schmidt orthonormalization process to retrieve of optical spherical and parabolic surfaces [115]. However, it was used only for surfaces whose form is either unchanged (flat surfaces) or changes monotonously and smoothly (spherical and parabolic surfaces).

Muravsky et al. [76, 78] developed the method of two-step interferometry with unknown phase shift (TSI UPS) of a reference beam allowing retrieving the phase map of the total surface relief area as well as the surface roughness and macrorelief phase maps. To this end, the phase spatial distribution of the nanorough surface, i.e.

the phase map $\varphi(k, l)$ of the surface relief area, is considered as the pixel-by-pixel summation of phase roughness (nanorelief) $\varphi_r(k, l)$ and phase macrorelief $\varphi_m(k, l)$ distributions, i.e.

$$\varphi(k, l) = \varphi_r(k, l) + \varphi_m(k, l). \tag{2.29}$$

Note, the macrorelief phase map $\varphi_m(k, l)$ consists of a form phase map $\varphi_F(k, l)$, waviness phase map $\varphi_W(k, l)$ and the slope of the quasi-flat surface at a small angle ω.

Taking into account (2.29), Eqs. (2.17) and (2.19) rewrite as follows:

$$\left.\begin{array}{l} I_1(k, l) = I_B(k, l) + I_M(k, l)\cos[\varphi_r(k, l) + \varphi_m(k, l)] \\ I_2(k, l) = I_B(k, l) + I_M(k, l)\cos[\varphi_r(k, l) + \varphi_m(k, l) + \alpha] \end{array}\right\}; \tag{2.30}$$

$$\left.\begin{array}{l} \tilde{I}_1(k, l) = I_M(k, l)\cos[\varphi_r(k, l) + \varphi_m(k, l)] \\ \tilde{I}_2(k, l) = I_M(k, l)\cos[\varphi_r(k, l) + \varphi_m(k, l) + \alpha] \end{array}\right\}. \tag{2.31}$$

2.5.1 *Method of Two-Step Interferometry with Unknown Phase Shift of a Reference Beam for Retrieving the Nanorough Surface*

The TSI UPS method was developed [76, 78] for retrieving a smooth or low-roughness surface relief of specimens or structural elements. In this method, in addition to recording two interferograms $I_1(x, y)$ and $I_2(x, y)$ described by Eq. (2.30), the object and reference wavefronts $I_o(x, y)$ and $I_r(x, y)$ should also be recorded. The normalized interferograms $\tilde{I}_1(x, y)$ and $\tilde{I}_2(x, y)$ [see Eq. (2.31)] are obtained by pixel-by-pixel subtraction of the background intensity distribution $I_B(x, y)$ from interferograms $I_1(x, y)$ and $I_2(x, y)$. To find the wrapped phase map $\varphi(k, l)$ in two quadrants, i.e. in the range $-\frac{\pi}{2} \le \varphi(k, l) \le \frac{\pi}{2}$, one can perform elementary trigonometric transformations in Eq. (2.31) and make the ratio

$$\frac{\tilde{I}_2(k, l)}{\tilde{I}_1(k, l)} = \cos\alpha - \mathrm{tg}[\varphi_r(k, l) + \varphi_m(k, l)]\sin\alpha, \tag{2.32}$$

from which the wrapped phase map $\varphi_w(k, l)$ of the total surface area is given by [76]

$$\varphi_w(k, l) = \varphi_r(k, l) + \tilde{\varphi}_m(k, l) = \arctan\left(\frac{\tilde{I}_1(k, l)\cos\alpha - \tilde{I}_2(k, l)}{\tilde{I}_1(k, l)\sin\alpha}\right), \tag{2.33}$$

where $\tilde{I}_{1,2}(k, l) = I_{1,2}(k, l) - I_B(k, l)$, $\tilde{\varphi}_m(k, l)$ is the wrapped macrorelief phase map.

Retrieval of a wrapped phase map $\varphi_w(k,l)$ in four quadrants, i.e. in the range $-\pi \le \varphi_n(k,l) + \tilde{\varphi}_m(k,l) \le \pi$, is performed using the "atan2" function [11] by converting Eq. (2.33) to the form

$$\begin{aligned}\varphi_w(k,l) &= \varphi_n(k,l) + \tilde{\varphi}_m(k,l) \\ &= \text{atan2}\Big[\tilde{I}_1(k,l)\cos\alpha - \tilde{I}_2(k,l), \tilde{I}_1(k,l)\sin\alpha\Big],\end{aligned} \tag{2.34}$$

where

$$\text{atan2}(y,x) = \begin{cases} \arctan\left(\frac{y}{x}\right) & \text{if } x > 0 \\ \arctan\left(\frac{y}{x}\right) + \pi & \text{if } x < 0 \text{ and } y \ge 0 \\ \arctan\left(\frac{y}{x}\right) - \pi & \text{if } x < 0 \text{ and } y < 0 \\ \frac{\pi}{2} & \text{if } x = 0 \text{ and } y > 0 \\ -\frac{\pi}{2} & \text{if } x = 0 \text{ and } y < 0 \\ \text{undefined} & \text{if } x = 0 \text{ and } y = 0. \end{cases}$$

To produce the macrorelief wrapped phase map $\tilde{\varphi}_m(k,l)$, only harmonic terms of the wrapped phase map $\varphi_w(k,l)$ that defined by Eq. (2.34) are extracted. To this end, the low-frequency spatial filtering of interferograms harmonic terms

$$I_{s0} = \sin\Big\{\text{atan2}\Big[\tilde{I}_1(k,l)\cos\alpha - \tilde{I}_2(k,l), \tilde{I}_1(k,l)\sin\alpha\Big]\Big\}, \tag{2.35}$$

$$I_{c0} = \cos\Big\{\text{atan2}\Big[\tilde{I}_1(k,l)\cos\alpha - \tilde{I}_2(k,l), \tilde{I}_1(k,l)\sin\alpha\Big]\Big\} \tag{2.36}$$

is fulfilled using well-known average sine/cosine method [2] and direct and inverse DFTs, that is

$$\begin{aligned}\sin\big[\tilde{\varphi}_m(k,l)\big] &\approx I_s\big[\tilde{\varphi}_m(k,l)\big] \\ &= \text{DFT}^{-1}\big\{\text{DFT}\big[\sin(\varphi_n(k,l) + \tilde{\varphi}_m(k,l))\big] \times LPCF\big\},\end{aligned} \tag{2.37}$$

$$\begin{aligned}\cos\big[\tilde{\varphi}_m(k,l)\big] &\approx I_c\big[\tilde{\varphi}_m(k,l)\big] \\ &= \text{DFT}^{-1}\big\{\text{DFT}\big[\cos(\varphi_n(k,l) + \tilde{\varphi}_m(k,l))\big] \times LPCF\big\},\end{aligned} \tag{2.38}$$

where *LPCF* is the low-pass cut-off filter.

Usually, the cut-off frequency f_c of this filter is chosen such that there is an inverse relationship between f_c and cut-off wavelength λ_c, that is $f_c = (\lambda_c)^{-1}$, where λ_c is the branch point between the surface roughness and waviness components and its values for relevant surface roughness parameters are given in [51]. As a result, the macrorelief wrapped phase map $\tilde{\varphi}_m(k,l)$ can be expressed as

$$\tilde{\varphi}_m(k,l) = \text{atan2}\big\{I_s\big[\tilde{\varphi}_m(k,l)\big], I_c\big[\tilde{\varphi}_m(k,l)\big]\big\}. \tag{2.39}$$

The whole procedure of low-pass filtering is shown in Fig. 2.5. In order to simplify the flow-charts of algorithms implementing two-step PSI methods, we will further replace the specified block diagram of low-frequency filtering with the process "Extraction $\tilde{\varphi}_m(k, l)$". After unwrapping, the wrapped phase map $\tilde{\varphi}_m(k, l)$ is transformed into the phase macrorelief of the test surface area $\varphi_m(k, l)$. The macrorelief height map $h_m(k, l)$ of the test surface area is given by

$$h_m(x, y) = \frac{\lambda \cdot \varphi_m(x, y)}{4\pi}, \tag{2.40}$$

if the incident and reflected beams in an interferometer are directed normally to the surface of the test object [see also (2.2)].

The surface nanorelief (roughness) phase map $\tilde{\varphi}_r(k, l)$ was defined as the difference between the wrapped phase map $\varphi_w(k, l)$ of the total surface relief, expressed by Eq. (2.34), and the wrapped phase map $\tilde{\varphi}_m(k, l)$ of the surface macrorelief, expressed by Eq. (2.39), that is

$$\begin{aligned} \tilde{\varphi}_r(k, l) &= \varphi_w(k, l) - \tilde{\varphi}_m(k, l) \\ &= \operatorname{atan2}\left[\tilde{I}_1(k, l)\cos\alpha - \tilde{I}_2(k, l), \tilde{I}_1(k, l)\sin\alpha\right] \\ &\quad - \operatorname{atan2}\left\{I_s\left[\tilde{\varphi}_m(k, l)\right], I_c\left[\tilde{\varphi}_m(k, l)\right]\right\}. \end{aligned} \tag{2.41}$$

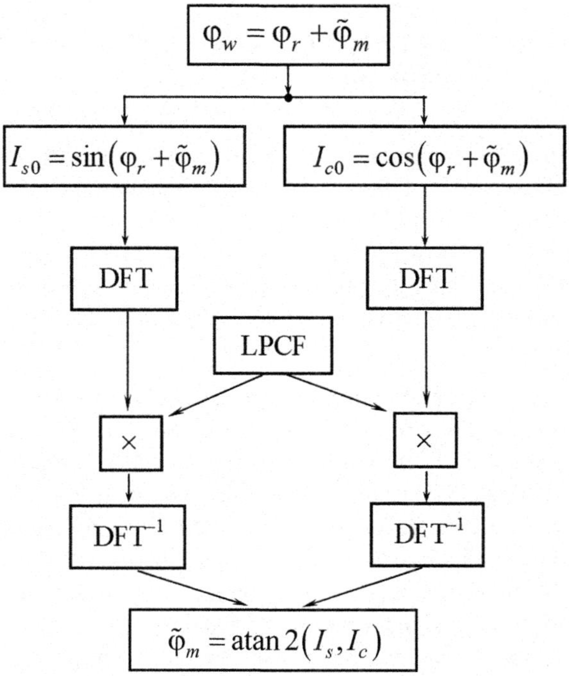

Fig. 2.5 Flow-chart of the process "Extraction $\tilde{\varphi}_m(k, l)$": LPCF is the low-pass cut-off filter; DFT is the digital Fourier transform; DFT^{-1} is the inverse digital Fourier transform

However, the second term of Eq. (2.41) is a cyclic function that has singularities. Therefore, the phase map $\tilde{\varphi}_r(k,l)$ differs significantly from the searched map $\varphi_r(k,l)$. To remove the singularities located on the map $\tilde{\varphi}_r(k,l)$ in loci with coordinates 2π and -2π, sine and cosine functions were used. Indeed, operations $I'_s[\varphi_r(k,l)] = \sin[\tilde{\varphi}_r(k,l)]$ and $I'_c[\varphi_r(k,l)] = \cos[\tilde{\varphi}_r(k,l)]$ remove singularities on the map $\tilde{\varphi}_r(k,l)$, since $\sin(2\pi) = \sin(-2\pi) = \sin(0)$ and $\cos(2\pi) = \cos(-2\pi) = \cos(0)$ [76]. Then the searched nanorelief phase map can be expressed as

$$\varphi_r(k,l) = \text{atan2}\left[I'_s(\varphi_r(k,l)), I'_c(\varphi_r(k,l))\right]. \tag{2.42}$$

The height map $h_r(k,l)$ of the studied surface nanorelief is obtained after the transition from the phase map $\varphi_r(k,l)$ to the metric one, using a ratio similar to (2.40), i.e.

$$h_r(k,l) = \frac{\varphi_r(k,l) \times \lambda}{4\pi}. \tag{2.43}$$

Since in the height map $h_r(k,l)$ the height discontinuities between adjacent samples are lesser than $\lambda/4$, there is no need to unwrap the obtained phase map $\varphi_r(k,l)$.

Flow-chart of the TSI UPS method is shown in Fig. 2.6.

2.5.2 *Assessment of Errors of Unknown Phase Shift Extraction*

In order to evaluate the effectiveness of the PPCA for definition of the unknown phase shift α of the reference beam, absolute and random errors of α extraction were calculated on test computer models of smooth and nanorough surfaces in the phase shift from 0 to 180°. For this purpose, Eq. (2.23) was used as a basis of the PPCA. First, errors of the phase shift α extraction were evaluated for optically smooth test surfaces, and further nanorough surfaces, which were created by introducing normal additive noise, were analyzed.

Absolute errors of unknown phase shift extraction for smooth surfaces. To evaluate the absolute errors of unknown phase shift extraction for smooth surfaces, a test bipolar phase object shown in Fig. 2.7 was generated by computer function "3 peaks 500" (in GNU Octave). This object will be further considered as a test bipolar phase surface. Interferograms $I_1(k,l)$ and $I_2(k,l)$ given by Eq. (2.17) were synthesized for the test bipolar surface. In this case, the flat reference beam $I_r(k,l)$ coincided with the object one $I_o(k,l)$ in the direction of propagation, and the interference patterns were focused in the recording plane.

To estimate the errors over the entire range of phase shifts, the given angle α_1 was chosen from 0 to π with a step of $\pi/18$. These values of α_1 were entered into

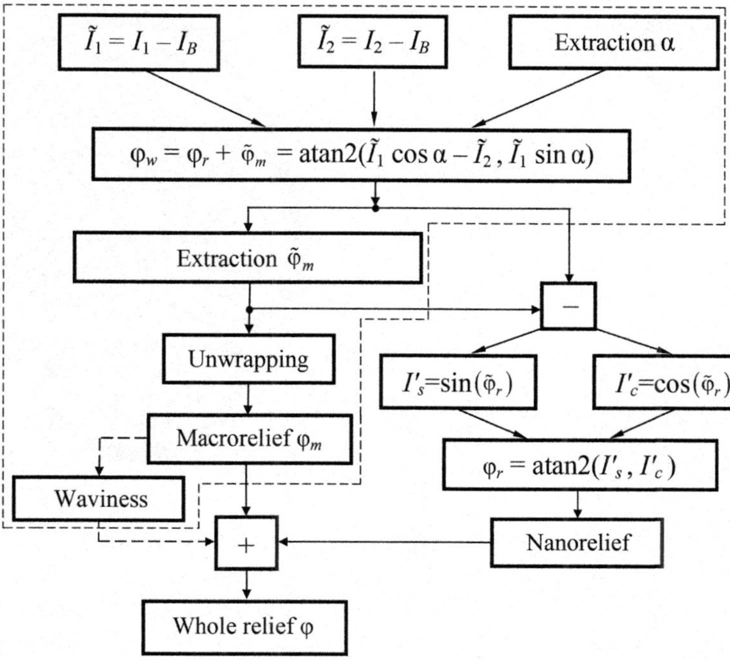

Fig. 2.6 Flow-chart of the TSI UPS method. The flow-chart part, surrounded by a dashed contour line, performs the retrieval of only the wrapped phase map of the surface macrorelief

Fig. 2.7 Test bipolar phase object

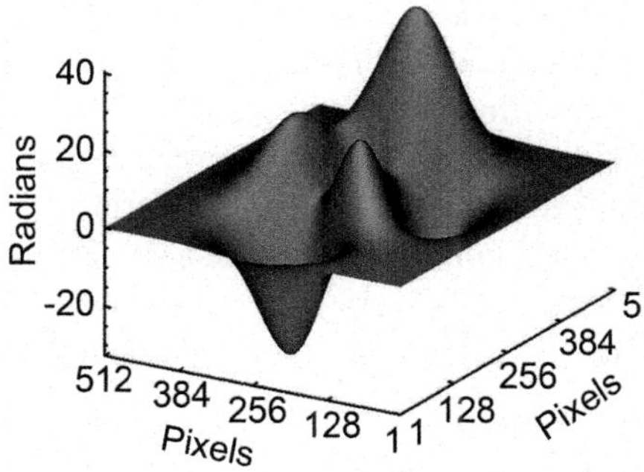

Eq. (2.17), and values of angles α_2 were calculated using Eq. (2.23). Each value of α_2 was evaluated for both the full format of the test bipolar phase surface and its central part. The calculation of the absolute error $|\Delta\alpha| = |\alpha_2 - \alpha_1|$ for each given phase shift was performed for the full format of the test phase surface ("Surface 1"), for the central square part of this surface ("Surface 1a"), and for the phase surface "Surface 1b". "Surface 1b" was synthesized by inclining of "Surface 1" relative to the rotation axis located along its left edge at an angle $\omega = 2\pi/(K-1)$ entailing the phase shift of the right edge on 2π rad. The interferogram of "Surface 1" and

Fig. 2.8 Interferogram of "Surface 1" and "Surface 1a"

"Surface 1a" is shown in Fig. 2.8. The interferogram of "Surface 1" is surrounded by solid lines, and interferogram of "Surface 1a" is surrounded by a dashed contour.

Dependencies "$|\Delta\alpha|$ versus α_1" in the range from 0 to π with a step of $\pi/18$ for "Surface 1", "Surface 1a" and "Surface 1b" are shown in Fig. 2.9. The analysis of these dependencies indicates that the absolute error $|\Delta\alpha| = |\alpha_2 - \alpha_1|$ of the phase shift α_1 for "Surface 1a" is less than the corresponding error for "Surface 1" almost all over the range from 0 to π, except for the subrange 1.43–1.68 rad. This behavior of $|\Delta\alpha|$ can be explained by a decrease in the contribution to the error of pixel values with equal phases and intensities, most of which are located on the outer edges of the phase field of "Surface 1". Therefore, cutting these edges reduces the error. At the same time, the error $|\Delta\alpha|$ for "Surface 1b" is less than the one for "Surface 1" and "Surface 1a" almost over the range from 0 to π, which is explained by the specifics of the phase field of "Surface 1b", since each pixel of this phase surface has an additional phase shift $\delta\omega_{k,l} = 2\pi/(K-1)$ relative to the corresponding pixel of "Surface 1". Averaging over the whole ensemble of intensities of this phase field using Eq. (2.23) leads to a decrease in the absolute error.

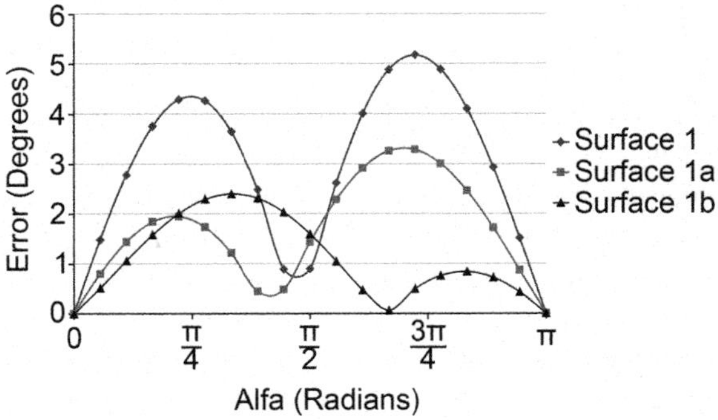

Fig. 2.9 Dependencies of the absolute error $|\Delta\alpha|$ on the phase shift α_1 for "Surface 1" (curve ◆), "Surface 1a" (curve ■) and "Surface 1b" (curve ▲)

Absolute errors of unknown phase shift extraction for nanorough surfaces. To estimate the absolute errors in extracting the unknown phase shift between interferograms recording the nanorough surface, the test bipolar phase surface shown in Fig. 2.7 was chosen. Additive Gaussian noise with different RMS values was introduced into this surface. To match computer simulation data with the experimental ones, it is convenient to convert the phase difference $\delta\varphi$ on the phase map into the height difference δh on the surface height map using Eq. (2.43). Similarly, test bipolar phase surface ("Surface 1") can be converted into the test bipolar height surface using the elementary transformation

$$\delta h = \frac{\lambda \cdot \delta\varphi}{4\pi}, \tag{2.44}$$

and the RMS phase error σ_φ can be converted into the RMS height error σ_h:

$$\sigma_h = \frac{\lambda \cdot \sigma_\varphi}{4\pi}. \tag{2.45}$$

The absolute error $|\Delta\alpha|$ of phase shift extraction for nanorough surfaces was determined in the same way as for the smooth surfaces, using the test smooth bipolar height surface with RMS roughness value $(Sq)_{sim0} = 0$, to which eight additive Gaussian noise values were added and eight rough test surfaces were synthesized with simulated RMS roughness values $(Sq)_{sim1} = \lambda/100$, $(Sq)_{sim2} = \lambda/50$, $(Sq)_{sim3} = \lambda/30$, $(Sq)_{sim4} = \lambda/20$, $(Sq)_{sim5} = \lambda/15$, $(Sq)_{sim6} = \lambda/10$, $(Sq)_{sim7} = \lambda/8$, $(Sq)_{sim8} = 2\lambda/15$. Note that such RMS roughness values approximately correspond to nanorough surfaces illuminated in interferometers by laser radiation in the visible range. For example, if the He–Ne laser is used ($\lambda = 633$ nm), then $(Sq)_{sim1} = 6.33$ nm and $(Sq)_{sim8} = 84.4$ nm. The absolute error $|\Delta\alpha|$ was calculated for nineteen values of the angle α_1 with a phase step equal to $\pi/18$, starting from 0.

Dependence of the absolute error $|\Delta\alpha|$ on α_1 for the test surface with RMS roughness $(Sq)_{sim4} = \lambda/20$ (curve 1) is shown in Fig. 2.10. In this figure, a similar dependence for the test smooth surface $(Sq)_{sim0} = 0$ is depicted as curve 2. Although the dependencies are similar, however, the range of curve 1 is approximately four times less than the range of curve 2, i.e. absolute error $|\Delta\alpha|$ for the rough surface $(Sq)_{sim4} = \lambda/20$ is about four times less than the same error for the test smooth surface. Analysis of dependencies of $|\Delta\alpha|$ on α_1 obtained for the initial test smooth surface and all eight test rough surfaces indicate a decrease of $|\Delta\alpha|$ in the range from 0 to π with increasing surface roughness. In particular, the dependencies $|\Delta\alpha|$ on the RMS roughness $(Sq)_{sim}$ for phase shifts $\alpha_{11} = \pi/6$ (solid curve) and $\alpha_{12} = 4\pi/9$ (dashed curve) are shown in Fig. 2.11. It is obvious that the absolute errors for $\alpha_{12} = 4\pi/9$ (dashed curve) are significantly smaller than errors for $\alpha_{11} = \pi/6$ (solid curve) in all range of surface roughness from $(Sq)_{sim0} = 0$ to $(Sq)_{sim8} = 2\lambda/15$.

Thus, a rather low level of absolute errors in the extraction of the unknown phase shift of the reference beam using synthesized test smooth and low-rough bipolar surfaces proved the validity of the PPCA for the TSI UPS method.

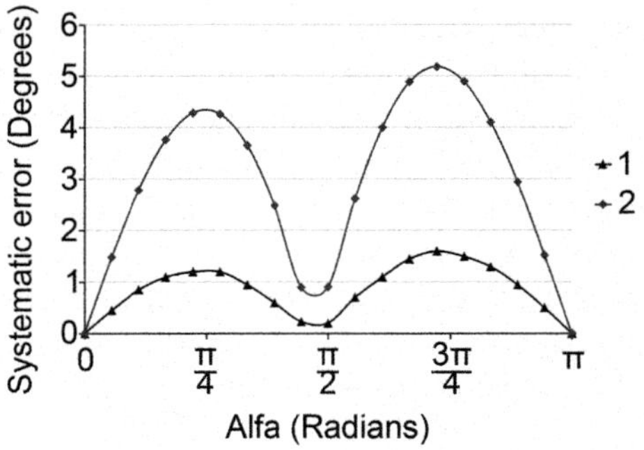

Fig. 2.10 Absolute error $|\Delta\alpha|$ as a function of given phase shift α_1 for test rough bipolar height surface $(Sq)_{sim4} = \lambda/20$ (curve 1) and the test smooth bipolar height surface $(Sq)_{sim0} = 0$ (curve 2)

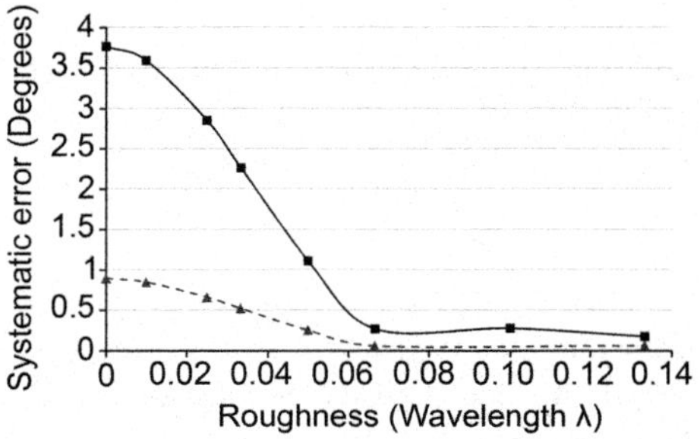

Fig. 2.11 Absolute error $\Delta\alpha$ as a function of $(Sq)_{sim}$ (wavelength λ is the measuring unit) at given $\alpha_{11} = \pi/6$ (solid curve) and $\alpha_{12} = 4\pi/9$ (dashed curve 2)

2.5.3 Computer Simulation of Test Nanorough Surfaces Retrieval

The effectiveness of the developed TSI UPS method was tested by reconstruction computer models of smooth and nanorough test bipolar phase surfaces. Nanorough phase surfaces were synthesized by introducing the additive Gaussian noise into the test smooth bipolar phase surface shown in Fig. 2.7.

An example of the implementation of the procedure to reconstruct the test bipolar phase surface with added Gaussian noise with RMS $Sq_\varphi = \frac{\pi}{5}$ for phase shift $\alpha_{11} = 30°$ is shown in Fig. 2.12. Such noise is used to imitate the RMS height of the scale-limited surface equal to $(Sq)_{sim4} = \lambda/20$. Initially, the interferograms $\tilde{I}_1(k,l)$ and $\tilde{I}_2(k,l)$ of "Surface 1" were synthesized. The interferogram $\tilde{I}_1(k,l)$ is shown in Fig. 2.12a. Further, the phase macrorelief of the surface is reconstructed using a low-pass filter with a normalized cut-off frequency $f_w = 0.2$ and the Flynn software [31, 32] for unwrapping the phase map $\tilde{\varphi}_m(k,l)$ [see (2.39)]. The macrorelief unwrapped phase map is shown in Fig. 2.12b. At the next step, according to Eq. (2.41), the phase map of the nanorelief $\tilde{\varphi}_r(k,l)$ containing singularities in pixels, the values of which are 2π and -2π, is calculated (see Fig. 2.12c). After the singularities elimination, the fine phase map of the nanorelief $\varphi_r(k,l)$ is obtained (see Fig. 2.12d, e). And finally the resultant phase map of the total surface relief area $\varphi(k,l)$ is synthesized [see Eq. (2.29) and Fig. 2.12f].

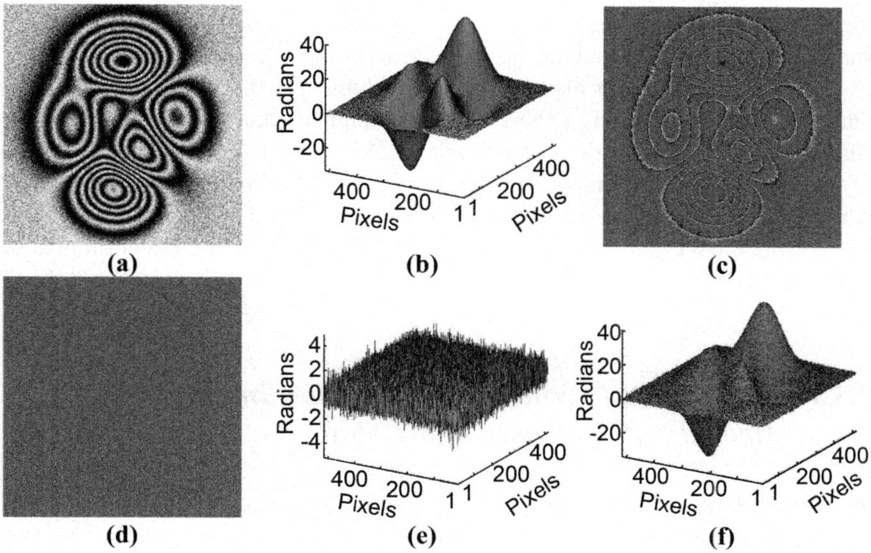

Fig. 2.12 Procedure to reconstruct the test bipolar phase surface with added Gaussian noise with RMS $\sigma_\varphi = \frac{\pi}{5}$ for phase shift $\alpha_{11} = \frac{\pi}{6}$ using the TSI UPS method: initial interferogram (**a**), retrieved phase macrorelief after unwrapping (**b**), retrieved phase microrelief containing singularities (**c**); retrieved phase microrelief after singularities elimination (**d**), isometric image of the retrieved phase microrelief (**e**), resultant retrieved phase surface relief (**f**)

Assessment of the reliability of the nanorough surface reconstruction using the proposed TSI UPS method was performed by analyzing random and absolute errors in the retrieving the test bipolar phase surface (see Fig. 2.7) corrupted by additive Gaussian noise with different RMS values in the range of phase shifts from 0 to π. Retrieved test phase maps were transformed into series of test height maps with different random noise RMS values $(Sq)_{sim}$ in the range $0 \leq (Sq)_{sim} \leq 0.07\lambda$ using Eq. (2.45). These maps were applied to obtain the dependencies of the RMS of macrorelief (σ_m), nanorelief (σ_r) and total relief (σ_T) on the RMS $(Sq)_{sim}$. As mentioned above, the RMS $(Sq)_{sim}$ can be approximately treated as the RMS height of the scale-limited surface. The obtained dependencies σ_m versus $(Sq)_{sim}$ increase monotonically with increase of $(Sq)_{sim}$. However, the dependence σ_m versus $(Sq)_{sim}$ for $\alpha_{11} = \pi/6$ increases from $4 \cdot 10^{-3}\lambda$, if $(Sq)_{sim} = 0$, to $16.5 \cdot 10^{-3}\lambda$, if $(Sq)_{sim} = 6.7 \cdot 10^{-2}\lambda$. The similar dependence for $\alpha_{12} = 4\pi/9$ is lower to the dependence for $\alpha_{11} = \pi/6$ up to $(Sq)_{sim} = 4.0 \cdot 10^{-2}\lambda$. At $(Sq)_{sim} > 4.0 \cdot 10^{-2}\lambda$ these dependencies merge and increase monotonically. The dependencies σ_T versus $(Sq)_{sim}$ for $\alpha_{11} = \pi/6$ monotonically decrease in the range $0 \leq (Sq)_{sim} \leq 0.05\lambda$ from $\sigma_T = 3.8 \cdot 10^{-3}\lambda$ to $\sigma_T = 1.1 \cdot 10^{-3}\lambda$ and for $\alpha_{12} = 4\pi/9$ from $\sigma_T = 0.55 \cdot 10^{-3}\lambda$ to $\sigma_T = 0.21 \cdot 10^{-3}\lambda$. If $(Sq)_{sim} > 0.05\lambda$, then σ_T begins to rise sharply for α_{11} and α_{12}.

Absolute errors of the reconstructed macro-, nano- and total surface reliefs are very low. The absolute error maximum values are reached at $(Sq)_{sim} = 0$ and equal to $3.4 \cdot 10^{-3}\lambda$ for $\alpha_{11} = \pi/6$ and to $8.0 \cdot 10^{-4}\lambda$ for $\alpha_{12} = 4\pi/9$. If the simulated surface roughness increased, the absolute errors tend to decrease.

Comparison with other methods and algorithms of two-step PSI with unknown phase shift shows that this method possesses higher accuracy of the test bipolar surface relief reconstruction. For example, if $(Sq)_{sim} = \lambda/16$, then the RMS error of the total surface relief reconstruction is $\sigma_T = 0.054$ using the TSI UPS method, while the RMS errors are $\sigma_T = 0.21$ when using the GSO algorithm, $\sigma_T = 0.32$ when using the ST algorithm, and $\sigma_T = 0.25$ when using the ROF algorithm at the phase shifts range $\pi/6 \le \alpha_1 \le 4\pi/9$.

2.5.4 Experimental Verification of Two-Step Interferometry Method with Unknown Phase Shift

The TSI UPS method was tested on the experimental setup based on the Twyman–Green interferometer (see Fig. 2.13). The setup contains a single-mode He–Ne laser (radiation wavelength $\lambda = 633$ nm, output power $P = 50$ mW), a PZT with a mirror attached to it to implement phase shift, a special jig plate for precision mounting of the test specimen with a given surface roughness, digital CCD camera SONY XCD-SX910 (1/2 " IT CCD-SXGA, 1280 $(H) \times 960$ (V), pixel size 4.65 μm × 4.65 μm and pixel pitch $p_x = p_y = 4.65$ μm), a control unit and PC for the implementation of the TSI UPS method. The gauge block (GB) with a metric nominal length $l_n = 30$ mm was chosen for experimental verification of the TSI UPS method. The arithmetic mean deviation of the roughness profile [see Eq. (2.21)] for this GB is $Ra < 0.020$ μm, and a metric nominal length for this roughness is $l_r = 0.08$ mm according to the International Standard ISO 4288:1996 [49].

The purpose of the experimental studies was to reconstruct the surface relief of the investigated GB measuring surface, including the reconstruction of both its macrorelief, i.e. the waviness and the surface slope, and the surface nanorelief (roughness). The optical scheme of the Twyman–Green interferometer was adjusted to register the GB measuring surface with such a spatial resolution that would be sufficient for a robust restoration of its relief while simultaneously satisfying the requirements of the Shannon sampling theorem. For this purpose, it is necessary to choose the acceptable sampling interval for interferograms during their registration. Most of the known methods for determining the sampling interval are based on calculating either the cumulative power spectrum [68] of the analyzed fringe pattern or its autocorrelation function (ACF) [88, 91].

To find the allowable image sampling interval and, accordingly, the linear (lateral) magnification M_l of the optical system generating fringe patterns on the matrix photodetector of a digital camera, a method based on calculating the correlation length τ_0 of the ACF of the surface roughness profile was used. This profile is approximated by the exponential function given by

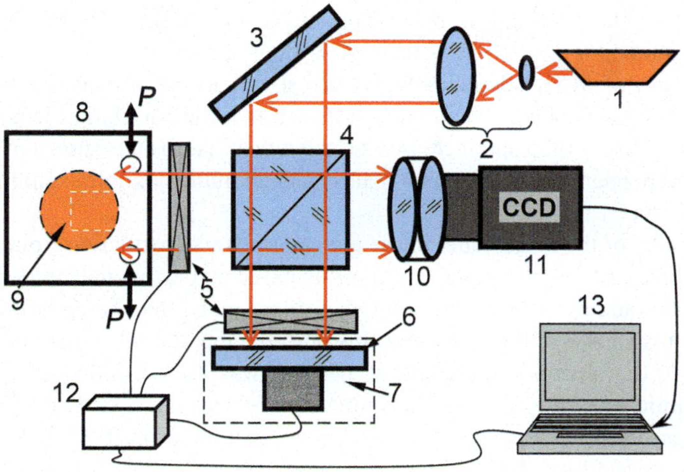

Fig. 2.13 Scheme of the experimental setup based on the Twyman–Green interferometer: He–Ne laser (1); laser beam collimator (2); rotating mirror (3); beam splitter (4); electromechanical shutters (5); movable mirror on a piezoelectric mount (6); piezoelectric transducer (7); specimen attached to a jig plate (8); region of interest (9); lens (10); CCD camera (11); control unit (12); computer (13)

$$\rho(\tau) = \exp\left(-\tau/\beta^*\right), \tag{2.46}$$

where β^* is the ACF spacing parameter characterizing the correlation length τ_0. Whitehouse [126] and later Archard and Onions [82] proposed to define the correlation length on the level of 10% concerning the ACF maximum. In this case, $\tau_0 = 2.3\beta^*$. To calculate the sampling interval τ_s, the approach proposed by Pavlus [88] was chosen, according to which this interval should not exceed (0.4–0.5)β^*, i.e.

$$\tau_s = (0.17 - 0.21)\tau_0. \tag{2.47}$$

Assessment of maximum accessible sampling interval τ_{sm} was performed using proposed by Poon and Bhushan approach [91]. According to this approach, the correlation length τ_0 for digital images does not change until the sampling interval exceeds the Nyquist interval. After exceeding the sampling interval, the correlation length starts increasing sharply. This approach was used to define the maximum accessible sampling interval τ_{sm}. To realize this approach, the interferograms of the surface area of the studied gauge block were recorded at different values of the linear magnification M_l of the optical system of the Twyman–Green interferometer shown in Fig. 2.13. The example of the obtained ACF along the chosen line of the retrieved phase map of the gauge block surface area is shown in Fig. 2.14. This ACF was obtained in optical system of the interferometer, which projected the surface area on the photodetector array with linear magnification $M_l = 2.325$. According to this

ACF, the correlation length is $\tau_0 = 6$ pixels, i.e. $\tau_0 = \frac{p_x}{M_l} \cdot 6 = 12\ \mu\text{m}$, and the sampling interval is $\tau_s = (0.17 - 0.21)\tau_0 = 2.0 - 2.5\ \mu\text{m}$.

The results of ACFs calculations obtained at various linear magnifications M_l of the interferometer optical scheme have testified that the correlation length τ_0 does not change if $M_l \geq 0.85$ and increases sharply otherwise, as it is shown in Fig. 2.15. Obtained dependence τ_0 versus τ_s indicates that maximum possible sampling interval $\tau_{sm} = 5.5\ \mu\text{m}$.

The results of the macrorelief phase map $\varphi_m(k, l)$ reconstruction from two interferograms recorded by the experimental setup based on the Twyman–Green interferometer of the studies surface area of the gauge block and differed only by the arbitrary phase shift is shown in Fig. 2.16. For this purpose, the TSI UPS method was used. The size of the recorded digital fringe patterns was 800 × 600 pixels, the linear magnification was $M_l = 2.325$, and the metric size was $1.6 \times 1.4\ \text{mm}^2$. Calculation of the phase shift between the obtained fringe patterns using Eq. (2.19) indicated that $\Delta\alpha = 0.644$ rad. The waviness phase map $\varphi_W(k, l)$, represented in Fig. 2.17,

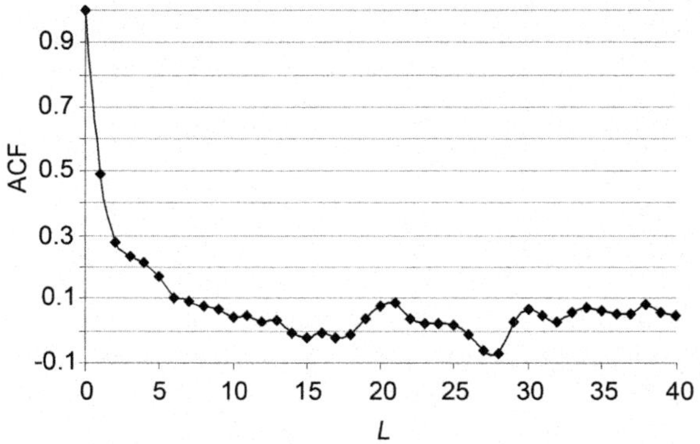

Fig. 2.14 Autocorrelation function of the phase map along the linear horizontal profile: L is the pixel number

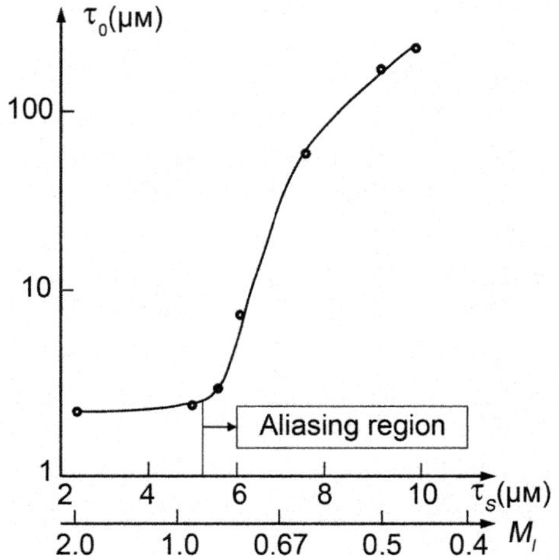

Fig. 2.15 Dependence of the ACF correlation length τ_0 on the sampling interval τ_s and linear magnification M_l of the Twyman–Green interferometer optical system

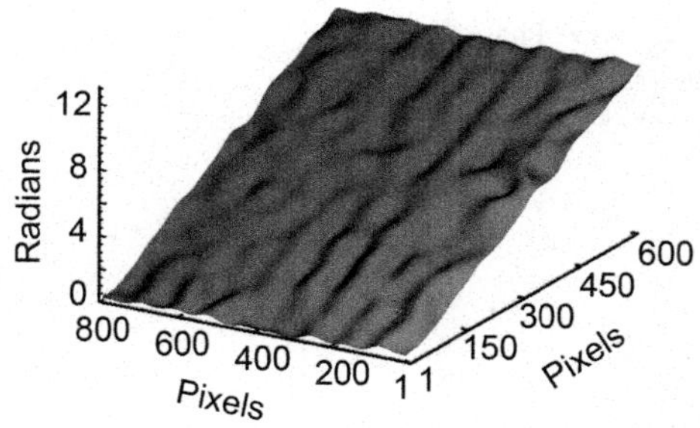

Fig. 2.16 Macrorelief phase map $\varphi_m(k, l)$ of the gauge block surface area

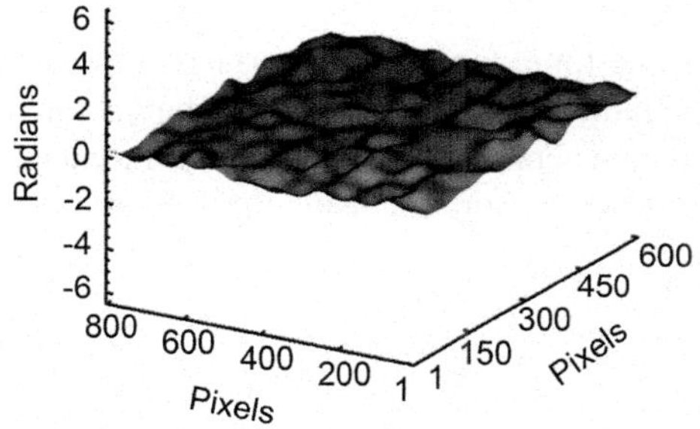

Fig. 2.17 Waviness phase map $\varphi_W(k, l)$ of the gauge block surface area

was extracted from the macrorelief phase map by eliminating both the slope of 5.5 $\cdot\ 10^{-4}$ rad and the small contribution of very low spatial frequencies corresponding to the form. The nanorelief phase maps with singularity points (pixels) and after their elimination are shown in Fig. 2.18. The retrieved phase map $\varphi(k, l)$ of the total surface relief area is shown in Fig. 2.19.

The surface roughness of the gauge block selected area was determined by calculating the arithmetical mean height of the scale-limited surface Sa. The areal height parameter is $Sa_r = 24.7$ nm for the nanorelief surface height map $h_r(k, l)$. At the

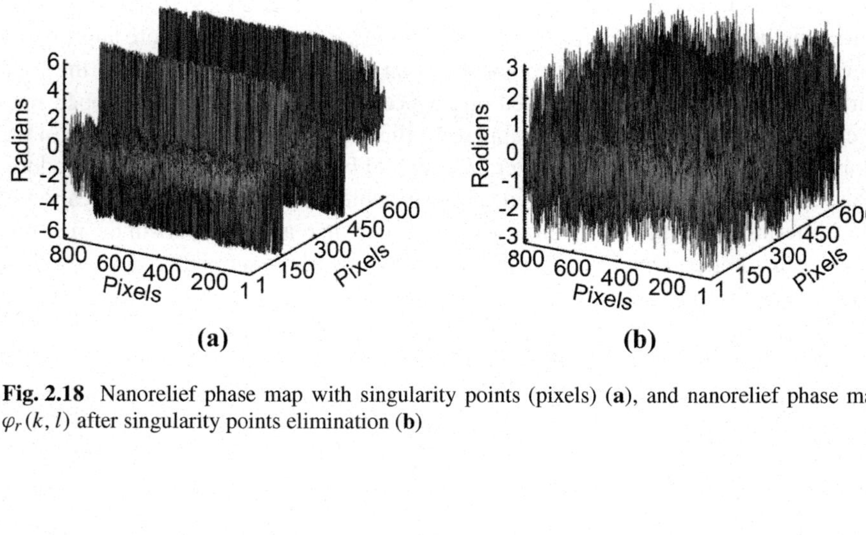

Fig. 2.18 Nanorelief phase map with singularity points (pixels) (**a**), and nanorelief phase map $\varphi_r(k, l)$ after singularity points elimination (**b**)

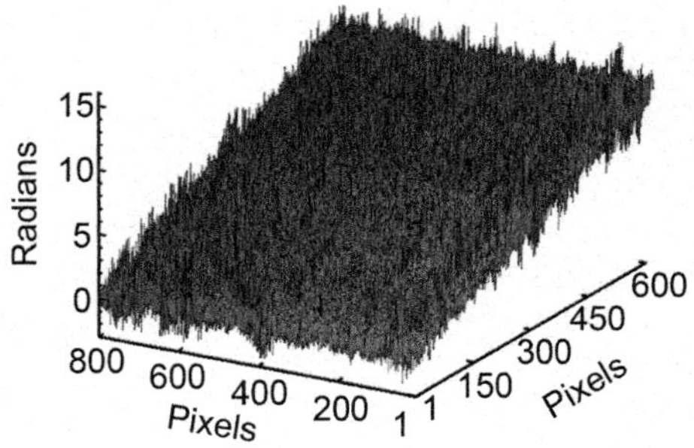

Fig. 2.19 Phase map $\varphi_r(k, l)$ of the total surface relief area

same time, for the total surface relief surface height map $h(k, l)$, the areal height parameter is $Sa = 26.6$ nm. The obtained results of calculations of the roughness parameters of the gauge block surface area testified to the high reliability of the real surfaces restoration using the TSI UPS method.

2.5.5 *Method of Iterative Two-Step Interferometry with Unknown Phase Shift for Retrieving the Nanorough Surface*

Many 2D techniques for extraction roughness, waviness and form along the surface profile have been developed [95]. The developed TSI UPS method allows extracting separately 3D surface nanorelief, i.e. 3D surface roughness, and 3D surface macrorelief including waviness and form from the retrieved total surface relief. However, during modeling and technical realization of the procedure for retrieval of the reliefs of smooth and nanorough surfaces using the TSI UPS method, two factors were identified that have a significant impact on the increase in errors in determining the relief components. First, during validation of this method on test and real surfaces, it was found that the error in the retrieving the full relief is much less than the errors in retrieving both the nanorelief and macrorelief [76]. Therefore, it is more appropriate to extract the macrorelief phase map from an unwrapped phase map of the total surface relief, than from a wrapped one, since errors caused by the wrapped map cyclic structure are absent in unwrapped. To this end, in order to reduce errors of the phase relief retrieval, the second iteration should be introduced into this procedure. The second factor is related to the edge effects, which in the case of the application of DFT filtering to extract the macrorelief phase map increase the errors in the restored total surface relief. One of the simplest approaches to reduce the errors caused by edge effects includes enlargement of the format of the restored phase map, spatial filtering of the extended map and returning to its previous format.

Taking into account these two factors, a new method of iterative two-step interferometry with unknown phase shift (ITSI UPS) of a reference beam for retrieving the nanorough surface has been developed, which initial version is described in [81]. A simplified flow-chart of the two-stage algorithm performing this method is shown in Fig. 2.20. Its first stage is finished by the operation of the coarse phase map $\varphi_w(k, l)$ generation, and the second stage is marked with a dashed contour.

At the first iterative step of the algorithm second stage, the initial coarse wrapped phase map $\varphi_w(k, l)$ of the total surface relief area is expanded around the perimeter in all four vertical and horizontal directions to suppress edge effects. Computer simulations have shown that for a 512 × 512 pixel phase map, it is sufficient to limit oneself to fifty columns of pixels from the left and right edges and fifty rows from the top and bottom. The expanded phase map $(\varphi_w)_{\exp 1}$ is filtered using the low-pass (LP) filtering procedure, the flow-chart of which is shown in Fig. 2.5. Due to this procedure, edge effects initiated by the direct and inverse DFTs are suppressed. Simultaneously, the LP filtering procedure allows avoiding iterations, because the LP spatial filtering is performed in spatial frequency domain [81]. To suppress a random noise in the phase map $(\varphi_w)_{\exp 1}$ and extract a coarse wrapped phase map $\tilde{\varphi}_{mc}(k, l)$ of the surface macrorelief, it is necessary to choose a cut-off frequency f_{c0}. This frequency can be

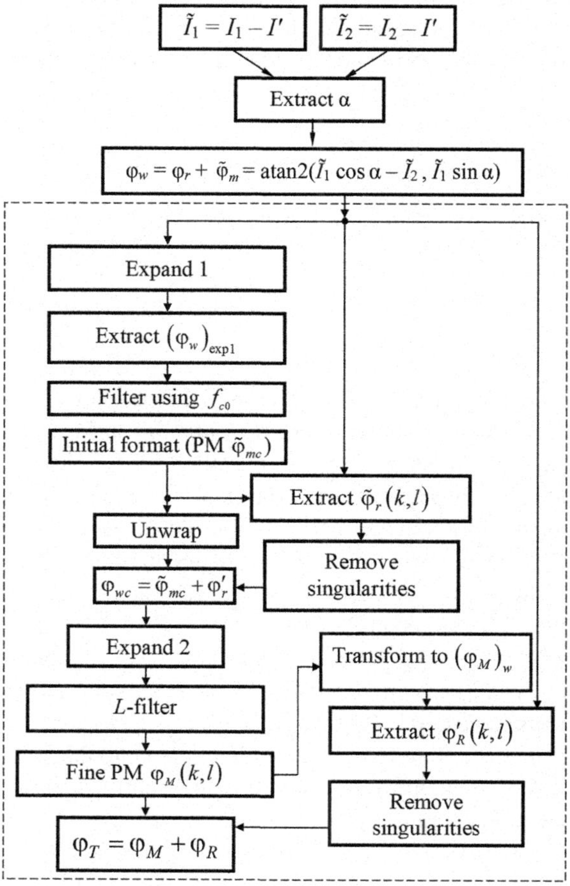

Fig. 2.20 Simplified flow-chart of the algorithm performing the ITSI UPS method. The second stage of the algorithm is highlighted by a dashed contour

determined by calculating the maximum frequency of the initial phase map $\varphi_w(k, l)$ saw-tooth fringe structure using the algorithm for estimation of a fringe orientation and fringe density developed by Yang et al. [137]. The coarse phase map $\tilde{\varphi}_r(k, l)$ of the nanorelief is produced by subtraction of a coarse wrapped phase map $\tilde{\varphi}_{mc}(k, l)$ of the surface macrorelief from the initial coarse wrapped phase map $\varphi_w(k, l)$. After unwrapping the coarse wrapped phase map $\tilde{\varphi}_{mc}(k, l)$ and removing singularities from $\tilde{\varphi}_r(k, l)$ (see Sect. 2.5.1 and Fig. 2.6), the coarse phase map $\varphi_{wc}(k, l)$ of the total surface relief is produced as the sum of $\tilde{\varphi}_{mc}(k, l)$ and $\tilde{\varphi}_r(k, l)$, that is $\varphi_{wc}(k, l) = \tilde{\varphi}_{mc}(k, l) + \tilde{\varphi}_r(k, l)$.

At the second iterative step highlighted by a dashed contour, errors of the coarse phase map $\varphi_{wc}(k, l)$ are suppressed and the final macrorelief and nanorelief phase masks are extracted with the help of a low-frequency linear L-filter [51]. This filter is equivalent to cut-off filters used in 2D profilometry [95], and its nesting index frequency f_c can be defined from a cut-off wavelength $\lambda_c = (f_c)^{-1}$, which is the branch point between surface roughness and waviness components. The cut-off wavelength λ_c values for corresponding surface roughness parameters are represented in several ISO and ASME standards. Functions of the L-filter can perform several rejection filters, for example, the low-pass Gaussian filter (LPGF) and Rk filter [50, 95]. In the TSI UPS method, the LPGF was used. According to the international standard ISO 16610-21:2011 [50], the transfer characteristic of this filter is given by

$$S(u) = \exp\left(-\pi^2\tilde{\alpha}^2\lambda_c^2u^2\right), \tag{2.48}$$

which is the Fourier transform of the weighting function

$$s(x) = (\tilde{\alpha}\lambda_c)^{-1}\exp\left[-\pi\left(\frac{x}{\tilde{\alpha}\lambda_c}\right)^2\right], \tag{2.49}$$

where $\tilde{\alpha} = \sqrt{\ln 2/\pi}$; u is the spatial frequency.

If we consider not a profile, but a 3D surface relief, then Eqs. (2.48) and (2.49) can also be used, provided that the transfer characteristic and weighting function have circular symmetry. Note that Gaussian spatial filtering can be implemented using both convolution [126] and DFT, including fast Fourier transform [95]. To increase the noise immunity of Gaussian filtering and, in particular, to get rid of such disturbances as deep valleys, the two-step Gaussian corrective filtering, or the Rk filter [50, 95] can be used.

So, the total coarse phase map $\varphi_{wc}(k, l)$ obtained after first iterative step is expanded once more and its expanded version is filtered by the cut-off LPGF. A fine phase map $\varphi_M(k, l)$ of the surface macrorelief area is produced after conversion of the extended filtered total coarse phase map into the initial format. Further, the macrorelief phase map $\varphi_M(k, l)$ should be transformed into a wrapped one $[\varphi_M(k, l)]_w$ using the next operations:

$$\Phi_{sM} = \sin[\varphi_M(k, l)],\ \Phi_{cM} = \cos[\varphi_M(k, l)],$$

$$[\varphi_M(k,l)]_w = \arctan(\Phi_{sM}, \Phi_{cM}). \tag{2.50}$$

The intermediate phase map $\varphi'_R(k,l)$ of the nanorelief (surface roughness) is retrieved by subtraction of the wrapped map $[\varphi_M(k,l)]_w$ from the initial coarse wrapped phase map $\varphi_w(k,l)$. After repeated removal of singularities in the intermediate phase map $\varphi'_R(k,l)$, a fine phase map $\varphi_R(k,l)$ of the nanorelief is obtained. Finally, a fine total phase map $\varphi_T(k,l)$ is generated as a result of pixel-by-pixel summation of the fine phase maps $\varphi_M(k,l)$ and $\varphi_R(k,l)$.

2.5.6 Retrieval of Test Surface Using Iterative Two-Step Interferometry Method with Unknown Phase Shift

Step-by-step retrieving of the total phase map $\varphi_T(k,l)$ and its components, namely the macrorelief $\varphi_M(k,l)$ and nanorelief (roughness) $\varphi_R(k,l)$ phase maps of the test nanorough bipolar phase surface using ITSI UPS method, is shown in Fig. 2.21. In this case, the macrorelief consists of the waviness and form, and the phase map $\varphi_M(k,l)$ of the macrorelief is the sum of the phase maps of waviness $\varphi_W(k,l)$ and form $\varphi_F(k,l)$, i.e. $\varphi_M(k,l) = \varphi_W(k,l) + \varphi_F(k,l)$. In order to obtain the test nanorough surface with the RMS height $Sq = \lambda/15$, the phase additive Gaussian noise with RMS phase $Sq_\varphi = 4\pi/15$ was introduced into the test bipolar phase surface shown in Fig. 2.7. Two interferograms of this surface differed by a given phase shift $\alpha_1 = \pi/5$ were synthesized. The initial interferogram $\tilde{I}_1(k,l)$ is shown in Fig. 2.21a. The wrapped phase map $[\varphi_M(k,l)]_w$ is shown in Fig. 2.21b, and the phase maps $\varphi_M(k,l)$, $\varphi_R(k,l)$ and $\varphi_T(k,l)$ are shown in Fig. 2.21c–e, respectively.

The test nanorough bipolar phase surface with a size of 512 × 512 pixels was chosen also to analyze the aliasing influence on reliability of the surface roughness reconstruction. To this aim, histograms of the initial test nanorough bipolar phase surface and the retrieved fine total phase map $\varphi_T(k,l)$ were calculated and studied at the range $\varphi_R(k,l) < \pi$. The histogram of phase distribution of the initial test nanorough bipolar phase surface with embedded phase additive Gaussian noise $\left(Sq_\varphi = 4\pi/15\right)$ and a given phase shift $\alpha_1 = \pi/5$ in the range $(-\pi, \pi)$ and the histogram of phase distribution of the retrieved total phase map $\varphi_T(k,l)$ of this initial test nanorough bipolar phase surface in the same range are depicted in Figs. 2.22 and 2.23, respectively. As can be seen, both histograms in this case are still localized within one phase cycle from $-\pi$ to π, which indicates the absence of aliasing.

To evaluate the effectiveness of the developed ITSI UPS method, random and absolute errors of the retrieved test surfaces with different roughnesses for given values of the phase shifts were calculated [78]. Obtained dependencies of these errors on the roughness of the test surfaces indicate a significant decrease in errors compared to the errors that occur when using the TSI UPS method. For example, the random errors of the test macrorelief phase maps are reduced by a factor of 1.6 for a surface with RMS height $Sq = \lambda/100$ and by a factor 5 for surfaces with RMS height

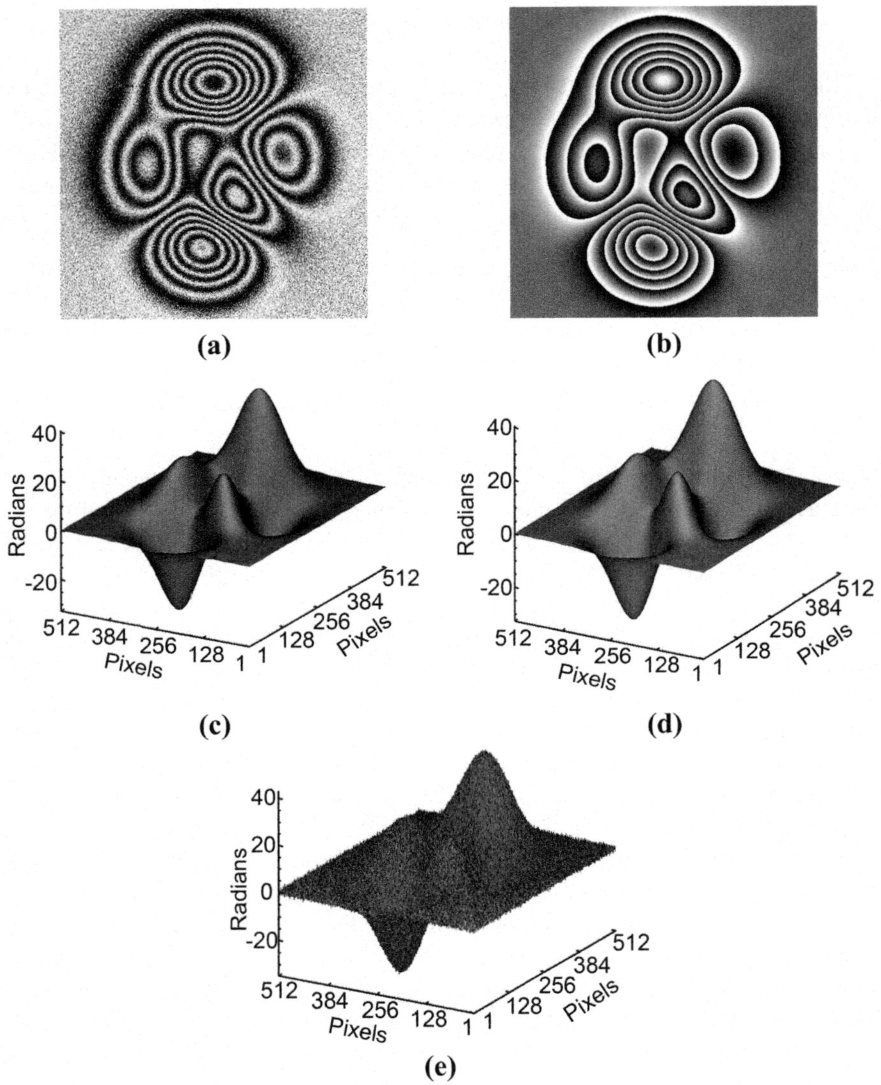

Fig. 2.21 Retrieving the test nanorough bipolar phase surface ($Sq = \lambda/15$, $Sq_{\varphi} = 4\pi/15$, $\alpha_1 = \pi/5$) with the help of the ITSI UPS method: initial interferogram $\tilde{I}_1(k, l)$ (**a**); wrapped fine phase map $[\varphi_M(k, l)]_w$ of the test phase surface macrorelief (**b**); unwrapped fine phase map $\varphi_M(k, l)$ of the macrorelief (**c**); fine phase map $\varphi_R(k, l)$ of the surface roughness (nanorelief) (**d**); retrieved total phase map $\varphi_T(k, l)$ of the test nanorough bipolar phase surface (**e**)

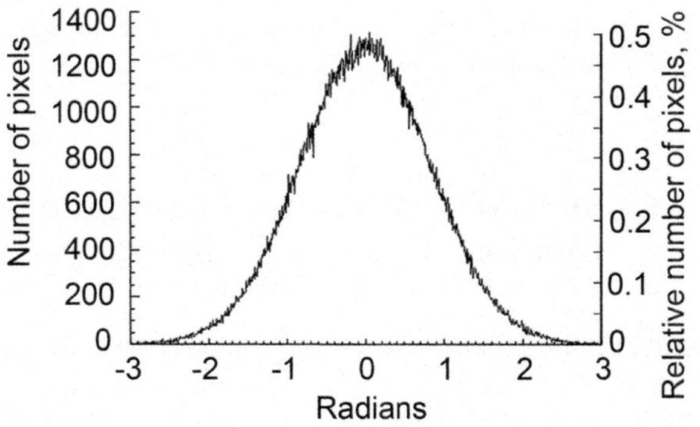

Fig. 2.22 Histogram of the phase distribution of the initial test nanorough bipolar phase surface with embedded phase additive Gaussian noise $(Sq_{\varphi} = 4\pi/15)$

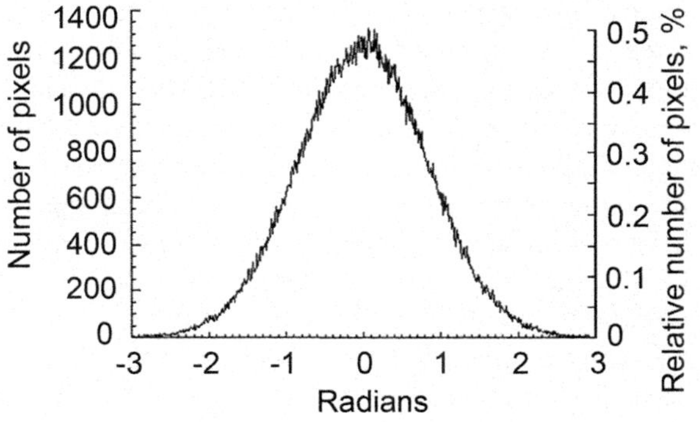

Fig. 2.23 Histogram of the phase distribution of the retrieved total phase map $\varphi_T(k, l)$ of the bipolar test surface with embedded phase additive Gaussian noise $(Sq_{\varphi} = 4\pi/15)$

$Sq = \lambda/10$. On average, over the entire roughness range, that is, from $Sq = \lambda/100$ to $Sq = \lambda/10$, the total errors decrease by about 4 times.

The obtained results confirm the high efficiency of the proposed ITSI UPS method during the reconstruction of smooth and nanorough surfaces. Computer modeling of test surfaces and experimental studies of the gauge block measuring surface have shown that the developed TSI UPS and ITSI UPS methods allow retrieving surfaces with roughness, which RMS does not exceed $\lambda/10$. Surfaces with higher roughness can also be retrieved up to $\lambda/8$, but at $Sq > \lambda/8$ random errors of restored surfaces increase rapidly. The simulation and experimental results also clearly show that surface reconstruction errors decrease with increasing laser wavelength in the interferometer. These methods have practical application for studying the fatigue properties of metals and alloys specimens near rounded and fractured stress concentrators, where it is especially important to determine the parameters of the fatigue process zone (FPZ) and estimate the nanorough surface evolution and change in roughness of the polished surface near the concentrator during cyclic loading. The possibility to extract the surface nanorelief from the total surface relief by these methods is also of great significance in studying the surface behavior near stress concentrators. Thanks to this fact, it is possible to study the surface roughness near the edges of the notches, which usually have an inevitably large curvature resulting from the notch drilling or milling.

2.6 Method of Three-Step Phase Shifting Interferometry with Unknown Phase Shifts Between Interferograms

To find the phase $\varphi(k,l)$ in the fundamental equation for two-step PSI (2.5), it is necessary to define the interferogram intensity $I(k,l)$, the background intensity $I_B(k,l)$ and the intensity (fringe) modulation (visibility function) $I_M(k,l)$ or the fringe visibility (modulation) $V(k,l) = \frac{I_M(k,l)}{I_B(k,l)}$. Therefore, at least three interferograms should be obtained for calculating the phase $\varphi(k,l)$. The basic conventional three-step algorithms for phase φ definition are considered, for example, by Creath [16] and Schreiber and Brunning [104]. Their advantage consisting in the comparative simplicity of the algorithm implementation is offset by large errors in the reconstruction of the phase map. Nevertheless, due to this advantage and the ability to reduce phase map retrieval errors, several three-step algorithms with unknown phase shifts have been created. For example, two blind self-calibrating algorithms to estimate two blind phase shifts have been developed by Guo et al. [39, 40]. In the first algorithm [40], to extract the searched phase shift, the minimum value of spatial correlation between the reference beam I_r and recovered phase φ was calculated. However, the second-order correlation initiated by the cross power spectrum is unable to describe the nonlinear correlation between these two signals because their spectra overlap very weakly. The third-order correlation between I_r and φ, estimated by the cross-bispectrum, made it possible to overcome this lack [39]. This blind self-calibration algorithm can estimate blind phase shifts between interferograms. However, the computational procedure of phase shifts extraction is rather complicated. In addition, computer simulations can only be implemented for fixed and equal phase shifts used in conventional three-step phase shifting algorithms, for example, for such phase shifts $\alpha_i = 0, 2\pi/3, 4\pi/3$. Therefore, the assessment of errors of retrieved phase maps can be carried out only for fixed and equal phase shifts. A noniterative three-step PSI algorithm based on geometric concept of volume enclosed by a surface using three unknown and unequal phase steps was proposed by Meneses Fabian [69]. The algorithm allows reaching the low errors in extraction of blind phase shifts and retrieval of phase objects. However, the recorded interferograms should have at least several fringes that considerably restrict the capabilities of this algorithm.

Another method is based on the modulation light estimation by applying the least squares iterative algorithm [70]. Note that the requirements for a large spread of illumination and the object phase limit its performance. The method of phase reconstruction based on normalized difference maps generated by three interferograms with unknown phase shifts was proposed by Wang et al. [124]. In this method, three interferograms were subtracted from each other, thus getting rid of the background intensity. Next, the obtained difference maps were reconstructed and normalized, eliminating their amplitude inhomogeneity. Difference phase maps were used to retrieve the desired phase map with the help of the two-step Gram–Schmidt algorithm [122].

TSI UPS and ITSI UPS methods considered in Sect. 2.5 use recording of the object and reference wavefronts $I_o(k,l)$ and $I_r(k,l)$ to suppress the background intensity $I_B(k,l)$. Therefore, these methods are time-consuming, since they demand separate recording of $I_o(k,l)$ and $I_r(k,l)$ except fringe patterns $I_1(k,l)$ and $I_2(k,l)$. To overcome this problem, the method of three-step PSI with unknown phase shifts (3SI UPS) between reference and object waves was developed. In this method, two unknown phase shifts are calculated immediately after recording the fringe patterns $I_1(k,l)$, $I_2(k,l)$ and $I_3(k,l)$. After determining the phase shift values, two difference phase maps are produced to exclude the background intensity [79]. The 3SI UPS method also retrieves separately the macrorelief and nanorelief surface areas and the total surface relief. This method, as well as the ITSI UPS method, includes two stages. At the first stage, a coarse phase map of the total surface relief is produced using three recorded fringe patterns $I_1(k,l)$, $I_2(k,l)$ and $I_3(k,l)$. The 3SI UPS method second stage is similar to the second stage of the ITSI UPS method.

2.6.1 Description of the Method of Three-Step Phase Shifting Interferometry with Unknown Phase Shifts

Three digital fringe patterns $I_1(k,l)$, $I_2(k,l)$ and $I_3(k,l)$ with respective unknown phase shifts α_1, α_2 and α_3 of the reference wave are recorded in the two-beam interferometer. The set of three fundamental equations of PSI can be given by

$$\left.\begin{aligned} I_1(k,l) &= I_B(k,l) + I_M(k,l)\cos[\varphi(k,l)+\alpha_1] \\ I_2(k,l) &= I_B(k,l) + I_M(k,l)\cos[\varphi(k,l)+\alpha_2] \\ I_3(k,l) &= I_B(k,l) + I_M(k,l)\cos[\varphi(k,l)+\alpha_3] \end{aligned}\right\}. \tag{2.51}$$

Using Eq. (2.23), the phase shift $\alpha_{21} = \alpha_2 - \alpha_1$ between interferograms $I_1(k,l)$ and $I_2(k,l)$ is expressed as

$$\alpha_{21} = \alpha_2 - \alpha_1 = \arccos[\rho(I_1, I_2)] = \arccos\frac{\langle(I_1 - \langle I_1\rangle)(I_2 - \langle I_2\rangle)\rangle}{\sigma_{I1}\sigma_{I2}}, \tag{2.52}$$

and the phase shift $\alpha_{31} = \alpha_3 - \alpha_1$ between interferograms $I_1(k,l)$ and $I_3(k,l)$ is expressed as

$$\alpha_{31} = \alpha_3 - \alpha_1 = \arccos[\rho(I_1, I_3)] = \arccos\frac{\langle(I_1 - \langle I_1\rangle)(I_3 - \langle I_3\rangle)\rangle}{\sigma_{I1}\sigma_{I3}}, \tag{2.53}$$

where $\sigma_{I_1(k,l)}$,$\sigma_{I_2(k,l)}$,$\sigma_{I_3(k,l)}$ are the RMS of intensities distributions in interferograms $I_1(k,l)$, $I_2(k,l)$ and $I_3(k,l)$, respectively.

In order to calculate the wrapped phase map $\varphi_w(k,l)$, the set (2.51) is reduced to the set of two equations expressed as

$$\left.\begin{array}{l}\frac{I_2(k,l)-I_B(k,l)}{I_1(k,l)-I_B(k,l)}=\frac{\cos[\varphi(k,l)+\alpha_2]}{\cos[\varphi(k,l)+\alpha_1]}=\frac{\cos\alpha_2-\tan\varphi(k,l)\sin\alpha_2}{\cos\alpha_1-\tan\varphi(k,l)\sin\alpha_1}\\ \frac{I_3(k,l)-I_B(k,l)}{I_1(k,l)-I_B(k,l)}=\frac{\cos[\varphi(k,l)+\alpha_3]}{\cos[\varphi(k,l)+\alpha_1]}=\frac{\cos\alpha_3-\tan\varphi(k,l)\sin\alpha_3}{\cos\alpha_1-\tan\varphi(k,l)\sin\alpha_1}\end{array}\right\}. \tag{2.54}$$

These equations are reduced to the common quadratic ones

$$t(k,l)\tan^2\varphi(k,l)+g(k,l)\tan\varphi(k,l)+r(k,l)=0, \tag{2.55}$$

where

$$\begin{aligned}t(k,l)=\;&0.5[I_3(k,l)-I_1(k,l)][\cos\alpha_{21}-\cos(2\alpha_1+\alpha_{21})]\\&+0.5[I_1(k,l)-I_2(k,l)][\cos\alpha_{31}-\cos(2\alpha_1+\alpha_{31})]\\&+[I_2(k,l)-I_3(k,l)]\sin^2\alpha_1;\end{aligned} \tag{2.56}$$

$$\begin{aligned}g(k,l)=\;&[I_3(k,l)-I_2(k,l)]\sin 2\alpha_1\\&+[I_1(k,l)-I_3(k,l)]\sin 2(\alpha_1+\alpha_{21})\\&+[I_2(k,l)-I_1(k,l)]\sin(2\alpha_1+\alpha_{31});\end{aligned} \tag{2.57}$$

$$\begin{aligned}r(k,l)=\;&[I_2(k,l)-I_3(k,l)]\cos^2\alpha_1\\&+0.5[I_3(k,l)-I_1(k,l)][\cos\alpha_{21}+\cos(2\alpha_1+\alpha_{21})]\\&+0.5[I_1(k,l)-I_2(k,l)][\cos\alpha_{31}+\cos(2\alpha_1+\alpha_{31})].\end{aligned} \tag{2.58}$$

Equations (2.56)–(2.58) contain only differences between the intensities of interferograms, and hence, the background $I_B(k,l)$ is excluded in all these equations. As a rule, the initial phase is considered equal to $\alpha_1=\pm\pi n$ $(n=0,1,2,\ldots)$. In this case $t(k,l)=0$, and Eq. (2.55) is reduced to a linear one, that is

$$g(k,l)\tan\varphi(k,l)+r(k,l)=0, \tag{2.59}$$

and

$$\tan\varphi(k,l)=-\frac{r(k,l)}{g(k,l)}, \tag{2.60}$$

where

$$g=[I_1-I_3]\sin\alpha_{21}+[I_2-I_1]\sin\alpha_{31}, \tag{2.61}$$

$$r=[I_2-I_3]+[I_3-I_1]\cos\alpha_{21}+[I_1-I_2]\cos\alpha_{31}, \tag{2.62}$$

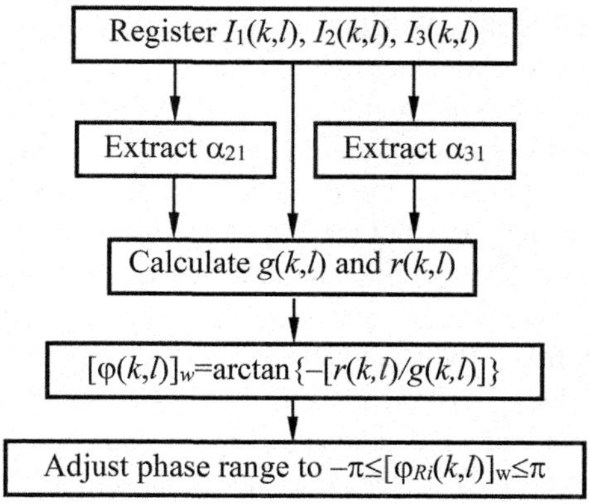

Fig. 2.24 Flow-chart of the 3SI UPS method first stage

and a coarse wrapped phase map $[\varphi(k, l)]_w$ is given by

$$[\varphi(k, l)]_w = \arctan\left[-\frac{r(k, l)}{g(k, l)}\right]. \tag{2.63}$$

The range of the coarse wrapped phase map $[\varphi(k, l)]_w$ is expanded twice, i.e. to $-\pi \leq \varphi_w(k, l) \leq \pi$ by using the "atan2" function [11]. A flow-chart of the first stage of the 3SI UPS method implementation displaying obtained theoretical results is depicted in Fig. 2.24.

The second stage of the 3SI UPS method is similar to the second stage of the ITSI UPS method, which is shown on the flow-chart in Fig. 2.20 after the operation of the coarse phase map $\varphi_w(k, l)$ generation ("$\varphi_w = \varphi_r + \tilde{\varphi}_m = \text{atan2}\left(\tilde{I}_1 \cos\alpha - \tilde{I}_2, \tilde{I}_1 \sin\alpha\right)$") and is highlighted by a dashed contour.

2.6.2 Estimation of Errors

Computer simulation of the developed 3SI UPS method was performed similarly to computer simulations of the TSI UPS and ITSI UPS methods. The test bipolar phase surface, depicted in Fig. 2.7, was used to analyze the 3SI UPS method reliability, as well as errors in both the unknown phase shifts extraction and the test surface relief retrieval.

Absolute errors of unknown phase shifts. The results of absolute errors of two unknown phase shifts α_{21} and $\alpha_{32} = \alpha_3 - \alpha_2 = \arccos[\rho(I_2, I_3)]$ extraction calculated by the 3SI UPS method [see Eq. (2.51)] showed that they have wavelike dependencies on the given phase shifts in the range from 0 to π. At the same time, the obtained dependencies indicate a slight increase in these errors, by a maximum of 2 times, in certain areas of given phase shifts from 0 to π, compared with the similar dependencies for such errors obtained by the TSI UPS method (see Sect. 2.5.2 and

Fig. 2.10). Note that the dependencies $|\Delta\alpha|$ versus α_1 shown in Fig. 2.10 are identical to dependencies of absolute errors $|\Delta\alpha_{21}|$ on the given phase shifts α_{21} calculated by the 3SI UPS method. For different combinations of phase shifts, these errors change. But in general, there is a slight increase up to 2 times in the absolute errors $|\Delta\alpha_{32}|$ calculated by the 3SI UPS method, compared with the absolute errors $|\Delta\alpha|$ calculated by the TSI UPS method, in certain intervals of phase shifts in the range from 0 to π. Nonetheless, there are phase intervals in which absolute errors $|\Delta\alpha_{32}|$ are less than errors $|\Delta\alpha|$. For example, the mean absolute errors (MAEs) $|\Delta\alpha_{\Sigma}|$ of the phase shift α_{21} between interferograms I_2 and I_1 using the TSI UPS method and the MAEs $|\Delta\alpha_{\Sigma,32}|$ of the phase shift α_{32} between interferograms I_3 and I_2 using the 3SI UPS method were analyzed [79]. To this end, the test bipolar phase surface (see Figs. 2.7 and 2.12) corrupted by additive Gaussian noise with RMS value $\sigma_{\varphi} = \pi/5$ radians was used. This procedure simulated a test nanorough phase surface with RMS roughness $Sq_{\varphi} = 4\pi/20$, which corresponds to RMS height $Sq = \lambda/20$. The results of the MAEs $|\Delta\alpha_{\Sigma}|$ and $|\Delta\alpha_{\Sigma,32}|$ calculations showed that $|\Delta\alpha_{\Sigma,32}| > |\Delta\alpha_{\Sigma}|$ in most of the phase shift range. Nevertheless $|\Delta\alpha_{\Sigma,32}| < |\Delta\alpha_{\Sigma}|$ in the range $18° < \alpha_{21} < 69°$, if $\alpha_{32} = 140°$, and in the range $96° < \alpha_{21} < 168°$, if $\alpha_{32} = 30°$. It is also shown that the MAE $|\Delta\alpha_{\Sigma,32}|$ decreases monotonically over the entire range of phase shifts with an increase in the phase roughness Sq_{φ} of the test bipolar phase surface.

Errors of phase maps retrieval. To assess the impact of errors on the total phase map $\varphi_T(k, l)$ retrieval with the help of the 3SI UPS method, the same bipolar phase test surface was used (see Fig. 2.7). Gaussian additive noise with different RMS values was added to the test for synthesis of test nanorough surfaces. An ideal cut-off filter, a Butterworth filter ($n = 1$) and a Gaussian filter were used to perform low-pass filtering during the first and second iterations. At the first iteration of the second stage of the algorithm performing the 3SI UPS method, the normalized cut-off frequency was equal $(f_{c0})_n = f_{c0}/f_s = 0.2$, where f_s is the spatial sampling frequency. This cut-off frequency was defined using the Yang algorithm [137]. At the second iteration of this algorithm at its second stage, the normalized cut-off frequency $f_{cn} = f_c/f_s = 3.33 \cdot 10^{-2}$ was chosen according to f_c-criterion proposed by Muravsky et al. [76, 78, 81]. The procedure of retrieval of the test bipolar nanorough phase surface with the RMS $Sq_{\varphi} = 4\pi/20$ (the RMS height $Sq_{\varphi} = \lambda/20$), as well as both the phase macrorelief and nanorelief, is shown in Fig. 2.25.

Histograms of the random errors distributions in the restored phase maps of the macrorelief φ_M, roughness (nanorelief) φ_R and total relief φ_T of the test bipolar nanorough surface with RMS $Sq_{\varphi} = 4\pi/20$ of the introduced Gaussian additive noise are shown in Fig. 2.26. The dimensions of all the phase maps are 512 × 512 pixels; i.e. each the map contains 2^{18} pixels. To extract the macrorelief and roughness phase maps, the ideal low-pass filter was used.

Retrieving the total relief phase map φ_T from the test bipolar nanorough surface at RMS $Sq_{\varphi} = 4\pi/20$ was performed using the Gaussian, Butterworth and ideal LP filters. If $\alpha_{21} = 30°$ and $\alpha_{31} = 47°$, and also if Butterworth and Gaussian filters are used, the MAE values are 4.60×10^{-2} rad and 5.41×10^{-2} rad, respectively, and the RMS values of random errors are 2.75×10^{-2} rad and 3.22×10^{-2} rad, respectively.

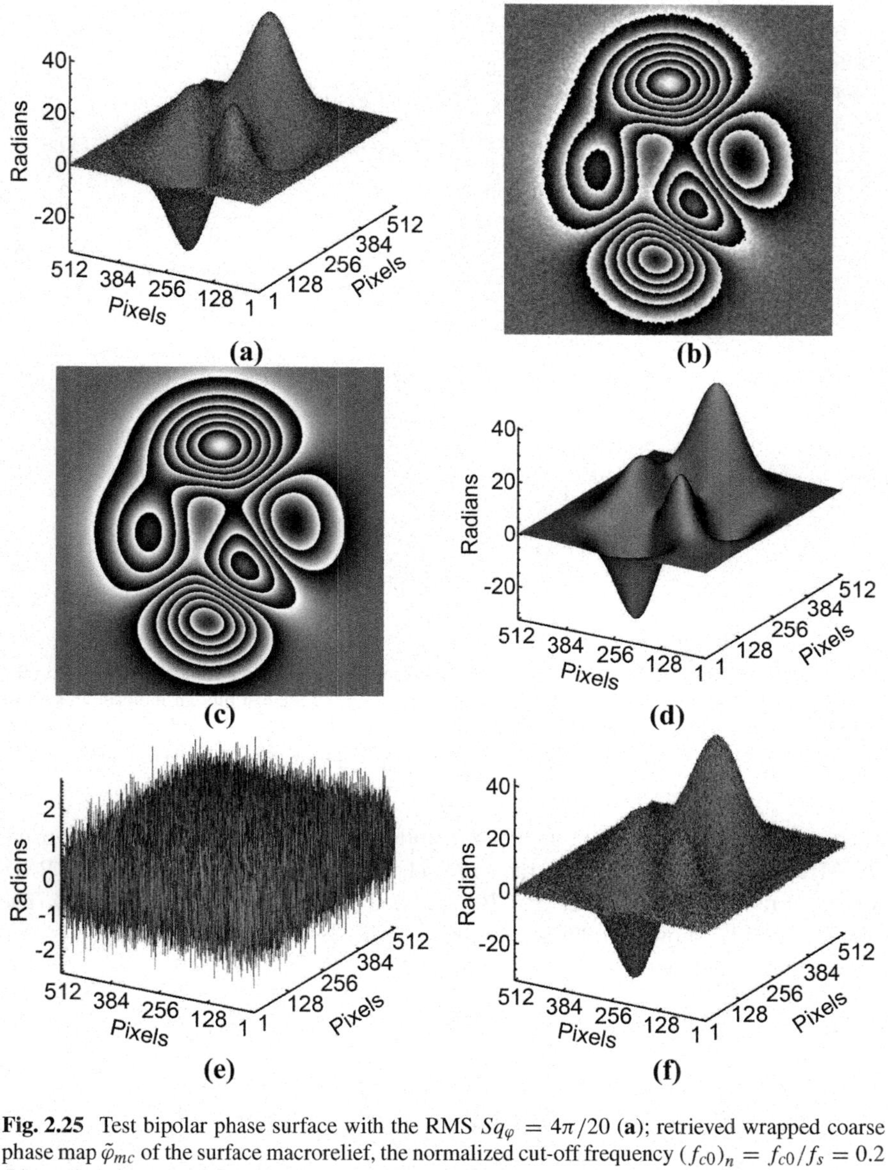

Fig. 2.25 Test bipolar phase surface with the RMS $Sq_\varphi = 4\pi/20$ (**a**); retrieved wrapped coarse phase map $\tilde{\varphi}_{mc}$ of the surface macrorelief, the normalized cut-off frequency $(f_{c0})_n = f_{c0}/f_s = 0.2$ (**b**); retrieved wrapped fine phase map $(\varphi_M)_w$, the normalized cut-off frequency $f_{cn} = f_c/f_s = 3.33 \cdot 10^{-2}$ (**c**); the retrieved macrorelief phase map φ_M (**d**); the retrieved nanorelief (roughness) phase map φ_R (**e**); the total phase map φ_T (**f**)

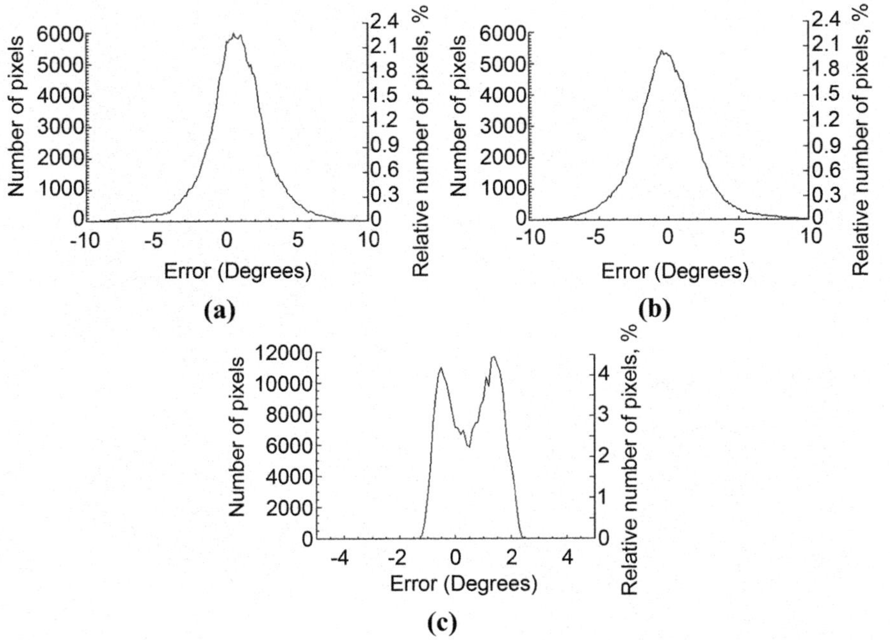

Fig. 2.26 Histograms of the random errors distributions in the restored phase maps of macrorelief φ_M (**a**), roughness (nanorelief) φ_R (**b**) and total relief φ_T (**c**) of the test bipolar nanorough surface with RMS $Sq_\varphi = 4\pi/20$ of introduced Gaussian additive noise and using the ideal low-pass filter with normalized cut-off frequency $f_{cn} = 3.33 \cdot 10^{-2}$. Each phase map contains 2^{18} pixels

If $\alpha_{21} = 30°$ and $\alpha_{31} = 140°$, and also if Butterworth and Gaussian filters are used, the MAE values are 4.60×10^{-2} rad and 5.41×10^{-2} rad, respectively, and the RMS values of random errors are 2.75×10^{-2} rad and 3.22×10^{-2} rad, respectively. The obtained results show that the MAEs decrease if $\alpha_{31} \rightarrow \pi$. A similar trend is also observed for random errors.

2.6.3 Implementation of the Method of Three-Step Interferometry with Unknown Phase Shifts

Experimental implementation of the 3SI UPS method was performed using the experimental setup based on the Twyman–Green interferometer, which scheme is depicted in Fig. 2.13. However, the control unit 12 in this setup has more functionality. The main difference of this control unit from the one mentioned in Sect. 2.5.4 is that it provides a smooth phase shift procedure, i.e. the integrating-bucket technique [37, 104, 130], and is used to implement the 3SI UPS method. The control unit generates a saw-tooth voltage that provides the smooth phase shift of the reference wave in the interferometer using the PZT. When a saw-tooth voltage from 0 to 48 V is applied, the mirror attached to the PSE in the PZT shifts by a distance δd, and the reference beam in the Twyman–Green interferometer shifts by a distance $2(\delta d)$; i.e. the smooth phase shift of the reference wavefront is equal to $\delta\varphi = 4\pi(\delta d)/\lambda$. The time of applying a smooth voltage to the PSE can be adjusted in the range from 0.2 to 90 s. The period of the reference phase change Δ_t during interferograms recording is chosen taking into account the MTF $\gamma = \mathrm{sinc}\left(\frac{\Delta_t}{2\pi}\right)$ and the fundamental equation for PSI adapted to the integrating-bucket technique [see Eq. (2.6)]. In Sect. 2.1, it is shown that if $\Delta_t = 0.06\pi$, then $\gamma = 0.9985$. Exposures providing such the period are quite real. For example, three interferograms of a flat nanorough surface are given in Fig. 2.27. The first of them was obtained without voltage (0 V), the second—for 12 V and the third—for 20 V. During the experiment, the time of smooth voltage rise is $T = 3.0$ s, the exposure time is $\tau_j = 0.001$ s, and $\gamma = 1.0$; i.e. MTF for smooth phase shift at such exposures remains the same as for stepwise shift.

For implementation of the 3SI UPS method, the gauge block was used. Its arithmetic mean deviation of the roughness profile is $Ra < 0.020$ μm, and a metric nominal length for this roughness is $l_r = 0.08$ mm [49]. In the experimental setup shown in Fig. 2.13, the sampling interval was chosen equal to $\tau_s = 2.0$ μm in the optical scheme of the Twyman–Green interferometer and the linear magnification was $M_l = 2.325$, because the pixel pitch of the matrix photodetector of the digital camera SONY XCD-SX910 (1/2" IT CCD-SXGA, 1280 $(H) \times 960$ (V) is $p_x = p_y = 4.65$ μm. This increase makes it possible to record the GB surface area equal to 2.048×1.536 mm^2.

The Gaussian LP filters with normalized spatial cut-off frequencies equal to $(f_{c0})_n = f_{c0}/f_s = 0.21$ and $f_{cn} = f_c{\cdot}f_s = 3.35 \cdot 10^{-2}$ for the first and second iterations, respectively, were used to implement the 3SI UPS method. Unknown phase shifts α_{21}, α_{31} and, if necessary, α_{32} were calculated using recorded interferograms I_1, I_2 and I_3. With the help of the developed software, phase shifts $\alpha_{21} = 23.4°$ and $\alpha_{31} = 79.5°$ were calculated. Entering their values in the software implementing the 3SI UPS method, the spatial phase distributions of intermediate phase maps, as well as final phase maps of the surface area macrorelief, nanorelief and total relief were retrieved.

The initial three interferograms of the GB surface area recorded in the experimental setup containing the Twyman–Green interferometer are shown in Fig. 2.28.

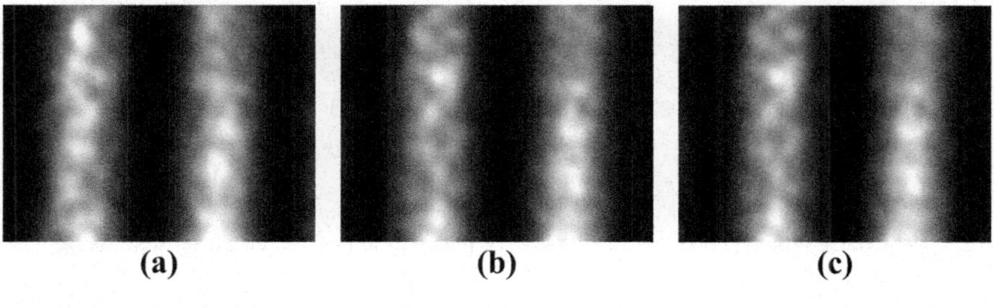

Fig. 2.27 Interferograms of a flat nanorough surface recording by the integrating-bucket technique: $U = 0$ V (**a**), $U = 12$ V (**b**), $U = 20$ V (**c**)

The coarse wrapped phase map $\tilde{\varphi}_{mc}(k, l)$ of the macrorelief surface area obtained after first iteration and the fine wrapped phase map $[\varphi_M(k, l)]_w$ of the macrorelief surface area obtained during second iteration are shown in Fig. 2.29.

The unwrapped fine phase map $\varphi_M(k, l)$ of the macrorelief, the fine phase map $\varphi_R(k, l)$ of the roughness (nanorelief) and the final fine total phase map $\varphi_T(k, l)$ of the GB surface area are shown in Fig. 2.30.

The fine total phase map $\varphi_T(k, l)$ of the GB surface area was transformed into total surface relief height map $h_T(k, l)$. According to the obtained spatial distributions of the heights of the surface area, the arithmetic mean height of the scale-limited surface was calculated, which in discrete form is given by

$$Sa = \frac{1}{N \cdot M} \sum_{m=0}^{M-1} \sum_{n=0}^{N-1} |z(x_m, y_n)|. \tag{2.64}$$

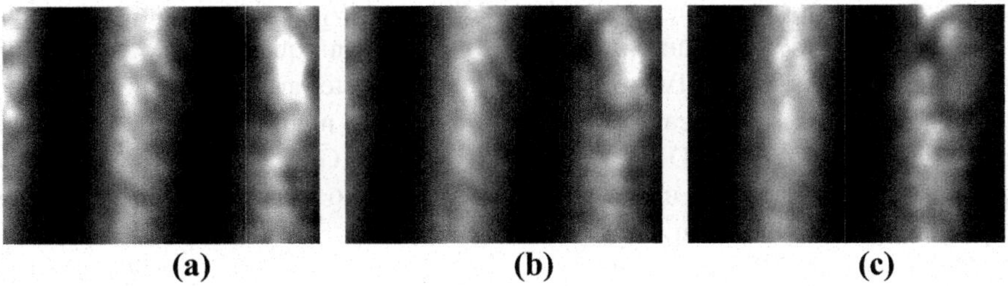

Fig. 2.28 Recorded fringe patterns of the GB surface area: initial fringe pattern $I_1(k, l)$ without phase shift of the reference beam (**a**); fringe pattern $I_2(k, l)$, if the reference beam phase is shifted by $\alpha_{21} = 23.4°$ (**b**); fringe pattern $I_3(k, l)$, if the reference beam phase is shifted by $\alpha_{31} = 79.5°$ (**c**)

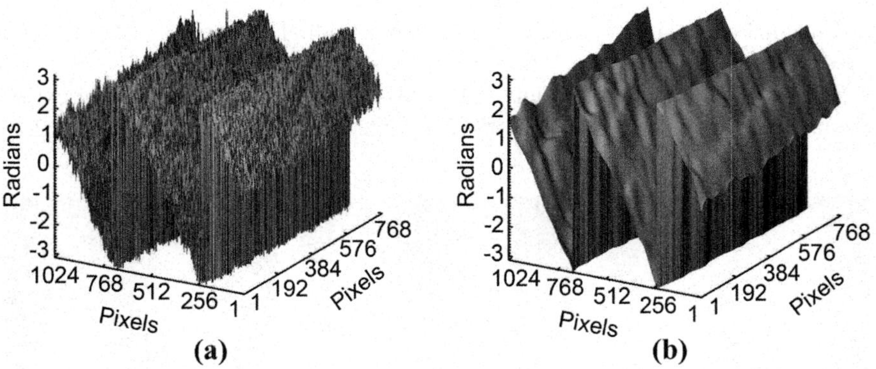

Fig. 2.29 Coarse wrapped phase map $\tilde{\varphi}_{mc}(k, l)$ of the macrorelief obtained after first iteration (**a**); the fine wrapped phase map $[\varphi_M(k, l)]_w$ of macrorelief obtained during second iteration (**b**)

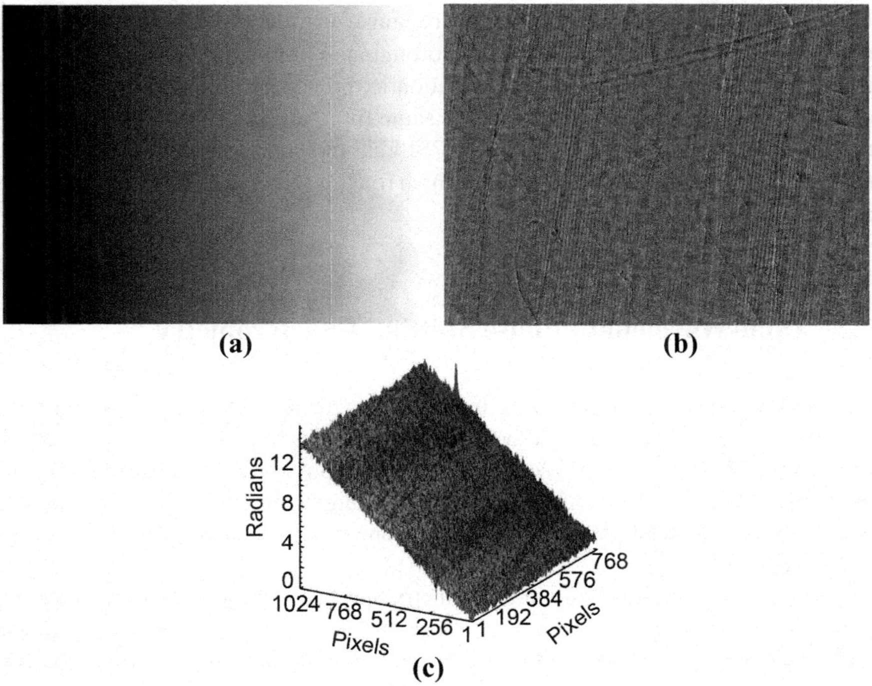

Fig. 2.30 Retrieved fine phase maps: phase map $\varphi_M(k, l)$ of macrorelief (**a**); phase map $\varphi_R(k, l)$ of roughness (nanorelief) (**b**); total phase map $\varphi_T(k, l)$ of surface relief area (**c**)

The results of calculations have shown that for the chosen surface area of the GB the arithmetic mean height $Sa = 14.7$ nm and the root mean square height $Sq = 18.8$ nm. These results good coincide with the nominal arithmetic mean deviation of this gauge block roughness profile equal to $Ra < 0.020$ μm [49].

The proposed 3SI UPS method has several common features and properties with the TSI UPS and ITSI UPS methods. In particular, it also uses a correlation approach to the determination of unknown phase shifts α_{21}, α_{31} and α_{32}. However, the method uses three interferograms instead of the two ones used in the TSI UPS and ITSI UPS methods. At the same time, an important feature of the 3SI UPS method is that it is not necessary to additionally record the intensity spatial distributions $I_o(k, l)$ and $I_r(k, l)$ of the object and reference beams. Therefore, the two-step PSI methods are time-consuming and not always can be used to study mechanical properties of constructive materials and structural elements, especially when they are studied under the influence of vibrations or aggressive environments. The 3SI UPS method can be used to record the kinetics of surface changes under the action of mechanical loads or aggressive environments, and the speed of the method will depend mainly on the capabilities of the control units for the registering fringe patterns. In addition, this method has the ability to extract both high-frequency and low-frequency components of the surface relief, including 3D roughness and 3D waviness. The 3SI

UPS method is based on the principles of single-wavelength PSI. Therefore, it is effective only for retrieving smooth, nanorough and partially microrough surfaces, if the laser source generates optical radiation in the visible or near-infrared regions of the electromagnetic spectrum. At the same time, the use of two-wavelength or multiwavelength PSI, which can use the 3SI UPS method, will significantly expand the range of rough surfaces retrieval, in which roughness parameters reach units, tens and even hundreds of microns.

2.7 Dual-Wavelength Phase Shifting Interferometry

In recent decades, new methods of multiwavelength laser interferometry with the simultaneous use of two, three or more coherent sources have been intensively developed [47, 103, 104, 125]. These methods make it possible to significantly expand the range of measurements of the surface relief heights by generating a synthesized wave, the beat wavelength of which is much longer than the real wavelengths of the laser radiation.

If in single-wavelength laser interferometry the phase difference between adjacent pixels of the retrieved phase mask of a test surface exceeds π, then, as mentioned above (see Sect. 2.1), there is a phase ambiguity modulo 2π [103]. In this case, it is impossible to restore the phase map site in these two pixels. That is, exceeding the phase interval π between adjacent pixels causes the so-called 2π ambiguity problem. In particular, if the interferogram is produced in a two-beam interferometer with a normal incidence of the beam on the test surface, the 2π ambiguity problem will not occur if the height difference δh_{adj} between adjacent samples of the surface relief does not exceed $\lambda/4$ [see Eqs. (2.2) and (2.14)]. For example, the 2π ambiguity in a two-beam single-wavelength interferometer containing a He–Ne laser ($\lambda = 633$ nm) does not arise, if $\delta h_{\mathrm{adj}} \leq 158$ nm.

Dual-wavelength interferometry methods make it possible to significantly expand the range of measurement of surface height differences Δh by using two wavelengths λ_1 and λ_2 in an optical interferometer, which leads to the formation of the synthesized beat wave. The principles of dual-wavelength interferometry were laid down in the method of dual-wavelength holography developed by Wyant and designed to control aspherical surfaces [129]. Later, the dual-wavelength interferometry techniques were updated and modified by Cheng et al. [13, 15, 19, 132, 133]. Dual-wavelength interferometry makes it possible to significantly expand the range of measuring the surface relief within 2π due to the synthetic beat wavelength

$$\Lambda_{12} = \frac{\lambda_1 \lambda_2}{|\lambda_1 - \lambda_2|}. \tag{2.65}$$

The wavelength Λ_{12} exceeds the wavelengths λ_1 and λ_2, which are used to generate and record series of interferograms. Equation (2.65) shows that the smaller the difference between the wavelengths λ_1 and λ_2, the greater the wavelength Λ_{12}. Due to the formation of Λ_{12}, the real height h_Λ of the surface relief in two-beam interferometers at normal incidence of the object wave to the test surface, taking into account Eq. (2.2), is given by

$$h_\Lambda = \frac{\Lambda_{12} \cdot (\Delta\varphi)}{4\pi}. \tag{2.66}$$

So, the retrieved surface relief heights can be as many times larger in comparison with the relief heights, which are retrieved using the single-wavelength PSI, how many times Λ_{12} is larger than λ_1 and λ_2. Hence, the surface height jumps that can be retrieved are increased by Λ_{12}/λ_1 or Λ_{12}/λ_2 times.

2.7.1 Basic Methods of Dual-Wavelength Interferometry

Two main methods of dual-wavelength interferometry, proposed by Cheng and Wayant [13], have been developed to solve the problem of 2π-uncertainty. In the first method, a series of interferograms with phase shifts $\Delta\delta_j = \alpha = 2\pi j/J$ is recorded at the wavelength λ_1 and a wrapped phase map modulo 2π is produced. Using second wavelength λ_2, a series of interferograms with the appropriate phase shifts is also recorded and a second wrapped phase map is produced. These two wrapped maps are combined in such a way as to produce a new wrapped map $\varphi_{12}(k, l)$ modulo 2π, the phase period of which already corresponds to the wavelength Λ_{12}. The phase corresponding to Λ_{12} can be given by [13]

$$\varphi_{12} = \varphi_1 - \varphi_2 = \frac{2\pi\,\Delta_{\mathrm{op}}}{\Lambda_{12}}, \tag{2.67}$$

where φ_1 and φ_2 are the phases of wavefronts for λ_1 and λ_2; Δ_{op} is the difference optical path.

After unwrapping of the phase map $\varphi_{12}(k, l)$, the height jumps of the surface relief between two consecutive samples (pixels) in a restored unwrapped phase map $\varphi_{12u}(k, l)$ are already reproduced within $\Delta h_{12} \le \Lambda_{12}/4$. This indicates that dual-wavelength interferometry provides retrieval of Λ_{12}/λ_1 or Λ_{12}/λ_2 times greater surface heights than is provided in single-wavelength interferometry with corresponding wavelengths λ_1 or λ_2. For example, if the wavelengths are $\lambda_1 = 640$ nm and $\lambda_2 = 660$ nm, then $\Lambda_{12} = 21.12$ μm. In this case, the allowable height differences for single-wavelength interferometry are $\Delta h_1 = \lambda_1/4 = 160$ nm and $\Delta h_2 = \lambda_2/4 = 165$ nm, and for dual-wavelength $\Delta h_{12} = \Lambda_{12}/4 = 5.28$ μm.

In the second method, two series of interferograms for λ_1 and λ_2 are recorded and the wrapped phase map $\varphi_{12}(k, l)$ is calculated directly from all obtained interferograms using the next equation [13]:

$$\varphi_{12} = \text{arctg}\left[\frac{\sin(\varphi_1 - \varphi_2)}{\cos(\varphi_1 - \varphi_2)}\right] = \text{arctg}\left[\frac{\sin\varphi_1\cos\varphi_2 - \cos\varphi_1\sin\varphi_2}{\cos\varphi_1\cos\varphi_2 + \sin\varphi_1\sin\varphi_2}\right]. \quad (2.68)$$

As shown in [103], any phase shifting algorithm can be used for dual-wavelength interferometry.

However, the increase in height differences by Λ_{12}/λ_1 or Λ_{12}/λ_2 times in the surface relief retrieved with the help of dual-wavelength interferometry is accompanied by approximately the same increase in noise. To eliminate these noises, Parshall and Kim [87] developed the algorithm containing three stages, namely (i) producing a coarse map that is free of discontinuities and possesses a longer range (by Λ_{12}/λ_1 or Λ_{12}/λ_2 times); (ii) suppression of amplified noise; (iii) the fine map synthesis [87]. The similar ideology for dual-wavelength digital speckle shearing interferometry using the algorithm of phase error correction combined with windowed Fourier filtering for noise suppression was proposed by Tay et al. [113]. The doubling of the number of recorded fringe patterns is another problem of dual-wavelength interferometry, especially when multistep methods are used. For example, the time-consuming seven-phase step algorithm for each wavelength to restore 3D surface profiles using the two-wavelength sequential illumination technique was proposed by Kumar et al. [57].

2.7.2 *Two-Step Dual-Wavelength Interferometry Method: Theoretical Model*

In contrast to multistep dual-wavelength PSI, a new two-step dual-wavelength interferometry method (TS DWIM), which uses only two pairs of fringe patterns for surface relief retrieval was developed at PMI [52]. Each pair is produced for one of two wavelengths λ_1 or λ_2, and two fringe patterns in each pair are differed only by an arbitrary phase shift of a reference wave. Simultaneously, this method effectively suppresses the noise of the synthesized interferogram with a wavelength Λ_{12}. In this method, the phase shift calibration procedure is absent and it is enough to use only one phase shift to record both pairs of fringe patterns.

When studying the temporal evolution of the surface under static or cyclic loads or under the influence of aggressive media, this method can be combined with the two-step PSI methods, in particular with TSI UPS and ITSI UPS ones (see Sect. 2.5). It is especially important to combine these methods when the maximum difference in height between adjacent samples begins to exceed $\lambda_1/4$ or $\lambda_2/4$. This combination of the TS DWIM with TSI UPS or ITSI UPS methods makes it possible to constantly monitor the surface nanorelief, the roughness of which is monotonically increasing.

Extraction of unknown phase shift. The algorithm for extraction an unknown phase shift based on the PPCC calculation considered in Sect. 2.4 can be used for the TS DWIM, taking into account the peculiarities of dual-wavelength PSI. For wavelength λ_1, Eq. (2.17) can be rewritten as

$$\left.\begin{aligned} I_{11}(k,l) &= I_{B1}(k,l) + I_{M1}(k,l)\cos[\varphi_1(k,l)] \\ I_{12}(k,l) &= I_{B1}(k,l) + I_{M1}(k,l)\cos[\varphi_1(k,l) + \alpha_1] \end{aligned}\right\} \tag{2.69}$$

and for wavelength λ_2 as

$$\left.\begin{aligned} I_{21}(k,l) &= I_{B2}(k,l) + I_{M2}(k,l)\cos[\varphi_2(k,l)] \\ I_{22}(k,l) &= I_{B2}(k,l) + I_{M2}(k,l)\cos[\varphi_2(k,l) + \alpha_2] \end{aligned}\right\}, \tag{2.70}$$

where I_{11} and I_{12} are the intensities of fringe patterns recorded at the wavelength λ_1; I_{21} and I_{22} are the intensities of fringe patterns recorded at the wavelength λ_2; $I_{B1} = I_{o1} + I_{r1}$ and $I_{B2} = I_{o2} + I_{r2}$ are the background intensities at the wavelengths λ_1 and λ_2; $I_{M1} = 2\sqrt{I_{o1}I_{r1}}$ and $I_{M2} = 2\sqrt{I_{o2}I_{r2}}$ are the visibility functions; I_{o1} and I_{o2} are the intensities distributions of the object wavefront at λ_1 and λ_2, respectively; I_{r1} and I_{r2} are the intensities distributions of the reference wavefront at λ_1 and λ_2, respectively.

Unknown phase shifts α_1 and α_2 are defined using PPCC and as well as Eq. (2.23) can be given by

$$\alpha_1 = \arccos[\rho(I_{11}, I_{12})] = \arccos\frac{\langle (I_{11} - \langle I_{11}\rangle)(I_{12} - \langle I_{12}\rangle)\rangle}{\sigma_{I11}\sigma_{I12}}, \tag{2.71}$$

$$\alpha_2 = \arccos[\rho(I_{21}, I_{22})] = \arccos\frac{\langle (I_{21} - \langle I_{21}\rangle)(I_{22} - \langle I_{22}\rangle)\rangle}{\sigma_{I21}\sigma_{I22}}, \tag{2.72}$$

where σ_{I11} and σ_{I12} are the standard deviations of I_{11} and I_{12}; σ_{I21} and σ_{I22} are the standard deviations of I_{21} and I_{22}.

Technically, the proposed TS DWIM is the easiest to implement without changing the phase shift α_2 when recording two interferograms I_{21} and I_{22} at wavelength λ_2. In this case, taking into account Eq. (2.2), the shift of the reference wavefront with the wavelength λ_1 by a distance Δl is given by

$$\Delta l = \frac{\lambda_1}{4\pi}\alpha_1. \tag{2.73}$$

The shift of the reference wavefront with the wavelength λ_2 by the same distance Δl is given by

$$\Delta l = \frac{\lambda_2}{4\pi}\alpha_2. \tag{2.74}$$

Therefore, if the displacement Δl is the same for two pairs of interferograms, the relationship between α_1 and α_2 using (2.73) and (2.74) is expressed as

$$\alpha_2 = \frac{\lambda_1}{\lambda_2}\alpha_1, \quad \alpha_1 = \frac{\lambda_2}{\lambda_1}\alpha_2. \tag{2.75}$$

Therefore, the only one phase shift, for example α_1, can be used to record two pairs of fringe patterns. In this case, Eq. (2.69) remains unchanged, and Eq. (2.70), taking into account (2.75), becomes

$$\left.\begin{array}{l} I_{21}(k,l) = I_{B2}(k,l) + I_{M2}(k,l)\cos[\varphi_2(k,l)] \\ I_{22}(k,l) = I_{B2}(k,l) + I_{M2}(k,l)\cos\left[\varphi_2(k,l) + \frac{\lambda_1}{\lambda_2}\alpha_1\right] \end{array}\right\}. \tag{2.76}$$

Algorithm for implementation two-step dual-wavelength interferometry method. Due to expanding the height range of the surface relief retrieved by the TS DWIM, a total phase map $\varphi_{T,\Lambda}(k,l)$ of the surface relief area can be considered as the pixel-by-pixel summation of the microrelief phase map $\varphi_{R,\Lambda}(k,l)$ or surface roughness phase map, as well as the macrorelief phase map $\varphi_{M,\Lambda}(k,l)$, i.e.

$$\varphi_{T,\Lambda}(k,l) = \varphi_{R,\Lambda}(k,l) + \varphi_{M,\Lambda}(k,l). \tag{2.77}$$

In Eq. (2.77), the subscripts "$_{12}$" in Λ_{12} are omitted for brevity. We will use this simplification for most of subsequent equations and formulas.

In order to retrieve these maps at the first stage of the algorithm for implementing the TS DWIM, elementary trigonometric transformations, similar to transformation of Eq. (2.31) into Eq. (2.32), are performed using Eqs. (2.69) and (2.70), that is

$$\left.\begin{array}{l} \frac{\tilde{I}_{12}(k,l)}{\tilde{I}_{11}(k,l)} = \cos\alpha_1 - \tan\varphi_{1,w}(k,l)\sin\alpha_1 \\ \frac{\tilde{I}_{22}(k,l)}{\tilde{I}_{21}(k,l)} = \cos\alpha_2 - \tan\varphi_{2,w}(k,l)\sin\alpha_2 \end{array}\right\}. \tag{2.78}$$

where $\tilde{I}_{11} = I_{11} - I_{B1}$, $\tilde{I}_{12} = I_{12} - I_{B1}$, $\tilde{I}_{21} = I_{21} - I_{B2}$, $\tilde{I}_{22} = I_{22} - I_{B2}$, $\varphi_{1,w}(k,l) = \tilde{\varphi}_{r1}(k,l) + \tilde{\varphi}_{m1}(k,l)$ is the coarse wrapped phase map of the surface relief area for wavelength λ_1, $\varphi_{2,w}(k,l) = \tilde{\varphi}_{r2}(k,l) + \tilde{\varphi}_{m2}(k,l)$ is the coarse wrapped phase map of the surface relief area for wavelength λ_2, $\tilde{\varphi}_{r1}$ and $\tilde{\varphi}_{r2}$ are the phase maps of surface roughness area for wavelengths λ_1 and λ_2, respectively; $\tilde{\varphi}_{m1}$ and $\tilde{\varphi}_{m2}$ are the wrapped phase maps of the surface macrorelief for wavelengths λ_1 and λ_2, respectively.

Two Eq. (2.78) which are being used to retrieve the coarse wrapped phase maps $\varphi_{1,w}(k,l)$ in range $-\pi \le \varphi_{1,w}(k,l) \le \pi$ and the coarse phase map $\varphi_{2,w}(k,l)$ in range $-\pi \le \varphi_{2,w}(k,l) \le \pi$ are given by

$$\begin{aligned} \varphi_{1,w}(k,l) &= \tilde{\varphi}_{r1}(k,l) + \tilde{\varphi}_{m1}(k,l) \\ &= \operatorname{atan2}\left[\tilde{I}_{11}(k,l)\cos\alpha_1 - \tilde{I}_{12}(k,l), \tilde{I}_{11}(k,l)\sin\alpha_1\right], \end{aligned} \tag{2.79}$$

$$\begin{aligned} \varphi_{2,w}(k,l) &= \tilde{\varphi}_{r2}(k,l) + \tilde{\varphi}_{m2}(k,l) \\ &= \operatorname{atan2}\left[\tilde{I}_{21}(k,l)\cos\alpha_2 - \tilde{I}_{22}(k,l), \tilde{I}_{21}(k,l)\sin\alpha_2\right]. \end{aligned} \tag{2.80}$$

After calculation of the phase maps $\varphi_{1,w}(k,l)$ and $\varphi_{2,w}(k,l)$ using Eqs. (2.79) and (2.80), the difference wrapped phase map $\left[\tilde{\varphi}_{\Lambda}(k,l)\right]_w$ for synthetic wavelength Λ_{12} is given by

$$(\tilde{\varphi}_{\Lambda})_w = \varphi_{1,w} - \varphi_{2,w} = \operatorname{atan2}\left(\tilde{I}_{11}\cos\alpha_1 - \tilde{I}_{12}, \tilde{I}_{11}\sin\alpha_1\right) \\ - \operatorname{atan2}\left(\tilde{I}_{21}\cos\alpha_2 - \tilde{I}_{22}, \tilde{I}_{21}\sin\alpha_2\right). \quad (2.81)$$

The obtained synthesized phase map

$$(\tilde{\varphi}_{\Lambda})_w = \varphi_{\Lambda r} + (\varphi_{\Lambda m})_w \quad (2.82)$$

provides the phase retrieval in four quadrants in range $-2\pi \le (\tilde{\varphi}_{\Lambda})_w \le 2\pi$, where $\varphi_{\Lambda r}(k,l)$ is the course microrelief phase map. To reduce this range twice to $-\pi \le (\varphi_{\Lambda})_w \le \pi$, the harmonic terms $I_{s,o}(k,l) = \sin\left[\varphi_{\Lambda r} + (\varphi_{\Lambda m})_w\right]$ and $I_{c,o}(k,l) = \cos\left[\varphi_{\Lambda r} + (\varphi_{\Lambda m})_w\right]$ are used. Therefore, the synthesized difference phase map $\left[\tilde{\varphi}_{\Lambda}(k,l)\right]_w$ is transformed into a new course wrapped phase map $[\varphi_{\Lambda}(k,l)]_w$ of the total surface relief, which is expressed as

$$(\varphi_{\Lambda})_w = \operatorname{atan2}\left(I_{s,o}, I_{c,o}\right) \\ = \operatorname{atan2}\left\{\sin\left[\varphi_{\Lambda r} + (\varphi_{\Lambda m})_w\right], \cos\left[\varphi_{\Lambda r} + (\varphi_{\Lambda m})_w\right]\right\}. \quad (2.83)$$

The flow-chart of the first stage of the algorithm for the TS DWIM realization is shown in Fig. 2.31.

The next stage of this algorithm implementation is similar to the second stage of the algorithm performing the ITSI UPS method (see Sect. 2.5.5), which is highlighted by a dashed contour. However, in this algorithm, the phase map $[\varphi_{\Lambda}(k,l)]_w$ is used

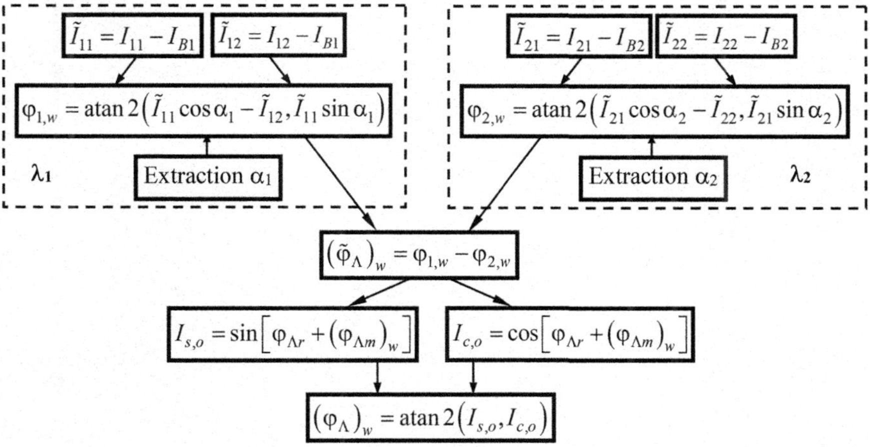

Fig. 2.31 Flow-chart of the first stage of the algorithm for the TS DWIM realization

instead of the map $\varphi_w(k, l)$. As a result, the fine phase map $\varphi_{T,\Lambda}(k, l)$ of the total relief given by Eq. (2.77) is obtained.

2.7.3 Computer Simulation of Two-Step Dual-Wavelength Interferometry Method

Computer simulations of the TS DWIM were performed with the aim to its verification using three test phase surfaces that are often used in dual-wavelength interferometry for developed algorithms reliability assessment. These test surfaces, namely phase slope and phase step, are shown in Figs. 2.32 and 2.33. The test bipolar phase surface depicted in Fig. 2.7 is also used to estimate random errors of retrieved synthesized final phase maps $\varphi_{R,\Lambda}(k, l)$, $\varphi_{M,\Lambda}(k, l)$ and $\varphi_{T,\Lambda}(k, l)$. These test phase maps can be easily transformed into the test height maps imitating the surface reliefs using Eq. (2.2). The additive Gaussian phase noise with different RMS σ_h was introduced into each test smooth surface shown in Figs. 2.7, 2.32 and 2.33 for estimation of the areal surface roughness influence on a phase relief retrieval errors.

Two pairs of fringe patterns were synthesized according to Eqs. (2.69) and (2.70) for two wavelengths $\lambda_1 = 532.0$ nm and $\lambda_2 = 632.8$ nm, respectively. Introducing phase shifts α_1 for the first pair of Eq. (2.69) and α_2 for the second pair of Eq. (2.70)

Fig. 2.32 Phase slope

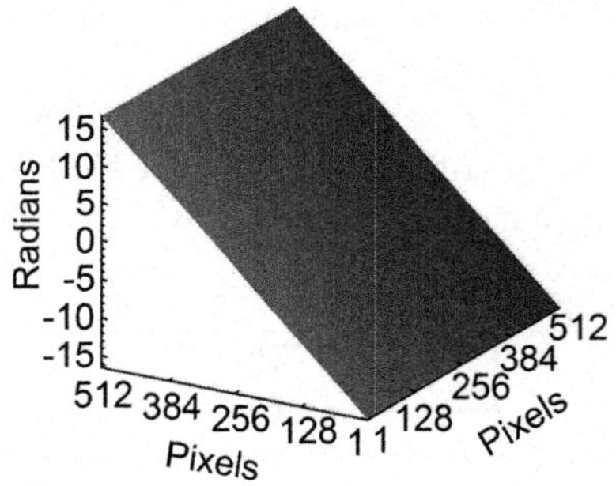

Fig. 2.33 Phase step

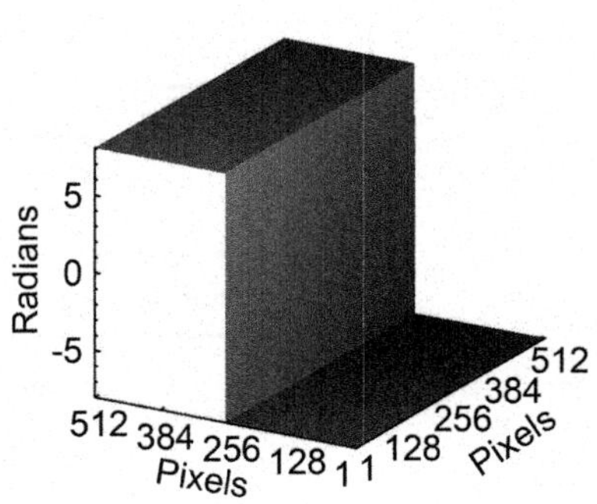

was performed by taking into account Eq. (2.76). Using Eqs. (2.69) and (2.76), the only one phase shift can be fulfilled during recording fringe patterns I_{11} and I_{12} for wavelength λ_1 and fringe patterns I_{21} and I_{22} for wavelength λ_2. In this case, Eq. (2.75) is used to calculate phase shift angle α_2, if α_1 is given, and vice versa. Such choice of phase shifts makes it possible to carry out the next procedure for recording two series of fringe patterns in a two-wavelength interferometer containing two lasers with radiation wavelengths λ_1 and λ_2:

(i) record the interferogram I_{11} by the laser 1 without phase shift ($\alpha_1 = 0$); record the reference I_{r1} and object I_{o1} beams to obtain the background $I_{B1} = I_{r1} + I_{o1}$;
(ii) perform a blind phase shift α_1 of the laser 1 reference beam and record the interferogram I_{12};
(iii) overlap the initial unsplit beam of the laser 1 and open the beam of the laser 2;
(iv) record the interferogram I_{22} by the laser 2 at the same phase shift α_1 corresponding to the phase shift $\frac{\lambda_2}{\lambda_1}\alpha_2$;
(v) return the reference beam of the laser 2 to the initial state (before its phase shift) and record the interferogram I_{21} in the absence of phase shift ($\alpha_2 = 0$); record the reference I_{r2} and object I_{o2} beams to obtain the background $I_{B2} = I_{r2} + I_{o2}$.

The same dependence as between α_1 and α_2 in Eq. (2.75) takes place between the phase random noise introduced into a test object, which RMS corresponds to the phase RMS height $Sq_{1\varphi}$ of the scale-limited surface for λ_1, and the phase random noise introduced into the same test object, which RMS corresponds to the phase RMS height $Sq_{2\varphi}$ for λ_2, that is

$$\frac{Sq_{1\varphi}}{Sq_{2\varphi}} = \frac{\alpha_2}{\alpha_1}, \tag{2.84}$$

besides,

$$Sq_{1\varphi} = \frac{4\pi(Sq)}{\lambda_1}, \quad Sq_{2\varphi} = \frac{4\pi(Sq)}{\lambda_2}, \tag{2.85}$$

where Sq is the RMS height of the scale-limited surface.

Figures 2.34 and 2.35 show the results of the TS DWIM performing on the example of the test phase slope surface with a size of 512 × 512 pixels (see Fig. 2.32). Two series of the phase slope fringe patterns described by Eqs. (2.69) and (2.70) were synthesized in accordance with Eqs. (2.75), (2.84) and (2.85) and using phase RMS heights $Sq_{1\varphi} = 0.373$ rad and $Sq_{2\varphi} = 0.314$ rad those correspond to the RMS height $Sq = \lambda/40 = 15.8$ nm, and phase shifts $\alpha_1 = 35.68°$ and $\alpha_2 = 30.0°$ for $\lambda_1 = 532.0$ nm and $\lambda_2 = 632.8$ nm, respectively. The phase slope height has been chosen so that to provide one period of wavelength Λ for the retrieved PM. In this case, the second iteration and unwrapping are omitted.

Similar results were obtained for the same test phase slope surface with phase RMS heights $Sq_{1\varphi} = 1.502$ rad and $Sq_{2\varphi} = 1.257$ rad those correspond to the RMS height $Sq = \lambda/10 = 63.3$ nm, and phase shifts $\alpha_1 = 35.68°$ and $\alpha_2 = 30.0°$ for

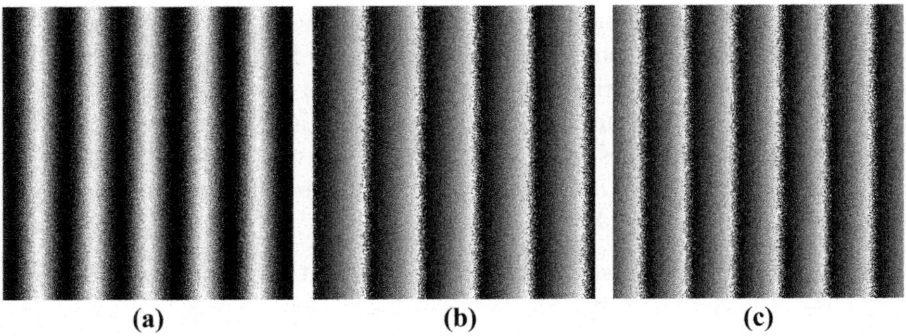

Fig. 2.34 Retrieval of the phase slope with phase RMS height $Sq_{2\varphi} = 0.314$ rad: initial interferogram $I_{21}(k, l)$ with phase RMS height $Sq_{2\varphi} = 0.314$ rad for wavelength $\lambda_2 = 632.8$ nm (**a**); retrieved phase map $\varphi_{2,w}(k, l)$ with phase RMS height $Sq_{2\varphi} = 0.314$ rad for $\alpha_2 = 30.0°$ and $\lambda_2 = 632.8$ nm (**b**); retrieved phase map $\varphi_{1,w}(k, l)$ with phase RMS height $Sq_{1\varphi} = 0.373$ rad for $\alpha_1 = 35.68°$ and $\lambda_1 = 532.0$ nm (**c**)

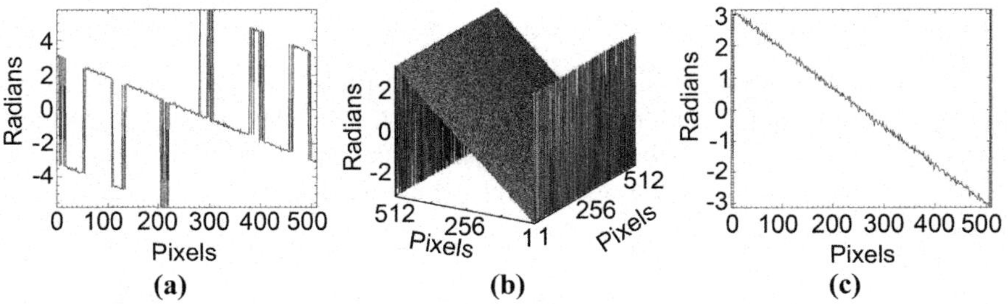

Fig. 2.35 Retrieval of the slope phase map $[\varphi_\Lambda(k, l)]_w$ with RMS height $Sq = \lambda/40 = 15.8$ nm obtained at the first stage of the algorithm for the TS DWIM realization (**a**); isometric image of retrieved total phase map $\varphi_{T,\Lambda}(k, l)$ for wavelength $\Lambda = 3340$ (**b**) and its profile (**c**)

$\lambda_1 = 532.0$ nm and $\lambda_2 = 632.8$ nm, respectively, as it is shown in Figs. 2.36 and 2.37.

Histograms of absolute and relative errors distributions of the phase slopes relief retrieval with added Gaussian phase noises emulating surface roughness $Sq_2(\lambda_2/40) = 15.8$ nm and $Sq_2(\lambda_2/10) = 63.3$ nm, corresponding to phase heights $Sq_{2\varphi}(\lambda_2/40) = 0.314$ rad and $Sq_{2\varphi}(\lambda_2/10) = 1.257$ rad, respectively, are shown in Fig. 2.38.

As it is demonstrated by obtained distributions, error of the total phase map retrieval is decreased, if the surface roughness is increased, since it was shown in Sects. 2.5.3 and 2.5.6. Decrease of the unknown phase shift retrieval error with increase of surface roughness [76] also influences on the decrease of total phase map retrieval error.

The developed method was also used for retrieval of phase step with a size of 512×512 pixels exceeding 2π for both λ_1 and λ_2. Example of retrieval of a phase step equal to 805.7 nm or 19.03 rad. for λ_1, 16.0 rad for λ_2 and 3.032 rad for Λ with

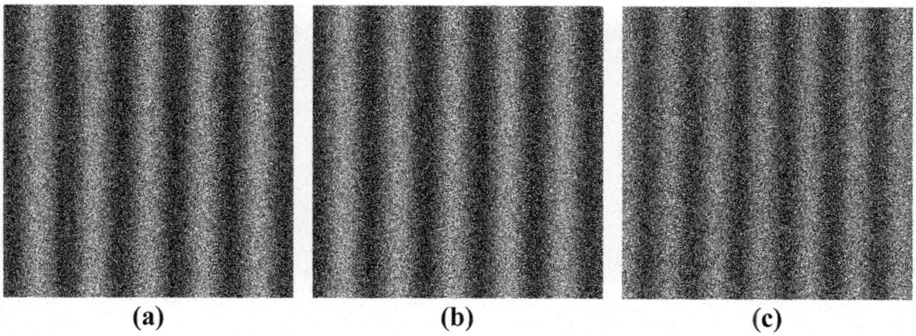

Fig. 2.36 Retrieval of the phase slope: initial interferogram $I_{21}(k, l)$ with phase RMS height $Sq_{2\varphi} = 1.257$ rad for wavelength $\lambda_2 = 632.8$ nm (**a**); retrieved phase map $\varphi_{2,w}(k, l)$ with phase RMS height $Sq_{2\varphi} = 1.257$ rad for $\alpha_2 = 30.0°$ and $\lambda_2 = 632.8$ nm (**b**); retrieved phase map $\varphi_{1,w}(k, l)$ with phase RMS heights $Sq_{1\varphi} = 1.502$ rad for $\alpha_1 = 35.68°$ and $\lambda_1 = 532.0$ nm (**c**)

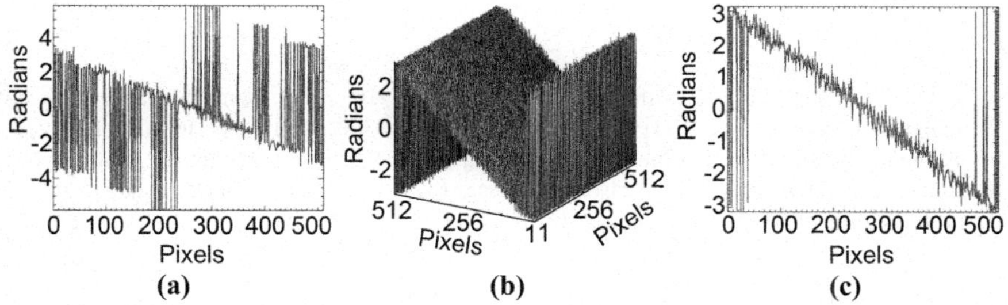

Fig. 2.37 Retrieval of the slope phase map $[\varphi_\Lambda(k, l)]_w$ with RMS height $Sq = \lambda/10 = 63.3$ nm obtained at the first stage of the algorithm for the TS DWIM realization (**a**); isometric image of retrieved total phase map $\varphi_{T,\Lambda}(k, l)$ for wavelength $\Lambda = 3340$ (**b**) and its profile (**c**)

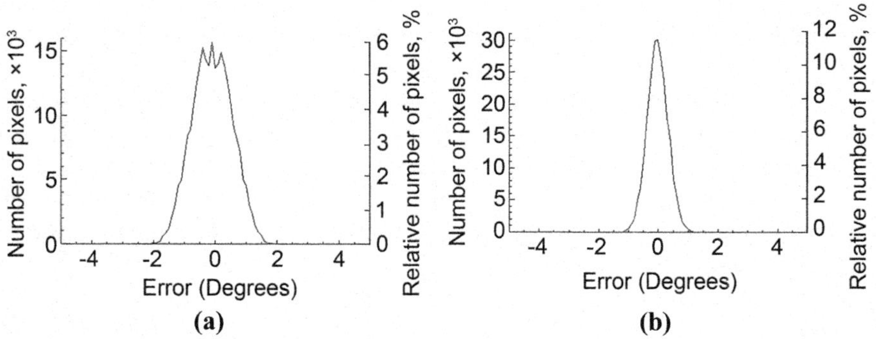

Fig. 2.38 Histograms of absolute and relative errors distributions of phase slopes retrieval with RMS heights $Sq_2(\lambda_2/40) = 15.8$ nm (**a**) and $Sq_2(\lambda/10) = 63.3$ nm (**b**). Absolute error distributions are computed in pixel units along the vertical axis

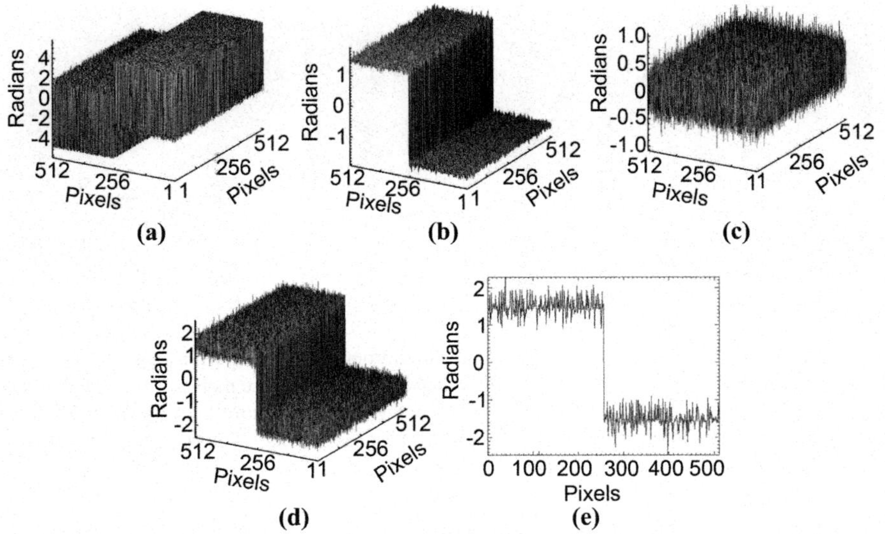

Fig. 2.39 Phase step retrieval with RMS height $Sq = \lambda/10 = 63.3$ nm: isometric image of the wrapped phase map $[\tilde{\varphi}_{\Lambda}(k,l)]_w$ for wavelength $\Lambda = 3340$ nm after performing Eq. (2.81) (**a**); isometric image of macrorelief phase map $\varphi_{M,\Lambda}(k,l)$ (**b**); isometric image of microrelief (roughness) phase map $\varphi_{R,\Lambda}(k,l)$ (**c**); isometric image of the unwrapped total phase map $\varphi_{T,\Lambda}(k,l)$ (**d**); profile of the total phase map $\varphi_{T,\Lambda}(k,l)$ (**e**)

added Gaussian phase noise emulating the surface roughness $Sq_2(\lambda_2/10) = 63.3$ nm is represented in Fig. 2.39.

The obtained results show the effectiveness of the proposed TS DWIM to reconstruct the test phase slope and phase step low-rough surfaces.

The test bipolar phase surface shown in Fig. 2.7 was used for more detailed estimation of surface roughness influence on errors of the test relief retrieval using the TS DWIM. For this purpose, different RMS values Sq_2 of Gaussian noise (from $\lambda_2/10$ to $\lambda_2/2$) emulating surface roughness were added to the test surface. Results of retrieval of the test phase surface relief with introduced phase noise $Sq_{2\varphi} = 0.314$ rad corresponding to $Sq_2 = \lambda_2/4 = 158.3$ nm for $\lambda_2 = 632.8$ nm are shown in Fig. 2.40.

Random error histograms of macrorelief, microrelief and total relief retrieval are depicted in Fig. 2.41. The absolute error distributions are computed in pixel units along the vertical axis.

These histograms indicate that the random errors of the retrieved total phase map are lesser than the random errors of retrieved macrorelief and nanorelief, which confirms a strict negative correlation relationship between random errors of the test macrorelief and microrelief, as shown in [76, 81]. Figure 2.42 indicates the dependencies of the surface relief random errors retrieval versus the RMS surface roughness Sq values of the macrorelief $h_{M,\Lambda} = \Lambda \cdot (\varphi_{M,\Lambda})/4\pi$, microrelief $h_{R,\Lambda} = \Lambda \cdot (\varphi_{R,\Lambda})/4\pi$ and total relief $h_{T,\Lambda} = \Lambda \cdot (\varphi_{T,\Lambda})/4\pi$ for wavelength $\Lambda = 3340$ nm. These dependencies are compared with the corresponding ones obtained using the algorithm

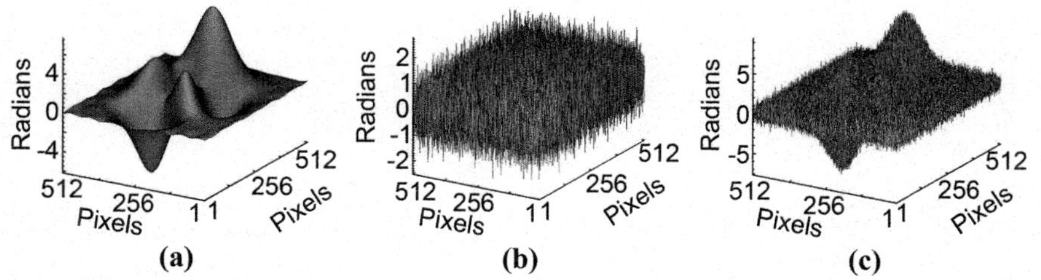

Fig. 2.40 Retrieved phase maps of the test bipolar phase surface with introduced phase noise $Sq_{2\varphi} = 0.314$ rad: macrorelief phase map $\varphi_{M,\Lambda}(k,l)$ (**a**), microrelief (roughness) phase map $\varphi_{R,\Lambda}(k,l)$ (**b**), total relief phase map $\varphi_{T,\Lambda}(k,l)$ (**c**)

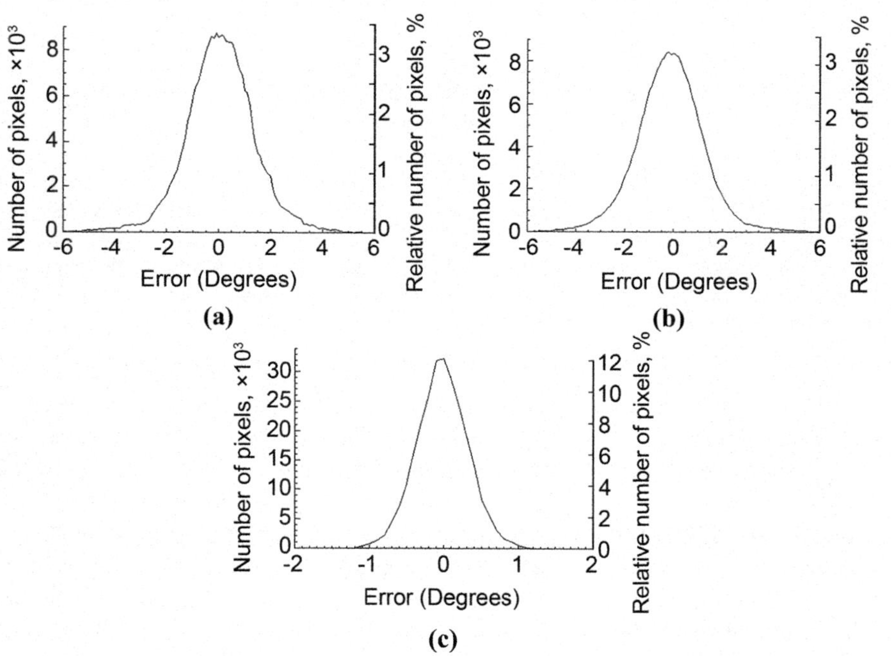

Fig. 2.41 Histograms of random error distributions of retrieved macrorelief $\varphi_{M,\Lambda}(k,l)$ (**a**), microrelief $\varphi_{R,\Lambda}(k,l)$ (**b**) and total relief $\varphi_{T,\Lambda}(k,l)$ (**c**) test phase maps ($Sq_{2\varphi} = 0.314$ rad)

performing the TSI UPS method for the same wavelength $\Lambda = 3340$ nm. As shown in all the obtained dependencies, the random errors in retrieving the total relief covering the range $Sq = 60$–160 nm also are smaller than the random errors in retrieving the macrorelief or microrelief. Random errors calculated using the TSI UPS method and TS DWIM for $Sq > 160$ are practically the same.

The results of systematic errors estimation show that their influence on the total error is very small. Therefore, total errors can be identified with random ones.

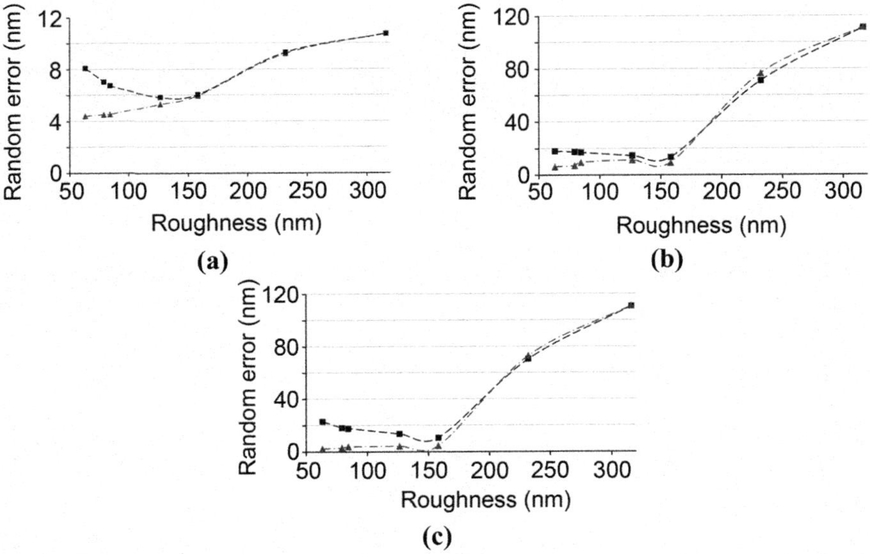

Fig. 2.42 Relationships between random errors of the surface relief retrieval and RMS roughness Sq of the retrieved macrorelief $h_{M,\Lambda} = \Lambda \cdot (\varphi_{M,\Lambda})/4\pi$ (**a**), microrelief $h_{R,\Lambda} = \Lambda \cdot (\varphi_{R,\Lambda})/4\pi$ (**b**) and total relief $h_{T,\Lambda} = \Lambda \cdot (\varphi_{T,\Lambda})/4\pi$ (**c**). Dashed curve (samples are marked as ■) is obtained by the TSI UPS method for $\Lambda = 3340$ nm, and dashed-dot one (samples are marked as ▲) is obtained by the TS DWIM for $\lambda_1 = 532.0$ nm, $\lambda_2 = 632.8$ nm and $\Lambda = 3340$ nm

The dependencies between random errors and RMS roughness of the retrieved total relief obtained by TSI UPS method are shown in Fig. 2.43 as solid curve (samples are marked as •). Comparison of this dependence with one obtained by the TS DWIM for $\Lambda = 3340$ μm (see Fig. 2.42c), which also is presented in Fig. 2.43, shows that the TSI UPS method can retrieve surface reliefs with RMS areal roughness up to $Sq \approx \lambda_2/8$ (or $Sq \approx \lambda_1/8$), and the proposed TS DWIM—up to $Sq \approx \Lambda/8$. So, the allowable random errors of these reliefs retrieval can be provided by the TSI UPS method for $\lambda_2 = 632.8$ nm, if $Sq_2 \leq 70$ nm, and by the TS DWIM, if $60 \leq Sq \leq 300$ nm.

2.7.4 Three-Frame Two-Wavelength Interferometry Method

In order to increase the speed of the fringe patterns recording, the three-frame two-wavelength interferometry method (3F TWIM) was developed. The 3F TWIM also performs unknown phase shifts between interferograms. In this method, only three interferograms recorded at wavelength λ_1 and three interferograms recorded at wavelength λ_2 are used. The 3F TWIM, like the TS DWIM, is mainly intended for the reconstruction of 3D surface relief areas in a much wider range of surface roughness compared to single-wavelength TSI UPS and ITSI UPS methods. However, in

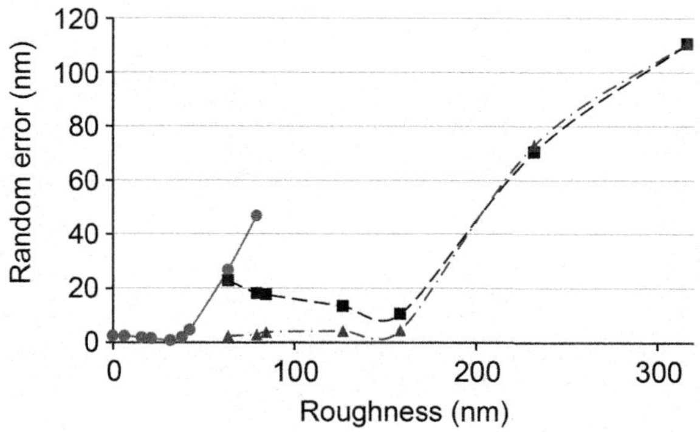

Fig. 2.43 Dependencies "random error versus roughness" for total relief retrieval. Solid curve (samples are marked as •) is obtained by TSI UPS method for $\lambda_2 = 632.8$ nm. Dashed curve (samples are marked as ■) is obtained by the same method for $\Lambda = 3340$ nm, and dashed-dot one (samples are marked as ▲) is obtained by the TS DWIM for $\Lambda = 3340$ nm

contrast to the TS DWIM, the 3F TWIM does not require to record additionally the intensities spatial distributions $I_{o1}(k, l)$ and $I_{o2}(k, l)$ of the object laser beam and $I_{r1}(k, l)$ and $I_{r2}(k, l)$ of the reference laser beam at wavelengths λ_1 and λ_2, which significantly increases its speed.

The need to develop the 3F TWIM is caused by the results of numerous fatigue experiments with polished smooth surfaces of CT-specimens with round or U-shaped notches and the study of the evolution of the surface topography directly at the notches during cyclic loads [74, 79]. Various metal and alloy materials were studied in PMI to determine the dimensions of the FPZ, including its structural mechanical parameter d^* [83–85], and the dimensions of the monotonic and cyclic plasticity zones. For this purpose, the ITSI UPS and 3SI UPS methods were used [76, 79, 81]. These methods make it possible to retrieve the 3D surface topography with the RMS roughness values $Sq \leq \lambda/10$ and even $Sq \leq \lambda/8$ and provide the mentioned above mechanical characteristics for materials with high cyclic plasticity, for example, low-carbon steels. These characteristics were also determined by the ITSI UPS method for some materials with moderate and even low cyclic plasticity during measuring the roughness parameters of the 3D relief in CT-specimens near the notch at the initial stage of cyclic loads [81]. However, the surface roughness increases sharply for most such materials in the intermediate and final stages of fatigue failure. Therefore, the surface topology of these materials cannot be retrieved near the notches by conventional methods of single-wavelength PSI due to the existence of the above-mentioned problem of "2π ambiguity". For this purpose, new interferometric two-wavelength methods, namely TS DWIM and 3F TWIM, have been proposed for retrieval of the 3D surface relief, whose roughness parameter $Sq \geq (\lambda/10 - \lambda/8)$.

Using Eq. (2.17), the first set of equations consisting of three interferograms recorded at wavelength λ_1 can be expressed as

$$\left.\begin{aligned} I_{11} &= I_{B1} + I_{M1}\cos\varphi_{1,w} \\ I_{12} &= I_{B1} + I_{M1}\cos(\varphi_{1,w} + \alpha_{21}) \\ I_{13} &= I_{B1} + I_{M1}\cos(\varphi_{1,w} + \alpha_{31}) \end{aligned}\right\}, \tag{2.86}$$

and the second set of equations consisting of three interferograms recorded at wavelength λ_2, as

$$\left.\begin{aligned} I_{21} &= I_{B2} + I_{M2}\cos\varphi_{2} \\ I_{22} &= I_{B2} + I_{M2}\cos(\varphi_{2,w} + \tilde{\alpha}_{21}) \\ I_{23} &= I_{B2} + I_{M2}\cos(\varphi_{2,w} + \tilde{\alpha}_{31}) \end{aligned}\right\}, \tag{2.87}$$

where α_{21} is the phase shift angle between the fringe patterns I_{12} and I_{11}, α_{31} is the phase shift angle between the fringe patterns I_{13} and I_{11}, $\tilde{\alpha}_{21}$ is the phase shift angle between the fringe patterns I_{22} and I_{21}, $\tilde{\alpha}_{31}$ is the phase shift angle between the fringe patterns I_{23} and I_{21}, $I_{B1,2} = I_{o1,2} + I_{r1,2}$ are the background intensities, $I_{M1,2} = 2\sqrt{I_{o1,2}I_{r1,2}}$ are the modulation intensities, $\varphi_{1,w}(k,l) = \tilde{\varphi}_{r1}(k,l) + \tilde{\varphi}_{m1}(k,l)$ is the coarse wrapped phase map of the surface relief area for wavelength λ_1, $\varphi_{2,w}(k,l) = \tilde{\varphi}_{r2}(k,l) + \tilde{\varphi}_{m2}(k,l)$ is the coarse wrapped phase map of the surface relief area for wavelength λ_2, $\tilde{\varphi}_{r1}$ and $\tilde{\varphi}_{r2}$ are the phase maps of surface roughness area for wavelengths λ_1 and λ_2, respectively; $\tilde{\varphi}_{m1}$ and $\tilde{\varphi}_{m2}$ are the wrapped phase maps of the surface macrorelief for wavelengths λ_1 and λ_2, respectively.

Phase shifts $\alpha_{21}, \alpha_{31}, \tilde{\alpha}_{21}$ and $\tilde{\alpha}_{31}$ are defined using Eq. (2.23), that is

$$\alpha_{21} = \arccos[\rho(I_{11}, I_{12})] = \arccos\frac{\langle(I_{11} - \langle I_{11}\rangle)(I_{12} - \langle I_{12}\rangle)\rangle}{\sigma_{I11}\sigma_{I12}}, \tag{2.88}$$

$$\alpha_{31} = \arccos[\rho(I_{11}, I_{13})] = \arccos\frac{\langle(I_{11} - \langle I_{11}\rangle)(I_{13} - \langle I_{13}\rangle)\rangle}{\sigma_{I11}\sigma_{I13}}, \tag{2.89}$$

$$\tilde{\alpha}_{21} = \arccos[\rho(I_{21}, I_{22})] = \arccos\frac{\langle(I_{21} - \langle I_{21}\rangle)(I_{22} - \langle I_{22}\rangle)\rangle}{\sigma_{I21}\sigma_{I22}}, \tag{2.90}$$

$$\tilde{\alpha}_{31} = \arccos[\rho(I_{21}, I_{23})] = \arccos\frac{\langle(I_{21} - \langle I_{21}\rangle)(I_{23} - \langle I_{23}\rangle)\rangle}{\sigma_{I21}\sigma_{I23}}, \tag{2.91}$$

where $\sigma_{I11}, \sigma_{I12}, \sigma_{I13}, \sigma_{I21}, \sigma_{I22}$ and σ_{I23} are the RMS values of $I_{11}, I_{12}, I_{13}, I_{21}, I_{22}$ and I_{23}, respectively.

Registration of each of the three interferograms occurs during a smooth phase shift of the reference beam. Therefore, such a basic mode for recording fringe patterns is proposed, although there may be additional modes depending on different technical circumstances and experimental conditions. The main mode involves the recording of the first pair of interferograms I_{11} and I_{21} at two wavelengths λ_1 and λ_2 at the beginning of the smooth phase shift under the action of linearly increasing saw-tooth voltage (see Fig. 2.44). Each wavelength has its own saw-tooth equal π. In Fig. 2.44, the shorter saw-tooth is generated for λ_1, since $\lambda_1 < \lambda_2$, and the longer one is generated for λ_2. The next two interferograms with arbitrary phase shifts for each wavelength are also recorded within its own saw-tooth, however during one cycle, which is not shorter than the temporal length of the longer saw-tooth generated by wavelength λ_2. In this figure, the saw-tooth generated for λ_1 is marked by dashed lines, and the saw-tooth generated for λ_2 is marked by solid lines. The moments of three interferograms recording at wavelength λ_1 are indicated by dashed columns, and the moments of three interferograms recording at wavelength λ_2 are marked by solid columns.

Consider the first stage of the 3F TWIM, which consists in the recording of three interferograms for each wavelength and the formation of a wrapped phase map for the synthetic beat wavelength Λ. Using Eqs. (2.86)–(2.91), let's find the total phase map $\varphi_{T,\Lambda 3}(k,l)$ of the surface relief area that is considered as the pixel-by-pixel summation of the microrelief phase map $\varphi_{R,\Lambda 3}(k,l)$ or surface roughness phase map, as well as the macrorelief phase map $\varphi_{M,\Lambda 3}(k,l)$. To exclude the background intensity I_{B1}, the set of three Eq. (2.86) should be reduced to

$$\left.\begin{array}{l} \frac{I_{12}-I_{B1}}{I_{11}-I_{B1}} = \cos\alpha_{21} - \tan\varphi_{1,w'}\cdot\sin\alpha_{21} \\ \frac{I_{13}-I_{B1}}{I_{11}-I_{B1}} = \cos\alpha_{31} - \tan\varphi_{1,w'}\cdot\sin\alpha_{13} \end{array}\right\}. \tag{2.92}$$

After elementary trigonometric transformations of (2.92), the course wrapped phase map for wavelength λ_1 is given by

$$\varphi_{1,w'} = \arctan\left[\frac{(I_{12}-I_{13}) + (I_{13}-I_{11})\cos\alpha_{21} + (I_{11}-I_{12})\cos\alpha_{13}}{(I_{13}-I_{11})\sin\alpha_{21} + (I_{11}-I_{12})\sin\alpha_{31}}\right]. \tag{2.93}$$

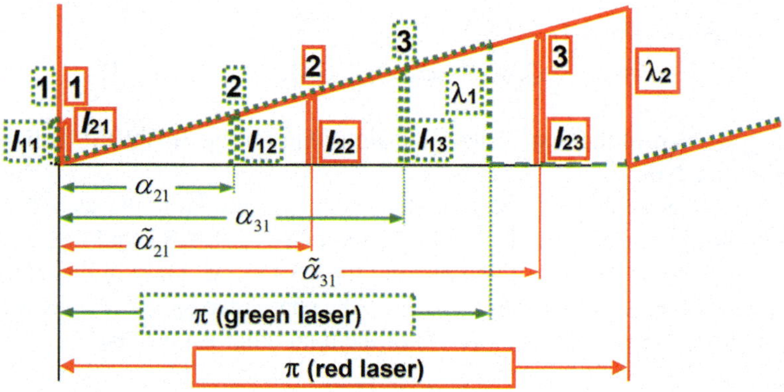

Fig. 2.44 Saw-tooth diagrams of smooth linear phase shifts at wavelengths λ_1 and λ_2 ($\lambda_2 > \lambda_1$), and moments of interferograms I_{11}, I_{12} and I_{13} recording (moments 1, 2, 3 are surrounded by dashed contours), as well as moments of interferograms I_{21}, I_{22} and I_{23} recording (moments 1, 2, 3 are surrounded by solid contours)

Excluding of the background intensity I_{B1} from the set of Eq. (2.87) leads to the set of two equations, that is

$$\left.\begin{aligned}\frac{I_{22}-I_{B2}}{I_{12}-I_{B2}} &= \cos\tilde{\alpha}_{21} - \tan\varphi_{2,w'}\cdot\sin\tilde{\alpha}_{21}\\ \frac{I_{23}-I_{B2}}{I_{21}-I_{B2}} &= \cos\tilde{\alpha}_{31} - \tan\varphi_{2,w'}\cdot\sin\tilde{\alpha}_{31}\end{aligned}\right\}, \tag{2.94}$$

and the coarse wrapped phase map for wavelength λ_2 is given by

$$\varphi_{2,w'} = \arctan\left[\frac{(I_{22}-I_{23}) + (I_{23}-I_{21})\cos\tilde{\alpha}_{21} + (I_{21}-I_{22})\cos\tilde{\alpha}_{31}}{(I_{23}-I_{21})\sin\tilde{\alpha}_{21} + (I_{21}-I_{22})\sin\tilde{\alpha}_{31}}\right]. \tag{2.95}$$

Obtained Eqs. (2.93) and (2.95) contain only differences of interferograms $(I_{12}-I_{13}), (I_{13}-I_{11}), (I_{11}-I_{12}), (I_{22}-I_{23}), (I_{23}-I_{21}), (I_{21}-I_{22})$, where background intensity is excluded indicating the automatic normalization of interferograms.

As in the TS DWIM, the coarse wrapped phase map $\left[\tilde{\varphi}_{\Lambda 3}(k,l)\right]_w$ for synthesized beat wavelength Λ is given by

$$(\tilde{\varphi}_{\Lambda 3})_w = \varphi_{\Lambda r'} + (\varphi_{\Lambda m'})_w = \varphi_{1,w'} - \varphi_{2,w'}. \tag{2.96}$$

This equation is similar to Eq. (2.82), only Eqs. (2.94) and (2.95) are inserted into Eq. (2.96) instead of Eqs. (2.79) and (2.80), which are inserted into Eq. (2.82). Using harmonic terms of the synthesized phase map $\left[\tilde{\varphi}_{\Lambda 3}(k,l)\right]_{w'}$, the new transformed synthesized course wrapped phase map $[\varphi_{\Lambda 3}(k,l)]_w$ of the total surface relief is expressed as

$$\begin{aligned}(\varphi_{\Lambda 3})_w &= \operatorname{atan2}\left(I_{s,o'}, I_{c,o'}\right)\\ &= \operatorname{atan2}\left\{\sin\left[\varphi_{\Lambda r'} + (\varphi_{\Lambda m'})_w\right], \cos\left[\varphi_{\Lambda r'} + (\varphi_{\Lambda m'})_w\right]\right\}.\end{aligned} \tag{2.97}$$

The flow-chart of the first stage of the algorithm for the 3F TWIM realization is shown in Fig. 2.45.

The next stage of this algorithm implementation is similar to the second stage of the algorithm performing the ITSI UPS method (see Sect. 2.5.5 and Fig. 2.20) and the second stage of the algorithm realizing the TS DWIM. Only in this algorithm, instead of the maps $\varphi_w(k,l)$ and $[\varphi_\Lambda(k,l)]_w$, the phase map $[\varphi_{\Lambda 3}(k,l)]_w$ is used, and as a result, a fine phase map $\varphi_{T,\Lambda 3}(k,l)$ of the total relief is given by

$$\varphi_{T,\Lambda 3}(k,l) = \varphi_{R,\Lambda 3}(k,l) + \varphi_{M,\Lambda 3}(k,l). \tag{2.98}$$

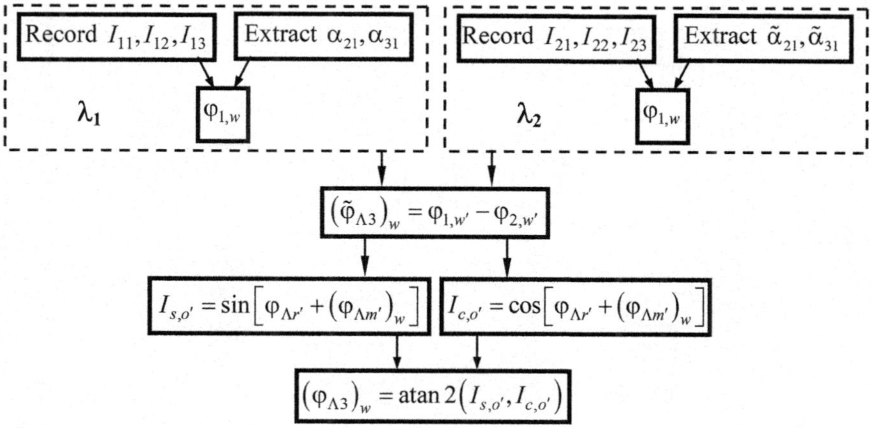

Fig. 2.45 Flow-chart of the first stage of the algorithm for the 3F TWIM realization

2.7.5 Experimental Verification of Two-Wavelength Phase Shifting Interferometry Methods

To verify experimentally the reliability of retrieving the test surface areas using the developed TS DWIM and 3F TWIM, the experimental setup for two-wavelength PSI with unknown phase shifts based on a Twyman–Green interferometer was constructed. During experiments, He–Ne laser ($\lambda_1 = 633$ nm) and laser diode ($\lambda_2 = 650$ nm) were used.

The laser diode coherence length measured in a Michelson interferometer is equal about 4.2 mm. Therefore, the optical scheme of the setup was carefully adjusted. The test cast iron specimen with flat grinding, which surface roughness corresponds to the 6th grade according to GOST 25142-82 [35], was chosen for experiments. The linear magnification of the Twyman–Green interferometer optical scheme was chosen according to the approach described in Sect. 2.5.4 and was equal to $M_l = 1.5$. An objective Yupiter 37A (focal length $f = 135$ mm, focal number $f_\# = 3.5$) projects interferograms onto the matrix photosensor 1/2 " IT CCD-SXGA (1280 (H) × 960 (V), pixel size 4.65 μm × 4.65 μm and pixel pitch $p_x = p_y = 4.65$ μm) of the digital CCD camera SONY XCD-SX910. The experimental setup scheme is shown in Fig. 2.46.

In this setup, two unexpanded beams from the He–Ne laser (1) and the diode laser (2) coinciding in the direction of propagation starts from the beam splitter (4). After the beam splitter (5), two laser beams are expanded and collimated in an optical collimating system (6) containing a spatial low-pass filter with a diameter of 15 μm. Two collimated beams are introduced into object and reference channels of the Twyman–Green interferometer, where the object beam falls on the surface area of the test cast iron specimen (10). The object beam reflected from the test surface area and the reference beam reflected from a movable mirror on a piezoelectric mount (8) interfere and the fringe pattern is projected on the matrix photosensor of the

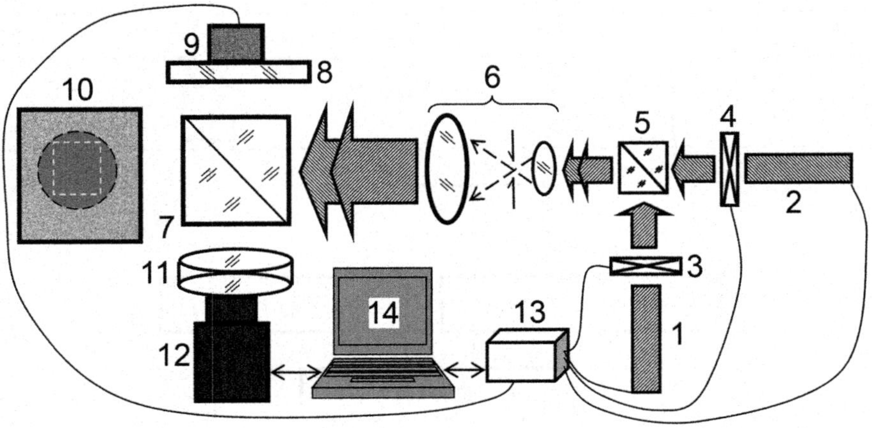

Fig. 2.46 Scheme of the experimental setup for verification of two-wavelength phase shifting interferometry methods: He–Ne laser ($\lambda_1 = 633$ nm) (1); laser diode ($\lambda_2 = 650$ nm) (2); shutters (3) and (4); beam splitters (5) and (7); optical collimator (6); movable mirror (8); piezoelectric transducer (9); test cast iron specimen (10); objective Yupiter 37A (11); digital camera SONY XCD-SX910 (12); controller (13); computer (14)

digital camera (12) using the objective Yupiter 37A (11). A controller (13) and a computer (14) control the experimental setup operating mode, including controlling shutters (3) and (4), piezoelectric transducer (9), as well as the digital camera (12). The controller includes an electronic control unit for smooth displacement of the movable mirror (8) to ensure the 3F TWIM fulfillment. In the computer (14), the unknown phase shifts are calculated and software implementing algorithms for the TS DWIM and 3F TWIM are run.

During the experimental verification of the 3F TWIM, the algorithm for this method realization was used. The test cast iron specimen (10), the arithmetic mean deviation of which roughness profile is $Ra = 1.6$ μm (see Fig. 2.47), was fastened on the special jig plate. Six interferograms I_{11}, I_{12}, I_{13}, I_{21}, I_{22} and I_{23} of the test specimen surface area with a size of 3.17×2.38 mm^2 were recorded during the smooth phase shift of the reference beam by π radians for a longer wavelength $\lambda_2 = 650$ nm. After calculation of unknown phase shifts $\alpha_{21}, \alpha_{31}, \tilde{\alpha}_{21}$ and $\tilde{\alpha}_{31}$ using Eqs. (2.88)–(2.91), the phase map $\varphi_{T,\Lambda 3}(k, l)$ of the total surface relief, as well as the phase maps $\varphi_{R,\Lambda 3}(k, l)$ and $\varphi_{M,\Lambda 3}(k, l)$ of the microrelief and macrorelief of the test specimen area, was retrieved. At the intermediate stages of the total surface relief retrieval of the test specimen area, the wrapped phase maps were extracted. The wrapped phase map of the macrorelief obtained after first iteration at the second stage of the algorithm for the 3F TWIM realization is shown in Fig. 2.48a; and the unwrapped phase map of the macrorelief obtained after second iteration at the second stage of this algorithm fulfillment is shown in Fig. 2.48b.

The final phase map $\varphi_{R,\Lambda 3}(k, l)$ of the surface microrelief area of the test cast iron specimen is shown in Fig. 2.49.

Fig. 2.47 Test cast iron specimen, $Ra = 1.6\ \mu m$ (6th grade of surfaceroughness)

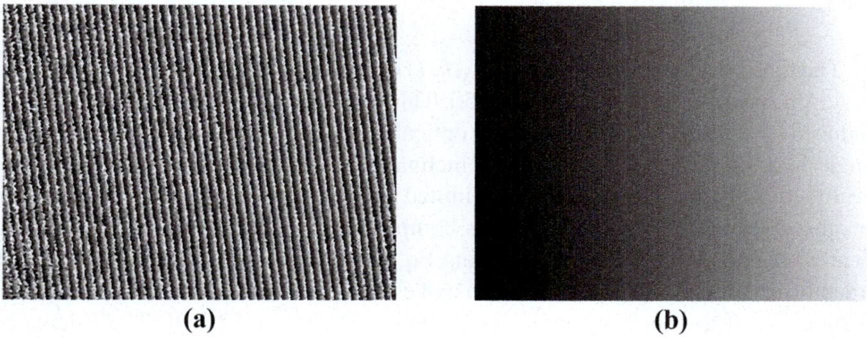

Fig. 2.48 Wrapped phase map of the surface macrorelief area obtained after first iteration at the second stage of the algorithm for the 3F TWIM realization (**a**); final unwrapped phase map $\varphi_{M,\Lambda 3}(k, l)$ of the macrorelief area of the test cast iron specimen (**b**)

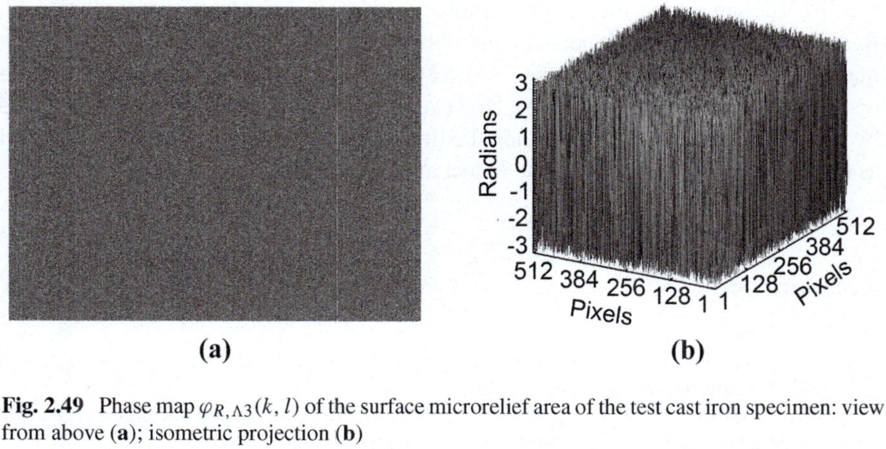

Fig. 2.49 Phase map $\varphi_{R,\Lambda 3}(k, l)$ of the surface microrelief area of the test cast iron specimen: view from above (**a**); isometric projection (**b**)

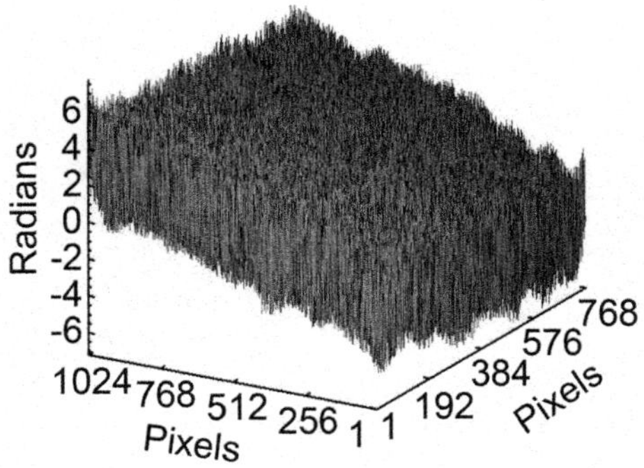

Fig. 2.50 Final retrieved phase mask $\varphi_{T,\Lambda 3}(k,l)$ of the total surface relief area

The final retrieved phase mask $\varphi_{T,\Lambda 3}(k,l)$ of the total surface relief area of the test cast iron specimen is shown in Fig. 2.50. This phase map has very small inclination induced by a little disadjustment of the optical scheme. The tangent of the inclination angle is equal to $2.63 \cdot 10^{-3}$ rad, and inclination angle is $1°30'$. Calculation of the arithmetical mean height of the scale-limited surface has shown that $Sa = 1.49$ μm. In this case, the sampling theorem is completely satisfied, since the synthesized beat wavelength is $\Lambda = 24.2$ μm. Using Eq. (2.66), the phase map $\varphi_{T,\Lambda 3}(k,l)$ was transformed into the height map, which makes it possible to calculate the areal height parameters of the surface roughness [see Eqs. (2.27) and (2.28)]. The calculated value of Sa was somewhat less than the nominal one, which is $Sa_n = 1.6$ μm. This circumstance is most likely related to the long-term use of the sample for practical purposes.

Similar investigations were performed for study the reliability of the developed TS DWIM. The same experimental setup and the same test cast iron specimen were used. Results of the arithmetical mean height of the scale-limited surface of the total height map $h_{T,\Lambda}(k,l)$ have shown that $Sa_n = 1.52$ μm. This result almost coincides with the result obtained using the 3F TWIM. However, if the time to record six interferograms by the 3F TWIM is about 1–3 s, then the time to record four interferograms, as well as the intensity distributions of the object and reference waves, is about 3 min.

References

1. 4D Technology. The leader in innovative optical metrology. Access mode http://www.4dtechnology.com/products/index.php
2. Aebischer HA, Waldner S (1999) A simple and effective method for filtering speckle interferometric phase fringe patterns. Opt Commun 162:205–210
3. Baer G, Schindler J, Pruss C, Siepmann J, Osten W (2014) Calibration of a non-null test interferometer for the measurement of aspheres and free-form surfaces. Opt Exp 22(25):31200–31211
4. Baldi P, Brunak S, Chauvin Y, Andersen CAF, Nielsen H (2000) Assessing the accuracy of prediction algorithms for classification: an overview. Bioinformatics 16(5):412–424

5. Bobkov YuN (1984) Korreliatsionnyye Metody i Ustrojstva Izmiereniya Sostavliayushchih Komponientov Signalov (Correlation methods and devices for measuring signal components). Vyshcha Shkola, Lviv
6. Bobkov YuN, Molodov VD (1969) Sposob izmierieniya fazovogo sdviga (Method for phase shift measurement). USSR Patent No. 256,867
7. Bone DA, Bachor HA, Sandeman RJ (1986) Fringe-pattern analysis using a 2-D Fourier transform. Appl Opt 25(10):1653–1660
8. Brock N, Hayes J, Kimbrough B, Millerd J, North-Morris M, Novak M, Wyant JC (2005) Dynamic interferometry. In: Sasian JM, Koshel RJ, Juergens RC (eds) Novel optical systems design and optimization VIII, vol 5875. SPIE, Bellingham, WA, pp 58750F
9. Bruning JH, Herriott DR, Gallagher JE, Rosenfeld DP, White AD, Brangaccio DJ (1974) Digital wavefront measuring interferometer for testing optical surfaces, lenses. Appl Opt 13(11):2693–2703
10. Buckland JR, Huntley JM, Turner SRE (1995) Unwrapping noisy phase maps by use of a minimum-cost-matching algorithm. Appl Opt 34(23):5100–5108
11. Burger W, Burge MJ (2013) Principles of digital image processing—advanced techniques, vol 3. Springer, London
12. Chen CW, Zebker HA (2001) Two-dimensional phase unwrapping with use of statistical models for cost functions in nonlinear optimization. J Opt Soc Amer A 18(2):338–351
13. Cheng Y, Wyant JC (1984) Two-wavelength phase shifting interferometry. Appl Opt 23(24):4539–4543
14. Ching NH, Rosenfeld D, Braun M (1992) Two-dimensional phase unwrapping using a minimum spanning tree algorithm. IEEE Trans Image Process 1(3):355–365
15. Creath K (1987) Step height measurement using two-wavelength phase-shifting interferometry. Appl Opt 26(14):2810–2816
16. Creath K (1988) V Phase-measurement interferometry techniques. In: Wolf E (ed) Progress in optics, vol 26. North-Holland, Amsterdam, pp 349–393
17. Creath K, Schmit J (2005) Phase-measurement interferometry. In: Guenther RD (ed) Encyclopedia of modern optics, vol 2. Elsevier Academic Press, San Diego, CA, pp 364–374
18. Creath K, Wyant JC (1990) Absolute measurement of surface roughness. Appl Opt 29(26):3823–3827
19. Creath K, Cheng Y-Y, Wyant JC (1985) Contouting aspheric surfaces using two-wavelength phase-shifting interferometry. Opt Acta 32(12):1455–1464
20. Dalmau O, Rivera M, Gonzalez A (2016) Phase shift estimation in interferograms with unknown phase step. Opt Commun 372:37–43
21. de Groot P (2002) Method and system for profiling objects having multiple reflective surfaces using wavelength-tuning phase-shifting interferometry. US Patent No. 6,359,692
22. de Groot P (2009) Design of error-compensating algorithms for sinusoidal phase shifting interferometry. Appl Opt 48(35):6788–6796
23. de Groot PJ (2019) A review of selected topics in interferometric optical metrology. Rep Prog Phys 82(5):056101
24. Deck LL (2014) Model-based phase shifting interferometry. Appl Opt 53(21):4628–4636
25. Deng J, Wang H, Zhang F, Zhang D, Zhong L, Lu X (2012) Two-step phase demodulation algorithm based on the extreme value of interference. Opt Lett 37(22):4669–4671
26. ESDI—surface and wavefront metrology. Access mode http://www.esdimetrology.com/
27. Estrada JC, Servin M, Vargas J (2012) 2D simultaneous phase unwrapping and filtering: a review and comparison. Opt Lasers Eng 50(8):1026–1029
28. Farrell C, Player M (1992) Phase step measurement and variable step algorithms in phase-shifting interferometry. Meas Sci Technol 3(10):953
29. Flores VH, Rivera M (2020) Robust two-step phase estimation using the Simplified Lissajous Ellipse Fitting method with Gabor filters bank preprocessing. Opt Commun 461:125286
30. Flores VH, Reyes-Figueroa A, Carrillo-Delgado C, Rivera M (2020) Two-step phase shifting algorithms: where are we? Opt Laser Technol 126:106105

31. Flynn TJ (1997) Two-dimensional phase unwrapping with minimum weighted discontinuity. J Opt Soc Amer A 14(10):2692–2701
32. Ghiglia DC, Pritt MD (1998) Two-dimensional phase unwrapping: theory, algorithms, and software. Wiley-Interscience, New York
33. Ghiglia DC, Romero LA (1996) Minimum *Lp*-norm two-dimensional phase unwrapping. J Opt Soc Am A 13(10):1999–2013
34. Goldestein RM, Zebker HA, Werner CL (1988) Satellite radar interferometry: two-dimensional phase unwrapping. Radio Sci 23(4):713–720
35. GOST 25.142-82 (1982) Sherohovatost' poverhnosti. Terminy i opredeleniya (Surface roughness. Terms and definitions). Izd Standartov, Moscow
36. Graham Optical Systems. Instruments for Science and Industry made in the USA. Extensive line of optical instruments offered by Graham Optical Systems. Access mode http://www.grahamoptical.com/index.html
37. Greivenkamp JE (1984) Generalized data reduction for heterodyne interferometry. Opt Eng 23(4):350–352
38. Greivenkamp JE, Brunning JH (1992) Phase shifting interferometry. In: Malakara D (ed) Optical shop testing, 2nd edn. Wiley, New York, pp 501–598
39. Guo H (2011) Blind self-calibrating algorithm for phase-shifting interferometry by the use of cross-bispectrum. Opt Exp 19(8):7807–7815
40. Guo H, Zhao Z, Chen M (2007) Efficient iterative algorithm for phase-shifting interferometry. Opt Lasers Eng 45(2):281–292
41. Hariharan P (2000) Phase-shifting interferometry: minimization of systematic errors. Opt Eng 39(4):967–969
42. Hariharan P, Oreb BF, Eiju T (1987) Digital phase-shifting interfeometry: a simple error-compensating phase calculation algorithm. Appl Opt 26(13):2504–2505
43. Hettwer A, Kranz J, Schwider J (2000) Three channel phase-shifting interferometer using polarization-optics and a diffraction grating. Opt Eng 39(4):960–966
44. Hibino K, Hanayama R, Burke J, Oreb B (2004) Tunable phase-extraction formulae for simultaneous shape measurement of multiple surfaces with wavelength-shifting interferometry. Opt Exp 12(23):5579–5594
45. Huntley JM (1989) Noise-immune phase unwrapping algorithm. Appl Opt 28(16):3268–3270
46. Huntley JM (1998) Automated fringe pattern analysis in experimental mechanics: a review. J Strain Anal Eng Des 33(2):105–125
47. Ibrahim DGA, Yasui T (2018) Multi-object investigation using two-wavelength phase-shift interferometry guided by an optical frequency comb. Appl Phys Lett 112(17):171101
48. ISO 4287:1997 (1997) Geometrical Product Specifications (GPS)—surface texture: profile method—terms, definitions and surface texture parameters. International Organization for Standardization, Geneva
49. ISO 4288:1996 (1996) Geometrical Product Specifications (GPS)—surface texture: profile method—rules and procedures for the assessment of surface texture. International Organization for Standardization, Geneva
50. ISO 16610-21:2011 (2011) Geometrical product specification (GPS)—Filtration—Part 21: Linear profile filters: Gaussian filters. International Organization for Standardization, Geneva
51. ISO 25178-2012 (2012) Geometrical Product Specification (GPS)—surface texture: areal. Surface texture indications (Part 1); Terms, definitions and surface texture parameters (Part 2); Specification operators (Part 3). International Organization for Standardization, Geneva
52. Kmet' AB, Muravsky LI, Voronyak TI, Stasyshyn IV (2019) Two-step dual-wavelength interferometry for surface relief retrieval. Opt Appl XLIX(2):331–343
53. Kreis T (2005) Handbook of holographic interferometry: optical and digital methods. WILEY–VCH Verlag GmbH & Co, Weinheim
54. Kreis TM, Jueptner WP (1992) Fourier transform evaluation of interference patterns: demodulation and sign ambiguity. In: Pryputniewicz RJ (ed) Laser interferometry IV: computer-aided interferometry, vol 1553. SPIE, Bellingham, WA, pp 263–274

55. Kuechel MF, Wiedmann W (1990) In-process metrology for large astronomical mirrors. In: Sander GM, Reid PB, Baker LR (eds) Advanced optical manufacturing and testing, vol 1333. SPIE, Bellingham, WA, pp 280–294
56. Kulkarni R, Rastogi P (2018) Two-step phase demodulation algorithm based on quadratic phase parameter estimation using state space analysis. Opt Lasers Eng 110:41–46
57. Kumar UP, Bhaduri B, Kothiyal MP, Mohan NK (2009) Two wavelength interferometry for 3-D surface profiling. Opt Lasers Eng 47(2):223–229
58. Lai G, Yatagai T (1991) Generalized phase-shifting interferometry. J Opt Soc Am A 8(5):822–827
59. Larkin KG, Oreb BF (1992) Design and assessment of symmetrical phase-shifting algorithms. J Opt Soc Am A 9(10):1740–1748
60. Larkin KG, Bone DJ, Oldfield MA (2001) Natural demodulation of two-dimensional fringe patterns. I. General background of the spiral phase quadrature transform. J Opt Soc Am A 18(8):1862–1870
61. Liu F, Wu Y, Wu F (2015) Phase shifting interferometry from two normalized interferograms with random tilt. Opt Exp 23(8):19332–19946
62. Liu Q, Wang Y, Ji F, He J (2013) A three-step least-squares iterative method for tilt phase-shift interferometry. Opt Exp 21(24):29505–29515
63. Luo C, Zhong L, Sun P, Wang H, Tian J, Lu X (2015) Two-step demodulation algorithm based on the orthogonality of diamond diagonal vectors. Appl Phys B 119(2):387–391
64. Macy WW (1983) Two-dimensional fringe-pattern analysis. Appl Opt 22(23):3898–3901
65. Malacara D (2007) Twyman-Green interferometer. In: Malacara D (ed) Optical shop testing, 3rd edn. Wiley, Hoboken, NJ, pp 46–96
66. Malacara D, Servín M, Malacara Z (2005) Interferogram analysis for optical testing, 2nd edn. Taylor & Francis, Boca Raton, FL
67. Mantravadi MV, Malacara D (2007) Newton, Fizeau, and Haidinger interferometers. In: Malacara D (ed) Optical shop testing, 3rd edn. Wiley, Hoboken, NJ, pp 1–45
68. Mathia TG, Pawlus P, Wieczorowski M (2011) Recent trends in surface metrology. Wear 271(3):494–508
69. Meneses-Fabian C (2016) Self-calibrating generalized phase-shifting interferometry of three phase-steps based on geometric concept of volume enclosed by a surface. J Opt 18(12):125703
70. Meneses-Fabian C (2017) Self-calibrating generalized phase-shifting interferometry of three steps by fitting modulation light to a polynomial of degree K. J Opt 19(11):115701
71. Meng XF, Cai LZ, Wang YR, Yang XL, Xu XF, Dong GY, Shen XX, Cheng XC (2008) Wavefront reconstruction by two-step generalized phase-shifting interferometry. Opt Commun 281(23):5701–5705
72. Millerd J, North-Morris M (2017) Dynamic interferometry: measurement of space optics and structures. In: Lehmann P (ed) Optical measurement systems for industrial inspection X, vol 10329. SPIE, Bellingham, WA, p 103291G
73. Millerd JE, Brock NJ, Hayes JB, North-Morris MB, Novak M, Wyant JC (2004) Pixelated phase-mask dynamic interferometer. In: Creath K, Schmit J (eds) Optical science and technology, vol 5531. SPIE, Bellingham, WA, pp 304–314
74. Muravsky LI (2011) Rozroblennia metodiv ta zasobiv spekl-metrologiyi ta fazozsuvnoyi interferometriyi dlia doslidzhennia napruzheno-deformovanogo stanu konstruktsiynyh materialiv (Development of speckle metrology and phase-shifting interferometry methods and means for studying the stress-strain state of structural materials). In: Panasyuk VV (ed) Fizyko-Mehanichnyy Instytut (Do 60-richchia z chasu zasnuvannia): monografiya (Physico-Mechanical Institute (To the 60th anniversary since the foundation): monograph). SPOLOM, Lviv, pp 213–231
75. Muravsky LI, Kmet AB, Voronyak TI (2011, September) Retrieval of diffusing surface by two-frame interferometric method with blind phase shift of a reference wave. In: Kashyap R, Têtu M, Kleiman RN (eds) Photonics North 2011, vol 8007. SPIE, Bellingham, WA, pp 385–391

76. Muravsky LI, Kmet' AB, Voronyak TI (2012) Retrieving the relief of a low-roughness surface using a two-step interferometric method with blind phase shift of a reference wave. Opt Lasers Eng 50(11):1508–1516
77. Muravsky L, Kmet' A, Voronyak T (2013) Two approaches to the blind phase shift extraction for two-step electronic speckle pattern interferometry. Opt Eng 52(10):101909
78. Muravsky LI, Voronyak TI, Kmet' AB (2014) Lazerna interferometriya poverhni dlia potreb tehnichnoyi diagnostyky (Laser interferometry of surface for needs of technical diagnostics). Spolom, Lviv
79. Muravsky LI, Kmet' AB, Stasyshyn IV, Voronyak TI, Bobitski YV (2018) Three-step interferometric method with blind phase shifts by use of interframe correlation between interferograms. Opt Lasers Eng 105:27–34
80. Muravsky LI, Ostash OP, Kmet' AB, Voronyak TI, Andreiko IM (2011) Two-frame phase shifting interferometry for retrieval of smooth surface and its displacements. Opt Lasers Eng 49(3):305–312
81. Muravsky LI, Picart P, Kmet' AB, Voronyak TI, Ostash OP, Stasyshyn IV (2016) Evaluation of fatigue process zone dimensions in notched specimens by two-step phase shifting interferometry technique. Opt Eng 55(10):104108
82. Onions RA, Archard JF (1973) The contact of surfaces having a random structure. J Phys D Appl Phys 6(3):289–304
83. Ostash OP (2006) New approaches in fatigue fracture mechanics. Mater Sci 42(1):5–19
84. Ostash OP, Panasyuk VV (2001) Fatigue process zone at notches. Int J Fatigue 23(7):627–636
85. Ostash OP, Panasyuk VV, Kostyk EM (1999) A phenomenological model of fatigue macrocrack initiation near stress concentrators. Fatigue Fract Eng Mater Struct 22(2):161–172
86. Ostash OP, Muravs'kyi LI, Voronyak TI, Kmet' AB, Andreiko IM, Vira VV (2011) Determination of the size of the fatigue prefracture zone by the method of phase-shifting interferometry. Mater Sci 46(6):781–788
87. Parshall D, Kim MK (2006) Digital holographic microscopy with dual-wavelength phase unwrapping. Appl Opt 45(3):451–459
88. Pawlus P (2007) Digitization of surface topography measurement results. Measurement 40(6):672–686
89. Pearson correlation coefficient. https://en.wikipedia.org/wiki/Pearson_correlation_coefficient. Last edited on 24 Aug. 2022
90. Picart P (2007) Interference and fringe pattern demodulation, SCIRN-PRN, Université du Maine. http://www.optique-ingenieur.org/en/courses/OPI_ang_M02_C06/co/Grain_OPI_ang_M02_C06_4.html
91. Poon CY, Bhushan B (1995) Comparison of surface roughness measurements by stylus profiler, AFM and non-contact optical profiler. Wear 190(1):76–88
92. QPSI^{TM} technology. Access mode https://www.zygo.com/support/technologies/qpsi-technology
93. Quiroga JA, Bernabeu E (1994) Phase-unwrapping algorithm for noisy phase-map processing. Appl Opt 33(29):6725–6731
94. Quiroga JA, Gómez-Pedrero JA, Garcia-Botella A (2001) Algorithm for fringe pattern normalization. Opt Commun 197(1–3):43–51
95. Raja J, Muralikrishnan B, Fu S (2002) Recent advances in separation of roughness, waviness and form. Precis Eng 26(2):222–235
96. Rastogi PK (ed) (2000). Wiley, Hoboken, NJ
97. Reyes-Figueroa A, Flores VH, Rivera M (2021) Deep neural network for fringe pattern filtering and normalisation. Appl Opt 60(7):2022–2036
98. Rivera M, Dalmau O, Gonzalez A, Hernandez-Lopez F (2016) Two-step fringe pattern analysis with a Gabor filter bank. Opt Lasers Eng 85:29–37
99. Rodgers JL, Nicewander WA (1988) Thirteen ways to look at the correlation coefficient. Am Stat 42:59–66
100. Saide D, Trusiak M, Patorski K (2017) Evaluation of adaptively enhanced twoshot fringe pattern phase and amplitude demodulation methods. Appl Opt 56(19):5489–5500

101. Sasaki O, Okazaki H, Sakai M (1987) Sinusoidal phase modulating interferometer using the integrating-bucket method. Appl Opt 26(6):1089–1093
102. Schmit J, Creath K (1995) Extended averaging technique for derivation of error-compensating algorithms in phase-shifting interferometry. Appl Opt 34(19):3610–3619
103. Schmit J, Creath K, Wyant JC (2007) Surface profilers, multiple wavelength, and white light intereferometry. In: Malacara D (ed) Optical shop testing, 3rd edn. Wiley, Hoboken, NJ, pp 667–755
104. Schreiber H, Brunning JH (2007) Phase shifting interferometry. In: Malacara D (ed) Optical shop testing, 3rd edn. Wiley, Hoboken, NJ, pp 547–666
105. Schwider J, Burow R, Elssner K-E, Grzanna J, Spolaczyk R, Merkel K (1983) Digital wavefront measuring interferometry: some systematic error sources. Appl Opt 22(21):3421–3432
106. Shough D, Kwon OY, Leary DF (1990) High-speed interferometric measurement of aerodynamic phenomena. In: Ulrich P, Wilson LE (eds) Propagation of high-energy laser beams through the earth's atmosphere, vol 1221. SPIE, Bellingham, WA, pp 394–403
107. Sirohi SR (2009) Optical methods of measurement: wholefield techniques, 2nd edn. Taylor & Francis, Boca Raton, FL
108. Smythe R, Moore R (1984) Instantaneous phase measuring interferometry. Opt Eng 23(3):361–364
109. Su X, Chen W (2004) Reliability-guided phase unwrapping algorithm: a review. Opt Lasers Eng 42(3):245–261
110. Surrel Y (1997) Design of phase-detection algorithms insensitive to bias modulation. Appl Opt 36(4):805–807
111. Sykora D, de Groot P (2011) Instantaneous measurement Fizeau interferometer with high spatial resolution. In: Burge JH, Fähnle OW, Williamson R (eds) Optical manufacturing and testing IX, vol 8126. SPIE, Bellingham, WA, pp 288–297
112. Takeda M, Ina H, Kobayashi S (1982) Fourier-transform method of fringe-pattern analysis for computer-based topography and interferometry. J Opt Soc Amer A 72(1):156–160
113. Tay CJ, Quan C, Niu H, Bhaduri B (2011) Phase retrieval in two-wavelength DSSI using a combined filtering method. Optik 122(23):2114–2118
114. Tian C, Liu S (2016) Demodulation of two-shot fringe patterns with random phase shifts by use of orthogonal polynomials and global optimization. Opt Exp 24(4):3202–3215
115. Tian C, Liu S (2016) Two-frame phase-shifting interferometry for testing optical surfaces. Opt Exp 24(16):18695–18708
116. Toyooka S, Tominaga M (1984) Spatial fringe scanning for optical phase measurement. Opt Commun 51(2):68–70
117. Trusiak M, Patorski K, Pokorski K (2013) Hilbert–Huang processing for single-exposure two-dimensional grating interferometry. Opt Exp 21(23):28359–28379
118. Trusiak M, Wielgus M, Patorski K (2014) Advanced processing of optical fringe patterns by automated selective reconstruction and enhanced fast empirical mode decomposition. Opt Lasers Eng 52:230–240
119. van Brug H (1999) Phase-step calibration for phase-stepped interferometry. Appl Opt 38(16):3549–3555
120. Vargas J, Quiroga JA, Belenguer T, Servin M, Estrada JC (2011) Two-step self-tuning phase-shifting interferometry. Opt Exp 19(2):638–648
121. Vargas J, Quiroga JA, Sorzano COS, Estrada JC, Carazo JM (2011) Two-step interferometry by a regularized optical flow algorithm. Opt Lett 36(17):3485–3487
122. Vargas J, Quiroga JA, Sorzano COS, Estrada JC, Carazo JM (2012) Two-step demodulation based on the Gram–Schmidt orthonormalization method. Opt Lett 37(3):443–445
123. Voronyak TI, Kmet' AB, Muravs'kyi LI (2009) Determination of the 3D fields of displacements by the method of phase-shifting speckle interferometry. Mater Sci 45(3):372–377
124. Wang H, Luo C, Zhong L, Ma S, Lu X (2014) Phase retrieval approach based on the normalized difference maps induced by three interferograms with unknown phase shifts. Opt Exp 22(5):5147–5154

125. Warnasooriya N, Kim MK (2010) Quantitative phase imaging using multi-wavelength optical phase unwrapping. In: Costa N, Cartaxo A (eds) Advances in lasers and electro optics. InTech, Rijeka, Shanghai, pp 769–786
126. Whitehouse DJ, Archard JF (1970) The properties of random surfaces of significance in their contact. Proc R Soc A Math Phys Eng Sci 316(1524):97–121
127. Wielgus M, Sunderland Z, Patorski K (2015) Two-frame tilt-shift error estimation and phase demodulation algorithm. Opt Lett 40(15):3460–3463
128. Womack KH (1984) Interferometric phase measurement using spatial synchronous detection. Opt Eng 23(4):391–395
129. Wyant JC (1971) Testing aspherics using two-wavelength holography. Appl Opt 10(9):2113–2118
130. Wyant JC (1975) Use of an ac heterodyne lateral shear interferometer with real–time wavefront correction systems. Appl Opt 14(11):2622–2626
131. Wyant JC (2013) Computerized interferometric surface measurements. Appl Opt 52(1):1–8
132. Wyant JC, Creath K (1989) Two-wavelength phase-shifting interferometer and method. US Patent No. 4,832,489
133. Wyant JC, Oreb BF, Hariharan P (1984) Testing aspherics using two-wavelength holography: use of digital electronic techniques. Appl Opt 23(22):4020–4023
134. Xu J, Li Y, Wang H, Chai L, Xu Q (2010) Phase-shift extraction for phase-shifting interferometry by histogram of phase difference. Opt Exp 18(23):24368–24378
135. Xu XF, Cai LZ, Wang YR, Meng XF, Zhang H, Dong Y, Chen XX (2007) Blind phase shift extraction and wavefront retrieval by two-frame phase-shifting interferometry with an unknown phase shift. Opt Commun 273(1):54–59
136. Xu XF, Cai LZ, Wang YR, Yang XL, Meng XF, Dong GY, Chen XX, Zhang H (2007) Generalized phase-shifting interferometry with arbitrary unknown phase shifts: direct wavefront reconstruction by blind phase shift extraction and its experimental verification. Appl Phys Lett 90(12):121124
137. Yang X, Yu Q, Fu S (2007) An algorithm for estimating both fringe orientation and fringe density. Opt Commun 274(2):286–292
138. Yi Z, Hongguang L (2013) A denoising scheme for DSPI fringes based on fast bi-dimensional ensemble empirical mode decomposition and BIMF energy estimation. Mech Syst Sig Process 35(1–2):369–382
139. Zappa E, Busca G (2008) Comparison of eight unwrapping algorithms applied to Fourier-transform profilometry. Opt Lasers Eng 46(2):106–116
140. Zeng F, Tan Q, He M, Kang Y, Gu H, Zhou Z, Jin G (2015) Interferometric spherical surface testing with unknown phase shifts. Opt Exp 23(11):14360–14379
141. Zhang S (2018) Absolute phase retrieval methods for digital fringe projection profilometry: a review. Opt Lasers Eng 107:28–37
142. Zhang Q, Han Y, Wu Y (2019) Comparison and combination of three spatial phase unwrapping algorithms. Opt Rev 26(4):380–390
143. Zhao M, Huang L, Zhang QC, Su XY, Asundi S, Qian KM (2011) Quality-guided phase unwrapping technique: comparison of quality maps and guiding strategies. Appl Opt 50(33):6214–6224
144. ZYGO Corporation. Home Page. Access mode http://www.zygo.com/

Chapter 3
Application of Phase Shifting Interferometry Methods for Diagnostics of Materials Surface

Abstract This chapter considers the application of the developed methods of single- and dual-wavelength phase shifting interferometry (PSI) for nondestructive evaluation of materials and structural elements. Special attention is paid to the application of PSI methods to study the fatigue life of compact tension specimens (CT-specimens) made of metals or alloys with round and U-shaped notches. The performed researches include defining the fatigue process zone (FPZ) size $d*$, evaluating the cyclic and monotonic plasticity zones geometric parameters as well as determining the site and time of fatigue crack initiation. In order to accurate determine the dimensions of the FPZ and the parameter $d*$ for various CT-specimens of metals and alloys, a criterion for achieving the maximum surface roughness at the FPZ boundary was proposed. Experimental verification of the proposed criterion was performed using the iterative method of two-step interferometry with an unknown phase shift. To determine the site and time of a fatigue macrocrack initiation in every pixel of the retrieved surface roughness height maps, the technique for definition of the roughness increments of the surface relief area near the notch tip during the CT-specimen fatigue was developed. The evolution of the surface roughness near the notch tip in metal and alloy CT-specimens during their fatigue was estimated by increasing the arithmetic mean roughness value Sa at each pixel of the surface roughness height maps obtained after each predetermined number of load cycles, until the appearance of a fatigue macrocrack on the surface. Results of experiments confirmed the hypothesis that the fatigue macrocracks initiation is most likely occurs within the zone $D*$, which covers the FPZ and narrow band outside it.

Described in Chap. 2 single- and two-wavelength methods of PSI with unknown phase shifts between fringe patterns were used for diagnostics of structural materials and constructional elements. Essential attention was paid to research the fatigue failure phenomena and fatigue life of compact tension (CT) specimens made of metals or alloys with round and U-shaped notches. The fulfilled researches include the determination of the geometric parameters of a fatigue process zone (FPZ), cyclic and monotonic plasticity zones as well as the definition of the location and time of the fatigue crack initiation. Let us dwell in more detail on the results of the conducted research.

Z. Nazarchuk et al., *Optical Metrology and Optoacoustics in Nondestructive Evaluation of Materials*, Springer Series in Optical Sciences 242,
https://doi.org/10.1007/978-981-99-1226-1_3

3.1 Determination of Fatigue Process Zone, Cyclic and Monotonic Plastic Zones Parameters

Cyclic loading of specimens with round notches or U-shaped stress concentrators leads to initiation and formation of the FPZ in the notch vicinity as well as the cyclic and monotonic plastic zones, schematically outlined in Fig. 3.1. Researches of geometric and structural parameters of these zones are important for assessing the fatigue life and residual resource of cyclically loaded structural elements [34]. Ostash et al. [28, 29, 33] developed the hypothesis of the FPZ formation near the round or U-shaped notches in CT-specimens. In this hypothesis, the FPZ can be considered as a subsurface prefracture mezozone within the cyclic plastic zone r_{pc} [28, 29]. During fatigue, the FPZ is the main physical barrier for all the fatigue microcracks, except one that overcome this barrier and is transformed into a macrocrack. The FPZ size $d*$ is the structural mechanical parameter of materials, which depends on the mechanical properties of the material and its microstructure [28, 29, 33]. This parameter can be treated as a material constant, which defines the maximum value of the local elastic–plastic stress range $\Delta\sigma_y$ at a critical distance $d*$ from the notch root. The equivalent distances were introduced earlier by Neuber (distance $\rho*$) [27], Peterson (distance a_p) [36] and Pluvinage (distance x_{eff}) [38]. The FPZ size $d*$ can be considered as the delimitation criterion of two main fatigue damage stages determined by periods N_i and N_p, being defined as the number of cycles N_i before nucleation of a minimum fatigue macrocrack in a material at the first stage and the number of cycles N_p during the macrocrack growth at the second one. This criterion can be used for calculation of fatigue life

$$N_f = N_i + N_p, \tag{3.1}$$

that is [29]

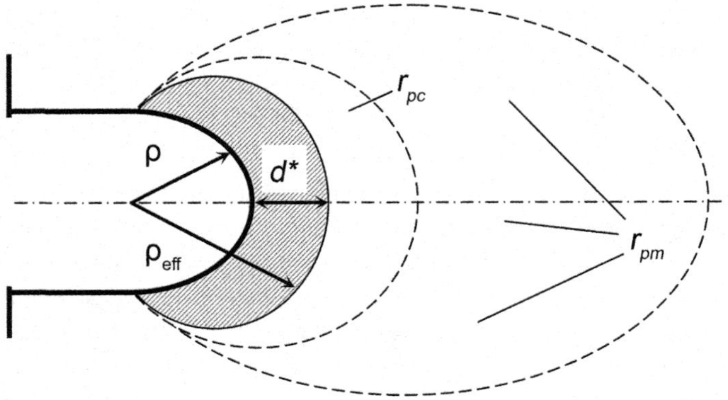

Fig. 3.1 Schematic representation of the fatigue process zone (shaded) as well as the cyclic (r_{pc}) and monotonic (r_{pm}) plastic zones near round notch with a radius ρ

$$N_f = N_i + N_p = N_{i|a_i=d^*} + \int\limits_{d^*}^{a_c} \frac{da}{F[\Delta K(\Delta P, a)]}, \tag{3.2}$$

where a_i is the macrocrack initial length, $a_i = d*$ is the criterion of a microcrack transition in the macrocrack, $a_c = f(\Delta K_{fc})$, K_{fc} is the cyclic fracture toughness, $F[\Delta K]$ is the function describing a diagram of the fatigue macrocrack growth rate, ΔK is the range of stress intensity factor.

The reliability of estimating the effective fatigue stress concentration factor K_f for cyclic stresses also depends on the accuracy of determining the size $d*$ according to the Ostash–Panasyuk formula [28, 29]

$$K_f = 1 + \frac{K_t - 1}{1 + \sqrt{d^*/\rho_{\text{eff}}}}, \tag{3.3}$$

where K_t is the theoretical stress concentration factor, $\rho_{\text{eff}} = \rho + d*$ is the effective radius of the notch, ρ is the radius of the round notch.

Known methods for determining the parameter $d*$ can be divided into nondestructive and destructive. Among the nondestructive ones, the method of microanalysis (in particular, X-ray diffraction) of the FPZ of cyclically loaded specimens [28, 33] and the method of reflected electron diffraction [32] should be noted. The method of reflected electron diffraction makes it possible to determine $d*$ as the size of a local zone, in which metal or alloy grains rotate and their crystallographic orientation changes relative to the load axis under the influence of plastic deformation caused by cyclic loads. Experimental confirmation of grain rotation under tensile and cyclic loads has also been demonstrated using a scanning laser confocal microscope [41] and scanning white light interferometry [39, 40, 44]. In particular, Wang et al. [44] and Sola et al. [40] showed that, after saturation of the growth heights of slip bands, the out-of-plane motion of grains plays the main role in the formation of the surface topography. However, the methods based on fixing grain rotation should be supplemented by additional machine vision and artificial intelligence algorithms, which would allow clearly determining the common level of grains inclinations for more accurate definition of the FPZ boundary.

Destructive is the method of establishing the zone of the initial macrocrack formation, in which the fracture surface is analyzed using microfractograms [33, 37]. This method gives satisfactory results when the fracture micromechanism abruptly changes during the transition from a microcrack to a macrocrack, for example, at low test temperatures [33]. However, known methods and techniques are either quite time consuming or give very approximate and unstable measurement results. Therefore, the problem of nondestructive, fairly accurate and rapid determination of the parameter $d*$ is quite relevant.

In this regard, two new nondestructive techniques for determining $d*$ based on the use of two-step single-wavelength PSI methods (i.e. TSI UPS and ITSI UPS methods) have been proposed. The first technique (Technique 1) is based on execution of that part of the TSI UPS method (see Sect. 2.5.1), which retrieves only unwrapped phase map $\varphi_m(k, l)$ of the macrorelief surface area. The part of the algorithm performing the TSI UPS method is highlighted with a dashed outline in Fig. 2.6. In Technique 1, two surface macrorelief areas are reconstructed near the notch edge before cyclic loading of the studied CT-specimen and after a certain number of cyclic loads. These macroreliefs are then superimposed and subtracted from each other. As a result, a field of surface deformations is obtained. The parameter $d*$ is determined using the resulting strain field as the distance from the notch edge to the point of maximum strain in the direction perpendicular to the direction of load application to the specimen [19, 30, 31]. The second technique (Technique 2) uses the ITSI UPS method (see Sect. 2.5.5) and the criterion to determine the FPZ size. This criterion is based on the assumption that the surface roughness of the specimen near the notch edge after applying cyclic loads reaches maximum values at the FPZ boundary.

The experimental setup based on the Twyman–Green interferometer that is described in Sect. 2.5.4 (see Fig. 2.13) was used for implementation of these techniques. Standard CT-specimens were used in the experiment; the cross-section of such a typical specimen is shown in Fig. 3.2. Before applying the cyclic loads to them, the specimens' surfaces were ground and polished near stress concentrators. To check the state of the polished surface, the specimens were fixed with a special conductor, placing them in one of the arms of the interferometer. The surfaces were polished until the operator began to observe a clear interference pattern with a minimum density of fringes at the output of the Twyman–Green interferometer. To fix the initial state of the smooth or nanorough surface relief near the stress concentrator, two interference patterns with a dimension of 800 × 600 pixels were recorded by the digital camera namely the initial $I_1(k, l)$ pattern and the pattern $I_2(k, l)$ differing from the first one by an unknown phase shift of the reference beam. After that, the specimen was subjected to cyclic loading. After a certain number of cycles, it was fixed again in the conductor and the next pair of interferograms was recorded, and the same surface relief area was reconstructed. Such pairs of interferograms were recorded several times after a predetermined number of cycles until the end of the planned experiment.

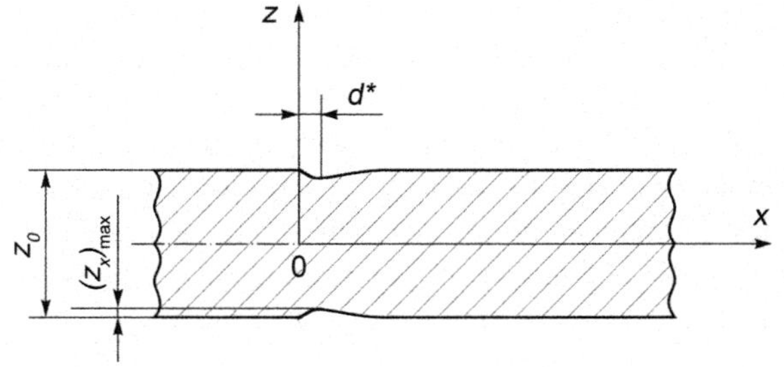

Fig. 3.2 Scheme of the plastic flow localization of the material in the section of the specimen near the notch edge

3.1.1 Determination of Fatigue Process Zone Size Using Two-Step Phase Shifting Interferometry Method and Technique 1

Determination of the fatigue process zone for steel 08kp specimens. The structural mechanical parameter $d*$ for a CT-specimen made of low-carbon steel 08kp, similar to steels 1008, 1010 and A619 with a size of $24 \times 24\ \text{mm}^2$ and a thickness of 4 mm was first determined using Technique 1. For this purpose, a characteristic point of the FPZ boundary was chosen within the zone of plastic deformations, where the cross-section of the specimen near the notch is most thinned due to plastic deformation along the axis z (see Fig. 3.2). The FPZ size was defined as the distance from the notch edge to the abscissa of the characteristic point, where the local deformation has a maximum, i.e. where the specimen cross-section reaches its minimum value. The value of local deformation at some point x of the specimen cross-section is $\varepsilon_{zx} = 2z_x \big/ z_0$, where z_0 is the specimen thickness. Thus, if we record two interferograms of the studied surface relief area before the loads of the specimen and then two interferograms of the same area after cyclic loads, the two surface areas can be retrieved using the TSI UPS method. After subtraction of these two surface relief areas, it is possible to find the point of the greatest thinning of the sample, the distance from which to the edge of the notch is equal to $d*$.

To implement Technique 1, standard CT-specimens made of steel 08kp were used. The example of such specimen and the image of its studied area near the notch are shown in Fig. 3.3. Since the problem of 2π ambiguity for macrorelief is not as critical as for the nanorelief, the linear magnification M_l of the interferometer optical scheme was chosen to cover not only the FPZ, but also the cyclic one and if possible the monotonic plastic zones. For these experiments, the chosen linear magnification was $M_l = 0.6$. The size of the investigated area of the polished surface is $10 \times 7.5\ \text{mm}^2$, the thickness of the specimen is 4 mm, and the diameter of the notch is 3 mm. Cyclic loads with a coefficient of load cycle asymmetry value $R = 0.1$ and a range of local stress $\Delta\sigma_y(0) = 890$ MPa were applied to the specimen. The combination of the position of the specimen before cyclic loads with its position after its cyclic loads was provided using Technique 1 for interactive matching of reference pixels in the computer. The alignment error was $\pm\,0.5$ px, i.e. $\pm\,2.33\ \mu\text{m}$ in the imaging plane using a SONY XCD-SX910 digital camera, where pixel pitch is $p_x = p_y = 4.65\ \mu\text{m}$.

Interferograms of the same polished surface area near the notch tip, obtained before cyclic loading and after application of a certain number of loading cycles, are represented in Fig. 3.4.

According to the sequence of operations for retrieval the surface macrorelief using TSI UPS method and algorithm performing this method (see the part of the represented in Fig. 2.6 flow-chart highlighted with a dashed outline), at first the wrapped phase map of the macrorelief area after 100 loading cycles shown in Fig. 3.5a was produced. Then, it was unwrapped using the software Flynn [13, 14]. The unwrapped

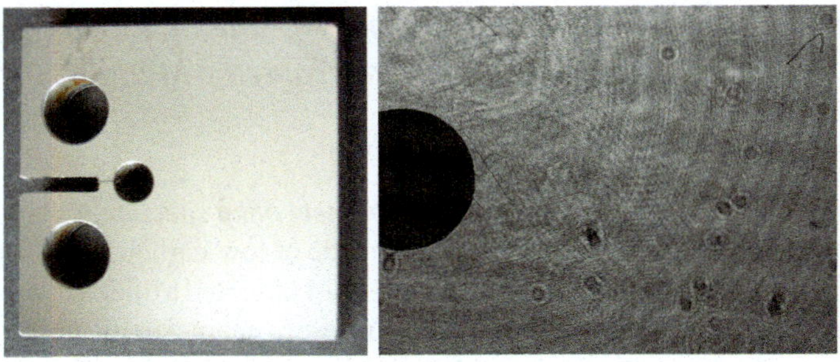

(a) (b)

Fig. 3.3 CT-specimen from steel 08kp with round notch 3 mm in diameter (**a**); studied surface area near the notch (**b**)

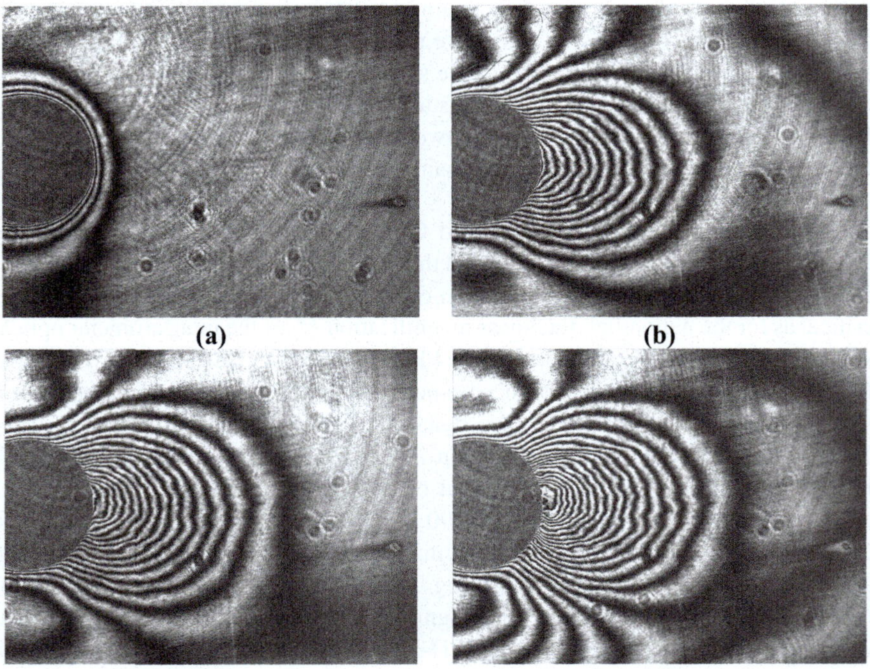

(c) (d)

Fig. 3.4 Fringe patterns of the steel 08kp CT-specimen surface area near the notch tip: before loading (**a**); after 100 loading cycles (**b**); after 10,000 cycles (**c**); after 40,000 cycles (**d**)

phase map of the macrorelief surface area is shown in Fig. 3.5b. The surface area profile obtained along the horizontal dashed line is shown in Fig. 3.5c.

Phase maps of the surface area macrorelief of the studied CT-specimen (see Fig. 3.3b) before its loading and after 100, 10,000 and 40,000 loading cycles were obtained, and profiles of these maps along the dashed line shown in Fig. 3.5a were

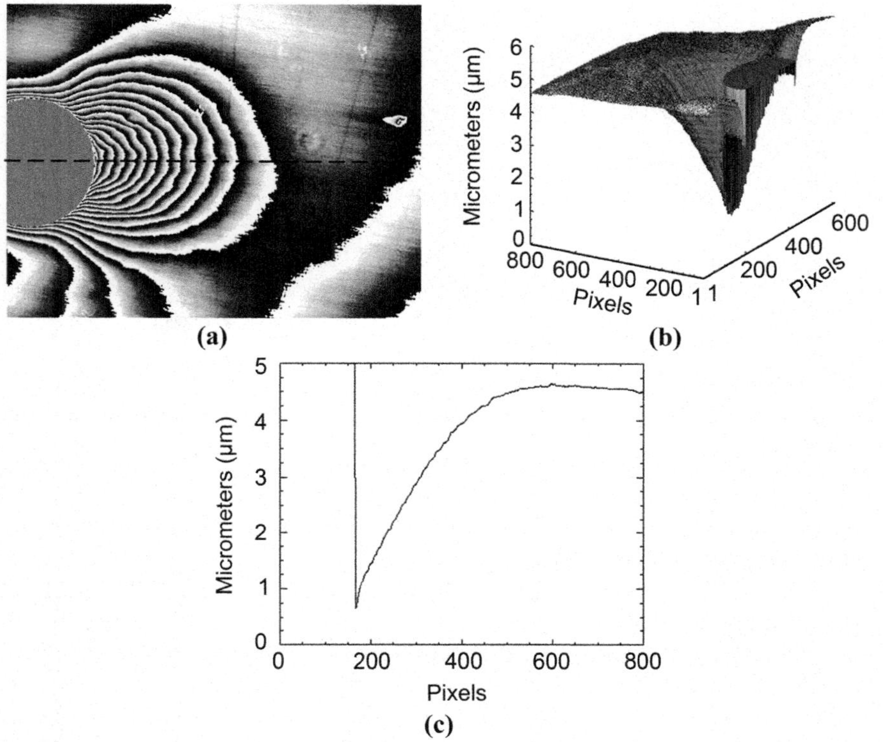

Fig. 3.5 Processing of the fringe patterns of the surface relief area shown in Fig. 3.3b obtained after 100 loading cycles: wrapped phase map (**a**); unwrapped phase map (**b**); surface profile along the dashed line depicted in Fig. 3.5a (**c**)

extracted. To find the value of the size of $d*$, extracted profiles after 100, 10,000 and 40,000 loading cycles were substracted from the profile extracted before the loading. Obtained difference profiles after 100 and 40,000 loading cycles are shown in Fig. 3.6. They can be considered as distributions of plastic deformations along the selected cross-sectional line of the displacement field. Figure 3.6b shows that the largest thinning of the studied CT-specimen is 5.0 μm or 0.25% after 40,000 loading cycles, which is calculated using Eq. (2.40), where $\lambda = 633$ nm. According to this profile, it is easy to determine the parameter $d*$, which is equal to 21 ± 1 pixels, i.e. for the steel 08kp CT-specimen, if $R = 0.1$, we obtain $d* = 262.5 \pm 12.5$ μm.

Except the FPZ size determination, the geometric parameters of the cyclic and monotonic plastic zones also can be defined. As it is shown in Figs. 3.4b–d and 3.5a, the monotonic plastic zone is defined by the boundaries of a system of closed annular interference fringes concentrated near the round notch.

Determination of fatigue plastic zones for aluminum alloys specimens. The geometric parameters of the FPZ, as well as the cyclic and monotonic plastic zones near the round notch, were also determined for CT-specimens made of D16T similar to alloy 2024-T4 and V95T similar to alloy 7075 under cyclic loads (see Fig. 3.7).

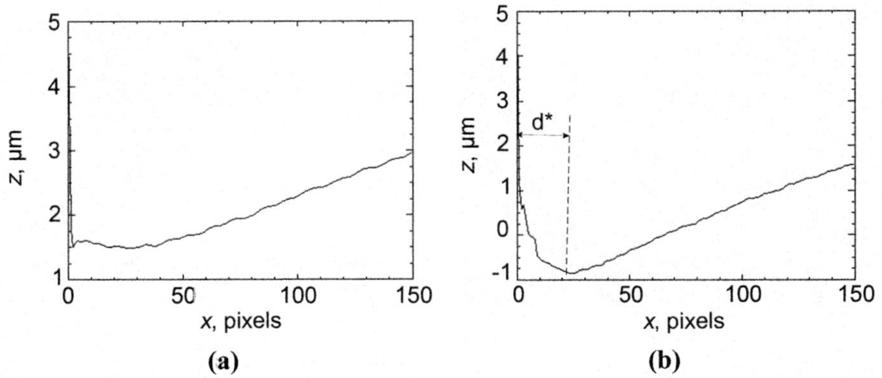

Fig. 3.6 Difference profiles of the deformed surface near the notch tip after 100 (**a**) and 40,000 (**b**) loading cycles

The specimens had dimensions of $48 \times 48 \times 4$ mm^3 and round or U-shaped notches with diameters of 1.5 or 3 mm.

Studies of CT-specimens made of aluminum alloy D16T using the developed experimental setup described in Sect. 2.5.4 (see Fig. 2.13) are represented in [19]. Retrieval of the surface areas of specimens near their notches was performed with the help of the TSI UPS method and the algorithm for extraction of the surface macrorelief phase map (see the part of the flowgraph highlighted with a dashed outline represented in Fig. 2.6). Measurement results have shown that $d* \approx 540$ µm for this alloy, if the load cycle asymmetry coefficient value (stress ratio) is $R = P_{\min} / P_{\max} = 0.5$.

By implementing the full algorithm performing the TSI UPS method, which flowchart is shown in Fig. 2.13, it is possible to observe the dynamics of changes in roughness of the studied surface area with an increase in the number of cycles. So,

Fig. 3.7 CT-specimens made of aluminum alloys D16T and V95T

the surface roughness reliefs (nanoreliefs) of the alloy D16T CT-specimen with the U-shaped notch of 1.5 mm in rounding radius obtained in the experimental setup based on the Twyman–Green interferometer (linear magnification is equal to $M_l = 0.85$) at different number of load cycles are depicted in Fig. 3.8. Here, the plastic zones and FPZs are outlined. The thick black line covers the surface macrorelief area that changes under cyclic loads. It can be identified with the monotonic plastic zone r_{pm}. The thin black line covers the FPZ. The thick dashed line covers the nanorelief surface area changing with numbers of cycles. This area can be identified with the cyclic plastic zone r_{pc}. As a result of the retrieved surfaces processing, it was found that in the nanorelief surface area, the surface roughness increases with the increase of loading cycle's quantity. However, before the appearance of a crack, the geometric parameters of these zones do not actually change with a change in the number of loading cycles.

Another situation in the cyclic and monotonic plastic zones changes and the FPZ formation took place while studying CT-specimens from the alloy V95T, which is characterized by a low level of plasticity [18]. Prior to the crack appearance in the alloy V95T specimen with the round notch of 1.5 mm in diameter, the monotonic plastic zone r_{pm} was not observed at all. This is evidenced by those shown in Fig. 3.9 interference patterns on the deformed CT-specimen surface, which were obtained with a linear magnification equal to $M_l = 0.85$ in the experimental setup based on the Twyman–Green interferometer under the specimen cyclic loading. At the same time, up to the appearance of a crack in Fig. 3.9d, interference fringes with increasing number of cyclic loads practically do not change their position. This behavior of the fringes indicates that the out-of-plane deformations of the surface before the appearance of a crack remain practically unchanged. This was confirmed by an attempt to determine $d*$ using Technique 1, which proved to be unsuccessful. Therefore, Technique 1 for determining $d*$ for this type of materials with low plasticity turned out to be unsuitable.

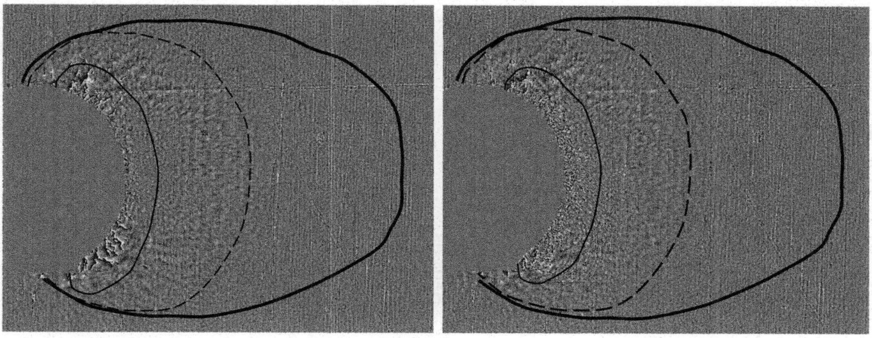

(a) **(b)**

Fig. 3.8 Fatigue plastic zones on the CT-specimen surface made of the alloy D16T at different numbers of loading cycles: 1000 (**a**), 5000 (**b**) cycles

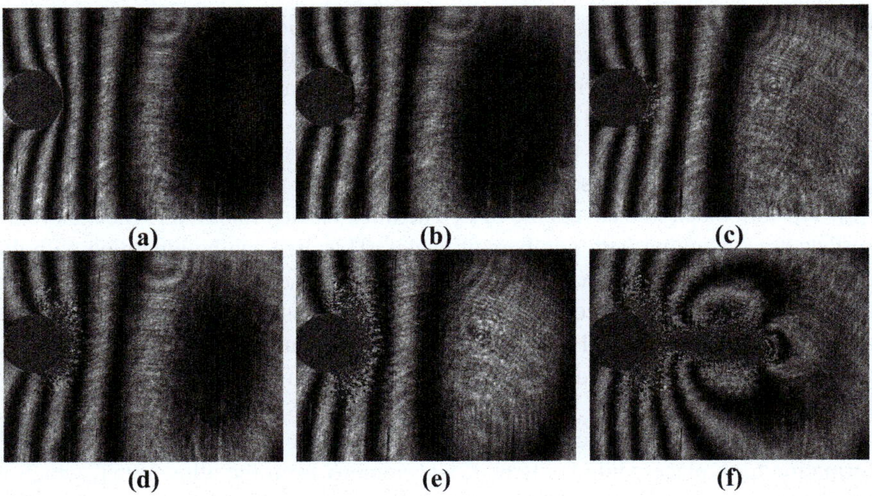

Fig. 3.9 Fringe patterns of the surface relief of the CT-specimen from alloy V95T at different numbers of loading cycles: before the loading, i.e. 0 cycles (**a**), after 1000 cycles (**b**), after 10,000 cycles (**c**), after 50,000 cycles (**d**), after 100,000 cycles (**e**), after 150,000 cycles (**f**)

On the other hand, the surface roughness increased monotonically with an increase in the number of load cycles. This is well seen in Fig. 3.9b–f directly near the round notch (see Fig. 3.9b–e) and the macrocrack edge (see Fig. 3.9f). In these places, the interference fringes are distorted by random spatial noise, the position of which corresponds to the increasing surface roughness with increasing number of cycles.

The reconstructed profile of the surface nanorelief for 50,000 loading cycles, which is shown in Fig. 3.10, is another clear confirmation that the surface roughness of the alloy V95T near the notch increases with increasing number of loading cycles. For clarity, an envelope dashed curve is superimposed on the roughness profile.

Phase maps of the alloy B95T CT-specimen surface relief at different numbers of load cycles are shown in Fig. 3.11. In this figure, the FPZ and plastic zones are

Fig. 3.10 Surface nanorelief profile of the alloy V95T CT-specimen near the notch after 50,000 cycles and its envelop

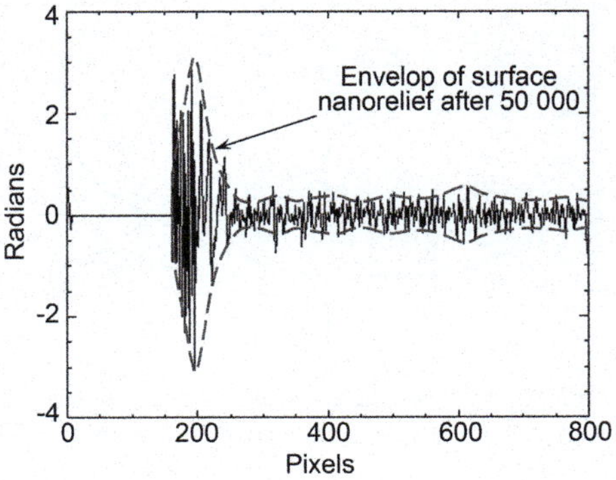

outlined in the same way as for the specimen made of D16T material (see Fig. 3.8), i.e. the static plastic zone is outlined by thick black curve, the FPZ—by thin black curve; and the zone of roughness change, that is, the cyclic plastic zone—by the thick dashed curve. Figure 3.11a–e shows that only the FPZ does not change with the number of loading cycles and before the crack appearance, the monotonic plastic zone developed only after the crack appearance, and the cyclic plastic zone increases with increasing number of loading cycles.

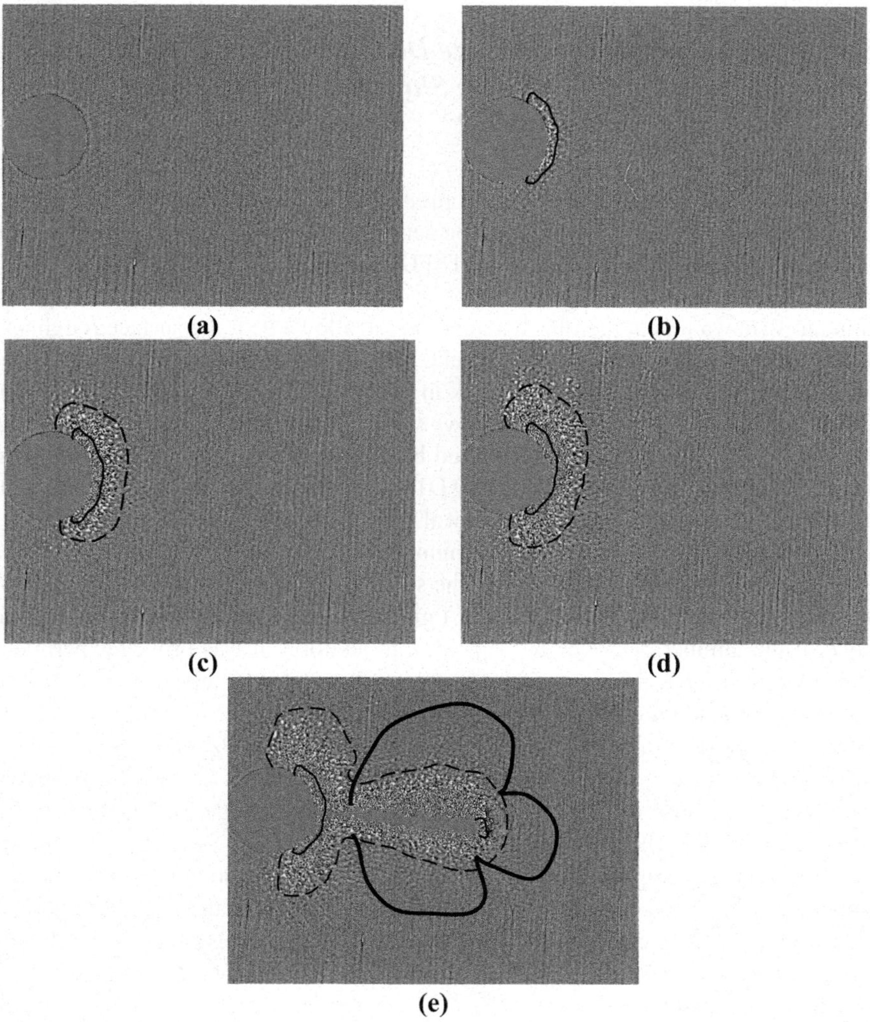

Fig. 3.11 Surface reliefs of the V95T CT-specimen for different numbers of cyclic loads: before loading, i.e. 0 cycles (**a**), after 10,000 cycles (**b**), after 50,000 cycles (**c**), after 100,000 cycles (**d**), after 150,000 cycles (**e**). Fatigue process zones are outlined by solid thin curve, cyclic plastic zones—by dashed curves, monotonic plastic zone—by solid thick curve

The results of the experiments indicate the different evolution nature of the cyclic and monotonic plastic zones for CT-specimens made of aluminum alloys D16T and V95T [18]. This behavior of the plastic zones can be explained by various mechanical characteristics of these materials, in particular the fact that the alloy D16T has moderate plasticity, and the alloy B95T has the low one. In addition, as shown in [12], the nature of the occurrence of the cyclic plastic zone also depends on the rate of strain hardening of the material under cyclic loads.

3.1.2 Fatigue Process Zone Size Determination by Using Iterative Two-Step Phase Shifting Interferometry Method and Technique 2

The results of the above studies of structural materials with different levels of plasticity showed that the FPZ size $d*$ can be determined only for CT-specimens of high or moderate material's plasticity, if the TSI UPS method is used. However, for materials with low plasticity, this zone cannot be determined by this method. At the same time, it is shown that at the FPZ boundary in the alloy V95T, the surface roughness grows and increases with increasing cyclic loads during the period N_i before nucleation of a minimum fatigue macrocrack in material. Known investigations confirm this behavior of various metals and alloys surfaces. In particular, surface roughness evolution at fatigue failure was observed by Chan et al. [7] for Ni200 specimens, Ignatovich and Yutskevich [15] for alloy D16AT, Buckner et al. [6], Davila et al. [10] and Diaz et al. [12]. For example, Chan et al. [7] showed that the roughness parameter Ra (see Eq. (2.25)) increases with the number of cycles under cyclic loading. Diaz et al. [12] experimentally showed that the surface roughness of the CT-specimen of stainless steel 304 with a notch radius of 1 mm also increases within the cyclic plastic zone, if the number of cycles N_i increases. In addition, it was proved experimentally that the CT-specimens surface roughness at the FPZ boundary increases with increasing of loading cycles N_i [18, 20].

In order to determine accurate the dimensions of the FPZ and essentially the parameter $d*$ for any metals and alloys, regardless of the level of their plasticity, the criterion of achieving the maximum surface roughness at the FPZ boundary was proposed [21, 22]. The scheme of the FPZ formation and stress distribution in the vicinity of the notch tip is shown in Fig. 3.12. The degree of material deformation and the range of local stresses $\Delta\sigma_y^*$ within the $D*$ zone, covering the FPZ and a narrow strip outside it and marked in this figure by a dashed curve, significantly exceed the same mechanical characteristics of the material outside the zone $D*$ (see Figs. 3.12 and 3.13). Moreover, the growth of the $\Delta\sigma_y^*$ is observed with disposal from the notch tip, and its maximum values are obtained at the FPZ boundary, that is, at the distance $d*$.

The reliability of this criterion is confirmed using results obtained by Arola and Williams [3], Ostash [29], Suraratchai et al. [42]. Arola and Williams [3] and

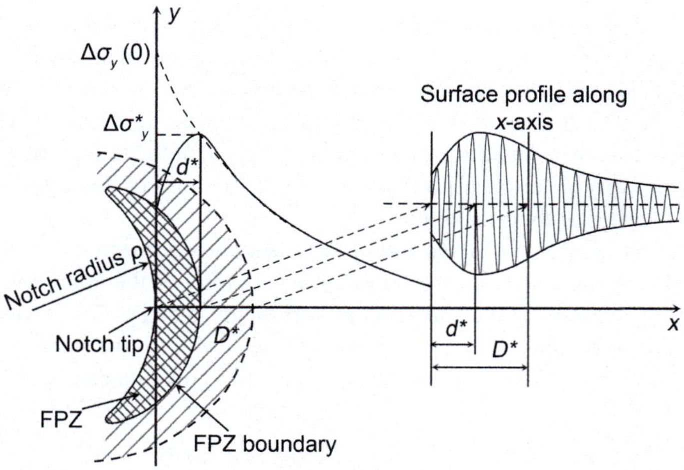

Fig. 3.12 Fatigue process zone (parameter $d*$) in specimen with notch and zone (parameter) $D*$, in which the local stress range $\Delta\sigma_y^*$ reaches its maximum values; $\Delta\sigma_y(0)$ is the maximum stress range at the notch tip ($x = 0$)

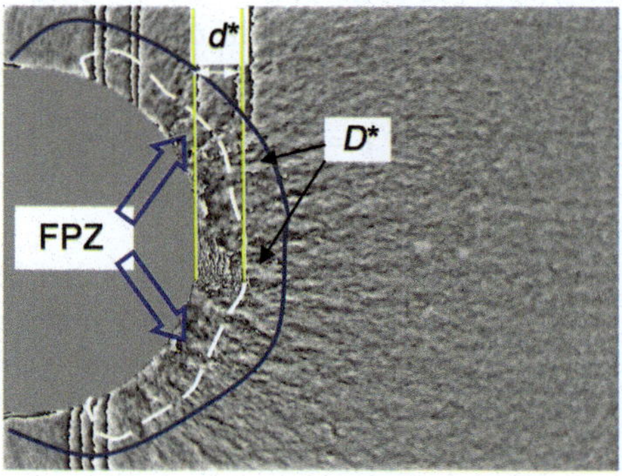

Fig. 3.13 Fatigue process zone (FPZ) $d*$ in the vicinity of the notch tip in 08kp steel CT-specimen outlined by a white dashed curve and zone $D*$ outlined by a black solid curve

Suraratchai et al. [42] showed that surface roughness is associated with the effect of stress concentration during fatigue of a material. Moreover, the dependencies "effective fatigue stress concentration factor K_f as a function of Ra" were calculated in [3] for abrasive waterjet machined surfaces of a rolled sheet steel AISI 4130CR specimens using the Neuber rule [26] and the Arola-Ramulu model [2]. These dependencies have shown that K_f increases strictly during the surface roughness evolution. They also obtained similar experimental dependencies.

On the other hand, Ostash et al. [28, 33] proved that the FPZ boundary corresponds to the maximum of the local stress range $\Delta\sigma_y^*$, which is proportional to K_f, if the nominal stress $\Delta\sigma_{\text{nom}}$ is constant, that is

$$\Delta\sigma_y^* = K_f \cdot \Delta\sigma_{\text{nom}}. \tag{3.4}$$

Consequently, the surface roughness evolution is accompanied by an increase in K_f and, respectively, an increase in $\Delta\sigma_y^*$. Since the maximum of the local stress range $\Delta\sigma_y^*$ is reached at the distance $d*$ from the notch tip [28, 29], it can be argued that the FPZ boundary corresponds to the maximum surface roughness. Thus, the FPZ size can be found by calculating the spatial distribution of the roughness parameters Ra or Sa in every pixel of the surface area in the vicinity of the notch tip and by definition the sites of their maximum values.

Calculation of spatial distribution of the roughness parameter Sa. To evaluate the surface roughness spatial distribution in the vicinity of the notch tip, the arithmetical mean height Sa of the scale-limited surface given by Eq. (2.27) was used. For a square definition area $A_s = A_x \times A_y$, where $A_x = A_y$, for the surface roughness height $h_r(x, y)$ in every point of the surface area, this equation can be rewritten as

$$Sa\left(x', y'\right) = \frac{1}{A_s} \iint\limits_{As} |h_r(x, y)| \mathrm{d}x\mathrm{d}y. \tag{3.5}$$

The retrieved digital height map $h_r(k, l)$, which produced from the nanorelief phase map $\varphi_r(k, l)$ using Eq. (2.43), make it possible to find the roughness parameter $Sa_M\left(k', l'\right)$ in every pixel corresponding to the center of each square subset $(A_s)_M$ of $A_{Mx} \times A_{My}$ pixels of the digital camera's matrix photodetector by the equation

$$Sa_M\left(k', l'\right) = \frac{1}{JQp_x p_y} \sum_{\left(k'=k-j+\frac{J-1}{2}\right)} \sum_{\left(l'=l-q+\frac{Q-1}{2}\right)} |h_{rM}(k, l)| p_x p_y, \tag{3.6}$$

where $h_{rM}(k, l)$ is the digital nanorelief height map, in which the pixel dimension coincides with the pixel dimension of the matrix photodetector; $j = 0, \ldots, J - 1$; $q = 0, \ldots, Q - 1$ are the pixel numbers in the square subset $(A_s)_M$ (J and Q are the odd numbers); p_x, p_y are the pixel pitches in the photodetector.

The resulting discrete spatial distribution of arithmetical mean heights $Sa\left(k', l'\right)$ is obtained by dividing of $Sa_M\left(k', l'\right)$ on the linear magnification M_l, that is

$$Sa\left(k', l'\right) = \frac{Sa_M\left(k', l'\right)}{M_l}. \tag{3.7}$$

The maximum values of this roughness parameter produce the desired FPZ boundary in the form of a thin strip near the notch tip. This boundary makes it possible to determine the size $d*$ and other geometric parameters of the FPZ.

Experimental results. Experimental validation of Technique 2 based on the proposed criterion to determine the FPZ using maximum surface roughness values at the FPZ boundary and the ITSI UPS method (see Sect. 2.5.5) was performed with the help of the experimental setup, which scheme is shown in Fig. 2.13. Several CT-specimens 4 mm thick and notch radii of 1.5 and 0.75 mm, made of low-carbon steel 08kp ($24 \times 24 \times 4$ mm^3) and aluminum alloys D16T and V95T ($55 \times 55 \times 4$ mm^3), were used for experiments. The specimen surfaces in the vicinity of the notch tip were carefully mechanically polished and further superfinished. The surfaces of CT-specimens made of steel 08kp were superfinished up to $Sa < 0.040$ μm (ISO Grade Scale Number 2). The surfaces of aluminum alloys D16T and V95T CT-specimens were superfinished up to $Sa < 0.065$ μm (ISO Grade Scale Number 3). Cyclic loading of all specimens was performed at the stress ratio $R = P_{\min}/P_{\max} = 0.1$ and at the maximum stress ranges equal to $\Delta\sigma_y(0) = 560$ MPa for steel 08kp specimens, $\Delta\sigma_y(0) = 460$ MPa for aluminum alloy D16T specimens and $\Delta\sigma_y(0) = 360$ MPa for aluminum alloy V95T specimens.

Optical scheme of the Twyman–Green interferometer was adjusted to reach the lateral magnification $M_l = 0.85$. Since the pixel pitch in the matrix photodetector of the SONY XCD-SX910 digital camera is equal to $p_x = p_y = 4.65$ μm, the sampling interval at such magnification is $\tau_s = 5.5$ μm. According to the one shown in Fig. 2.15, the influence of the autocorrelation function with correlation length τ_0 on the sampling interval τ_s, in such a sampling interval (i.e. $\tau_s = 5.5$ μm), aliasing is still practically absent. So, the noise caused by aliasing is still insignificant. This interval was chosen to record the studied surface areas fringe patterns both before and after cyclic loading.

To determine the dimensions of the FPZ, the obtained height maps $h_r(k, l)$ of the surface macrorelief area were transformed into the spatial distribution of the arithmetic mean surface roughness $Sa(k', l')$ using Eqs. (3.6) and (3.7). The dimensions of the square subset A_s were $A_s = A_x \times A_y = (A_s)_M / M_l = 15p_x \times 15p_y$. Such dimensions were chosen assuming that the roughness parameter $Sa \leq 0.025$ μm corresponds to minimum roughness of any sampling area $l_x \times l_y = 80 \times 80$ μm outside the cyclic plastic zone. To eliminate sharp height differences in the distributions of the roughness parameter $Sa(k', l')$, a Gaussian smoothing filter was used. Surface height maps of the spatial distribution of the roughness parameter $Sa(k', l')$ for the steel 08kp CT-specimen surface area near the notch before loading and after 10,000 cycles are shown in Fig. 3.14. Similar maps for the aluminum alloy V95T CT-specimen surface area are shown in Fig. 3.15. Maps of the spatial distribution of the roughness parameter $Sa(k', l')$ for the aluminum alloy D16T CT-specimen surface area near the notch before loading and after 5 000 cycles are shown in Fig. 3.16.

The obtained experimental results confirmed the proposed hypothesis of achieving maximum surface roughness at the FPZ boundary. In particular, in Figs. 3.14–3.16, the FPZ boundary is clearly distinguished in the form of a narrow strip, on which the surface roughness increases sharply. Moreover, the distance $d*$ depending on the direction angle of the radius vector $\vec{\rho}$, starting from the notch center, does not change. Measurements of the parameter $d*$ along the profile intersecting the center of the notch gave the following results: $d* = 213$ μm for 08kp CT-specimen (Fig. 3.14), $d* = 310$ μm for V95T CT-specimen (Fig. 3.15) and $d* = 540$ μm for D16T CT-specimen (Fig. 3.16).

Thus, proposed Technique 1 based on the TSI UPS method, make it possible to define the FPZ geometric parameters and determine the FPZ size $d*$, as well as the geometric parameters of the static and monotonic plastic zones for materials with high plasticity. To define the FPZ dimensions in materials with high, moderate and

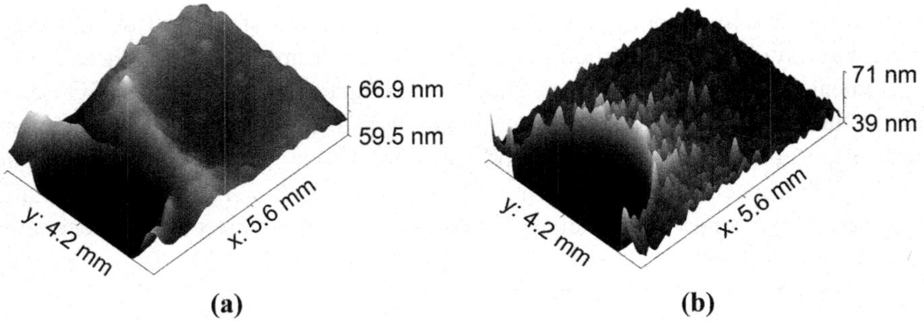

Fig. 3.14 Surface height maps of the roughness parameter $Sa(k', l')$ spatial distribution for the surface area of the steel 08kp CT-specimen near the notch of radius $\rho = 1.5$ mm before fatigue (**a**) and after 10,000 loading cycles (**b**)

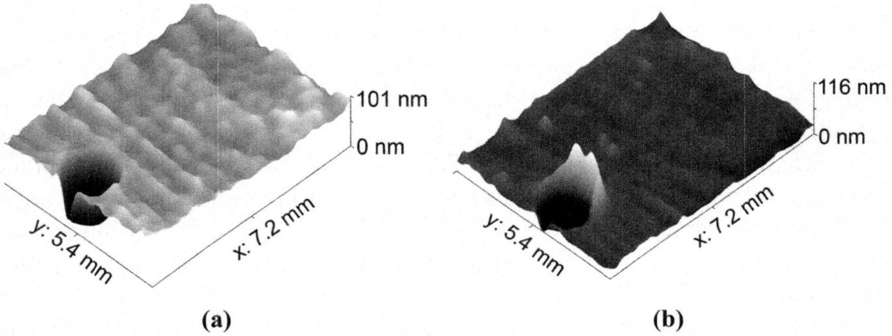

Fig. 3.15 Surface height maps of the roughness parameter $Sa(k', l')$ spatial distribution for the surface area of the V95T CT-specimen near the notch of radius $\rho = 0.75$ mm before fatigue (**a**) and after 10,000 loading cycles (**b**)

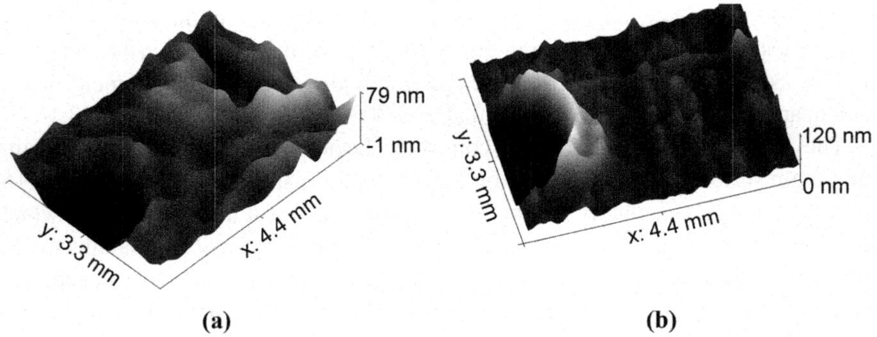

Fig. 3.16 Surface height maps of the roughness parameter $Sa(k', l')$ spatial distribution for the surface area of the D16T CT-specimen near the notch of radius $\rho = 0.75$ mm before fatigue (**a**) and after 5000 loading cycles (**b**)

low plasticity, Technique 2, based on the ITSI UPS method can be used. Technique 2 is more effective, since the FPZ sizes are determined using the surface roughness spatial distribution on the extracted surface nanorelief height map for a wide range of metal and alloy specimens.

3.2 Study of Fatigue Macrocrack Initiation in Thin Compact Tension Specimens

One of the urgent directions of the fatigue durability evaluation of structural materials is the prediction of the site and moment of a fatigue macrocrack initiation in metal structural elements with rounded notches. Usually, the fatigue durability is divided into three stages consisting of crack incubation or nucleation, crack initiation and crack propagation [4, 40]. A somewhat different approach assumes the existence of two main periods of fatigue failure, namely the period of origin of the minimum fatigue macrocrack in the material and the period of its growth [28, 29, 33, 34]. As mentioned above (see Sect. 3.1), these periods can be expressed by the number of cycles N_i and N_p, respectively, and the fatigue life N_f is given by Eq. (3.3). The N_i period can be divided into two substages, namely (i) the macrocrack nucleation period, which is characterized by a change in the substructure in the entire volume of the loaded material; (ii) the period of initiation and growth of microstructurally short cracks. The period N_p can also be divided into two stages, namely (i) the period of the macrocrack formation from one dominant microcrack; (ii) the period of the macrocrack growth to the critical sizes.

The process of the macrocrack initiation is closely connected with the formation and development of the FPZ during fatigue. The FPZ size $d*$ can be considered as the delimitation criterion of two main fatigue damage periods N_i and N_p. In this interpretation, the key criterion for assessing the fatigue life of structural elements with round and U-shaped notches is the moment of the fatigue macrocrack initiation at the end of the period N_i and at the beginning of the period N_p.

The effect of the surface roughness increase in various materials during fatigue in sites of macrocracks probable initiation was marked by several investigators. In particular, Suraratchai et al. [42] considered the roughness of the treated surface as a source of generating a local stress concentration that causes the propagation or nonpropagation of cracks. Using the theoretical stress concentration factor K_t, they determined the first fatigue damage stage N_i based on Basquin–Wöhler type power law, i.e. $S-N$ relation for solid material [9], and the second one N_p using the crack propagation law proposed by Paris and Erdogan [35]. However, they believed that the stress concentration causes only the propagation of surface cracks. The conducted researches allowed estimating the influence of roughness of the processed surfaces on fatigue durability of the aluminum alloy 7010. Buckner et al. [6] showed that the surface area of the aluminum specimen with high roughness may indicate the place where cyclic deformations and more developed slip bands are prevalent. They also experimentally demonstrated an increase of surface roughness during material fatigue and concluded that the evolution of surface roughness under the action of cyclic loads may be an indicator of fatigue crack formation. Computational results simulating the evolution of slip steps and surface roughness on a fatigue surface using dislocation dynamics and a cohesive zone approach [3] were presented by Brinckmann and Van der Giessen [5]. Ural et al. [43] obtained results indicating that a fatigue crack initiation on a free surface is predicted as the cooperative effect of the build-up of local stress by evolution of the dislocation distribution and development of surface roughness. Chan et al. [7, 8] showed that the surface roughness arithmetic mean Ra of the material Ni200 is proportional to the ratio of the slip band width to its size. During cyclic loads, this ratio increases with increasing N_i, and, consequently, Ra also increases. Based on these studies, as well as on studies conducted by Buckner [6] and Brinckmann and van der Giessen [5], Chan [8] argues that an area with high surface roughness may indicate fatigue crack initiation.

Optical methods, in combination with digital image processing techniques, have also been used to investigate the effect of fatigue loads on the evolution of surface roughness near round notches. Diaz et al. [12] and Armas et al. [1] used white light scattering from the surface of a stainless steel 304 specimen to estimate the expansion of the plastic zone around a round notch during its cyclic loading. Using the correlation contrast method [11], they showed that the cyclic plastic zone of this material is constantly growing with an increase in the number of cycles. Prior to the application of loads, the correlation contrast at the notch was unchanged and equal to unity, since the surface roughness was the same and very small throughout the studied area. With an increase in the number of load cycles, a cyclic plastic zone was produced, which boundaries were determined by a higher surface roughness compared to the roughness outside this zone. In this case, a decrease in the correlation contrast within the plastic zone corresponded to an increase in the contribution of the light scattered by the surface and, consequently, to an increase in its roughness. The experiment showed a direct relationship between the surface roughness and the level of plastic deformation of the material.

A new stage of experimental research to identify the site and time of fatigue macrocrack initiation and the relationship between changes in surface roughness and the processes of a crack nucleation, initiation and propagation was started by Kelton, Sola et al. [16, 17, 39, 40]. In all fatigue experiments, they used 3D Optical Profiler BRUCKER NPFLEX which provided high accuracy of the restored surface relief of polished pure nickel specimens. Kelton et al. [16, 17] studied the fatigue macrocrack propagation after its initiation in nickel sheet specimens. In particular, the dimensions of cyclic and monotonic plastic zones were accessed using measurements of arithmetic mean surface roughness Sa of a studied surface area. Authors have analyzed areal surface roughness changes of monotonic and cyclic plastic zones near the crack tip during fatigue with the help of the developed image registration and processing procedure. They have determined the boundary of monotonic plastic zone near the propagating crack tip and boundaries of monotonic and cyclic zones near the arresting crack tip. In addition, it was shown that the roughness parameter Sa increases near the crack tip arrested at triple point, if the number of cycles increase.

They also showed that the correlation between the increase in roughness and the stress intensity factor is observed at the triple point. Substantial topography changes near the triple points allowed them to introduce the surface roughness parameters as damage indices.

Research to predict the macrocrack initiation site and time in polycrystalline pure nickel through surface topography changes was fulfilled by Sola et al. [40]. As a criterion to predict the macrocrack initiation, they proposed to choose such damage indices as the arithmetic mean areal surface roughness ΔSa and the maximum areal surface height ΔS_t, which were calculated for difference images. These images were obtained by pixel-by-pixel subtraction of two 3D maps of surface heights restored after given fatigue cycles numbers N_i. The roughness parameters were calculated for square subsets, which dimensions were chosen to be commensurable with the grain size of pure nickel. Performed experimental results have shown that the crack initiated at the subset with the highest damage indices ΔSa and ΔS_t, as well as arithmetic mean surface roughness Sa in this subset increases steadily before the crack nucleation and stabilizes after its initiation. The obtained experimental results confirmed the authors' assumption that the main factors initiating the fatigue cracks appearance are a high increase in roughness in the site where a crack is likely to appear as well as the sites at the triple points and grain boundaries.

A detailed analysis of publications in this direction allowed the authors [40] to state that no one has conducted similar researches and they were the first to predict the site and time of a fatigue macrocrack initiation, using indices ΔSa and ΔS_t. This statement is valid for English-language publications, however, in Ukrainian-language publications, such studies were carried out before them. Similar researches were performed in PMI beginning from 2010 year. Results were printed in [18, 20]. Similar research results were also briefly published in English-language works by Muravsky et al. in 2019 [23] and by Nazarchuk et al. in 2021 [25].

3.2.1 Determination of Site and Time of Fatigue Macrocrack Initiation Along Given Profiles

Muravsky et al. [20] showed that in CT-specimens made of steel 08kp and aluminum alloy D16T, a macrocrack initiates most likely in the FPZ boundary area where the largest increase of the nanorelief roughness occurs during cyclic loading. According to the results of 08kp and D16T CT-specimens surfaces studies, the macrocracks initiation was observed after a certain number of cycles not in the places of their largest thinning. Indeed, Fig. 3.17b and c clearly shows the place of the largest thinning of the D16T CT-specimen, which corresponds to the concentric rings center on the interferograms of this specimen surface. This center is located at the distance d^*. However, the crack initiated and propagated not in the place of the largest thinning, but at some angle to the round notch axis.

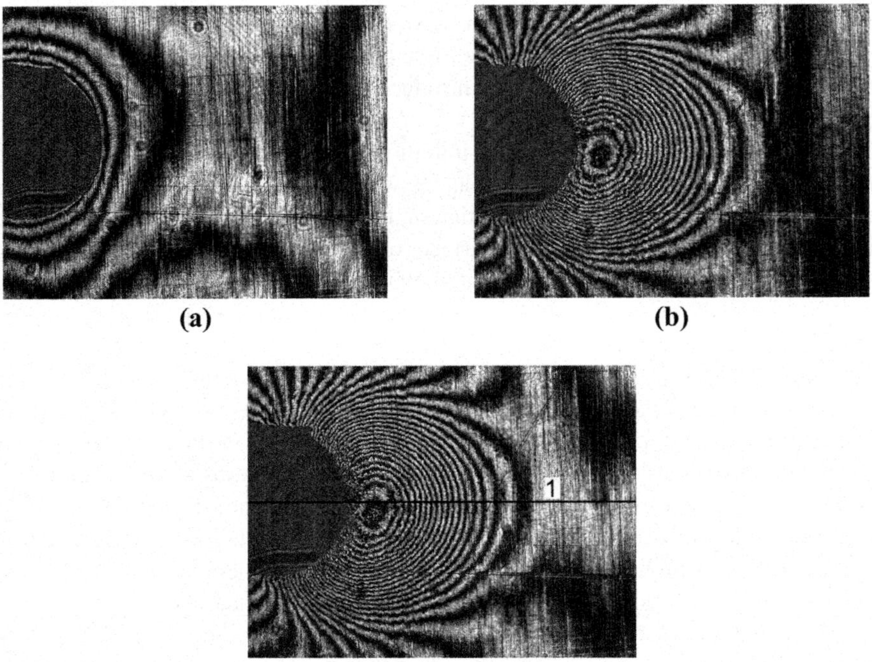

(a) (b)

(c)

Fig. 3.17 Interferograms of the D16T CT-specimen surface area with different number of load cycles: **a**—0 cycles; **b**—1000 cycles; **c**—5000 cycles

Quantitative evaluation of the roughness parameters of this surface near the round notch showed that the macrocrack arose exactly in the place where the largest increase in the nanorelief roughness occurred during the cyclic loads before its initiation. Confirmation of this statement is shown in Figs. 3.18 and 3.19. Figure 3.18 shows the profile of the surface nanorelief retrieved after 5000 cycles along line 1 of the round notch axis passing from the notch center (see Fig. 3.17c). Figure 3.19 shows the profile of the same surface nanorelief retrieved after 8000 cycles along the line passing from the notch center at the angle $\xi \approx 11.7°$ to the notch axis through the place on the notch edge from which the crack started. If before cyclic loads the arithmetic mean surface roughness Ra along the two selected profiles had approximately the same values, then with increasing number of loads it increased near the notch root compared to all other sections of the profile. Moreover, a larger Ra increase is observed along the line, where the macrocrack initiated.

Another confirmation of this assumption is shown in Fig. 3.20a–d. This figure depicts phase maps of nanoreliefs for a CT-specimen made of steel 08kp (see Fig. 3.3a) for various numbers of loading cycles before and after a macrocrack arising. This crack occurred between 24,000 and 27,000 cycles (see Fig. 3.20d) precisely from the site where the largest changes in surface roughness were observed for 10,000 and 20,000 cycles, as can be seen from Fig. 3.20a–c. Figure 3.20b shows

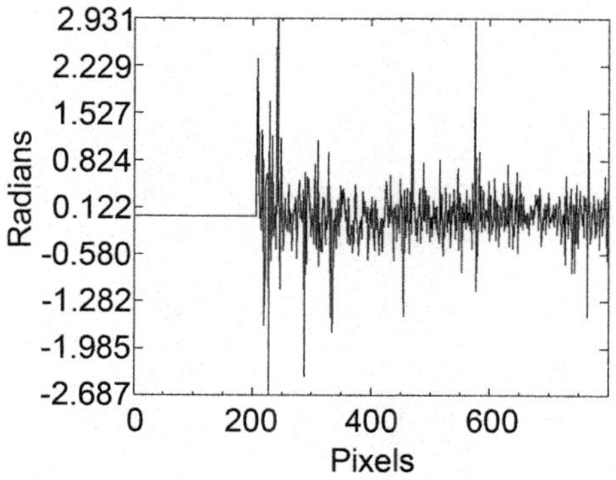

Fig. 3.18 Profile of the retrieved surface nanorelief (material D16T) after 5000 cycles along line 1 of the round notch axis passing from the notch center (see Fig. 3.17c)

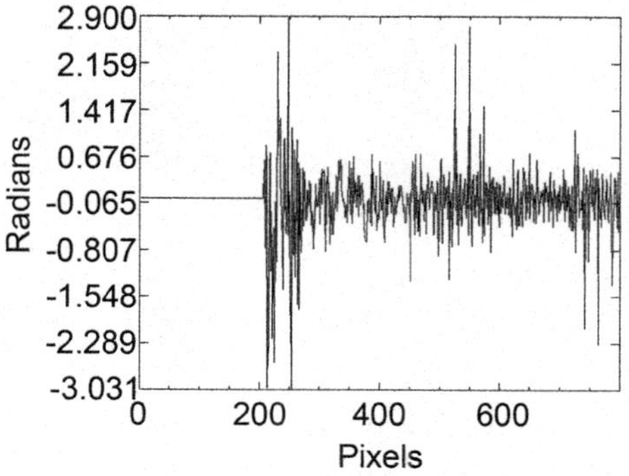

Fig. 3.19 Profile of the retrieved surface nanorelief (material D16T) after 8000 cycles along the line passing from the round notch center at the angle to the notch axis through the place on the notch edge from which the crack started

three surface relief radial profiles along which the RMS surface roughness Rq was measured. The maximum value of Rq for each profile was obtained at the point located at the FPZ boundary. The dynamics of the parameter Rq change from the number N_i of cycles for these three points is shown in Fig. 3.21. As it is demonstrated by the represented in this figure dependencies of RMS roughness Rq on the number of load cycles, the surface roughness at the point corresponding to the FPZ boundary for the two profiles 1 and 3 is practically independent of the number of cycles and varies slightly. At the same time, the surface roughness at the point corresponding to the FPZ boundary for profile 2, which passes exactly at the site where the crack originates, monotonically increases with the increase in the number of cycles until a macrocrack is formed at this site. At this point, the roughness increases almost three times after 20,000 cycles compared to its roughness before the start of the experiment (for the 0th cycle). However, the surface roughness at the points of intersection of profiles 1 and 3 with the FPZ boundary did not increase after the same number of cycles (see Fig. 3.21a, c). After the fatigue crack initiation at $N_i \approx 27{,}000$, the growth of roughness in the point of intersection of profile 2 with the FPZ boundary stopped, which became an important indicator of the moment of crack initiation. Thus, the roughness increment, which practically corresponds to the damage indices

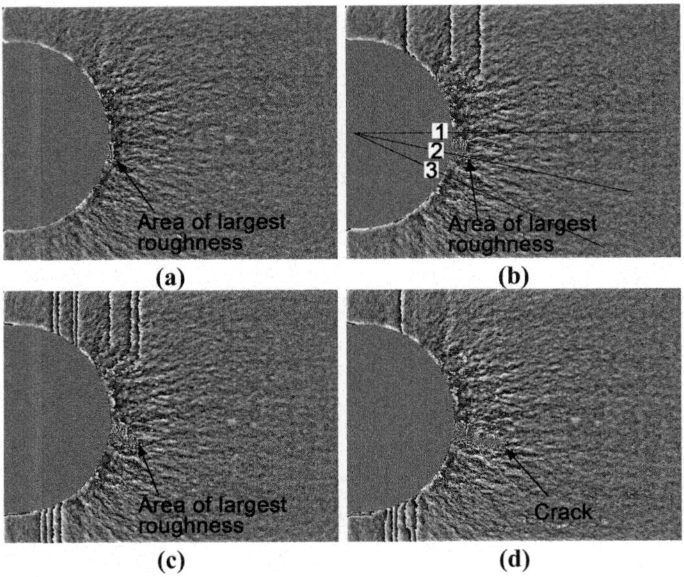

Fig. 3.20 Phase maps of the steel 08kp specimen surface nanoreliefs near the round notch for different number of load cycles: **a** 10,000, **b** 20,000, **c** 24,000, **d** 27,000

ΔSa and ΔS_t introduced by Sola et al. [40], was chosen as the indicator of the place and moment of crack formation.

Thus, the new technique (Technique 3) for determining the site and time of the fatigue macrocrack initiation along given profiles is proposed. The use of Technique 3 makes it possible to predict the site of a fatigue macrocrack initiation and estimate the period its origin N_i, establishing for this purpose the critical values of one of the roughness parameters (Ra or Rq).

3.2.2 New Technique for Determining the Site and Time of Fatigue Macrocrack Initiation in Every Pixel of Surface Roughness Height Map

The obtained results showed that the area of interest on the surface of the steel 08kp CT-specimen, where fatigue macrocracks are most likely to initiate and arise, is a narrow band surrounding the FPZ boundary. According to our hypothesis, which will be confirmed below, such an area of interest on the surface of any CT-specimen containing a round notch is the zone $D*$ shown in Fig. 3.12. The range of local stresses $\Delta\sigma_y^*$ reaches its maximum values in this zone, which covers the FPZ and the narrow strip outside it. Note that Sola et al. [40] do not indicate the area of interest, where fatigue macrocracks are most likely to occur. In this paper, it is pointed out

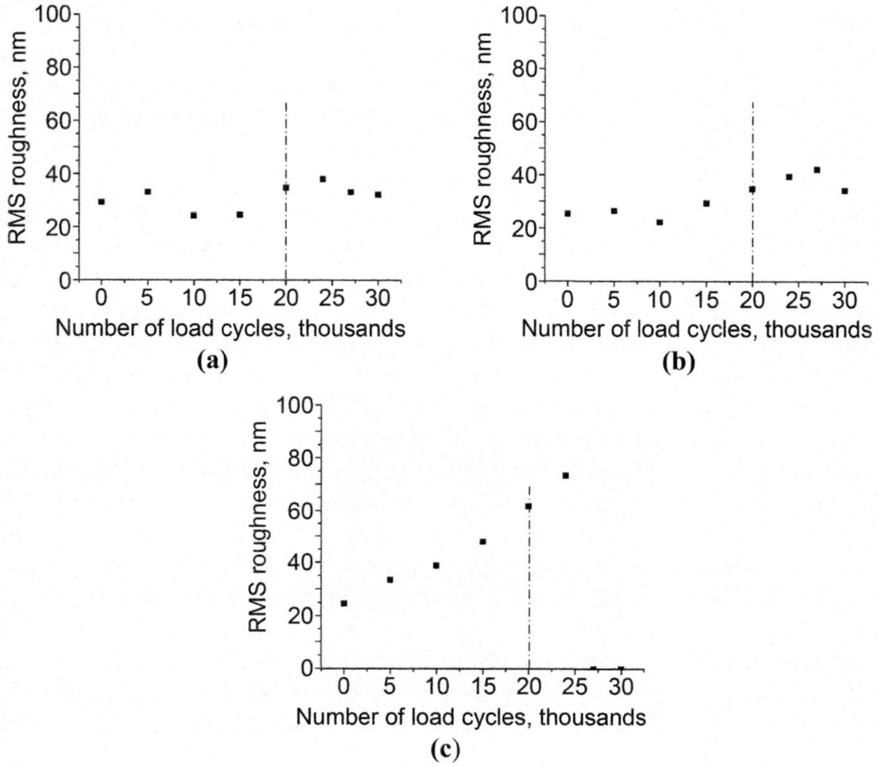

Fig. 3.21 Dependencies of Rq on the number of load cycles at the FPZ boundary of the steel 08kp CT-specimen surface at points on profiles 1 (**a**), 3 (**b**) and 2 (**c**)

that the most probable places of origin of fatigue macrocracks are triple points and grain boundaries. However, within the zone $D*$, there is a sufficient number of both triple points and grain boundaries. So, our hypothesis does not contradict the results of the study obtained in [40].

To improve developed Technique 3 for determination of the site and time of a fatigue macrocrack initiation along given profiles during cyclic loading of notched CT-specimens, the new technique for determining the site and time of a fatigue macrocrack initiation in every pixel of the retrieved surface roughness height map using the roughness increments of the surface relief area near the notch tip (Technique 4) was proposed. Technique 4 uses the 3F TWIM (see Sect. 2.7.4) for retrieving the phase maps $\varphi_{R,\Lambda 3}(k,l)$, $\varphi_{M,\Lambda 3}(k,l)$, $\varphi_{T,\Lambda 3}(k,l)$ of the nanorelief or surface roughness, macrorelief and total relief, respectively. To implement Technique 4, two experimental setups containing single-wavelength and two-wavelength Twyman–Green interferometers were used. The first one is similar to the setup described in Sect. 2.5.4 and shown in Fig. 2.13. However, the second setup, used to perform Technique 4, differs from that shown in Fig. 2.47 by application of a Nd:YAG laser generating a coherent beam at a wavelength $\lambda_1 = 532$ nm and a He–Ne laser generating the a coherent beam at a wavelength $\lambda_2 = 632.8$ nm. Such combination of laser sources in the setup makes it possible to produce the synthesized beat wavelength $\Lambda_{12} = 3340$ nm. Besides, two shutters are added to the interferometer optical scheme. The scheme of this experimental setup is shown in Fig. 3.22.

Necessity to use two setups to perform Technique 4 was caused by a monotonic increase in surface roughness near the notch tip in the surface zone $D*$ (see Fig. 3.12) of CT-specimens made of steel 08kp, as well as of aluminum alloys D16T and V95T, during their cyclic loading within the period N_i of fatigue failure. As shown above (e.g. in Sect. 2.5.6), the allowable error of the surface area phase map retrieval is occur, if $Sa < \lambda/8$. For instance, if $\lambda = 632.8$ nm, then $Sa < 80$ nm. However, the initiation of fatigue macrocracks for a number of materials with moderate and low plasticity usually begins at higher values of the roughness parameter ($Sa \geq 80$ nm) at later stages of the period N_i of the fatigue macrocrack initiation. Therefore, the two-wavelength Twyman–Green interferometer should be used in combination with the single-wavelength interferometer to retrieve reliably the surface relief of the studied specimens. Note that the single-wavelength version can be used by the setup containing the two-wavelength Twyman–Green interferometer. For example, the Nd:YAG laser 1 is working in this setup, but He–Ne laser 2 is not working (see Fig. 3.22).

Preparation of CT-specimens and conditions of their cyclic loading. Several thin CT-specimens were prepared for experimental studies. The specimens were made of low-carbon steel 08kp (dimensions 30 × 31 × 3 mm^3), as well as of aluminum alloys

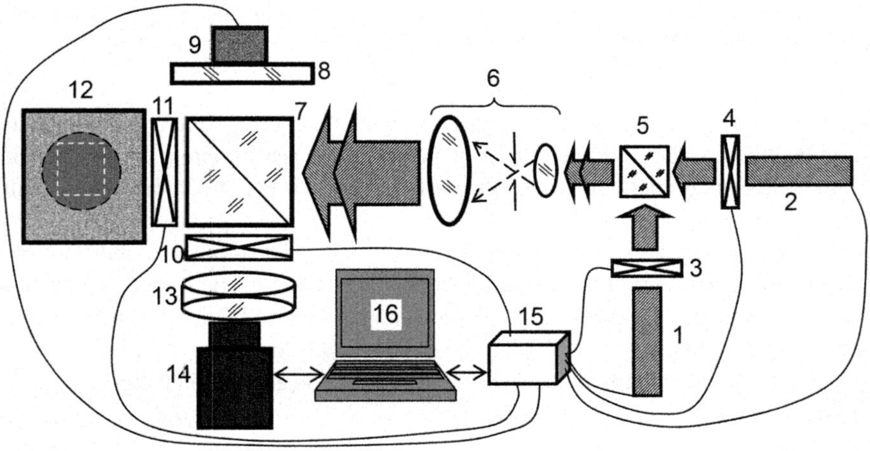

Fig. 3.22 Scheme of the experimental setup containing the two-wavelength Twyman–Green interferometer and performing Technique 4: Nd:YAG laser ($\lambda_1 = 532$ nm) (1); He–Ne laser ($\lambda_2 = 632.8$ nm) (2); shutters (3), (4), (10) and (11); beam splitters (5) and (7); optical collimator (6); movable mirror (8); piezoelectric transducer (9); test cast iron specimen attached to a jig plate (12); objective Yupiter 37A (13); digital camera SONY XCD-SX910 (14); controller (15); computer (16)

D16T (dimensions 48 × 48 × 4 mm^3) and V95T (dimensions 48 × 48 × 4 mm^3) and contained round notches with diameters 1.5 and 3 mm. The specimens surfaces were ground and then polished. The surfaces of steel 08kp specimens were polished to $Sa < 0.030$ μm (ISO Grade Scale Number 2, roughness class 14 according to GOST 25142–82). The surfaces of the aluminum alloys D16T and V95T specimens were polished to $Sa < 0.060$ μm (ISO Grade Scale Number 3, roughness class 13 according to GOST 25142–82). Experiments on cyclic loading of CT-specimens were carried out on a servo-hydraulic machine for fatigue testing "IMUR". Cyclic loads on all the specimens were carried out at a stress ratio $R = 0.1$. The local stress ranges were $\Delta\sigma_y(0) = 560$ MPa for low-carbon steel 08kp specimens, $\Delta\sigma_y(0) = 460$ MPa for aluminum alloy D16T specimens and $\Delta\sigma_y(0) = 210$ MPa for aluminum alloy V95T specimens. To record the interferograms of the studied area near the notch, cyclic loads were stopped after the 1st, 10th, 100th, 1000th, etc., cycles until the appearance of a fatigue macrocrack on the specimen surface.

Modes of interferograms recording. To record interferograms (fringe patterns) of the CT-specimens tested surface areas, the 3SI UPS method and 3F TWIM were used. The 3SI UPS method was performed at the initial stage of the material cyclic loading by using the experimental setup containing single-wavelength Twyman–Green interferometer, which is described in Sect. 2.5.4 and shown in Fig. 2.13. Another version of this interferometer can be realized in the experimental setup shown in Fig. 3.22, if one beam in the two-wavelength Twyman–Green interferometer is blocked by the shutters 3 or 4. This stage is completed when reaching a maximum surface roughness arithmetic mean $Sa < \lambda/8$, or $Sa < 80$ nm for a He–Ne laser ($\lambda = 632.8$ nm). The next final stage of the material cyclic loading realizing by the 3F TWIM is fulfilled using the experimental setup containing the two-wavelength Twyman–Green interferometer and performing Technique 4. Due to this technique, the surface relief phase maps can be reliably retrieved up to the reaching of the maximum values of the parameter Sa equal to $\Lambda_{12}/8$, that is $Sa_{\max} = 417$ nm, since $\Lambda_{12} = 3340$ nm in this setup.

Before cyclic loading CT-specimens were attached to the jig plate and three interferograms of the studied surface area were recorded by using the 3SI UPS method and the integrating-bucket technique providing a smooth phase shift procedure and driven by the controller 12 in Fig. 2.13 or the controller 15 in Fig. 3.22. The studied CT-specimen is then placed in the servo-hydraulic machine "IMUR", and cyclic loads are applied to the specimen. After a specified number of cycles, the machine is stopped and the specimen is installed into the jig plate in the experimental setup to record the next three interferograms in the single-wavelength interferometer or six interferograms in the two-wavelength interferometer. The work of the entire experimental setup is controlled by the computer 14 in the experimental setup shown in Fig. 2.13 or by the computer 16 in the setup shown in Fig. 3.22.

Digital processing of retrieved surface roughness height maps. Developed methods 3SI UPS and 3F TWIM make it possible to restore the fine nanorelief (surface roughness) phase maps $\varphi_{R,\Lambda}(k, l)$ (see Eq. (2.77)), which can be easy transformed to the surface roughness height maps $h_r(k, l)$ if $Sa < \lambda/8$, since $h = \lambda\varphi/4\pi$ in a single-wavelength Twyman–Green interferometer. If $75 < Sa < 417$ nm, the surface roughness phase map $\varphi_{R,\Lambda}(k,l)$ is transformed into the surface roughness height map (SRHM) by equation similar to (2.66), that is

$$h_r(k,l) = \frac{\Lambda_{12} \cdot \varphi_{R,\Lambda}(k,l)}{4\pi}. \tag{3.8}$$

Due to the developed methods ITSI UPS (see Sect. 2.5.5), 3SI UPS (see Sect. 2.6) and 3F TWIM (see Sect. 2.7.4), the procedure of coincidence of SRHMs obtained at different numbers of cycles is somewhat simplifies. This fact is explained by suppression of errors, which are caused by changes in the surface form and waviness and arise as a result of surface displacements in the out-of-plane direction thanks to the extraction of the SRHM from the total surface height map and the removal of the form and waviness. Since the servo-hydraulic machine for fatigue testing was stopped quite often to record interferograms of surface areas, the contour of the notch edge and characteristic points outside the cyclic plastic zone were chosen for the approximate coarse alignment of the SRHMs obtained for different number of cycles. To coincide the two height maps more accurately, characteristic points were used on the FPZ boundary after its detection and at the boundaries of the detected grains on the material surface. This approach can be considered close to the approach used by Sola et al. [40], if the aforementioned procedures for the SRHMs approximate coarse alignment are not taking into account.

The evolution of the surface roughness relief near the notch tip in metal and alloy CT-specimens during their fatigue can be estimated by increasing the arithmetic mean roughness value $Sa_M(k',l')$ within each square subset $(A_s)_M$ of $A_{Mx} \times A_{My}$ at each pixel of the SRHM $h_{rM}(k,l)$. Since $Sa_M(k',l')$ is defined from Eq. (3.6), its increment $\Delta Sa_{M,j}(k',l')$, which is the difference between $Sa_{M,j}(k',l')$ value obtained from the SRHM $h_{rM,j}(k,l)$ after $N_{i,j}$ cycles and $Sa_{M,j-1}(k',l')$ value obtained from the SRHM $h_{rM1,j-1}(k,l)$ after $N_{i,j-1}$ cycles ($N_{i,j-1} < N_{i,j} < N_i$), can be expressed as

$$\begin{aligned} &\Delta Sa_{M,j}(k',l') \\ &= \frac{1}{JQp_x p_y} \sum_{(k'=k-j+\frac{J-1}{2})} \sum_{(l'=l-q+\frac{Q-1}{2})} \left| h_{rM,j}(k,l) - h_{rM,j-1}(k,l) \right| p_x p_y. \end{aligned} \tag{3.9}$$

Since this increment is formed in the matrix photodetector plane, the real increment $\Delta Sa_j(k',l')$ characterizing the real surface relief is obtained by dividing of $\Delta Sa_{M,j}(k',l')$ on the linear magnification M_l (see also Eq. (3.7)), that is

$$\Delta Sa_j(k',l') = \frac{\Delta Sa_{M,j}(k',l')}{M_l}. \tag{3.10}$$

3.2.3 Surface Evolution of the Steel 08kp

The steel 08kp CT-specimen with a size of 30 × 31 × 3 mm^3 and a round notch 1.5 mm in diameter (see Fig. 3.23) was used to study its surface evolution during cyclic loading. Before loading and after its intermediate stages, the linear magnification in the optical interferometers of the experimental setups (see Figs. 2.13 and 3.22) was equal to $M_l = 2.3$. The retrieved spatial distribution of arithmetic mean values of heights, i.e. the spatial distribution of the roughness parameter $Sa(k', l')$ on the 3D surface relief area in the vicinity of the notch tip after 10,000 load cycles, is shown in Fig. 3.24. This distribution was obtained from a retrieved height map of this area, after which it was digitally processed according to the algorithm described by Eqs. (3.6) and (3.7). Three points are selected in the neighborhood of this region. At Point 1, the maximum value of the roughness parameter $Sa(k', l')$ is reached within the zone $D*$ (see Figs. 3.12 and 3.13), and the next two points (Point 2 and Point 3) are chosen in those sites of this region where the local maxima of $Sa(k', l')$ are reached.

The spatial distribution of the 3D roughness parameter $Sa(k', l')$ over the entire square of the studied area was determined. Due to this operation, there is no need

Fig. 3.23 Steel 08kp CT-specimen with a size of 30 × 31 × 3 mm^3 and a round notch 1.5 mm in diameter

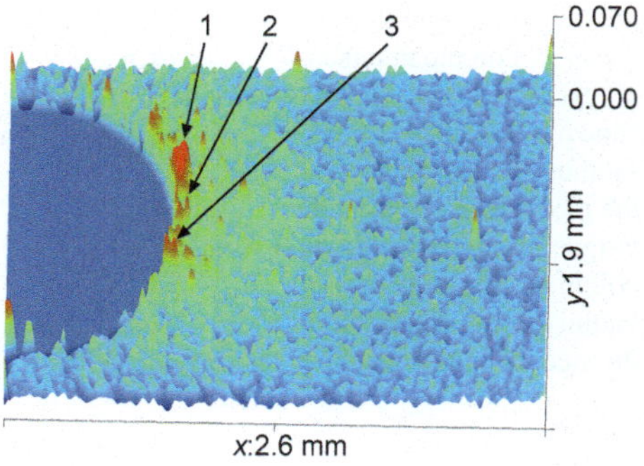

Fig. 3.24 Map of the spatial distribution of the 3D roughness parameter $Sa(k', l')$ on the steel 08kp CT-specimen surface area near the notch after 10,000 load cycles (top view)

to look for the place of the highest values of roughness in the vicinity of the FPZ boundary, as it was done in previous studies [18, 20]. Using the algorithm described by Eqs. (3.6) and (3.7), the maximum value of $Sa(k', l')$ was found automatically. In this figure, the maximum values of $Sa(k', l')$ do not always fall exactly on the FPZ boundary, but are within the zone $D*$ depicted in Figs. 3.12 and 3.13. This suggests that the most probable place of the fatigue macrocrack initiation should be sought within the zone $D*$, taking into account the influence of triple points and grain edges on its location.

The evolution of the roughness parameter $Sa(k', l')$ (see Eq. 3.7) defined within the square subset A_{s1} in its center, i.e. in the pixel (k_1, l_1) (Point 1 marked in Fig. 3.24) with increasing number of cycles, is shown in Fig. 3.25. The graph of $Sa(k_1', l_1')$ versus N_i shows that this roughness parameter increases monotonically and reaches its maximum $Sa_{\max}(k_1', l_1') \approx 140$ nm at the moment of a crack initiation at $N_i \approx 21{,}000$, when the crack arises at Point 1. After 10,000 cycles, the maximum roughness value at Point 1 is exactly traceable in comparison with marked in Fig. 3.24 local surface roughness maxima Point 2 (pixel (k_2, l_2)), which corresponds to the surface parameter $Sa(k_2', l_2')$), and Point 3 (pixel (k_3, l_3)), which corresponds to the surface parameter $Sa(k_3', l_3')$). Really, if at Point 1 the surface roughness arithmetic mean value is $Sa(k_1', l_1') = 70$ nm, then at Points 2 and 3 these roughness parameters are only equal to $Sa(k_2', l_2') = 43$ nm and $S_a(k_3', l_3') = 42$ nm, respectively, after 10,000 cycles. The dependencies of the surface roughness parameters $Sa(k_2', l_2')$ and $Sa(k_3', l_3')$ on the cycle's numbers N_i are shown in Fig. 3.26. These dependencies show that the values of $Sa(k_2', l_2')$ and $Sa(k_3', l_3')$ at Points 2 and 3 increase monotonically and nonlinearly with an increase of N_i, however, they do not reach the critical values at which the fatigue crack is initiated. Therefore, even at $N_{i,j} = 10{,}000$ cycles, it is safe to say that the fatigue macrocrack will nucleate and initiate at Point 1, i.e. in the pixel (k_1, l_1). After the initiation of the fatigue macrocrack, the roughness growth at Point 1 stopped, which is an indicator of the time of the crack initiation. Therefore, if the growth of roughness stops at the site of the fatigue macrocrack initiation, then this fact indicates that after approximately $N_i = 21{,}000$ cycles, the fatigue macrocrack initiated in the near-surface layer or on the surface.

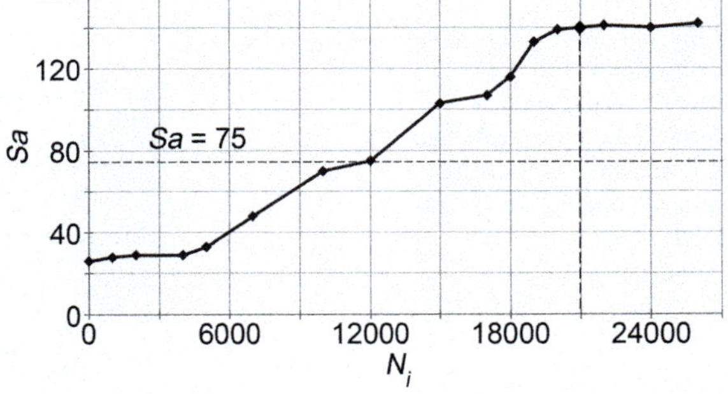

Fig. 3.25 Dependence of the roughness parameter $Sa(k_1', l_1')$ on the number of N_i cycles for Point (pixel) 1 shown in Fig. 3.24. Large rhombus on the curve shows the moment of origin of the fatigue macrocrack initiation within the zone $D*$ in the vicinity of the notch tip of the steel 08kp CT-specimen

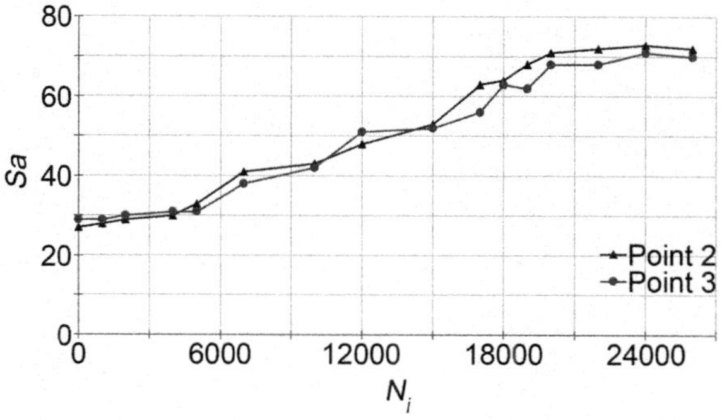

Fig. 3.26 Dependencies of the roughness parameters $Sa(k_2', l_2')$ (Point 2) and $Sa(k_3', l_3')$ (Point 3) on the number of N_i cycles for Points (pixels) 2 and 3 shown in Fig. 3.24

The dashed horizontal line at the level of $Sa_B = 75$ nm shows the boundary, below which the interferograms of the studied surface relief area were recorded using the single-wavelength interferometer in the experimental setup shown in Fig. 2.13. Above this line interferograms were recorded using the two-wavelength interferometer in the experimental setup shown in Fig. 3.22, since the upper limit of the retrieved surface nanorelief (surface roughness) height map from interferograms obtained using the single-wavelength interferometer ($\lambda = 633$ nm) is $Sa = \lambda/8 \approx 80$ nm. Because in subsets A_{s1}, A_{s2} and A_{s3}, the maximum surface roughness reached in the centers of subsets, i.e. in pixels (k_1, l_1), (k_2, l_2) and (k_3, l_3), respectively, the mentioned boundary Sa_B is chosen slightly below $Sa = \lambda/8$.

Dependencies of the roughness parameter $Sa(k', l')$ increments $\Delta Sa(k', l')$ for three Points 1, 2 and 3 on the number of N_i cycles, i.e. the dependencies of $\Delta Sa_j(k_1', l_1')$ (Point 1), $\Delta Sa_j(k_2', l_2')$ (Point 2) and $\Delta Sa_j(k_3', l_3')$ (Point 3) on N_i, are shown in Fig. 3.27. To construct such dependencies, the surface height increment maps obtained by the algorithm described by Eqs. (3.9) and (3.10) were produced from the nanorelief and microrelief surface height maps retrieved after $N_{i,1}, N_{i,2}, \ldots, N_{i,J}$ loading cycles. After completion of each of these sequences, the loading installation was stopped, the CT sample was removed from the servo-hydraulic machine for fatigue testing "IMUR" into one of the interferometers in the experimental setups shown in Figs. 2.13 and 3.22, to record interferograms and retrieve the SRHMs. Increment maps were constructed by subtracting from the SRHM, obtained after the $N_{i,j}$ sequence of cyclic loads, the SRHM, obtained after the $N_{i,j-1}$ adjacent sequence of cyclic loads. According to the obtained increment maps, the values of increments $\Delta Sa_j(k_1', l_1')$, $\Delta Sa_j(k_2', l_2')$ and $\Delta Sa_j(k_3', l_3')$ were calculated at Points 1, 2 and 3, as it is shown in Fig. 3.27. This figure shows that the largest increase is observed for Point 1 and the smallest one for Point 3. After the initiation of the fatigue macrocrack, the surface roughness growth at Point 1 stops. The surface roughness growth at Points 2 and 3 is also imperceptible.

Fig. 3.27 Dependencies of increments $\Delta Sa_j(k_1', l_1')$ (Point 1), $\Delta Sa_j(k_2', l_2')$ (Point 2) and $\Delta Sa_j(k_3', l_3')$ (Point 3) on the number of load cycles for the steel 08kp CT-specimen

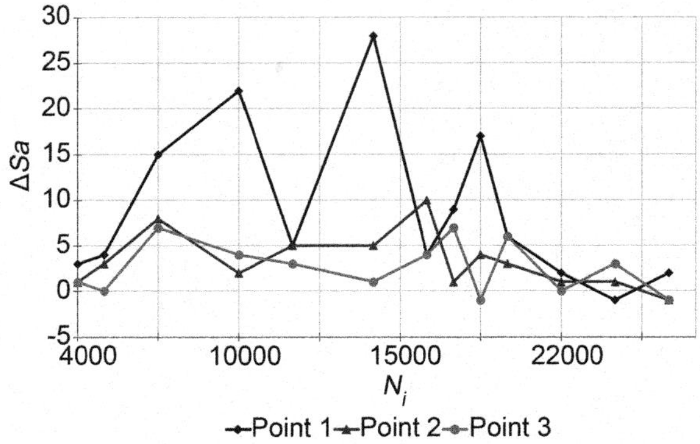

3.2.4 Surface Evolution of the Aluminum Alloy D16T

The aluminum alloy D16T CT-specimen with size of 48 × 48 × 4 mm^3 and a round notch 1.5 mm in diameter (see Fig. 3.28) was used to study the surface evolution during cyclic loading.

A retrieved total surface relief area of the CT-specimen near the round notch after 47,000 cycles is shown in Fig. 3.29. This figure illustrates a clear deflection of the surface in the vicinity of the notch, which is associated with the mechanical processing of the specimen, in particular the drilling of the notch and the subsequent grinding and polishing of the specimen surface. Due to the developed ITSI UPS (see Sect. 2.5.5) and 3SI UPS (see Sect. 2.6) methods, as well as 3F TWIM (see Sect. 2.7.4), we can separate the surface nanorelief or surface microrelief from waviness and form. Therefore, to study the surface relief evolution, it is possible to use only an extracted nanorelief or microrelief height maps retrieved by these methods, and not maps of the total surface relief, which were obtained in 3D Optical Profiler based on the white light scanning interferometry (WLSI). Note that Sola et al. [40] estimated the increase in roughness of a thin polycrystalline pure nickel specimen using the total surface relief height maps, cutting off the part of the FPZ adjacent directly to the edge of the notch or even entire FPZ part of the surface in the vicinity of the notch tip and, thereby, not taking into account the entire FPZ or its part. Such obstacle is caused by peculiarities of WLSI 3D Optical Profilers that cannot record

Fig. 3.28 CT-specimens from aluminum alloy D16T with round notch 1.5 mm in diameter

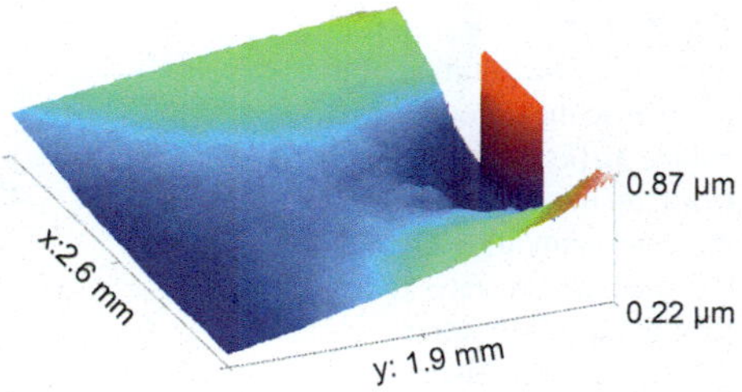

Fig. 3.29 Map of the aluminum alloy D16T CT-specimen total surface relief heights near the round notch of diameter 1.5 mm after 47,000 loading cycles retrieved by the 3F TWIM; $M_l = 2.3$

the studied surface located under given angle to the plane perpendicular to the optical axis of the profiler. However, analysis of the entire FPZ is very important for such studies; since the fatigue macrocrack initiation can be occur in any site within the zone $D*$ containing the FPZ.

The surface microrelief height maps of the arithmetic mean roughness values $Sa(k', l')$, obtained from the interferograms of the aluminum alloy D16T CT-specimen surface area near the notch tip after 25,000 and 43,000 cycles, are shown in Figs. 3.30 and 3.31. These maps were calculated according to the algorithm described by Eqs. (3.6) and (3.7). In Figs. 3.30 and 3.31, the largest arithmetic mean roughness values $Sa(k_1', l_1')$ and $Sa(k_2', l_2')$ at Points 1 and 2, respectively, are marked. These points were tracked starting from 1,100 cycles. And although Points 1 and 2 are outside the FPZ, but within the zone $D*$ (see Fig. 3.12).

The evolution of the roughness parameter values $Sa(k_1', l_1')$ and $Sa(k_2', l_2')$ defined within the square subsets A_{s1} and A_{s2} in their centers, i.e. in the pixels (k_1, l_1) and (k_2, l_2) (Points 1 and 2, marked in Figs. 3.30 and 3.31) with increasing the number of cycles, are shown in Fig. 3.32. As these dependencies show, the roughness at two points monotonously increases with an increase in the number of cycles up to 47,000 cycles. Note that the recording of the CT-specimen surface area fringe patterns before and during cyclic loads was performed in the experimental setup containing two-wavelength Twyman–Green interferometer (see Fig. 3.22). The fatigue macrocrack initiation occurred after 47,000 cycles. This fact was confirmed by the maximum value of the roughness parameter $Sa(k_1', l_1') = 225$ nm reached at Point 1. After

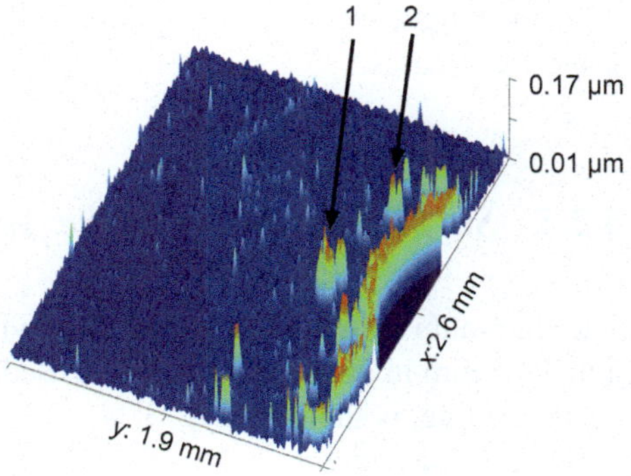

Fig. 3.30 Surface microrelief height map of the arithmetic mean roughness values $Sa(k', l')$, obtained from the interferograms of the aluminum alloy D16T CT-specimen surface area near the notch tip after 25,000 cyclic loads; $M_l = 2.3$

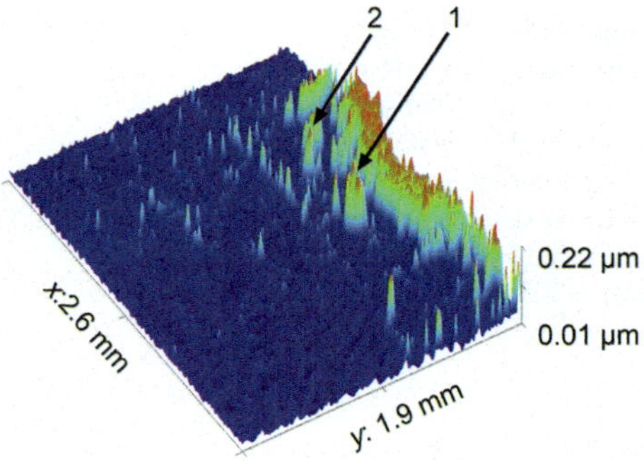

Fig. 3.31 Surface microrelief height map of the arithmetic mean roughness values $Sa(k', l')$, obtained from the interferograms of the aluminum alloy D16T CT-specimen surface area near the notch tip after 43,000 cyclic loads; $M_l = 2.3$

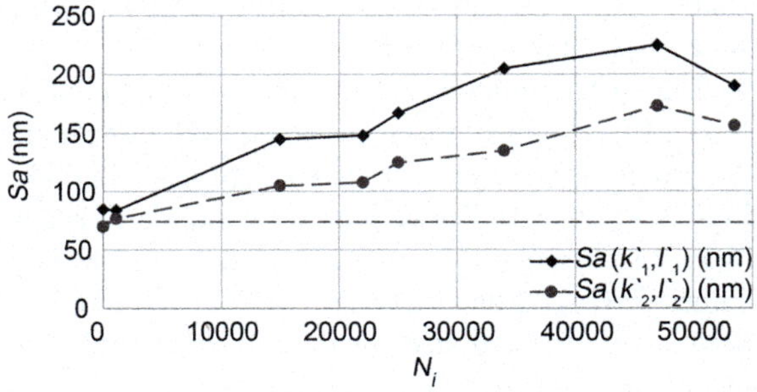

Fig. 3.32 Dependencies of $Sa(k_1', l_1')$ and $Sa(k_2', l_2')$ on number of cycles for Points 1 and 2 shown in Figs. 3.30 and 3.31. Points 1 and 2 are placed within the zone $D*$

47,000 cycles, the surface parameter $Sa(k_1', l_1')$ decreased due to the macrocrack formation. It should be noted that when studying the fatigue properties of the steel 08kp CT-specimen, the fatigue macrocrack initiated at the FPZ boundary. In this study, the macrocrack initiated outside the FPZ at Point 1, but within the zone $D*$ (see Fig. 3.12).

Using the algorithm for increments of the arithmetic mean roughness values $Sa(k', l')$ formation, described by Eqs. (3.9) and (3.10), the dependencies of increments $\Delta Sa_j(k_1', l_1')$ at Point 1 and $\Delta Sa_j(k_2', l_2')$ at Point 2 on N_i were constructed. These dependencies are shown in Fig. 3.33. The obtained dependencies show that their largest increase is observed for Points 1 and 2 from 15,000 to 22,000 cycles. After the fatigue macrocrack initiation at Point 1, the increase in roughness at Points 1 and 2 decreases.

3.2.5 Surface Evolution of the Aluminum Alloy V95T

The aluminum alloy V95T CT-specimen with a size of $48 \times 48 \times 4$ mm^3 and a round notch 1.5 mm in diameter, which was used to study its surface evolution during the cyclic loading, is shown in Fig. 3.34.

The surface microrelief height map of the arithmetic mean roughness values $Sa(k', l')$ over the studied surface area near the notch tip after 50,000 cyclic loads is

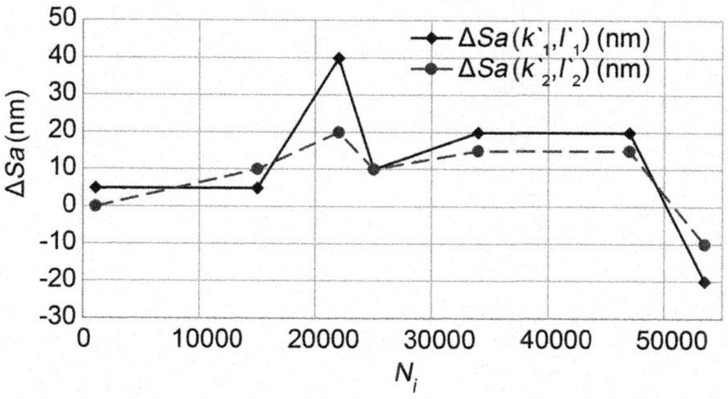

Fig. 3.33 Dependencies of increments $\Delta Sa_j(k_1', l_1')$ and $\Delta Sa_j(k_2', l_2')$ at Points 1 and 2 on the number of load cycles for the aluminum alloy D16T CT-specimen

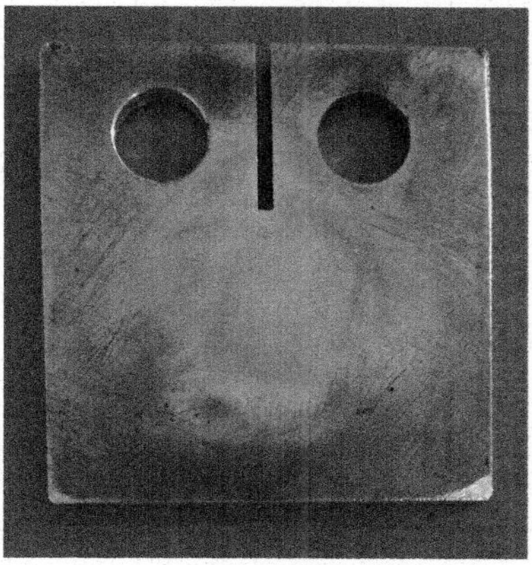

Fig. 3.34 Example of the aluminum alloy V95T CT-specimen. The round notch (Ø1.5 mm) is almost invisible because it is pinned to reduce the surface curvature at the edge of the notch when polished

shown in Fig. 3.35. Points (pixels) 1 and 2 with the highest height were selected on this surface area. These points were monitored from 10,000 cycles.

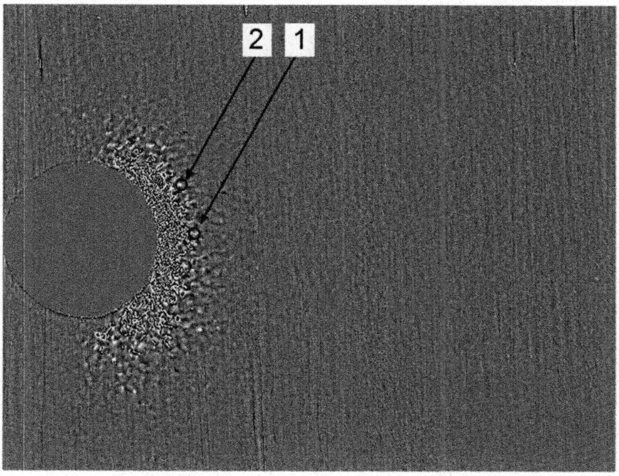

Fig. 3.35 Aluminum alloy V95T surface microrelief height map of the arithmetic mean roughness values $S_a(k', l')$ after 5000 cyclic loads; $M_l = 1.0$

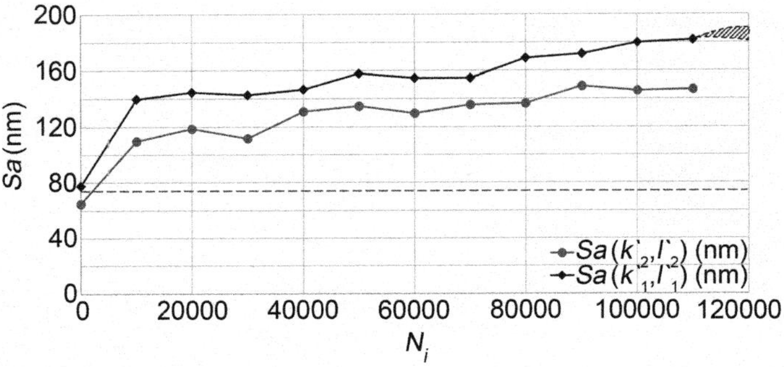

Fig. 3.36 Dependencies $Sa\left(k_1', l_1'\right)$ (broken line with diamonds) and $Sa\left(k_2', l_2'\right)$ (broken line with circles) on the number of cycles for Points 1 and 2, located near the FPZ boundary and within the zone $D*$

Dependencies of the roughness parameter values $Sa\left(k_1', l_1'\right)$ and $Sa\left(k_2', l_2'\right)$ defined within the square subsets A_{s1} and A_{s2} in their centers, i.e. in pixels (k_1, l_1) and (k_2, l_2) (Points 1 and 2, marked in Fig. 3.35) on the number of cycles, are shown in Fig. 3.36. These dependencies show that the roughness at two points monotonously increases with an increase in the number of cycles. At Point 1, after N_i=110,000 cycles, the arithmetic mean roughness value $Sa\left(k_1', l_1'\right)$ within the square subset A_{s1} reaches its maximum, equal to 180 nm. Unfortunately, the moment of crack formation could not be registered due to technical reasons. At the same time, the site of the macrocrack nucleation could already be determined after 10,000 cycles and its initiation could be clearly determined after 80,000 cycles, as it evidenced by the sharp rises of the broken line, which describes the increase in roughness at Point 1. At this point, a fatigue macrocrack arose. All the fringe patterns of the studied surface area were recorded using two-wavelength Twyman–Green interferometer included in the experimental setup shown in Fig. 3.22.

3.2.6 Discussion and Conclusion

Based on the developed concept of the FPZ formation [21, 29, 33] and taking into account the obtained experimental results, it can be stated that the fatigue macrocracks initiation is most likely occurs within the zone $D*$, which covers the FPZ and narrow band outside it (see Figs. 3.12 and 3.13). This assumption does not contradict the hypothesis proposed by Sola et al. [40] that fatigue macrocracks occur at triple points and grain boundaries. Indeed, the size $d*$ is much larger than the grain size of the material under study, for example, for steel 08kp CT-specimens $d* = 210-287$ μm [29], for aluminum alloy D16T CT-specimens $d* = 500-600$ μm [21] and for aluminum alloy V95T CT-specimens $d* = 300-350$ μm [21, 24] (see also Sect. 3.1.2). Therefore, the presence of a significant number of triple points and grain boundaries within the zone $D*$ is not in doubt.

Compared with the technique to predict the location of the fatigue crack initiation in pure nickel developed by Sola et al. [40], Technique 4 proposed in this Section has the next advantages:

(i) The most probable location of the fatigue macrocrack initiation is within the zone $D*$ (see Fig. 3.12), which completely covers the FPZ and narrow band outside it. Within this zone, the maximum range of local stresses $\Delta\sigma_y^*$ is reached. In [40], it is not indicated in which zone the most probable location of the fatigue crack initiation, but only in the triple points and grain boundaries.
(ii) From the total height map of the surface relief area, which includes the form, waviness and surface roughness, we extract only the surface roughness height map including the distribution of surface roughness in the entire FPZ. In [40], the relief increment is determined from the total height map of the surface relief, without taking into account the surface deflection at the notch edge and the lack of data on the surface roughness at the deflection site (see Fig. 3.29).
(iii) We determine the spatial distributions of the 3D roughness parameters $Sa\left(k', l'\right)$ and $\Delta Sa\left(k', l'\right)$ in each pixel of the 3D roughness distribution of the retrieved surface roughness height map, while in [40], these roughness parameters are determined in each subset of 50×50 pixels of the retrieved total height maps or their increments.

The main lack of Technique 4 proposed in this section consists in absence of proper software for coinciding height maps of the 3D roughness parameters $Sa_{j-1}\left(k', l'\right)$ and $Sa_j\left(k', l'\right)$ to obtain the maps of increments $\Delta Sa_j\left(k', l'\right)$ using the algorithm given by Eqs. (3.9) and (3.10).

References

1. Armas AF, Kaufmann GH, Galizzi GE (2004) Nondestructive evaluation of the fatigue damage accumulation process around a notch using a digital image measurement system. Opt Lasers Eng 41(3):477–487
2. Arola D, Ramulu M (1999) An examination of the effects from surface texture on the strength of fiber-reinforced plastics. J Compos Mater 33(2):102–123
3. Arola D, Williams CL (2002) Estimating the fatigue stress concentration factor of machined surfaces. Int J Fatigue 24(9):923–930
4. Boeff M, Hassan HU, Hartmaier A (2017) Micromechanical modeling of fatigue crack initiation in polycrystals. J Mater Res 32(23):4375–4386
5. Brinckmann S, van der Giessen E (2007) A fatigue crack initiation model incorporating discrete dislocation plasticity and surface roughness. Int J Fract 148(2):155–167
6. Buckner BD, Markov V, Lai LC, Earthman JC (2008) Laser-scanning structural health monitoring with wireless sensor motes. Opt Eng 47(5):054402
7. Chan KS, Tian JW, Yang B, Liaw PK (2009) Evolution of slip morphology and fatigue crack initiation in surface grains of Ni200. Metall Mater Trans A 40(11):2545–2556
8. Chan KS (2010) Roles of microstructure in fatigue crack initiation. Int J Fatigue 32(9):1428–1447
9. Ciavarella M, Monno F (2006) On the possible generalizations of the Kitagawa-Takahashi diagram and of the El-Haddad equation to finite life. Int J Fatigue 28(12):1826–1837

10. Davila A, Garnica G, Lopez JA, Carrion FJ (2009) Fatigue damage detection using a speckle-contrast technique. Opt Lasers Eng 47(3):398–402
11. Diaz FV, Kaufmann GH, Armas AF, Galizzi GE (2001) Optical measurement of the plastic zone size in a notched metal specimen subjected to low-cycle fatigue. Opt Lasers Eng 35(6):325–333
12. Díaz FV, Armas AF, Kaufmann GH, Galizzi GE (2004) Fatigue damage accumulation around a notch using a digital image measurement system. Exp Mech 44(3):241–246
13. Flynn TJ (1997) Two-dimensional phase unwrapping with minimum weighted discontinuity. J Opt Soc Amer A 14(10):2692–2701
14. Ghiglia DC, Pritt MD (1998) Two-dimensional phase unwrapping: theory, algorithms, and software. Wiley, New York
15. Ignatovich SR, Yutskevich SS (2012) Monitoring of the fatigue of d16aт alloy according to the characteristics of deformation surface pattern. Mater Sci 47(5):636–643
16. Kelton R, Fathi Sola J, Meletis EI, Huang H (2017) Study of the surface roughness evolution of pinned fatigue cracks, and its relation to crack pinning duration and crack propagation rate between pinning points. In: ASME International mechanical engineering congress and exposition, vol 58448, Nov 2017. American Society of Mechanical Engineers, pp V009T12A007
17. Kelton R, Sola JF, Meletis EI, Huang H (2018) Visualization and quantitative analysis of crack-tip plastic zone in pure nickel. JOM 70(7):1175–1181
18. Lobanov LM, Muravsky LI, Pivtorak VA, Voronyak TI (2017) Monitoryng napruzhenogo stanu elementiv konstruktsiy z vykorystanniam elektromagnitnyh hvyl' optychnogo diapazonu (Monitoring of Structural Elements Stress State with the Use of Electromagnetic Waves in the Optical Range), vol 3. In: Nazarchuk ZT (ed) Tehnichna diagnostyka materialiv i konstruktsiy (Technical Diagnostics of Materials and Structures), Reference manual in 8 volumes. Publishing House "Prostir-M", Lviv
19. Muravsky LI, Ostash OP, Kmet' AB, Voronyak TI, Andreiko IM (2011) Two-frame phase shifting interferometry for retrieval of smooth surface and its displacements. Opt Lasers Eng 49(3):305–312
20. Muravsky LI, Voronyak TI, Kmet' AB (2014) Lazerna Interferometriya Poverhni dlia Potreb Tehnichnoyi Diagnostyky (Laser Interferometry of Surface for Needs of Technical Diagnostics). Spolom, Lviv
21. Muravsky LI, Picart P, Kmet' AB, Voronyak TI, Ostash OP, Stasyshyn IV (2016) Evaluation of fatigue process zone dimensions in notched specimens by two-step phase shifting interferometry technique. Opt Eng 55(10):104108
22. Muravsky LI, Picart P, Kmet' AB, Voronyak TI, Ostash OP, Stasyshyn IV (2016) Two-step phase shifting interferometry technique for evaluation of fatigue process zone parameters in notched specimens. In: Creath K, Burke J, Gonsalves AA (eds) Interferometry XVIII, vol 9960. SPIE, Bellingham, WA, p 996011
23. Muravsky L, Voronyak T, Stasyshyn I, Fitio V (2019) Use of three-step interferometry with arbitrary phase shifts for retrieval and monitoring of nanoscale rough surfaces. In: 2019 IEEE 2nd Ukraine conference on electrical and computer engineering (UKRCON). IEEE, Lviv, July 2017, pp 792–796
24. Nazarchuk ZT, Muravsky LI (2018) Novel digital interferometry and image correlation techniques for nondestructive testing application In: Lobanov LM (ed) Nauka pro materialy: dosiagnennia ta perspektyvy (Materials Science: Achievements and Prospects), vol 2. Akademperiodyka, Kyiv, pp 374–392
25. Nazarchuk ZT, Voronyak TI, Muravsky LI (2021) Development of the optical-digital methods of monitoring of the surfaces of structural elements for the purposes of technical diagnostics. Mater Sci 57(3):344–354
26. Neuber H (1958) Theory of notch stresses: principles for exact calculation of strength with reference to structural form and material, 2nd edn. Springer Verlag, Berlin
27. Neuber H (1961) Theory of notch stresses: principles for exact calculation of strength with reference to structural form and material, vol 4547. USAEC Office of Technical Information
28. Ostash OP, Panasyuk VV (2001) Fatigue process zone at notches. Int J Fatigue 23(7):627–636

29. Ostash OP (2006) New approaches in fatigue fracture mechanics. Mater Sci 42(1):5–19
30. Ostash OP, Muravsky LI, Kuts OG, Andreyko IM Voronyak TI, Kmet AB, Vira VV (2009) Sposib vyznachennia rozmiriv vtomnoyi zony peredruynuvannia (Method for determination of the fracture process zone) (Ukrainian patent #42549). Ministry of Economy of Ukraine
31. Ostash OP, Muravs'kyi LI, Voronyak TI, Kmet' AB, Andreiko IM, Vira VV (2011) Determination of the size of the fatigue prefracture zone by the method of phase-shifting interferometry. Mater Sci 46(6):781–788
32. Ostash O, Muravsky L, Voroniak T, Chepil R, Tkach V, Vira V (2014) Struktura polia deformatsiy v okoli vershyny vyriziv i makrotrishchyn za tsyklichnogo navantazhennia (Strain field structure near the tip of notches and macrocracks at cyclic loading). In: Panasyuk VV (ed) Mekhanika ruynuvannya i mitsnist' konstruktsiy (Fracture Mechanics of Materials and Strength of Structures). Proceedings of the 5th international conference, Lviv, pp 71–78
33. Ostash OP, Panasyuk VV, Kostyk EM (1999) A phenomenological model of fatigue macrocrack initiation near stress concentrators. Fatigue Fract Eng Mater Struct 22(2):161–172
34. Panasyuk VV (2009) Some urgent problems of the strength of materials and durability of structures. Mater Sci 45(2):141–161
35. Paris PC, Erdogan F (1963) A critical analysis of crack propagation laws. J Fluids Eng 85(4):528–533
36. Peterson RE (1974) Stress concentration factor. Wiley, New York
37. Pluvinage G, Azari Z, Kadi N, Dlouhý I, Kozak V (1999) Effect of ferritic microstructure on local damage zone distance associated with fracture near notch. Theor Appl Fract Mech 31(2):149–156
38. Pluvinage G (1998) Fatigue and fracture emanating from notch; the use of the notch stress intensity factor. Nucl Eng Des 185(2):173–184
39. Sola JF, Kelton R, Meletis EI, Huang H (2019) A surface roughness based damage index for predicting future propagation path of microstructure-sensitive crack in pure nickel. Int J Fatigue 122:164–172
40. Sola JF, Kelton R, Meletis EI, Huang H (2019) Predicting crack initiation site in polycrystalline nickel through surface topography changes. Int J Fatigue 124:70–81
41. Stoudt MR, Levine LE, Creuziger A, Hubbard JB (2011) The fundamental relationships between grain orientation, deformation-induced surface roughness and strain localization in an aluminum alloy. Mater Sci Eng A 530:107–116
42. Suraratchai M, Limido J, Marbu C, Chieragatti R (2008) Modelling the influence of machined surface roughness on the fatigue life of aluminium alloy. Int J Fatigue 30(12):2119–2126
43. Ural A, Krishnan VR, Papoulia KD (2009) A cohesive zone model for fatigue crack growth allowing for crack retardation. Int J Solids Struct 46(11–12):2453–2462
44. Wang Y, Meletis EI, Huang H (2013) Quantitative study of surface roughness evolution during low-cycle fatigue of 316L stainless steel using Scanning Whitelight Interferometric (SWLI) Microscopy. Int J Fatigue 48:280–288

Chapter 4
Digital Speckle Pattern Interferometry for Studying Surface Deformation and Fracture of Materials

Abstract In this chapter, a two-step digital speckle pattern interferometry (DSPI) method with an unknown phase shift of a reference wave is described. The method implements the recording of two speckle fringe patterns (speckle interferograms (SIs)) before deformation of the studied surface differing in an arbitrary phase shift between the reference waves and two SIs after the surface deformation differing in the same arbitrary phase shift. In this case, the phases of the two initial SIs recorded before and after surface deformation coincide. As with the phase shifting interferometry methods considered in Chap. 2, the population Pearson correlation coefficient is used to calculate the unknown phase shift. Computer simulation and experimental verification of this method confirm its reliability to restore in-plane and out-of-plane surface displacements. Beam and sheet metal specimens, as well as composite specimens containing various defects and damages were studied by using this method performed with the help of 2D and 3D speckle interferometers. The displacement fields of the studied specimens were retrieved and the hidden defects were detected. The subtractive synchronized DSPI method for detection of subsurface defects in laminated composites is also represented in this Chapter. Subsurface defects were detected using an optical-digital speckle interferometer (ODSI) and ultrasonic excitation of the studied specimen in the frequency range from 10 to 150 kHz. Experiments performed demonstrate the ability of the ODSI to detect subsurface delaminations and disbonds in laminated composite panels.

Phase Shifting Interferometry and Digital Speckle Pattern Interferometry techniques have many common features. In particular, similar optical schemes of interferometers and speckle interferometers, the same phase shifting methods and phase unwrapping algorithms are used for their implementation. However, while the surface of an object is restored using PSI, then the surface displacement field is restored using the DSPI. In addition, if interferograms or fringe patterns of a studied object or surface are produced with the help of PSI, then speckle interferograms (SI) or speckle fringe patterns of an optically rough surface or a scattering medium are produced using DSPI. A SI is an intensity-modulated random spatial distribution of speckles (light spots) that carry information about the structure of the surface or scattering medium.

Z. Nazarchuk et al., *Optical Metrology and Optoacoustics in Nondestructive Evaluation of Materials*, Springer Series in Optical Sciences 242,
https://doi.org/10.1007/978-981-99-1226-1_4

In other words, SI are spatial distributions of speckles or speckle patterns focused in an optical system and modulated by the reference wavefront.

Recently, DSPI has become a powerful mean to study the stress–strain state of materials, structural elements and other urgent problems of experimental mechanics, NDT and TD [16, 43, 45, 56, 70, 89, 112]. The rapid development of DSPI is due to many factors, among which it is worth highlighting:

- precise determination of spatial surface displacement fields and scattering media movements;
- high sensitivity to in-plane and out-of-plane displacements and deformations of material surfaces and scattering media;
- presence of effective technical means for high-speed recording of speckle fringe patterns and their digital processing.

For a better understanding of the optical wavefronts transformations in DSPI, let us consider the main properties of speckles and SIs.

4.1 Formation of Speckles and Speckle Interferograms in Optical Systems

Let's consider some peculiarities of optical speckles and SIs formation in optical and optical-digital systems.

4.1.1 Optical Speckles Formation

Speckles are generated due to the multiple-beam interference of optical rays. As a result, a lot of sinusoidal interference patterns with different spatial frequencies and phase shifts produce a speckle pattern, some examples of that are represented in Fig. 4.1.

As a rule, speckles are associated with laser light [28, 29]. However, they can be produced by non-monochromatic light [30], as well as by thermal radiation [57] and acoustic waves [1, 79]. When the speckles propagate in free space, that is, in the field of objective speckles, their sizes increase with distance from the scatterer. Since the dimensions of the speckles grow in the longitudinal direction faster than in the transverse ones, the speckles are elongated in the direction of light propagation and acquire a cigar-like shape. In free space, the maximum spatial frequency of objective speckles corresponds to the interference of waves scattered at the most distant points in the illuminated surface area of the studied object, and is defined as α/λ, where α is the angle that contracts the illuminated area with the apex at the center of the objective speckle in free space, λ is the wavelength. The minimum speckle lateral size in free space corresponds to the value λ/α, and the longitudinal

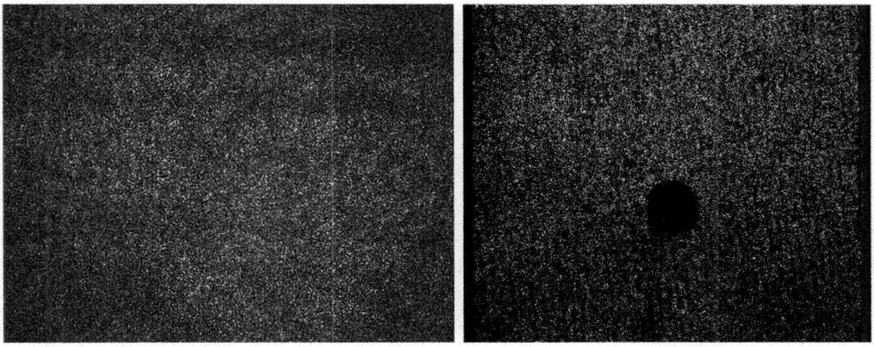

(a) (b)

Fig. 4.1 Speckle patterns of composites surfaces: the speckle pattern of a fiberglass painted surface (**a**); the speckle pattern of the surface of a polymer composite with a round hole (**b**)

one corresponds to the value λ / α^2. In subjective speckle field, the speckle pattern is generated due to the coherent superposition of diffraction patterns with random phase delays formed by the lens aperture. If the surface roughness is so low that it cannot be distinguished by the lens, then a significant number of diffraction patterns are superimposed. This leads to a speckle size distribution independent of surface roughness [111]. Under such conditions, the minimum lateral speckle size, which corresponds to the maximum spatial frequency of the speckle pattern equal to β / λ, is defined as λ / β, where β is the angle contracting the lens aperture with the apex at the center of the subjective speckle. So, the minimum lateral speckle size corresponds to the maximum resolution of the aberration-free optical imaging system. In the direction along the light propagation from the object surface, the minimum speckle size is defined as λ / β^2, which corresponds to the focal depth of the optical system [111] . If the optical system is limited by a circular exit pupil of a lens that forms a speckle image at a distance z from its back principal plane, then the average speckle size is determined as the diameter of the central lobe of Fourier transform of this pupil with a diameter of D, i.e. [28]

$$S_p = 1,22\frac{\lambda z}{D}. \tag{4.1}$$

Thus, the speckle average size depends linearly on the factor $\frac{z}{D}$ for given wavelength, that is, on the lens focal number $f_\# = \frac{f}{D}$, if $z = f$, where f is the lens focal length.

Speckle Metrology techniques are based on the use of matrix photodetectors, in particular, charge-coupled device (CCD) [37] and complementary metal–oxide–semiconductor (CMOS) image sensors [5]. Digital cameras with such sensors convert continuous images into discrete signals that are input to a computer and stored as digital images. In this case, the sizes of speckles must satisfy the Nyquist criterion, according to which [91]

$$\max\left|p_x, p_y\right| \leq \frac{\lambda z}{2D}, \tag{4.2}$$

or, taking into account (4.1),

$$\max\left|p_x, p_y\right| \leq 0,41S_p, \tag{4.3}$$

where p_x, p_y are the pixel pitches in a photodetector matrix Eq. (4.3) shows that the minimum subjective speckle diameter is 2.4 pixels.

4.1.2 Formation of Speckle Interferograms in Optical System

The sensitivity of measurements of speckles and speckle patterns parameters increases with the addition of a reference wave to the object wave that carries information about speckles. The resulting interference pattern contains dark and light fringes, the distance between which is determined by known interference relationships. If the studied object moves, then the SI brightness distribution changes, which corresponds to certain changes in the optical path of the rays. The movement of an object in DSPI is encoded as contrast variations in the speckle fringe pattern.

At the initial stage of development, the speckle interferometry techniques were based on direct visual observations or on recording variations in the speckle contrast on photographic film [24], which did not allow the SI formation in real time. Speckle interferometry techniques have been significantly improved, using video systems for recording, processing and visualization of SIs [10, 4551, 59]. Note that in DSPI schemes using the reference wave, the interference fringes are identical to those observed in holographic interferometry schemes [50], but in DSPI schemes they are of poorer quality. At the same time, in DSPI the sensitivity of the fringes to the direction and movement of the studied object surface can vary in wider ranges than in classical holographic interferometry [45, 50].

In DSPI, to match the spatial frequency of microinterference fringes formed as a result of mutual interference of the reference and object beams with the resolution of a digital camera, it is necessary to focus the rough surface speckle pattern formed by the object beam in the plane of the matrix photodetector. An example of a typical scheme of a digital speckle pattern interferometer, where the object beam forms a focused speckle pattern of the surface, is shown in Fig. 4.2. If in the optical scheme of a speckle interferometer the dimensions of the speckles are commensurate with the dimensions of the matrix photodetector pixels and satisfy Eqs. (4.2) and (4.3), then these speckles are distinguished by a digital camera, but worsen the recorded interference pattern. Consequently, SIs are identical to holograms of a focused image of an object, while in digital holography, as a rule, an unfocused wavefront from an object is recorded in the Fresnel or Fraunhofer approximation, interfering in the recording plane with the reference beam [82].

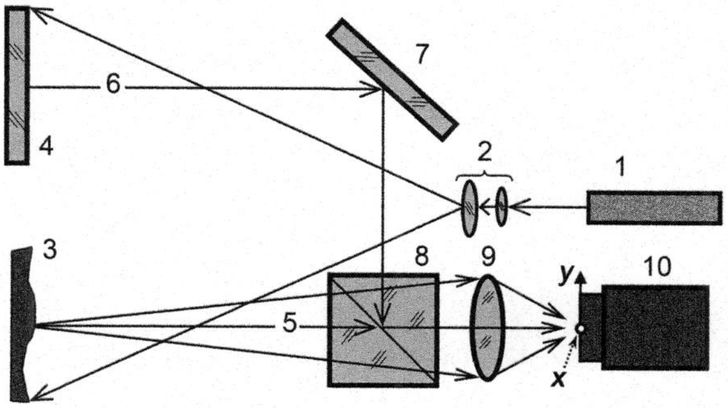

Fig. 4.2 Typical arrangement for digital speckle pattern interferometer: laser (1); laser beam collimator (2); object (3); reference mirror (4); object beam (5); reference beam (6); mirror (7); beam splitter (8); lens (9); digital camera (10)

Consider shown in Fig. 4.2 the plane (x, y) in which the SI of the undeformed studied object surface is generated. In this plane, where the matrix photodetector of the digital camera is located, the distributions of the complex amplitudes of the optical fields from the object and reference beams are expressed, respectively, as

$$E_o(x, y) = |E_o(x, y)| \exp[-j\varphi_o(x, y)], \tag{4.4}$$

$$E_r(x, y) = |E_r(x, y)| \exp[-j\varphi_r(x, y)], \tag{4.5}$$

where $|E_o(x, y)|$ and $|E_r(x, y)|$ are the real amplitude distributions, $\varphi_o(x, y)$ is the random phase distribution from the object rough surface, $\varphi_r(x, y)$ is the reference wavefront phase distribution.

Then the SI intensity distribution recorded by the photosensor is given by [50]

$$\begin{aligned} i_1(x, y) &= |E_o(x, y) + E_r(x, y)|^2 \\ &= i_o(x, y) + i_r(x, y) + 2\sqrt{i_o(x, y) i_r(x, y)} \cos \varphi(x, y), \end{aligned} \tag{4.6}$$

where $i_o(x, y)$ and $i_r(x, y)$ are the object and reference wavefronts intensities, $\varphi(x, y) = \varphi_o(x, y) - \varphi_r(x, y)$ is the random function that describes the phase difference between the object and reference wavefronts in every point of the plane (x, y).

For brevity, we omit the coordinates in the Eq. (4.6) and get

$$i_1 = |E_o + E_r|^2 = i_o + i_r + 2\sqrt{i_o i_r} \cos \varphi. \tag{4.7}$$

After applying the load to the studied object, its surface is deformed, and the intensity distribution of the second SI in the plane (x, y) is given by

$$i_2 = \left|E'_o + E_r\right|^2 = i'_o + i_r + 2\sqrt{i_o i_r} \cos(\varphi + \delta\varphi), \tag{4.8}$$

where $\delta\varphi(x, y)$ is the phase increment caused by the surface deformation in every point of the plane (x, y).

Since in DSPI rather small displacements are recorded, which are commensurate with the size of the speckles, it can be assumed that changes in the surface deformation only lead to a change in the phase of the object wave, while its amplitude remains unchanged, i.e. $E_o' = E_o$, whence $i_o' = i_o$. Then, after replacing E_o' with E_o and i_o' with i_o, Eq. (4.8) becomes

$$i_2 = |E_o + E_r|^2 = i_o + i_r + 2\sqrt{i_o i_r}\cos(\varphi + \delta\varphi). \tag{4.9}$$

Equations (4.7) and (4.9) are the basic ones in DSPI. They are suitable for describing any surface displacements within limits not exceeding the average speckle sizes in SIs.

There are two types of microdisplacements that are recorded using DSPI, namely, in-plane and out-of-plane ones concerning the object surface plane. To register the in-plane displacements of the rough surface, the optical illumination scheme with two rays $\vec{a}$ and $\vec{b}$ is used, which are usually two flat beams with the same angles of incidence to the observation axis [45]. The Leendertz interferometer, which provides sensitivity to a change in the phase difference between two beams for in-plane displacements and a constant phase difference for out-of-plane ones, is built according to this arrangement [10]. The interferometer scheme and the in-plane displacements of the surface are shown in Fig. 4.3a. To record the out-of-plane displacements of the rough surface using DSPI, a two-beam optical scheme is used, namely, the object beam $\vec{a}$ illuminates the surface and the reference beam $\vec{b}$ illuminates the photodetector. This scheme is most sensitive to changes in the phase difference between the object and reference beams for out-of-plane displacements and insensitive to in-plane ones. The interferometer scheme and the surface out-of-plane displacements are shown in Fig. 4.3b. Twyman–Green [60, 61], Mach–Zehnder [60] and Fizeao [62] interferometers are most often used to implement out-of-plane displacements in DSPI.

DSPI techniques are highly sensitive to the surface displacements. For each optical scheme, their sensitivity is different, but for any scheme it can be determined using the so-called sensitivity vector [50]

$$\vec{\eta}(x, y, z) = \left(\frac{2\pi}{\lambda}\right)\left[\vec{a}(x, y, z) - \vec{b}(x, y, z)\right]. \tag{4.10}$$

Using the sensitivity vector, the relationship between the phase change $\delta\varphi(x, y, z)$ and the displacement vector $d(x, y, z)$ can be expressed as

$$\begin{aligned}\delta\varphi(x, y, z) &= \vec{d}(x, y, z) \cdot \vec{\eta}(x, y, z) \\ &= \vec{d}(x, y, z) \cdot \left(\frac{2\pi}{\lambda}\right)\left[\vec{a}(x, y, z) - \vec{b}(x, y, z)\right],\end{aligned} \tag{4.11}$$

that is make possible to determine the displacement $d(x, y, z)$, if the phase change $\delta\varphi(x, y, z)$ in every point of the surface is known.

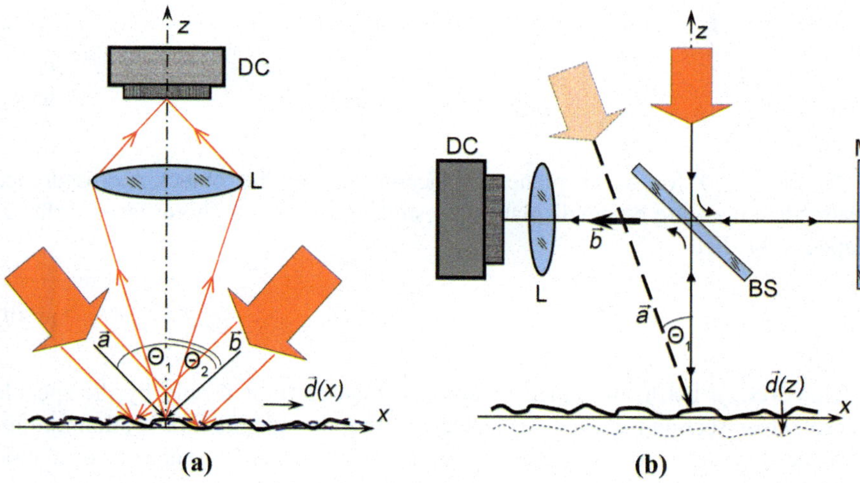

Fig. 4.3 The course of rays $\vec{a}$ and $\vec{b}$ in the interferometers to determine the surface in-plane (**a**) and out-of-plane (**b**) displacements: (L) lens; (DC) digital camera; (BS) beam splitter; $\vec{d}(x)$, $\vec{d}(z)$ are the surface displacement vectors; Θ_1 and Θ_2 are the angles of incidence of laser beams

From Eqs. (4.10) and (4.11) it is easy to determine the sensitivity of the optical system used to form SIs. For example, the sensitivity along the axis x for in-plane surface displacements according to Fig. 4.3a is given by

$$\eta(x) = \left(\frac{2\pi}{\lambda}\right)(\sin\Theta_1 + \sin\Theta_2), \tag{4.12}$$

and the phase change is

$$\delta\varphi(x) = d(x)\left(\frac{2\pi}{\lambda}\right)(\sin\Theta_1 + \sin\Theta_2), \tag{4.13}$$

where Θ_1 and Θ_2 are the angles of incidence of laser beams $\vec{a}$ and $\vec{b}$.

It is clear that along the axis y the sensitivity is zero, i.e. $\eta(y) = 0$. The sensitivity along the axis z for out-of-plane surface displacements according to Fig. 4.3b is given by

$$\eta(z) = \left(\frac{2\pi}{\lambda}\right)(\cos\Theta_1 + 1), \tag{4.14}$$

and the phase change is

$$\delta\varphi(z) = d(z)\left(\frac{2\pi}{\lambda}\right)(\cos\Theta_1 + 1). \tag{4.15}$$

If $\Theta_1 = -\Theta_2 = \Theta$ for in-plane displacements, then

$$\eta(x) = \left(\frac{4\pi}{\lambda}\right)\sin\Theta. \tag{4.16}$$

If $\Theta_1 = 0$ for out-of-plane displacements, i.e. the object and reference beams directions are perpendicular to the studied object and photosensor surfaces, respectively, then

$$\eta(z) = \frac{4\pi}{\lambda}. \tag{4.17}$$

However, in DSPI, the range of displacements is limited only by the speckle sizes in cross-section for in-plane displacements and in longitudinal section for out-of-plane displacements. In-plane displacements cause much greater decorrelation than out-of-plane ones, which is explained by the cigar-shaped speckles along the direction of their propagation. At large displacements, the level of decorrelation of SIs increases sharply, which makes it impossible to record them.

4.1.3 Some Restrictions Affecting the Efficiency of Digital Speckle Pattern Interferometry

In real conditions, the registration and processing of SIs is accompanied by a number of negative factors that can significantly affect the quality of speckle fringes and the reliability of retrieved displacement and deformation fields. In order to select the optimal modes for recording and processing of SIs, one should also take into account the peculiarities of the construction of optical schemes for SIs formation, as well as operational and dynamic characteristics of phase shifting elements and digital cameras.

Among the negative factors affecting the efficiency of the DPSI and, accordingly, the quality of SIs, there are discretization of the interference field, speckle decorrelation and noise of the spatiotemporal instability of the optoelectronic system for SI recording. The discretization of the interference field occurs in the cells of the photosensitive elements of the digital camera matrix photosensor. Speckle decorrelation arises from a change in the position of speckles on the SI due to the rough surface displacement. It depends on the type of displacements (in-plane or out-of-plane) and the parameters of the observation system, namely, sensitivity along the directions of displacements, aperture, optical magnification, spatial and amplitude resolution. The impact of discretization and decorrelation is minimized through the optimal combination of the above parameters [44, 81, 114]. Spatial and temporal instabilities of speckle interferometers are minimized by optimal choice of speckle sizes [44, 81, 114], stabilization of the optical system and optoelectronic units [44, 119], as well as statistical methods of the data recording and processing [26, 81, 118].

The considered properties of speckles and SIs are common for all speckle interferometry techniques, including correlation and temporal phase shifting DSPI. All attention will be paid to these techniques, since they are being developed at the Karpenko Physico-Mechanical Institute (PMI) of the National Academy of Sciences of Ukraine and are used to solve many problems of fracture mechanics, NDT and TD of constructive materials and structural elements.

4.2 Correlation Digital Speckle Pattern Interferometry

Correlation DSPI can be considered the simplest of all the other speckle interferometry techniques. Therefore, it is the most common and widely used for NDT and TD of materials. Correlation DSPI is based on the determination of the cross-correlation between two SIs, one of which was recorded before the application to the studied object a load or excitation, and the second one after the final or intermediate stages of the deformed object surface recording [51]. Mechanical loading, as well as ultrasonic, thermal or other type of excitation can be applied to the studied object. In correlation DSPI, the cumbersome cross-correlation operation can very often be replaced by a simpler pixel-by-pixel subtraction operation. Pixel-by-pixel subtraction of two SIs can be replaced by adding them. Then the maxima of the obtained correlation fringes correspond to the minima of the correlation fringes obtained by SIs subtraction. However, the speckle fringe patterns or difference SIs are more contrast, because their minimums correspond to zero intensity of the modulated signal [45].

Correlation DSPI is based on the implementation of the cross-correlation operation between two SIs and the formation the pattern of correlation fringes, the spatial distribution of which describes the movement of the surface relative to its initial state. As a measure of the correlation between two SI, the correlation coefficient can be chosen, the pixel-by-pixel calculation of which leads to the formation of the correlation fringe pattern. For SIs $i_1(x, y)$ and $i_2(x, y)$ (see Eqs. (4.7), (4.9)), the correlation coefficient is given by [45]

$$\rho_{12}(\delta\varphi) = \frac{\langle i_1 i_2\rangle - \langle i_1\rangle\langle i_2\rangle}{\sqrt{\langle i_1^2\rangle - \langle i_1\rangle^2}\sqrt{\langle i_2^2\rangle - \langle i_2\rangle^2}} = \frac{\langle i_1 i_2\rangle - \langle i_1\rangle\langle i_2\rangle}{\sigma_1\sigma_2}, \tag{4.18}$$

where σ_1 and σ_2 are the root mean squares (RMSs) of the random functions $i_1(x, y)$ and $i_2(x, y)$, which can be averaged separately because they are independent of each other.

According to [29], the following averaging conditions can be used in Eqs. (4.7), (4.9):

$$\langle\cos\varphi\rangle = \langle\cos(\varphi + \delta\varphi)\rangle = 0, \langle i_1^2\rangle = 2\langle i_1\rangle^2, \langle i_2^2\rangle = 2\langle i_2\rangle^2. \tag{4.19}$$

After substituting Eqs. (4.19) in (4.18), the correlation coefficient is given by [45]

$$\rho_{12}(\delta\varphi) = \frac{\langle i_1^2 \rangle + \langle i_2^2 \rangle + 2\langle i_1 \rangle \langle i_2 \rangle \cos(\delta\varphi)}{(\langle i_1 \rangle + \langle i_2 \rangle)^2}. \tag{4.20}$$

The correlation coefficient depends on the intensity distributions $i_1(x, y)$ and $i_2(x, y)$ and on the phase change $\delta\varphi$. If to assume that $\langle i_1 \rangle = \gamma \langle i_2 \rangle$, then Eq. (4.20) becomes simpler and reduces to a form as [45]

$$\rho_{12}(\delta\varphi) = \frac{1 + \gamma^2 + 2\gamma \cos(\delta\varphi)}{(1 + \gamma)^2}. \tag{4.21}$$

According to (4.21), the maximum value of ρ_{12} is equal to 1 and it is reached if $\delta\varphi = 2m\pi$. The minimum value is equal to $(1 - \gamma)^2 / (1 + \gamma)^2$, if $\delta\varphi = 2(m + 1)\pi$. The minimum value of ρ_{12} is equal to 0, if $\gamma = 1$, i.e. if $\langle i_1 \rangle = \langle i_2 \rangle$. In this case $\rho_{12}(\varphi) = 0.5 \cdot [1 + \cos(\delta\varphi)]$.

To reduce the correlation coefficient ρ_{12} calculations, the difference between i_1 and i_2 is used by pixel-by-pixel subtraction of them, and a subtraction fringe pattern (a subtraction specklegram [84]) is expressed as [45]

$$|i_{1-2}| = |i_1 - i_2| = 4\sqrt{i_1 i_2}|\sin(\varphi + \delta\varphi)|\left|\sin\frac{\delta\varphi}{2}\right| = 4W\sqrt{i_1 i_2}\left|\sin\frac{\delta\varphi}{2}\right|, \tag{4.22}$$

where $W = |\sin(\varphi + \delta\varphi)|$ is the noise variable component.

By calculating ρ_{12} or $|i_{1-2}|$, the phase displacement $\delta\varphi(x, y)$ can be approximately determined from Eqs. (4.21) or (4.22), respectively.

Although the correlation approach (see Eq. 4.21) more reliably reflects the cross-correlation of two SIs, the subtraction approach (see Eq. 4.22) is easier to implement and makes it possible to observe the displacement phase field in the first approximation after the SIs subtraction operation. Both approaches require the calculation of the averaged values of relatively large pixel arrays with the same phase shifts $\delta\varphi(k, l)$, i.e. fulfillment of the averaging conditions (4.19). In practice, these conditions are not met, since the averaging bases are limited to several tens of pixels with $\delta\varphi(k, l) \approx$ const. For digital subtraction specklegrams, the averaging base is formed as a two-dimensional array of pixels within a rectangular window. The problem of choosing the optimal window for the averaging low-pass (LP) filter is largely related from the shape and size of the array with $\delta\varphi \approx$ const in the subtraction speckle fringe pattern. In most cases, one has to deal with pixel arrays that are not isotropic in shape, size, or phase. Given these circumstances, it is common to use LP filtering with a fixed mask in the frequency domain or LP filtering in the spatial domain with the square averaging window. Its dimensions are determined on the basis of statistical analysis data, as well as a compromise between noise immunity and resolution of the resulting subtraction specklegram. In this case, as the window size increases, the noise immunity increases, but, accordingly, the resolution decreases. Therefore, it is not possible to obtain both low-noise and high-resolution correlation fringes in a specklegram, which limits the possibilities of correlation DSPI for quantitative analysis at high spatial frequencies. For low and medium spatial frequencies corresponding to several periods of correlation fringes (from 1 up to 10), the error in estimating the phases of the displacement modulo, i.e. without taking into account the displacement direction, is $\lambda/2$.

Due to large errors in measuring the position of correlation fringes, correlation DSPI is preferably used for NDT of various materials, constructional elements and scattering media, when there is no need for precise calculation of displacement fields and surface deformations. There are many problems of experimental mechanics and NDT of surfaces of materials and products, if it is sufficient to perform only a qualitative topological analysis of displacements with an error commensurate with half the laser source wavelength. Many of these problems can be solved by correlation DSPI. For example, correlation DSPI is used to control residual stresses near round holes in specimens [40, 119] and in areas of welded joints [53, 54], to determine bending strains in constructive elements [65, 109], as well as to estimate the strain rate of test specimens under tensile loads [27, 33, 38, 87].

4.3 Temporal Phase Shifting Digital Speckle Pattern Interferometry

Modern research in experimental mechanics, mechanics of dispersed media, TD and NDT, which are associated with the evaluation of displacements and deformations of materials surfaces and scattering media spatial fields, makes quite high demands on the accuracy of these fields' retrieval. Therefore, in recent years, temporal phase shifting (PS) DSPI [25, 46, 50, 80, 108] have been intensively developed, which can significantly improve the accuracy of measurements compared to the correlation DSPI. Indeed, the resolution of retrieved displacement fields in temporal PS DSPI can reach $\lambda/100$ [89] or less, while in correlation DSPI the resolution does not exceed $\lambda/2$.

4.3.1 Basic Algorithms of Temporal Phase Shifting Digital Speckle Pattern Interferometry

In temporal PS DSPI, phase shifting algorithms are used that are identical to those used in temporal PSI (see Chap. 2). These algorithms determine the phases of each point on the surface before deformation, i.e. $\varphi_1(x, y) = \varphi(x, y)$, and after deformation, i.e. $\varphi_2(x, y) = \varphi(x, y) + \delta\varphi(x, y)$. Each phase value in digital SIs depends on the argument sign, since it is calculated using the odd *arctan* function for every k, lth pixel of two phase maps $\varphi_1(k, l)$ and $\varphi_2(k, l)$ of the initial and deformed surfaces, respectively. A displacement or deformation phase map with sign indicating the displacement direction is calculated as the difference between the phase maps $\varphi_2(k, l)$ and $\varphi_1(k, l)$ [13], that is

$$\delta\varphi(k,l) = \varphi_2(k,l) - \varphi_1(k,l). \tag{4.23}$$

Therefore, in PS DSPI, it is possible to restore only the displacement phase map $\delta\varphi(k,l)$ by subtracting the phase maps $\varphi_1(k,l)$ and $\varphi_2(k,l)$. Direct calculation of these phase maps is impossible, because the surface is optically rough and the phase difference is greater than π between almost all neighboring pixels in SIs, i.e. the "2π ambiguity problem" occurs at almost every pixel of the SI [13]. However, the calculation of the displacement phase map $\delta\varphi(k,l)$ at each pixel is correct if the phase difference between adjacent pixels throughout the all map is less than π. This condition is fulfilled in those problems of NDT, TD and experimental mechanics, for which the maximum allowable range of the surface nano- and microdisplacements does not exceed the some lower threshold level of speckle decorrelation. The displacement phase map $\delta\varphi(k,l)$ in this case is comparable to a smooth or nanorough surface, and the phase difference between arbitrary two adjacent pixels on this map does not exceed π. For example, PS DSPI is a very effective tool for studying the deflected mode of constructive materials with flat rough surfaces, as well as surfaces with stress concentrators of the correct shape (holes, cutouts, notches, fillets, etc.). However, when studying the fracture processes at the crack tip and other singularities, PS DSPI is not always appropriate, because a sharp gradient of displacement is formed in these places, and the phase difference between adjacent pixels may exceed π.

Phase shifts in optical schemes of speckle interferometers are performed using various specialized devices. The most popular are piezoelectric transducers (PZTs) with mirrors attached to them, plane-parallel plates that change the tilt angle in the optical scheme, and acousto-optic Bragg cells [32]. In temporal PS DSPI, as in temporal PSI, two approaches are used to implement the phase shifts of beams in speckle interferometers. According to the first approach, phase shifts occur step by step both for the initial and for the final fixed positions of the studied surface area (for example, before and after the load application). In this case, the intensity distributions in SIs for these positions, considering Eqs. (4.7), (4.9), are given by

$$i_{1j}(k,l) = i_B(k,l) + i_M(k,l)\cos\left[\varphi(k,l) + \alpha_j\right], \tag{4.24}$$

$$i_{2j}(k,l) = i_B(k,l) + i_M(k,l)\cos\left[\varphi(k,l) + \delta\varphi(k,l) + \alpha_j\right], \tag{4.25}$$

where $i_B = i_o + i_r$ is the background intensity distribution, $i_M = 2\sqrt{i_o i_r}$ is the modulation intensity distribution.

In the second approach [31, 110], the phase shift $\alpha(t)$ of the beam occurs continuously in time by applying a constantly increasing saw-tooth voltage to the PZT. In this case, each SI is recorded for a certain time interval Δ_t. For this approach, Eqs. (4.24), (4.25), taking into account (2.6), can be expressed as

$$i_1(k,l,t) = \frac{1}{\Delta_t}\int\limits_{\alpha-0,5\Delta_t}^{\alpha+0,5\Delta_t} [i_B + i_M\cos(\varphi + \delta\varphi)]d\alpha$$

$$= i_B + i_M \sin c\left(\frac{\Delta_t}{2\pi}\right)\cos(\varphi + \delta\varphi), \tag{4.26}$$

$$i_2(k, l, t) = \frac{1}{\Delta_t}\int\limits_{\alpha-0,5\Delta_t}^{\alpha+0,5\Delta_t}[i_B + i_M\cos(\varphi + \delta\varphi + \alpha)]d\alpha$$

$$= i_B + i_M \sin c\left(\frac{\Delta_t}{2\pi}\right)\cos(\varphi + \delta\varphi + \alpha). \tag{4.27}$$

In Eqs. (4.26), (4.27), the pixel numbers k, l are omitted for brevity. We will use this simplification for most of the subsequent equations.

In temporal PS DSPI, the same phase shifting algorithms are used as in temporal PSI; however, J SIs should be generated for each nth state of the surface. For example, in a three-step phase shifting algorithm [14], in which $\alpha_j = -\alpha, 0, \alpha$, the surface phase fields before ($n = 1$) and after loading or excitation of the studied object ($n = 2$) can be expressed as

$$\varphi_{nj} = \arctan\left[\left(\frac{1 - \cos\alpha_j}{\sin\alpha_j}\right)\frac{i_{n1} - i_{n3}}{2i_{n2} - i_{n1} - i_{n3}}\right]. \tag{4.28}$$

Another three-step algorithm, in which $\alpha_j = 0, 2\pi/3, 4\pi/3$, is given by [43, 75]

$$\varphi_{nj} = \arctan\left(\sqrt{3}\frac{i_{n3} - i_{n2}}{2i_{n1} - i_{n2} - i_{n3}}\right). \tag{4.29}$$

The Carré algorithm [11] that was used in temporal PS DSPI for phase steps $\alpha_j = -3\alpha, -\alpha, \alpha, 3\alpha$ can be represented as [13, 17, 97, 98]

$$\varphi_{nj} = \arctan\left(\frac{\sqrt{(i_{n1} - i_{n4}) - (i_{n2} - i_{n3})[3(i_{n2} - i_{n3}) - (i_{n1} - i_{n4})]}}{[(i_{n2} + i_{n3}) - (i_{n1} + i_{n4})]}\right). \tag{4.30}$$

On the other hand, Mohan et al. [64], as well as Borza and Nistea [9] used the standard four-step phase shifting algorithm with phase steps $\alpha_j = 0, \pi/2, \pi, 3\pi/2$. The searched phase displacement field $\delta\varphi(k, l)$ for all these algorithms is determined from Eq. (4.23).

The obtained solutions for three- and four-step algorithms are quite sensitive to systematic errors caused by even small deviations of phase steps α_j from their nominal values [50, 61, 83]. Five-step algorithms, in particular, the Hariharan algorithm [35], are more resistant to such errors [7].

After finding the values of the phases $\delta\varphi(k, l)$ in each pixel, a wrapped phase map of the surface displacements is generated. The next operation is the phase unwrapping of the wrapped phase map. To transform the wrapped phase map into an unwrapped one, the same unwrapping algorithms are used as in temporal PS PSI (see Sect. 2.1).

4.3.2 Some Applications of Temporal Phase Shifting Digital Speckle Pattern Interferometry

Temporal PS DSPI is used to solve various problems in experimental mechanics. In particular, Dai et al. [15] used the temporal and spatial PS DSPI to measure the failure parameters of a beam sample of concrete reinforced with steel fibers under a three-point bending. Authors have been developed new techniques for interpreting the values of crack opening and its tip for different load levels using obtained interference fringes. Huang et al. [41] studied the characteristic of the fracture process zone in concrete pre-notched specimens using four-step PS electronic speckle pattern interferometry (ESPI) technique performed by the 3D ESPI System Q300. They showed that when concrete is tested for three-point bending, a crack initiates in concrete up to a load P, reaching approximately 40% of the peak load $P_{\max}$, and after the peak load the fracture process zone is move forward, if the load dropped to 25% of $P_{\max}$.

The technology of multistep displacement increments is often used to determine the parameters of local fracture of structural materials using PS DSPI. This technology provides small increments of the applied load to the studied samples to minimize decorrelation between the adjacent SIs. In this case, a series of phase shifted SIs is recorded at each of the fixed steps of the applied load. The technology was used to determine the stress intensity factor K_1 in homogeneous structural materials with one-sided edge crack [12] and in aluminum sample 1050A with a thickness of 6 mm using the universal machine INSTRON@ [75]. Displacement increments were also used to evaluate strain distributions at the edge of the carbon fiber plate under four-point bending [34] and to measure the crack opening in concrete beam specimens with a central notch (notch depth 30 mm, width 3 mm) with the help of a 3D electronic speckle interferometer Q300 [95]. Huang et al. [41] developed the incremental displacements collocation method to study concrete specimens.

Methods and means of temporal PS DSPI are very suitable for NDT and TD of structural elements, machine components and other products. In particular, Ettemeyer et al. [20] measured in-plane and out-of-plane strains and stresses on profiles of complex shape, car components and welded joints of structural elements, while combined measurements of the shape and strains of specimens, including honeycomb structures were carried out by Pfeifer et al. [76]. These authors compare the results of measurements performed by traditional methods that use strain gauges and temporal PS DSPI techniques, indicating that the advantages of the latter lie in the ability to build stress and strain distribution fields on a surface of arbitrary shape. Temporal PS DSPI also was used for detection of subsurface cracks in structural elements. For example, Kim et al. [49] have identified the internal cracks in pipelines using this technique.

Temporal PS DSPI is successfully used in the automotive and aviation industry. In particular, Yang and Ettemeier [112] determined the distribution of 3D strains on the surface of a DaimlerChrysler AG four-wheel drive transmission based on the created MicroStar 3D speckle interferometer. Lobanov et al. [55] investigated residual stresses in welded joints of heat-resistant magnesium alloy ML10 used in aircraft construction.

4.4 Two-Step Phase Shifting Digital Speckle Pattern Interferometry with Unknown Phase Shift of Reference Beam

To retrieve the surface displacement fields using temporal PS DSPI, it is necessary to implement multistep phase shifting algorithms. For their performing, at least three SIs of the studied surface area are recorded for each of its fixed states. A necessary condition for the implementation of these algorithms is to record the SI at each jth step for a fixed state of the studied surface area, during the registration of which one of the beams (reference or object) is shifted by an angle of $2\pi(j-1)/J$ relative to the same beam in the initial SI recorded at the first step ($j = 1, 2, ..., J$) at the same surface area fixed state. Such requirements significantly limit the speed of the temporal PS DSPI methods. In addition, to implement these algorithms, it is necessary to perform sufficiently accurate phase shifts of one of the beams (usually the reference one) on given angles. Therefore, the development of simpler and faster methods and algorithms for temporal PS DSPI is one of the urgent problems of optical NDT. One of promising directions for solving these problems consists in the development of two-step PS DPSI methods that are simpler than conventional multistep ones, since it is enough to record only two SIs for each state of the studied surface. Some two-step DPSI methods use a phase shift equal to $\pi/2$ [3, 48, 86, 115, 116]. However, a high-precision calibrated phase shifter must be used to perform them. The two-step ESPI method proposed by Sesselman and Gonzalves [85] differs from the previous ones, because it can carry out a phase shift at a given angle α within $0 < \alpha < \pi$. However, a calibrated phase shifter should also be used in this method, as it is necessary to know the phase shift value in order to calculate the surface displacements phase map.

4.4.1 *Method of Two-Step Digital Speckle Pattern Interferometry with Unknown Phase Shift of Reference Beam*

A new method of two-step digital speckle pattern interferometry (TS DSPI) with unknown phase shift of the reference beam was proposed by Voronyak et al. [101]. It was later used for 3D temporal PS DSPI [103] and modified by application of a population Pearson correlation coefficient (PPCC) to calculate the unknown phase shift [69]. The method restores the surface displacement field using only two SIs of the initial state of the surface, which differ only in the unknown (arbitrary) phase shift α of the reference beam, and two similar SIs of the surface deformed state also differing the same phase shift. In this case, the reference beam can be shifted to any angle in the entire range from 0 to π inclusive, i.e. $0 \le \alpha \le \pi$, since the systematic error of its extraction in this range does not exceed 1.3°. Note that a similar method, in which an arbitrary phase shift of the beam is implemented, was developed later [39]. The last one is inferior to the mentioned above in that it provides sufficient accuracy of unknown phase shift extraction only within small angles not exceeding $\pi/3$.

In the TS DSPI method, two digital SIs are recorded before loading or excitation of the studied surface, i.e. the initial SI $i_{11}(k, l)$ and the SI $i_{12}(k, l)$ differed from the initial one due to the phase shift of the reference beam by an unknown angle α_{21}. Intensity distributions of these SIs can be expressed as

$$i_{11}(k, l) = i_B(k, l) + i_M(k, l)\cos[\varphi(k, l)], \tag{4.31}$$

$$i_{12}(k, l) = i_B(k, l) + i_M(k, l)\cos[\varphi(k, l) + \alpha_{21}]. \tag{4.32}$$

After loading or excitation, the digital SIs $i_{21}(k, l)$ and $i_{22}(k, l)$ are recorded, and their intensity distributions are given by

$$i_{21}(k, l) = i_B(k, l) + i_M(k, l)\cos[\varphi(k, l) + \delta\varphi(k, l)], \tag{4.33}$$

$$i_{22}(k, l) = i_B(k, l) + i_M(k, l)\cos\left[\varphi(k, l) + \delta\varphi(k, l) + \tilde{\alpha}_{21}\right], \tag{4.34}$$

where $\tilde{\alpha}_{21}$ is the phase shift of the reference beam used to record the SI $i_{22}(k, l)$ concerning the reference beam used to record the SI $i_{21}(k, l)$, or, in short, the phase shift between SIs i_{21} and i_{22}.

To find the phase shift, an algorithm based on the calculation of the PPCC between two SIs is used. This algorithm is almost identical to the algorithm for calculating the phase shift in the TSI UPS method (see Sect. 2.5.1). The PPCC ρ can be considered as a normalized scalar product of two centered multidimensional vectors [6]. It can also be interpreted as the cosine of the phase shift between two temporary signals [8] or between two interferograms [66–68]. As it is shown by Muravsky et al. [69], it is also suitable for extracting the unknown phase shift between two SIs. So, the searched phase shift α_{21} between two SIs i_{11} and i_{12} is given by

$$\alpha_{21} = \arccos \rho(i_{11}, i_{12}) = \arccos \frac{\langle (i_{11} - \langle i_{11} \rangle)(i_{12} - \langle i_{12} \rangle) \rangle}{\sigma_{11}\sigma_{12}}, \tag{4.35}$$

and the phase shift $\tilde{\alpha}_{21}$ between SIs i_{21} and i_{22} is given by

$$\tilde{\alpha}_{21} = \arccos \rho(i_{21}, i_{22}) = \arccos \frac{\langle (i_{21} - \langle i_{21} \rangle)(i_{22} - \langle i_{22} \rangle) \rangle}{\sigma_{21}\sigma_{22}}, \tag{4.36}$$

where σ_{11} and σ_{12} are the RMS values of the SIs i_{11} and i_{12}, σ_{21} and σ_{22} are the RMS values of the SIs i_{21} and i_{22}.

One can also extract unknown phase shift using the subtraction specklegrams (difference SIs) $(i_{11} - i_{21})$ and $(i_{12} - i_{22})$, if $\alpha_{21} = \tilde{\alpha}_{21}$. In this case, the phase shift $\widehat{\alpha}_{21}$ is given by

$$\begin{aligned}\widehat{\alpha}_{21} &= \arccos \rho[(i_{11} - i_{21}, i_{12} - i_{22})] \\ &= \arccos \frac{\langle[(i_{11} - i_{21}) - \langle i_{11} - i_{21}\rangle][(i_{12} - i_{22}) - \langle i_{12} - i_{22}\rangle]\rangle}{\sigma_{(11-21)}\sigma_{(12-22)}}, \end{aligned} \tag{4.37}$$

where $\sigma_{(i_{11}-i_{21})}$ is the RMS value of the subtraction specklegram $(i_{11} - i_{21})$, $\sigma_{(i_{12}-i_{22})}$ is the RMS value of the subtraction specklegram $(i_{12} - i_{22})$.

To simplify the calculation of $\widehat{\alpha}_{21}$ in Eq. (4.37), the subtraction specklegrams $(i_{11} - i_{21})$ and $(i_{12} - i_{22})$ can be expressed as

$$(i_{11} - i_{21}) = 2i_M \sin\left(\varphi + \frac{\delta\varphi}{2}\right) \sin\left(\frac{\delta\varphi}{2}\right), \tag{4.38}$$

$$(i_{12} - i_{22}) = 2i_M \sin\left(\varphi + \frac{\delta\varphi}{2} + \alpha_{21}\right) \sin\left(\frac{\delta\varphi}{2}\right). \tag{4.39}$$

If $\alpha_{21} = \tilde{\alpha}_{21}$, then after the unknown phase shift α_{21} extraction, first the coarse displacement phase map $\delta\varphi_c(k, l)$ is determined using the next formula [69, 70, 85, 101]:

$$\begin{aligned}\delta\varphi_c &= 2\arctan\left[\left(\frac{i_{11} + i_{12} - i_{21} - i_{22}}{i_{11} - i_{12} + i_{21} - i_{22}}\right) \tan\frac{\alpha_{21}}{2}\right] \\ &= 2\arctan\left[\left(\frac{i_s}{i_c}\right) \tan\frac{\alpha_{21}}{2}\right], \end{aligned} \tag{4.40}$$

where $i_s = i_{11} + i_{12} - i_{21} - i_{22}$, $i_c = i_{11} - i_{12} + i_{21} - i_{22}$. Since $\alpha_{21} = \tilde{\alpha}_{21}$, phase shift $\tilde{\alpha}_{21}$ can also be used instead of α_{21} in Eq. (4.40).

Further, a sine/cosine average filter (SCAF) [2] is applied to generate a sine and cosine components of the coarse phase map $\delta\varphi_c(k, l)$. As a result of such the spatial filtering, only $i_{\sin} = \sin(\delta\varphi_c)$ and $i_{\cos} = \cos(\delta\varphi_c)$ harmonic terms are restored and the displacement wrapped phase map $\delta\varphi_w(k, l)$ is generated. An unwrapped fine displacement phase map $\delta\varphi(k, l)$ is obtained after a standard unwrapping procedure using, for example, the Flynn program [21, 26]. Flow-chart of the algorithm that performs the retrieval of the displacement phase map is shown in Fig. 4.4.

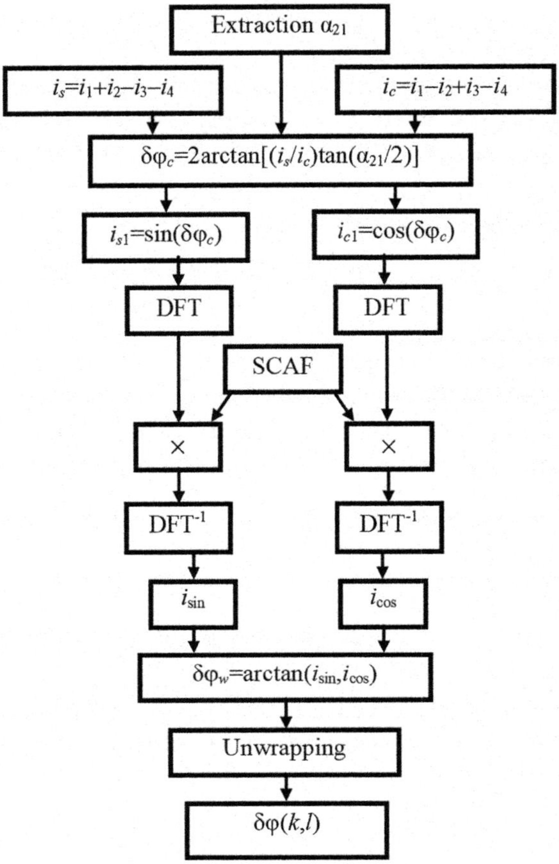

Fig. 4.4 Flow-chart of the algorithm performing the TS DSPI method: (DFT) digital Fourier transform; (DFT^{-1}) inverse digital Fourier transform; (SCAF) sine/cosine average filter

4.4.2 *Simulation of Speckle Interferograms of Test Surface in the Initial and Deformed States*

In order to assess the reliability to restore the surface displacement phase field using the TS DSPI method, it is necessary to generate the SIs in the initial and deformed states of the test surface and use them to verify this method. To this end, additive Gaussian noise corresponding to the RMS height $Sq = \lambda$ (see Eq. 2.28) was used for modeling the test flat rough surface. To simulate the object beam reflected from the test surface, it was assumed that the direction of laser illumination with a wavelength λ and the direction of observation in the optical scheme for SIs recording are perpendicular to this surface. Such conditions can be provided, for example, by the optical scheme of Twyman–Green interferometer. In this case, the RMS phase height $(Sq)_\varphi$ of the object wavefront reflected from the test surface in its immediate vicinity can be expressed as

$$(Sq)_\varphi = \frac{4\pi \cdot Sq}{\lambda}. \tag{4.41}$$

So, if $Sq = \lambda$, then $(Sq)_\varphi = 4\pi$.

The test deformed surface was simulated on the basis of the standard 3D bipolar phase test (see Fig. 2.7) modulated by the additive Gaussian phase noise with RMS value $(Sq)_\varphi = 4\pi$. To simulate SIs of the initial and deformed states of the test rough surface, a random reference wavefront was generated by a plane wave modulated by the same additive Gaussian phase noise. The initial flat phase rough surface and the deformed one, modulated by additive Gaussian phase noise with RMS phase height $(Sq)_\varphi = 4\pi$, are shown in Fig. 4.5.

The test random rough surfaces shown in Fig. 4.5 were used to generate speckle patterns. Synthesis of speckle patterns was performed by using the known methods described by Huntley [42] and modified by Sjödahl and Benckert [90], as well as by Dunkan and Kirkpatrick [18]. Simulation of SIs has been performed with taking into account the method proposed by Höfling and Osten [36], and adopted to the optical scheme of Twyman–Green interferometer, while using the random spatial phase distribution of the reference wavefront [72, 73]. The procedure of the simulated speckle patterns and SIs formation in the optical scheme of Twyman–Green interferometer is shown in Fig. 4.6.

To obtain the speckle pattern due to aforementioned methods, it is necessary to perform the 2D Fourier transform (FT) of the input surface, multiply the FT result by a circular mask of diameter D and fulfill the inverse FT. The circular mask is synthesized in Fourier domain and plays the role of the imaging pupil. The obtained complex amplitude distribution of the speckle pattern is squared modulo, since it is recorded by a matrix sensor. To provide the RMS phase height $(Sq)_\varphi = 4\pi$ for the object and reference wavefronts, the optical scheme of Twyman–Green interferometer was used to synthesize SIs. In Fig. 4.6, the synthesized rectangular flat rough surface is placed in Plane 1. The synthesized surface size is $K\tilde{p}_x \times L\tilde{p}_y$,

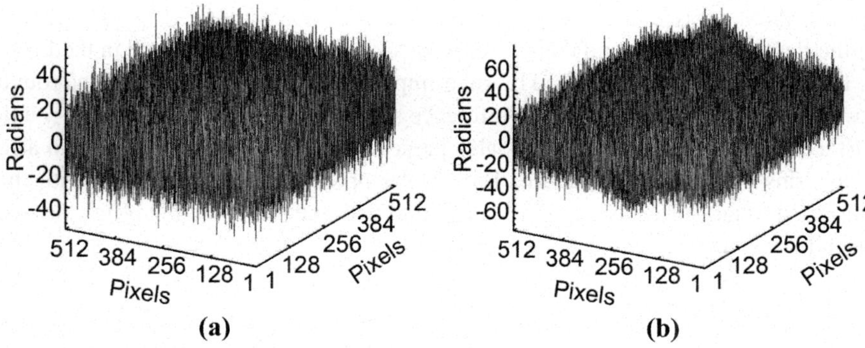

Fig. 4.5 Test phase rough surfaces with RMS phase height $(Sq)_\varphi = 4\pi$ for simulation of SIs: initial flat surface (**a**); deformed phase surface simulated using the standard 3D bipolar phase test (see Fig. 2.7) (**b**)

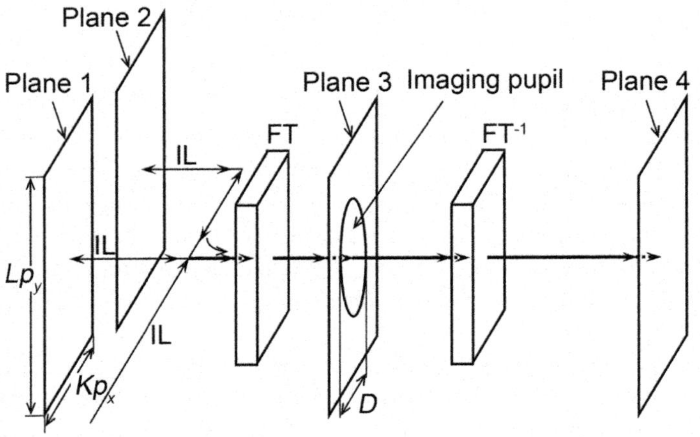

Fig. 4.6 Diagram of the simulated speckle patterns and speckle interferograms formation based on the optical scheme of Twyman–Green interferometer: (IL) incident light; (FT) Fourier transform; (FT^{-1}) inverse Fourier transform

where K, P are the number of pixels in two orthogonal directions along the x and y axes, respectively, and $\tilde{p}_x$, $\tilde{p}_y$ are the pixel pitches equal to pixel size. To generate the reference wavefront, the same flat rough surface is placed in Plane 2 that is tilted at an angle $\Theta_1/2$ relative to Plane 1 in order to generate the reference wavefront at an angle Θ_1 to the object wavefront (see Fig. 4.3b). The optical paths of these two wavefronts are the same. When simulating SIs in such the optical scheme, it is often convenient to choose the angle $\Theta_1 = 0$. Plane 1 and Plane 2 are illuminated by a parallel laser beam incident on the surface areas and perpendicular to Plane 1. The circular mask of diameter D placed in Plane 3 implements the function of the cut-off low-pass filter, that is, the imaging pupil in an optical system. This mask must satisfy Shannon sampling theorem, that is the maximum mask diameter is $D = K\tilde{p}_x/2$, if $K\tilde{p}_x \le L\tilde{p}_y$, or $D = L\tilde{p}_y/2$, if $L\tilde{p}_y \le K\tilde{p}_x$. In this case, a minimum subjective speckle diameter is 2.4 pixels [42, 90] (see Eq. 4.3). The simulated speckle pattern of the flat rough surface and the reference speckle pattern are generated in the Plane 4. In this Plane, a SI is generated by superimposing the speckle pattern of the rough surface placed in Plane 1 with the reference speckle pattern of the reference rough surface placed in Plane 2, which is also passes through the imaging pupil and also transforms into a speckle pattern. The similar procedure was performed with the deformed rough surface placed in Plane 1 and with the reference flat rough surface placed in Plane 2.

4.4.3 Extraction of Unknown Phase Shift Using Population Pearson Correlation Coefficient

In order to verify the reliability of the PPCC to extract the unknown phase shift between two SIs, a series of deformed bipolar phase surfaces with RMS phase heights $\pi/2 \leq (Sq)_\varphi \leq 4\pi$ were simulated using the standard 3D bipolar phase test (see Fig. 2.7). Speckle patterns of these surfaces were synthesized according to the procedure described in previous Sect. 4.4.2. One of the simulated deformed bipolar surfaces with RMS phase height equal to $(Sq)_\varphi = 4\pi$, intended to generate the object wavefront, is shown in Fig. 4.5b. Each reference flat surface intended to generate the reference wavefront had the same roughness and size as the rough deformed bipolar surface. To produce two series of SIs $i_{21}(k, l)$ and $i_{22}(k, l)$ (see Eqs. (4.33), (4.34)) differing by the given phase shift α_0, the first SI series was synthesized without phase shift, and the second one with phase shift equal to $\alpha_0 = \pi/6$. The phase shift between each pair of the SIs i_{21} and i_{22} with the same roughness was calculated using Eq. (4.36), and absolute error $|\Delta\tilde{\alpha}_{21}|$ between the calculated and given phase shifts was defined as

$$|\Delta\tilde{\alpha}_{21}| = |\tilde{\alpha}_{21} - \alpha_0| = |\arccos[\rho(i_{21}, i_{22})] - \alpha_0|. \tag{4.42}$$

Results of the absolute error $|\Delta\tilde{\alpha}_{21}|$ calculations represented in Table 4.1 indicate that if the surface roughness increases, the error $|\Delta\tilde{\alpha}_{21}|$ decrease, i.e. this error is smaller than the errors of the unknown phase shift extraction for smooth or low-rough surfaces considered in Chaps. 2 and 3.

Thus, the simulation results revealed that the correlation approach based on the application of the PPCC for calculation of the unknown phase shift and originally developed for two- and three-step temporal phase shifting PSI methods with a blind phase shifts of the reference beam considered in Chap. 2 is even more efficient for rough surfaces than for the smooth ones. That is, for the TS DSPI method this approach gives a smaller error in extracting an unknown phase shift.

To estimate the absolute errors of phase shifts α_{21}, $\tilde{\alpha}_{21}$ and $\widehat{\alpha}_{21}$ extracted from Eqs. (4.35)–(4.37) in the range from 0 to π, the series of SIs i_{11} and i_{12}, i_{21} and i_{22}, as well as the series of subtraction specklegrams (difference SIs) $(i_{11} - i_{21})$ and $(i_{12} - i_{22})$ were synthesized. For synthesis of these SIs, the simulated initial and deformed surfaces with RMS phase height equal to $(Sq)_\varphi = 4\pi$, generating the object wavefront, as well as the reference flat surface with the same RMS phase

Table 4.1 Absolute errors $|\Delta\tilde{\alpha}_{21}|$ calculated by Eq. (4.42)

$(Sq)_\varphi$	Sq	$\|\Delta\tilde{\alpha}_{21}\|$(deg)	$\|\Delta\tilde{\alpha}_{21}\|$(rad)
$\pi/2$	$\lambda/8$	0.209	$3.65 \cdot 10^{-3}$
$\pi/1.875$	$0.1(3)\lambda$	0.1745	$3.05 \cdot 10^{-3}$
2π	$\lambda/2$	0.0563	$9.83 \cdot 10^{-4}$
4π	λ	0.0261	$4.56 \cdot 10^{-4}$

height, generating the reference wavefront, were used. In this case, the phase shifts between SIs i_{11} and i_{12}, i_{21} and i_{22}, as well as between difference SIs were 10°, 20°,..., 170°, 180°. The absolute errors between the phase shift $\tilde{\alpha}_{21}$ evaluated via Eq. (4.36) and the given one α_0 were calculated by Eq. (4.42). The absolute errors between the phase shift α_{21} evaluated via Eq. (4.35) and the given one α_0, as well as between the phase shift $\widehat{\alpha}_{21}$ evaluated via Eq. (4.37) and the given one α_0 were calculated using the following equations:

$$|\Delta\alpha_{21}| = |\alpha_{21} - \alpha_0| = |\arccos[\rho(i_{11}, i_{12})] - \alpha_0|, \tag{4.43}$$

$$\left|\Delta\widehat{\alpha}_{21}\right| = \left|\widehat{\alpha}_{21} - \alpha_0\right| = |\arccos \rho[(i_{11} - i_{21}, i_{12} - i_{22})] - \alpha_0|. \tag{4.44}$$

The dependences of absolute errors $|\Delta\alpha_{21}|$, $|\Delta\tilde{\alpha}_{21}|$ and $\left|\Delta\widehat{\alpha}_{21}\right|$ as functions of the given phase shifts α_0 for the RMS phase height $(Sq)_\varphi = 4\pi$ and, consequently, for the RMS metric height $Sq = \lambda$ of the simulated initial and deformed surfaces are represented in Tables 4.2, 4.3 and 4.4, respectively.

The obtained results of calculations indicate a low level of absolute errors. The maximum error values are $|\Delta\alpha_{21}|_{max} = 58.2 \cdot 10^{-3}$ deg at $\alpha_0 = 70°$, $|\Delta\tilde{\alpha}_{21}|_{max} = 63.0 \cdot 10^{-3}$ at $\alpha_0 = 120°$ and $\left|\Delta\widehat{\alpha}_{21}\right|_{max} = 32.8 \cdot 10^{-3}$ deg at $\alpha_0 = 70°$. So, in order to achieve a minimum error in extracting the unknown phase shift, it is preferable to use Eq. (4.37) for calculation of $\widehat{\alpha}_{21}$.

Table 4.2 Absolute error $|\Delta\alpha_{21}|$ as a function of the given phase shift α_0, calculated using Eq. (4.43) $(Sq = \lambda)$

α_0(deg)	$\|\Delta\alpha_{21}\|$(deg)	α_0(deg)	$\|\Delta\alpha_{21}\|$(deg)
0	–	100	$46.2 \cdot 10^{-3}$
10	$3.50 \cdot 10^{-3}$	110	$36.4 \cdot 10^{-3}$
20	$14.4 \cdot 10^{-3}$	120	$25.7 \cdot 10^{-3}$
30	$27.4 \cdot 10^{-3}$	130	$14.5 \cdot 10^{-3}$
40	$37.8 \cdot 10^{-3}$	140	$5.37 \cdot 10^{-3}$
50	$47.7 \cdot 10^{-3}$	150	$1.23 \cdot 10^{-3}$
60	$54.7 \cdot 10^{-3}$	160	$2.88 \cdot 10^{-3}$
70	$58.2 \cdot 10^{-3}$	170	$1.53 \cdot 10^{-3}$
80	$57.9 \cdot 10^{-3}$	180	$1.22 \cdot 10^{-3}$
90	$53.9 \cdot 10^{-3}$	–	–

Table 4.3 Absolute error $|\Delta\tilde{\alpha}_{21}|$ as a function of the given phase shift α_0, calculated with the help of Eq. (4.42) $(Sq = \lambda)$

α_0(deg)	$\|\Delta\tilde{\alpha}_{21}\|$(deg)	α_0(deg)	$\|\Delta\tilde{\alpha}_{21}\|$(deg)
0	$8.5 \cdot 10^{-7}$	100	$48.6 \cdot 10^{-3}$
10	$15.4 \cdot 10^{-3}$	110	$57.8 \cdot 10^{-3}$
20	$23.6 \cdot 10^{-3}$	120	$63.0 \cdot 10^{-3}$
30	$27.5 \cdot 10^{-3}$	130	$61.8 \cdot 10^{-3}$
40	$26.3 \cdot 10^{-3}$	140	$56.7 \cdot 10^{-3}$
50	$20.4 \cdot 10^{-3}$	150	$45.7 \cdot 10^{-3}$
60	$9.07 \cdot 10^{-3}$	160	$32.9 \cdot 10^{-3}$
70	$5.36 \cdot 10^{-3}$	170	$18.7 \cdot 10^{-3}$
80	$20.1 \cdot 10^{-3}$	180	$1.75 \cdot 10^{-3}$
90	$35.4 \cdot 10^{-3}$	–	–

Table 4.4 Absolute error $\left|\Delta\widehat{\alpha}_{21}\right|$ as a function of the given phase shift α_0, calculated using Eq. (4.44) $(Sq = \lambda)$

α_0(deg)	$\Delta\widehat{\alpha}_{21}$(deg)	α_0(deg)	$\Delta\widehat{\alpha}_{21}$(deg)
0	$8.5 \cdot 10^{-7}$	100	$17.6 \cdot 10^{-3}$
10	$11.35 \cdot 10^{-3}$	110	$10.81 \cdot 10^{-3}$
20	$17.1 \cdot 10^{-3}$	120	$2.35 \cdot 10^{-3}$
30	$22.2 \cdot 10^{-3}$	130	$2.97 \cdot 10^{-3}$
40	$27.9 \cdot 10^{-3}$	140	$7.92 \cdot 10^{-3}$
50	$31.9 \cdot 10^{-3}$	150	$9.75 \cdot 10^{-3}$
60	$32.7 \cdot 10^{-3}$	160	$11.24 \cdot 10^{-3}$
70	$32.8 \cdot 10^{-3}$	170	$9.16 \cdot 10^{-3}$
80	$29.5 \cdot 10^{-3}$	180	$1.84 \cdot 10^{-3}$
90	$23.9 \cdot 10^{-3}$	–	–

4.4.4 Simulation of Test Phase Field of Surface Displacements

Verification of the proposed TS DSPI method was carried out with the help of two simulated SIs i_{11} and i_{12} of the initial test phase rough flat surface with a RMS phase height $(Sq)_\varphi = 4\pi$ (see Fig. 4.5a) and two SIs i_{21} and i_{22} of a deformed test phase rough surface of the same RMS phase height, which was simulated using the standard 3D bipolar phase test shown in Fig. 2.7. The given phase shift α_0 was equal to $\pi/3$. The test displacement phase map $\delta\varphi_t(k, l)$ was retrieved using Eq. (4.40) and the algorithm that implements the TS DSPI method and is shown in Fig. 4.4. The retrieved map $\delta\varphi_t(k, l)$ was compared with the standard 3D bipolar phase test, which was used for synthesis of the test deformed phase rough surface (see Fig. 2.7). Phase distribution of the error of the retrieved phase map $\delta\varphi_t(k, l)$ was obtained by pixel-by-pixel subtraction of the 3D bipolar test from this map. The histogram of

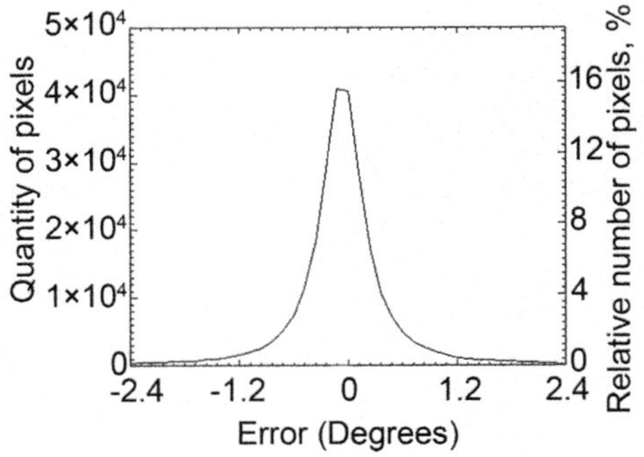

Fig. 4.7 Histogram of error distribution of the retrieved displacement phase map $\delta\varphi_t(k, l)$ with dimensions of 512×512 pixels

this distribution is shown in Fig. 4.7. The mean value of this error is $(-3.28) \cdot 10^{-3}$ deg and the values of 95% of errors are within $\pm 1.2°$.

4.4.5 Experimental Verification of Two-Step Digital Speckle Pattern Interferometry Method

Experimental verification of the TS DSPI method was performed with the help of an optical-digital experimental setup based on the Lindeertz interferometer [10, 45, 51] to study in-plane surface displacements and deformations, as well as the Mach–Zehnder interferometer [60] for research out-of-plane displacements and deformations. The optical schemes of the Lindeeretz and Mach–Zehnder interferometers are shown in Fig. 4.8.

For all the experiments realizing the TS DSPI method, the next multistep procedure to record SIs i_{11}, i_{12}, i_{21} and i_{22} was implemented: (1) recording the initial SI

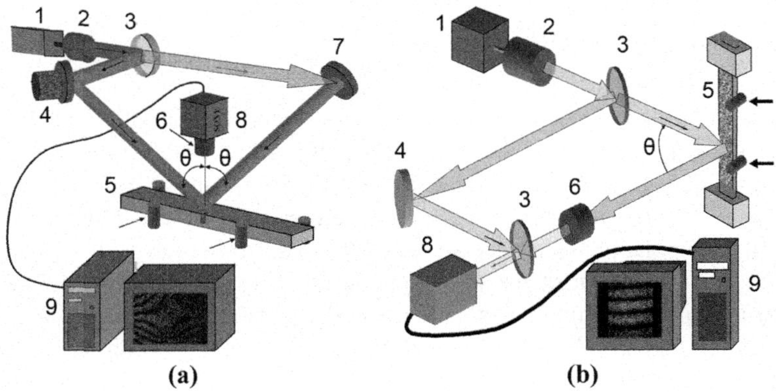

Fig. 4.8 Optical schemes of Lindeeretz (**a**) and Mach–Zehnder (**b**) interferometers: laser (1); laser beam expander and collimator (2) beam splitter (3); piezoelectric transducer with mirror (4); specimen fixed in load device (5); lens (6); mirror (7); digital camera (8); computer (9)

i_{11} at step 1; (2) performing an unknown (blind) phase shift of the reference beam at step 2; (3) recording the SI i_{12} at step 3; (4) loading of the studied specimen of constructive material or structural element at step 4; (5) recording the SI i_{22} at step 5; (6) turn off the phase shifter and return the reference beam to its initial state at step 6; (7) recording of the SI i_{21} at step 7.

Studying of in-plane surface displacements and deformations. Spatial fields of surface in-plane displacements and deformations were studied using the experimental setup containing the Lindeeretz interferometer (see Fig. 4.8a), a four-point bending device, a digital camera SONY XCD-SX910. The setup was placed on an optical table to suppress the influence of vibrations. Phase shift was implemented with the help of the piezoelectric transducer driven by the electronic unit. Beam metal specimens with dimensions 220 × 20 × 10 mm^3 were prepared for experiments.

The obtained and analyzed experimental data have indicated that there are two main negative error sources which should be suppressed, namely: (i) high level of noises and vibrations; (ii) fluidity of a metal under a fixed load. The level of noise and vibrations can be controlled by analyzing histograms of subtraction specklegrams at different time points. Using Eq. (4.31), the differences between two SIs at the same initial state of the studied surface at different time moments t_1, t_2, t_3 and t_4 can be given as $\delta i_{11}(t_1, t_2) = i_{11}(t_1) - i_{11}(t_2)$, $\delta i_{11}(t_1, t_3) = i_{11}(t_1) - i_{11}(t_3)$, $\delta i_{11}(t_1, t_4) = i_{11}(t_1) - i_{11}(t_4)$. In the absence of any temporal changes both in the state of the surface and in the experimental setup, these differences would be equal to zero. In reality, there are always interferences caused by these circumstances, which lead to the fact that the subtraction specklegrams (difference SIs) are always characterized by some noise level. An example of a histogram of the difference SIs for low noise, which is considered acceptable to record SIs, is shown in Fig. 4.9a, and an example of a histogram for high noise, which is considered unacceptable to record SIs, is shown in Fig. 4.9b. It was found that for an acceptable noise level, the relative systematic and random errors are $E_1 = 5.8 \cdot 10^{-5}$ and $\sigma_1 = 7.57 \cdot 10^{-3}$, respectively, and for a high one, $E_2 = -1.13 \cdot 10^{-4}$ and $\sigma_2 = 4.05 \cdot 10^{-2}$, where σ_1 and σ_2 are the RMS of the relative random error. According to the results of numerous experiments, the maximum allowable threshold level for relative random errors was chosen equal to $\sigma_{\text{thr}} = 2^{-7} = 7.81 \cdot 10^{-3}$, and for systematic errors, equal to $E_{thr} = 1 \cdot 10^{-3}$. Therefore, it is possible to filter SI by noise level using histograms of SI differences. Such filtering allows selecting SI with low noise.

If a constant load is applied to the specimen, the fluidity of material is observed during its surface recording, which is easy to detect by the SIs differences recorded at different times for the same load. An example of such a difference between two SIs recorded immediately after loading the sample with an interval of 15 s is shown in Fig. 4.10. If these SIs would be identical, these fringes would not appear. However, they are present and their number increases with increasing time intervals between the registration of two SI, i.e. the specimen surface moves, although the load does not change. This effect of material fluidity can be prevented by pausing after the specimen loading, the duration of which Δt is determined experimentally and depends on the mechanical characteristics of the material and the sensitivity threshold of the speckle interferometer. In experiments with steel specimens, the pausing duration is usually

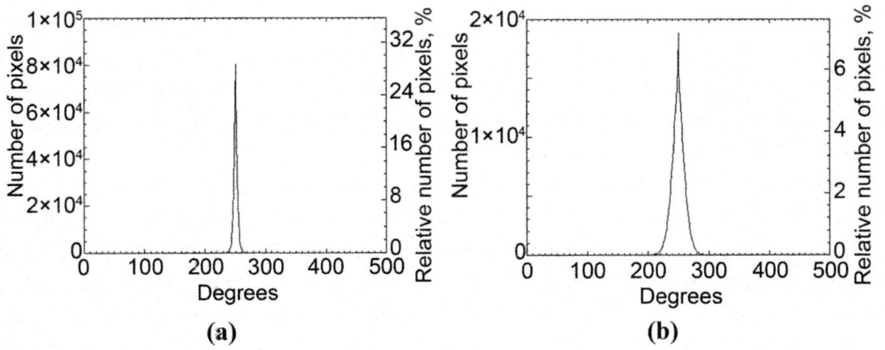

Fig. 4.9 Histograms of subtraction specklegrams $\delta i_{11}(t_1, t_2) = i_{11}(t_1) - i_{11}(t_2)$: acceptable noise level (**a**), unacceptable noise level (**b**). Dimensions of phase maps are 512×512 pixels

longer than 60 s and shorter than 180 s. It is advisable to take the same pause after all changes in the experimental conditions, i.e. after the depolarization of the PZT and after the shift of the reference beam phase. This prevents the PZT mirror from oscillating.

The proposed procedure for recording phase shifted SIs of real surfaces made it possible to effectively use the developed TS DSPI method to retrieve the in-plane displacement maps of structural materials surfaces. Figure 4.11 shows the intermediate and final results of the processing of the SIs and displacement phase maps of the surface area of the steel 20 beam with dimensions $220 \times 20 \times 10\ \text{mm}^3$ containing a V-shaped notch (steel 20 is similar to steels 1020, 1023, 1024 and G10200). The chosen area dimensions were $15.0 \times 11.25\ \text{mm}^2$, and the size of recorded SIs and the obtained phase maps is 800×600 pixels. The phase shift between SIs i_{21} and i_{22} was $\tilde{\alpha}_{21} = 44°$. The duration of the time pause after steps 2, 4 and 6 during the experiment was equal to 1 min. The minimum value of the unwrapped displacement phase map shown in Fig. 4.11d is -10.0 rad, and the maximum one is 21.2 rad.

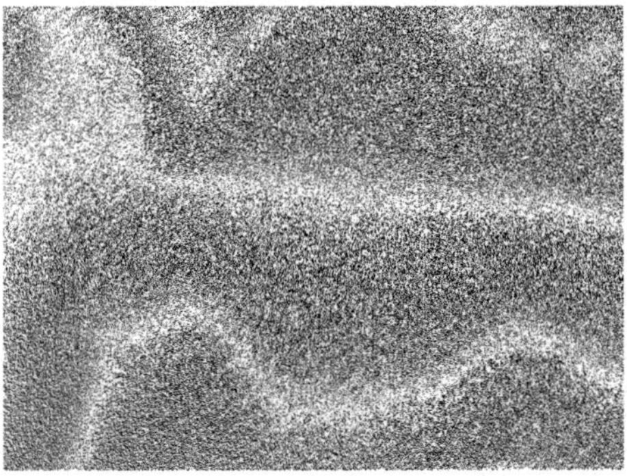

Fig. 4.10 Subtraction specklegram, obtained by recording two SIs at constant load with a time difference of 15 s

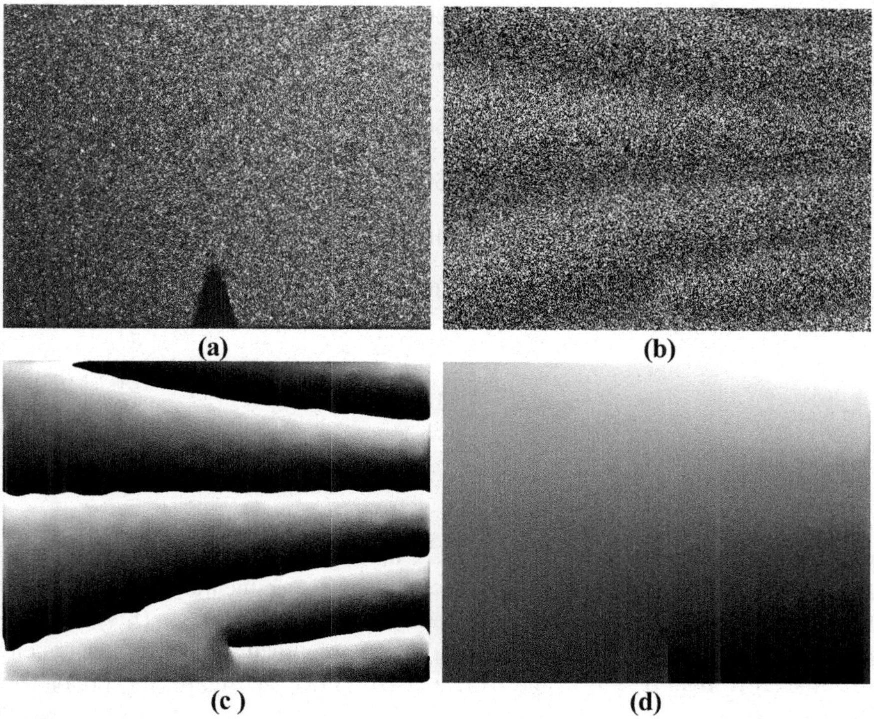

Fig. 4.11 Results of the processing of SIs and displacement phase maps of the surface area of the steel 20 beam containing a V-shaped notch according to the algorithm performing the TS DSPI method: recorded SI i_{22} (**a**); extracted coarse in-plane displacement phase map $\delta\varphi_c(k, l)$ (**b**); wrapped phase map $\delta\varphi_w(k, l)$ (**c**); unwrapped in-plane displacement phase map $\delta\varphi(k, l)$ (**d**)

Figure 4.12 illustrates the displacement phase map represented in Fig. 4.11d, divided by isolines of equal in-plane surface displacements at 30 levels (1 level is 50.4 nm in terms of the in-plane surface displacement).

Studying out-of-plane surface displacements and deformations. Out-of-plane displacement fields were studied using the TS DSPI method and the described above procedure to record SIs i_{11}, i_{12}, i_{21} and i_{22}. The experimental setup containing a Mach–Zehnder interferometer (see Fig. 4.8b), as well as the same four-point bending device (position 5 in Fig. 4.8b) and the digital camera SONY XCD-SX910 (position 8 in Fig. 4.8b). The experimental setup, as well as the previous one for the study of in-plane displacement fields, was mounted on the optical table to suppress the influence of vibrations. Metal beam specimens were tested under four-point bending, including beams with notches and transverse welds.

In one of the experiments, the steel beam specimen with dimensions 400 × 20 × 10 mm^3 consisting of two welded beams made of steel 20 and steel 45 was used (steel 45 is similar to steels 1042,1045, 1045H and G10450). In this specimen, a narrow cut 1 mm thick and deep from 2 mm up to 9 mm along the transverse weld was made on the side opposite to the side of the studied rough surface. The

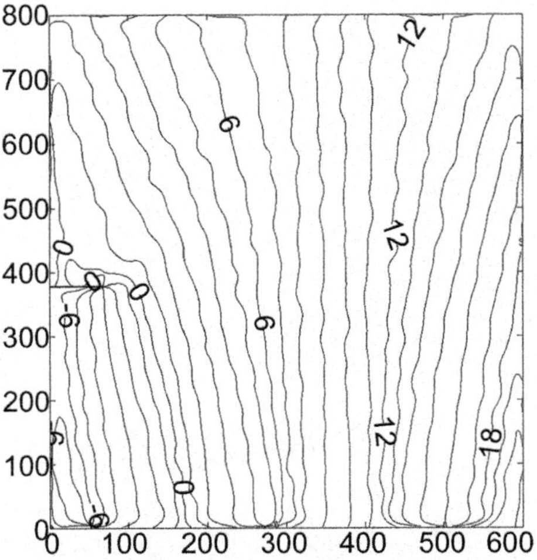

Fig. 4.12 Unwrapped in-plane surface displacement phase map of the steel 20 beam specimen surface area, shown in Fig. 4.11d. Map is divided by isolines of equal in-plane surface displacements at 30 levels (1 level is 1 rad or 50.4 nm in terms of the in-plane displacement)

chosen area dimensions were 15.0 × 11.25 mm^2, and the size of obtained SIs and phase maps is 800 × 600 pixels. The section of the specimen along the cut 1 mm thick is shown in Fig. 4.13. First, the SI series $i_{11}(t_{1,1})$, $i_{11}(t_{1,2})$, $i_{11}(t_{1,3})$, $i_{11}(t_{1,4})$ and $i_{12}(\tilde{t}_{1,1})$, $i_{12}(\tilde{t}_{1,2})$, $i_{12}(\tilde{t}_{1,3})$, $i_{12}(\tilde{t}_{1,4})$ were recorded at different moments of time $t_{1,1}$, $t_{1,2}$, $t_{1,3}$, $t_{1,4}$ and $\tilde{t}_{1,1}$, $\tilde{t}_{1,2}$, $\tilde{t}_{1,3}$, $\tilde{t}_{1,4}$ at significant preload of the specimen; then the load was slightly increased and the SI series $i_{22}(\tilde{t}_{2,1})$, $i_{22}(\tilde{t}_{2,2})$, $i_{22}(\tilde{t}_{2,3})$, $i_{22}(\tilde{t}_{2,4})$ and $i_{21}(t_{2,1})$, $i_{21}(t_{2,2})$, $i_{21}(t_{2,3})$, $i_{21}(t_{2,4})$ were recorded also at different moments of time $\tilde{t}_{2,1}$, $\tilde{t}_{2,2}$, $\tilde{t}_{2,3}$, $\tilde{t}_{2,4}$ and $t_{2,1}$, $t_{2,2}$, $t_{2,3}$, $t_{2,4}$, respectively. After each change in the experimental conditions, i.e. after steps 2, 4 and 6, there were pauses of about three minutes. The results of SIs processing in this experiment are illustrated in Fig. 4.14. The extracted unknown phase shift of the reference beam was $\alpha_{21} = 108.28°$; the minimum value of the displacement phase map was –42.89 rad, and the maximum one was + 9.64 rad (see Fig. 4.14d).

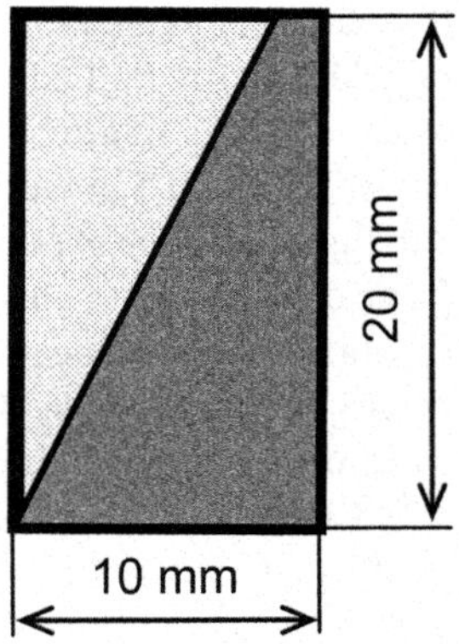

Fig. 4.13 Section of the specimen with dimensions 400 × 20 × 10 mm^3 along the narrow cut 1 mm thick and deep from 2 mm up to 9 mm along the transverse weld

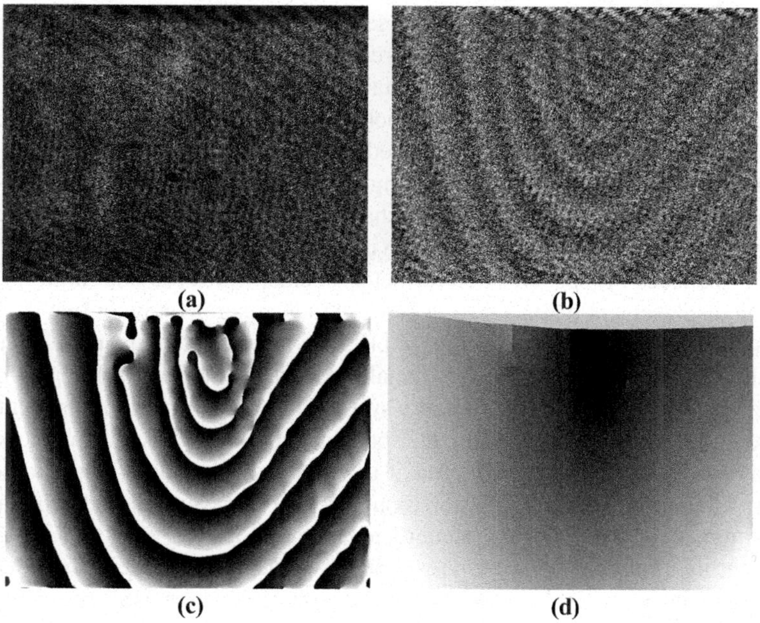

Fig. 4.14 Digital processing stages of SIs of the welded steel beam specimen surface area according to the algorithm performing the TS DSPI method: recorded SI i_{22} (**a**); extracted coarse out-of-plane displacement phase map $\delta\varphi_c(k, l)$ (**b**); wrapped phase map $\delta\varphi_w(k, l)$ (**c**); unwrapped out-of-plane displacement phase map $\delta\varphi(k, l)$ (**d**)

Thus, it is experimentally confirmed that the new TS DSPI method, which requires registration of only four SIs and performing a single phase shift of the reference beam to an unknown (arbitrary) angle α_{21}, provides precise retrieval of spatial displacement fields of materials and structural elements rough surfaces. The proposed method does not require any precision phase shifters and appropriate hardware required for their operation. Therefore, the cost of equipment for conducting experiments is significantly reduced.

4.5 Determination of 3D Displacement Fields by Two-Step Digital Speckle Pattern Interferometry

The schemes of digital speckle interferometers considered above can provide the determination and measurement of in-plane and out-of-plane displacements in one direction. To measure 3D displacement fields, more complex optical schemes are needed than, for example, the Leendertz [51], Mach–Zehnder or Twyman–Green [60] interferometers schemes. If, however, simplified schemes are used to generate 3D displacement fields, then a number of limitations arise. They negatively affect the providing of the interferometer necessary sensitivity and other operational parameters over the entire displacement field of the recorded surface area.

For example, in [19, 88, 96, 117], a fairly simple scheme of a speckle interferometer with divergent laser beams, which is simultaneously sensitive to in-plane and out-of-plane displacements, was used to obtain a complete description of the displacement and deformation fields along three mutual directions. However, in this scheme, the sensitivity to displacement changes at each point of the surface in these directions, which makes it problematic to use it for precision metrological purposes. A more advanced 3D speckle interferometer scheme, which provides uniform sensitivity of displacement fields, contains three digital cameras for recording the movements in three orthogonal directions [113]. In last years, more popular are 3D speckle interferometers that use the spatial carrier phase shifting technique [52, 58]. However, the developed interferometric systems increase the temporal resolution of the measurement results, but reduce the spatial resolution. In order to simultaneously record several speckle fringe patterns, it is necessary to use high-resolution digital camera containing a high-dimensional photosensor or several digital cameras and, in addition, a complex optical system to generate all SIs.

To ensure uniformity of sensitivity to out-of-plane and in-plane surface displacements over the entire field, a 3D digital speckle interferometer (DSI) was developed, in which a Mach–Zehnder interferometer for measuring out-of-plane displacements and two Leendertz interferometers for measuring in-plane displacements in two mutually perpendicular directions were used [103]. In contrast to similar schemes of a 3D speckle interferometer [107], where it is necessary to generate a significant number of SIs (usually 24 Sis or more) with phase shifts between adjacent SIs equal to $\pi/2$, in this 3D DSI, the SIs are generated using only one arbitrary phase shift $\alpha_{21} = \tilde{\alpha}_{21}$ between two SIs (see Eqs. 4.31–4.34)) recorded at the same undeformed or deformed state of the studied specimen surface [103]. Only one digital camera was used to record SIs, since under this condition the same scale of the studied surface area is guaranteed for all three interferometers. During the experiment, the digital camera sequentially records the SIs generated in the optical schemes of these three interferometers. The resulting digital speckle interferograms (12 in total) are used to determine the displacement fields in three orthogonal directions.

4.5.1 Experimental Setup of 3D Digital Speckle Interferometer

As mentioned above, the created experimental setup of the 3D DSI includes one Mach–Zehnder and two Leendertz interferometers. The scheme of the 3D DSI is shown in Fig. 4.15. Each of the interferometers contains a phase shifting element made of a piezoceramic cylinder to which a mirror is attached. Two single-mode He–Ne lasers LGN-215 with a wavelength of $\lambda = 0.6328$ μm and a power of 50 mW were used in this setup. The experimental setup also includes a digital CCD camera SONY XCD-S910 (resolution 1280 (H) × 960 (V)), a four-point bending load device and a computer. The studied metal and composite beam specimens, which were loaded with the help of the four-point bending device, are shown in Fig. 4.16. Speckle interferograms of the studied specimen surface areas in the initial and in the deformed states were recorded to restore surface displacement fields in three orthogonal directions.

The surface displacements caused by the specimen loading lead to a change in the phase difference of the interfering laser beams, i.e. to the formation of a phase displacement field $\delta\varphi$, which depends on both the surface displacement $d(x, y, z)$ and the angle Θ between normal to the specimen surface and illuminating beams. Since the Leendertz speckle interferometer is sensitive to in-plane surface displacements only in one direction transverse to the observer, its sensitivity according to Eq. (4.10)

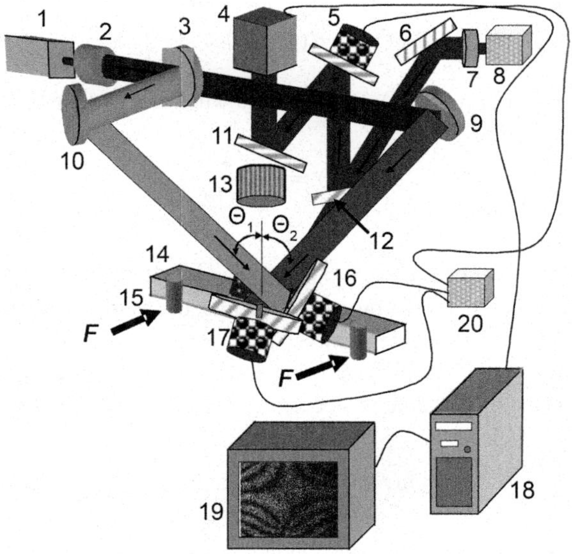

Fig. 4.15 Optical scheme of 3D DSI experimental setup: lasers (1, 8); laser beam collimators (2, 7); beam splitters (3, 11, 12); digital camera (4); piezoelectric transducers (5, 16, 17); deflecting mirrors (6, 9, 10); lens (13); specimen (14) fixed in load device (15); computer with monitor (18, 19); controller driving piezoelectric transducers (20)

(a) **(b)**

Fig. 4.16 Studied specimens: metal beams (**a**); laminated composite beams (**b**) and Fig. 4.3a along the x and y axes is respectively

$$\left.\begin{aligned}\eta(x) &= \tfrac{2\pi}{\lambda}(\sin\Theta_1 + \sin\Theta_2)\\ \eta(y) &= \tfrac{2\pi}{\lambda}(\sin\Theta_1 + \sin\Theta_2)\end{aligned}\right\}, \tag{4.45}$$

and the phase change according to (4.11) is

$$\left.\begin{aligned}\delta\varphi(x) &= d(x)\left(\tfrac{2\pi}{\lambda}\right)(\sin\Theta_1 + \sin\Theta_2)\\ \delta\varphi(y) &= d(y)\left(\tfrac{2\pi}{\lambda}\right)(\sin\Theta_1 + \sin\Theta_2)\end{aligned}\right\}. \tag{4.46}$$

If the angles between the normal to the specimen surface and the illuminating beams are the same, i.e. $\Theta_1 = -\Theta_2 = \Theta$, then

$$\left.\begin{aligned}\delta\varphi(x) &= d(x)\left(\tfrac{4\pi}{\lambda}\right)\sin\Theta\\ \delta\varphi(y) &= d(y)\left(\tfrac{4\pi}{\lambda}\right)\sin\Theta\end{aligned}\right\}. \tag{4.47}$$

Phase change $\delta\varphi(x) = \delta\varphi(y) = 2\pi$ occurs when the surface moves a distance

$$d_1(x, y) = \frac{\lambda}{2\sin\Theta_{x,y}}, \tag{4.48}$$

where $d_1(x, y) = \sqrt{d_1^2(x) + d_1^2(y)}$, $\Theta_{x,y} = \Theta$. This displacement is also called the sensitivity of the speckle interferometer per fringe, because it corresponds to one full fringe on the subtraction specklegram.

The sensitivity of the Mach–Zehnder speckle interferometer along the z axis for out-of-plane surface displacements according to (4.10) and Fig. 4.3b is

$$\eta(z) = \frac{2\pi}{\lambda}(1 + \cos\Theta_1), \tag{4.49}$$

and the phase change according to (4.11) is

$$\delta\varphi(z) = d(z)\left(\frac{2\pi}{\lambda}\right)(1 + \cos\Theta_1). \tag{4.50}$$

The sensitivity of the Mach–Zehnder speckle interferometer per fringe is

$$d_1(z) = \frac{\lambda}{1 + \cos\Theta_1}. \tag{4.51}$$

To determine the surface displacement fields in three directions, the phase surface displacement field $\delta\varphi(x, y, z)$ was first calculated by digital processing of speckle interferograms recorded before and after bending of the metal and composite specimens shown in Fig. 4.16. The calculated phase field is then converted into the surface displacement field $d(x, y, z)$ according to Eqs. (4.47), (4.50) using the TS DSPI method or according to Eqs. (4.48), (4.51) for correlation DSPI (see Sect. 4.2).

Phase displacement fields of surface areas of metal and composite specimens were generated using the TS DSPI method. To this end, in every direction along x, y and z axes recorded four SIs, namely, two SIs i_{11} and i_{12} before loading and two SIs i_{21} and i_{22} after loading. Before recording SIs i_{12} and i_{22}, the reference beam was shifted on arbitrary angle $\alpha_{21} = \tilde{\alpha}_{21} = \alpha$ concerning its initial position.

The developed 3D DSI experimental setup was used to study the stress–strain state of a series of metal and composite beam specimens. To create rough surfaces on some specimens, thin layers of fine-dispersed aluminum paint were applied to the studied surface areas, in particular, to surfaces of carbon fiber reinforced polymer (CFRP) laminate beam specimens.

4.5.2 Research Results on Retrieval of 3D Surface Displacement Fields of Steel and Composite Beam Specimens

In the course of experiments with steel beam specimens without and with stress concentrators, surface areas dimensions of $15 \times 11.25\,\text{mm}^2$ were studied. The dimensions of the recorded SIs are 800×600 pixels. The specimens' sizes are $200 \times 20 \times 10\,\text{mm}^3$, and $235 \times 20 \times 10\,\text{mm}^3$. The sensitivities per fringe of the Leendertz speckle interferometer along the x and y axes were $d_1(x) = 0.55\,\mu\text{m}$ and $d_1(y) = 0.47\,\mu\text{m}$, respectively; and sensitivity per fringe of the Mach–Zehnder speckle interferometer was $d_1(z) = 0.37\,\mu\text{m}$. Results of experiments with steel beam specimens, performed using the experimental setup of the 3D DSI were presented in [103]. These results showed that the configurations of surface displacement fields for specimens with and without stress concentrators differ only slightly for the y-component directed along the width of the specimens. In this direction, the influence of the concentrator is almost imperceptible. For the x-component directed along the specimen length, the field configurations, especially in the vicinity of the concentrator, were already noticeably different. And for the z-component, completely different configurations for surface displacement fields were observed.

The surface displacement fields of beam specimens made of CFRP laminate (see Fig. 2.44b) were investigated on the same experimental model of a three-coordinate electronic speckle interferometer, the scheme of which is shown in Fig. 4.15. The specimens with sizes of $200 \times 20 \times 3\ \text{mm}^3$ were made of laminated composite containing 7 layers, in which the surface layers are reinforced with fiber mesh. The working sizes of the specimen surface are $140 \times 20\ \text{mm}^2$ and the sizes of two plates to capture specimens are $30 \times 20\ \text{mm}^2$. Small bending loads were applied to the specimens according to the four-point bending scheme. The results of the experiments are shown in Fig. 4.17 at the specimen load of 150 N and in Fig. 4.18 at the specimen load of 200 N.

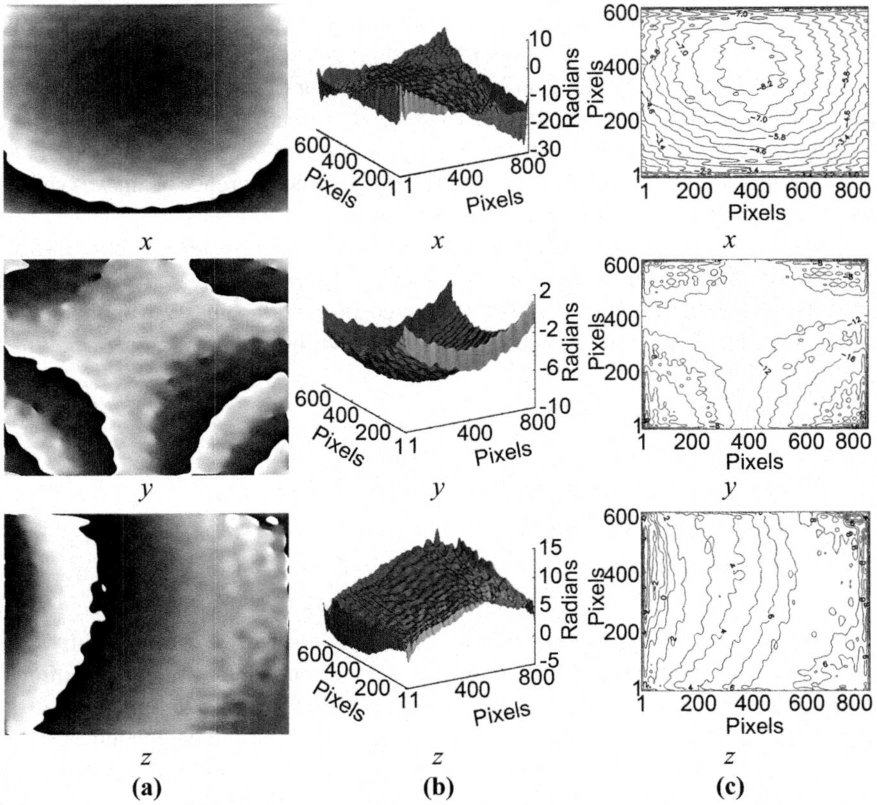

Fig. 4.17 Results of processing of the SIs obtained by recording the CFRP laminate beam specimen surface area with sizes of 15 × 11.25 mm^2 at the specimen load of 150 N. Retrieval of surface displacement maps was performed with the help of the algorithm that implements the TS DSPI method (see Fig. 4.4): along axis x (first line); along axis y (second line); along axis z (third line); wrapped phase maps of surface area displacements (column **a**); isometric images of the wrapped phase maps (column **b**); isolines of the displacement fields in μm (column **c**)

Obtained in Figs. 4.17b and 4.18b displacements phase maps reflect the matrix cell structure of the reinforcing fiber mesh located in the inner layers of the composite. However, rarely located on hundreds of nanometers isolines of surface displacement fields do not give a clear picture of such a structure (see Figs. 4.17c and 4.18c). If isolines of displacement fields are constructed at shorter intervals, the features of displacement fields on the composite specimen surface immediately become noticeable, which is clearly shown in Figs. 4.19 and 4.20. Thus, the reinforcing fiber mesh has a significant effect on the structure of the displacement fields of the composite beam specimen.

If to compare Figs. 4.19b and 4.20b, it is easy to conclude that the grid effect caused by the fiber meshes becomes more noticeable with an increase the load applied to the specimen. Since the ranges of displacements on the restored phase maps obtained

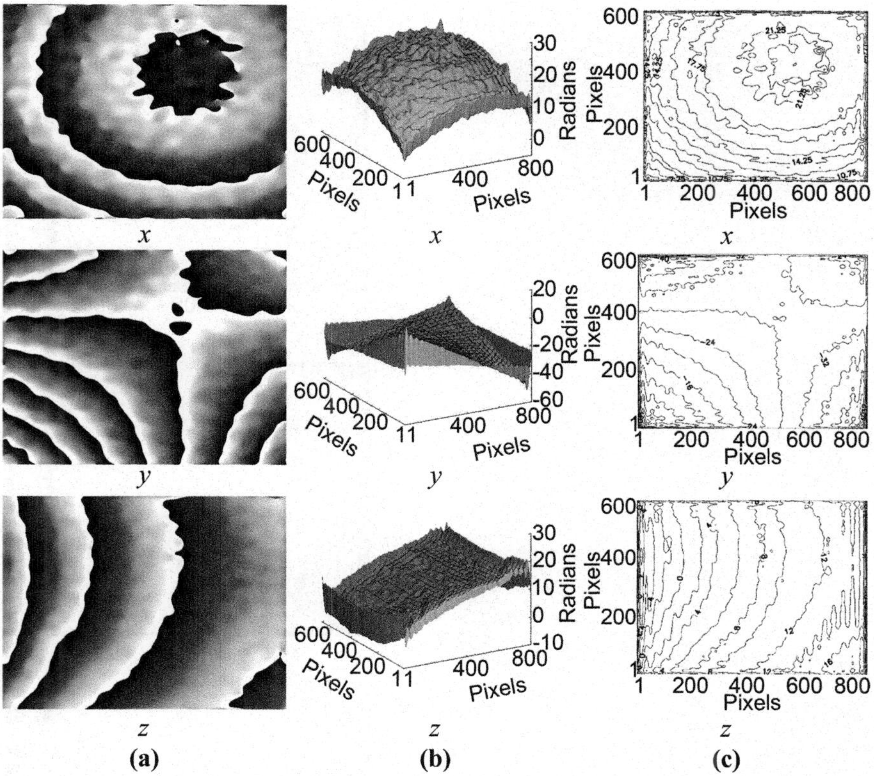

Fig. 4.18 Results of processing of the SIs obtained by recording the CFRP laminate beam specimen surface area with sizes of 15 × 11.25 mm^2 at the specimen load of 200 N. Retrieval of surface displacement maps was performed with the help of the algorithm that implements the TS DSPI method (see Fig. 4.4): along axis x (first line); along axis y (second line); along axis z (third line); wrapped phase maps of surface area displacements (column **a**); isometric images of the wrapped phase maps (column **b**); isolines of the displacement fields in μm (column **c**)

at the different loads are different for the x, y and z axes, the distances between the isolines on these maps are also different. Therefore, the pitches between adjacent isolines are 4 nm along x axis, 12 nm along y axis and 4 nm along z axis on the surface displacement phase maps shown in Fig. 4.19b, and the pitches between adjacent isolines are 12 nm along x axis, 24 nm along y axis and 7 nm along z axis on the surface displacement phase maps shown in Fig. 4.20b. The fiber mesh that reinforces the composite material appears more noticeably on the phase maps shown in Fig. 4.20b, although the distances between adjacent isolines are almost twice as large as compared to the phase maps in Fig. 4.19b. This effect is explained by the fact that the surface displacement fields shown in Fig. 4.20 are obtained with a higher specimen load. Consequently, the fiber mesh reinforcing the composite material is the more visible in the images of the surface displacement fields, the greater the load on the specimen.

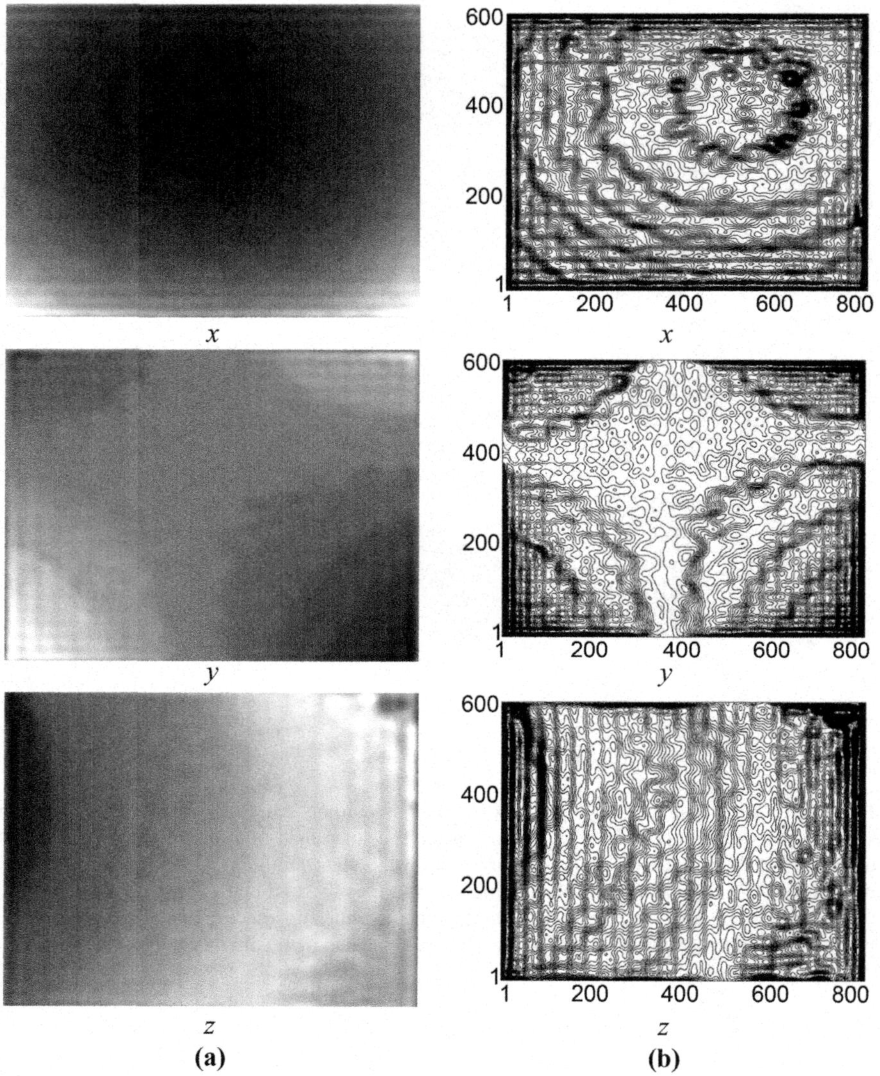

Fig. 4.19 Results of processing of the SIs obtained by recording the CFRP laminate beam specimen surface area with sizes of 15 × 11.25 mm^2 at the specimen load of 150 N. Retrieval of surface displacement maps was performed with the help of the algorithm that implements the TS DSPI method (see Fig. 4.4): along axis x (first line); along axis y (second line); along axis z (third line); wrapped phase maps of surface area displacements (column **a**); isolines of the displacement fields (column **b**). The isolines pitches are 4 nm for x axis, 12 nm for y axis and 4 nm for z axis

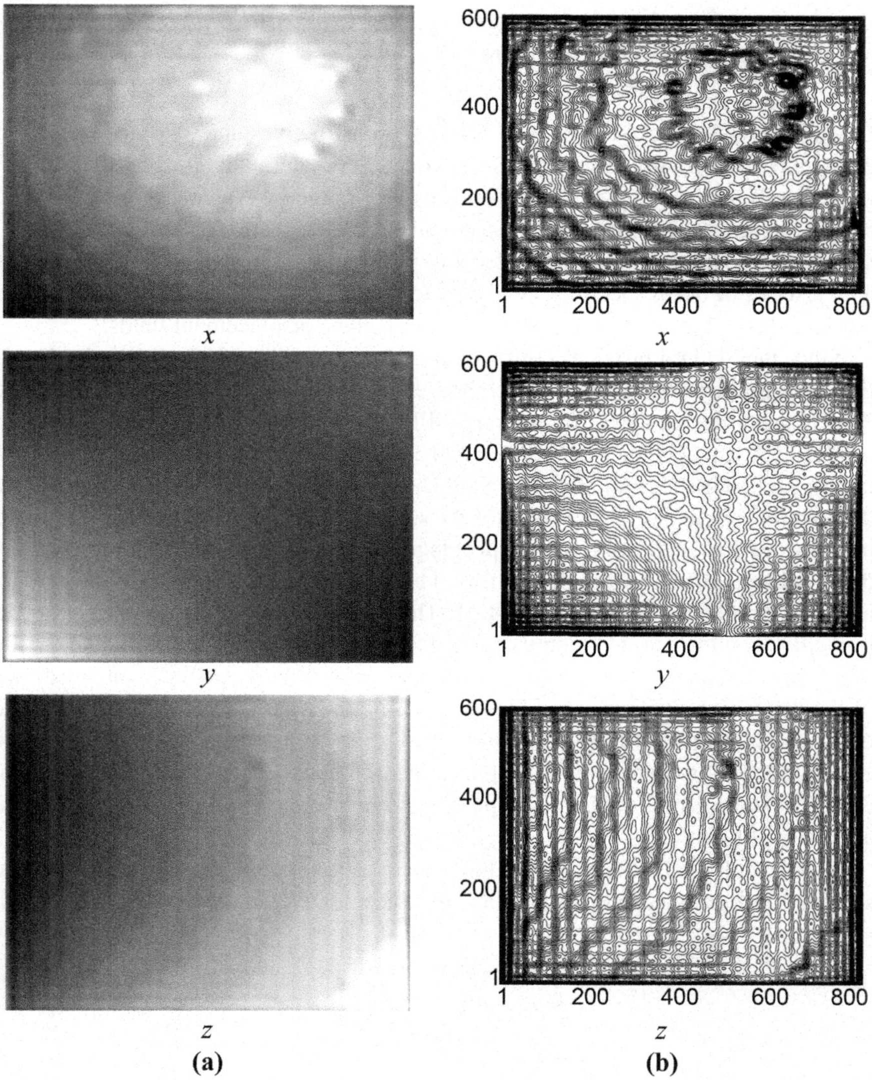

Fig. 4.20 Results of processing of the SIs obtained by recording the CFRP laminate beam specimen surface area with sizes of 15 × 11.25 mm^2 at the specimen load of 200 N. Retrieval of surface displacement maps was performed with the help of the algorithm that implements the TS DSPI method (see Fig. 4.4): along axis x (first line); along axis y (second line); along axis z (third line); wrapped phase maps of surface area displacements (column **a**); isolines of the displacement fields (column **b**). The isolines pitches are 12 nm along x axis, 24 nm along y axis and 7 nm along z axis

4.5.3 Comparative Assessment of In-Plane and Out-of-Plane Surface Displacements

One of the important problems of NDT of constructive materials and structural elements is the identification of internal defects for material diagnostics. The developed experimental setup of 3D DSI makes it possible to evaluate the influence of both in-plane and out-of-plane displacements of the same surface on the detection of surface, subsurface and internal defects in metal and composite specimens. To this end, a series of beam specimens with artificial and real defects were investigated to find and detect them using in-plane and out-of-plane displacement fields.

Since the 3D DSI provides registration both phase shifted SIs series and subtraction specklegrams, first there was compared the detection efficiency of subsurface defects using the standard correlation DSPI technique and the TS DSPI method on a steel 45 beam specimen with sizes 200 × 20 × 10 mm^3 containing an artificial internal defect in the form of a blind round hole of Ø3.5 mm and depths of 8 mm. The displacement fields were evaluated for the surface opposite to that in which the blind hole was drilled. The subtraction specklegram and the phase map of out-of-plane displacements of the surface area, from the opposite side of which the blind round hole was drilled, are shown in Fig. 4.21. The subtraction specklegram in Fig. 4.21a immediately allows you to identify the defect circled by a rectangular frame. On the other hand, the out-of-plane displacement phase map obtained in the form of isolines of the displacements field (see Fig. 4.21b) using the Mach–Zehnder interferometer optical scheme also give a noticeable effect. Therefore, a comparative analysis of the effectiveness of in-plane and out-of-plane displacements to identify defects using the 3D DSI experimental setup can be performed both the correlation DSPI technique and TS DSPI method.

Beams with blind cuts and blind holes were used for comparative analysis of the sensitivity of out-of-plane and in-plane displacements to defects by the correlation

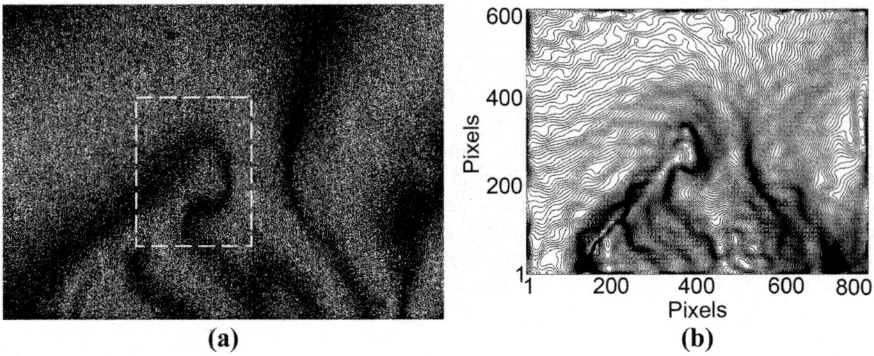

Fig. 4.21 Result of study of out-of-plane surface displacements of steel 45 beam specimen with a blind round hole on the opposite side: subtraction specklegram (**a**); displacement phase map obtained in the form of isolines of the displacements field (**b**)

DSPI technique using the 3D DSI. Figure 4.22 shows the correlation fringes on subtraction specklegrams corresponding to the in-plane displacements of the surface area of the beam with dimensions 400 × 20 × 10 mm^3 consisting of two welded beams made of steel 20 and steel 45 and considered above in Sect. 4.4.5. As show Fig. 4.22a, b, the specklegrams describing the in-plane surface displacements have the correlation fringes that characterize the monolithic defect-free steel beam, i.e. it is problematic to detect such a defect at the in-plane surface displacements along the *x* or *y* axes. However, the shown in Fig. 4.22c, d specklegrams, which describe the out-of-plane surface displacements, make it possible to find the location of the defect.

A similar experiment was performed with a steel 45 beam that has a round blind hole Ø3.5 mm and 6 mm deep. SIs of the surface area from the side of the blind hole were observed and recorded. Figure 4.23a, b shows the correlation fringes on subtraction specklegrams corresponding to the in-plane displacements of the surface area of the beam. As can be seen from these figures, the hole has almost no effect on the correlation fringes. However, the correlation fringes in the specklegram, which

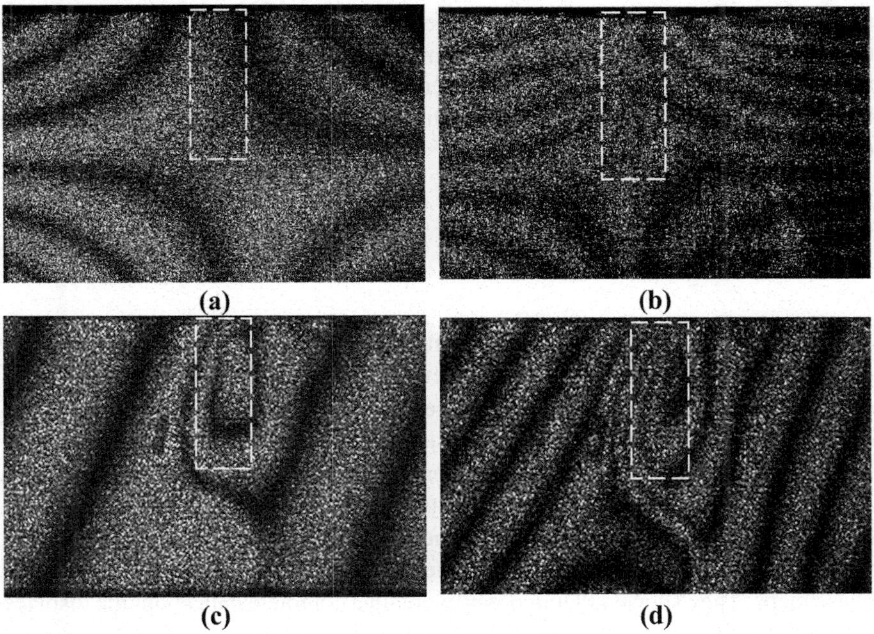

Fig. 4.22 Specklegrams corresponding to surface displacements of the beam specimen consisting of two welded beams made of steel 20 and steel 45 and containing the narrow cut 1 mm thick and deep from 2 mm up to 9 mm along the transverse weld on the side opposite to the side of the studied rough surface: specklegrams corresponding to in-plane surface displacements of the specimen with its loads equal to 100 N (**a**) and 300 N (**b**); specklegrams corresponding to out-of-plane surface displacements of the specimen with its loads equal to 100 N (**c**) and 300 N (**d**)

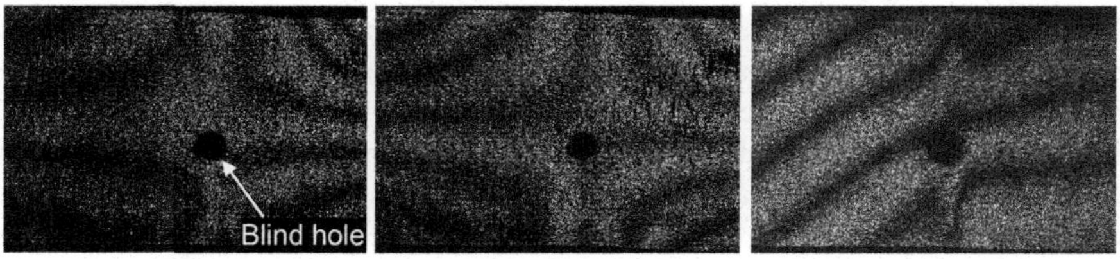

Fig. 4.23 Subtraction specklegrams of the surface area of the beam specimen with the round blind hole Ø3.5 mm and 6 mm deep: specklegrams corresponding to in-plane displacements at loads of 100 N (**a**) and 200 N (**b**); specklegram corresponding to out-of-plane displacements at a load of 200 N (**c**)

describes the out-of-plane displacements of the specimen surface area, possess a sharp curvature around the hole, as it is shown in Fig. 4.23c.

The results of the performed experiments indicated that only the analysis of the out-of-plane surface displacement fields is effective for the detecting subsurface and internal defects in the structural material specimens. Therefore, there is no need to use a 3D DSI, since defects cannot be detected in that part of the interferometer that is sensitive to in-plane displacements. The digital speckle interferometer, built according to the Twyman–Green interferometer scheme, is even more efficient, because its sensitivity reaches the maximum possible value for out-of-plane surface displacements.

4.5.4 Conclusion

The proposed TS DSPI method with arbitrary phase shift of the reference beam is modified to generate 3D surface displacement fields. The developed experimental setup of the 3D DSI makes it possible to generate separate displacement fields along x, y and z directions with the same sensitivity per fringe across the displacement field, which is provided by parallel reference and object beams in all three speckle interferometers. The results of experiments to determine the 3D displacement fields of the structural materials surfaces using the TS DSPI method with an unknown phase shift of the reference beam indicated its effectiveness for generating 3D displacement phase maps. They showed that in order to obtain a complete data about the displacement fields along and across the direction of observation, one can limit oneself to only twelve SIs.

4.6 Application of Digital Speckle Pattern Interferometry in Technical Diagnostics and Nondestructive Testing of Constructive Materials and Structural Elements

DSPI is one of the prominent directions of Optical Metrology widely used for NDT and TD of constructive materials and structural elements. Correlation and phase shifting DSPI techniques made it possible to analyze the spatial fields of surface displacements and deformations, which favorably distinguish them from the well-known methods for determining the surface strains [4]. The results of studies of metal and composite materials and structural elements by TS DPSI and correlation DSPI methods, given below, showed the effectiveness and feasibility of their use and further improvement to solve problems of experimental mechanics and TD. Let us dwell on some results of materials and structural elements TD using the aforementioned methods.

4.6.1 Detection of Surface and Hidden Defects in Metal and Alloy Specimens

In the course of experimental studies, beam and sheet metal and alloy specimens were used. The specimens contained certain defects. Defects were identified by local in homogeneities in the displacement field of the specimen surface under the action of mechanical load, as well as thermal or ultrasonic (US) excitation. Both artificial (blind cylindrical holes, cuts, etc.) and real defects (surface cracks, delamination of protective and restorative coatings, internal defects in welds) were considered. Among the internal (hidden) defects, the artificial and real ones were also distinguished. Artificial hidden defects were created by drilling blind holes or cutting oblique narrow cuts on one side of the specimen surface to register displacement fields on the other. As a result, hidden defects were formed at depths of 1–8 mm from the studied surface. To analyze real subsurface defects, specimens with protective coatings, as well as specimens with weld joints, were taken.

Specimens, some of which are presented in Fig. 4.24, in most cases were made in the form of beams 20 mm wide and 2–10 mm thick or cut steel rings with protective or restorative coatings. The observation surface areas ranged from 15×11.25 mm^2 to 30×20 mm^2. During the experiments, the specimens were loaded using the scheme of four-point bending or heated. Since it was established that out-of-plane displacements are more effective than in-plane ones for detecting defects (see Sect. 4.5.3), a Twyman–Green interferometer was used to achieve the maximum possible sensitivity to out-of-plane displacements. In the interferometer optical scheme, a flat plate with an optically rough surface was used to increase the contrast of fringe patterns. Figure 4.25 shows the Twyman–Green interferometer scheme, containing a flat plate with a rough surface 4 instead of a reference mirror, as well as two shutters 3, which make it possible to obtain separate images of the observation area and the reference rough plate on the matrix photosensor of a digital camera 9. Therefore, it is possible to adjust the optical paths of the reference and object beams with an error that does not exceed the depth of field of the digital camera's lens 9. In addition, it is possible to separately record optical noise resulting from the scattering of the expanded laser beam 1 on the surfaces of the beam splitter 2, for their subsequent compensation. In this case, the optical noise is recorded with simultaneously closed shutters 3.

Detection of hidden cuts and blind holes. Figure 4.26 shows the correlation fringes on the subtraction specklegram of the defect-free steel 20 beam specimen. The fringes are straight, even and of the same width. This indicates the uniformity of the surface displacement. Figures 4.27 and 4.28 show the results of similar studies of the steel beam specimen with dimensions 400 × 20 × 10 mm^3 consisting of two welded beams made of steel 20 and steel 45 with the hidden oblique narrow cut at the site of the weld (see Sect. 4.4.5 and Fig. 4.13). The upper part of these figures

(a) **(b)**

Fig. 4.24 Beam and cut ring specimens: beam specimens with protective coatings and welded beam specimens (**a**); beam and cut ring specimens with restorative coatings (**b**)

Fig. 4.25 Scheme of the Twyman–Green interferometer for studying the out-of-plane surface displacement fields: expanded and collimated laser beam generated by a He–Ne laser (1); beam splitter (2); two shutters (3); flat plate with rough surface (4); PZT (5); controller (6); test specimen (7); lens (8); digital camera (9); computer (10)

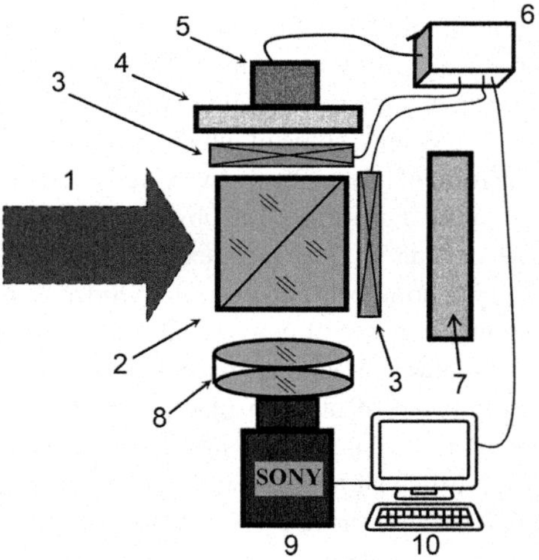

corresponds to a place with a greater cat's depth equal to 9 mm. Correlation fringes, as well as isolines of the displacement field, have the form of parabolas, the vertices of which are located at the oblique cut. In addition, in the upper part to the left of the location of the cat in Figs. 4.27 and 4.28a, there is a local fringe curvature (highlighted by the dashed frame), caused by a defect in the material, which was found only during experiments. The displacement phase map of this surface was also retrieved using the TS DSPI method with the unknown phase shift of the reference beam. At a load of 200 N, the range of phase displacements of the specimen surface in the observation area with dimensions of 15.0 × 11.25 mm^2 is $\delta\varphi_{\max}(z) = 52.53$ rad, which according to Eq. (4.15) at $\Theta_1 = 0$ corresponds to an out-of-plane displacement range $d_{\max}(z) = 2.51$ μm (see Fig. 4.28a). The displacement range along the profile of the oblique narrow cut is $d_p(z) = 1.76$ μm (see Fig. 4.28b).

Similar studies were carried out with a steel 20 beam specimen containing three blind cylindrical holes with a diameter of 3.5 mm and a depth of 8 mm shown in Fig. 4.29a. To increase the reliability of detecting hidden holes from the opposite side with drilled holes, a preload of up to 2000 N was applied to the specimen. The subtraction specklegram of the studied surface area out-of-plane displacements after additional 300 N load is shown in Fig. 4.29b.

During the study of the out-of-plane displacement field of the same surface area by the TS DSPI method, the range of phase displacements in the observation area after the preload 2000 N and additional load 300 N is $\delta\varphi_{\max}(z) = 55.91$ rad, which

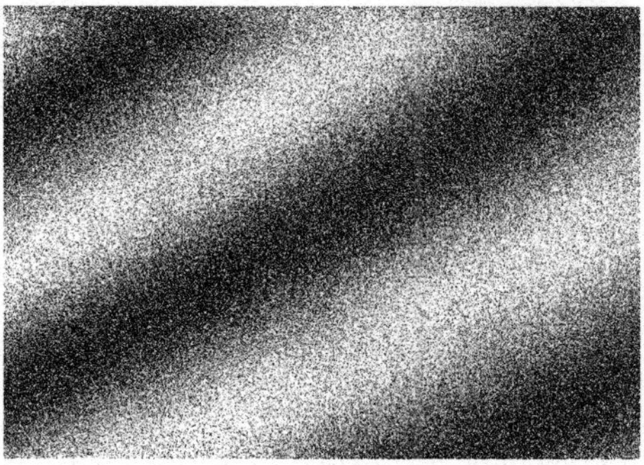

Fig. 4.26 Subtraction specklegram of the defect-free 20 steel beam specimen at load of 100 N containing straight and even fringes of the same width

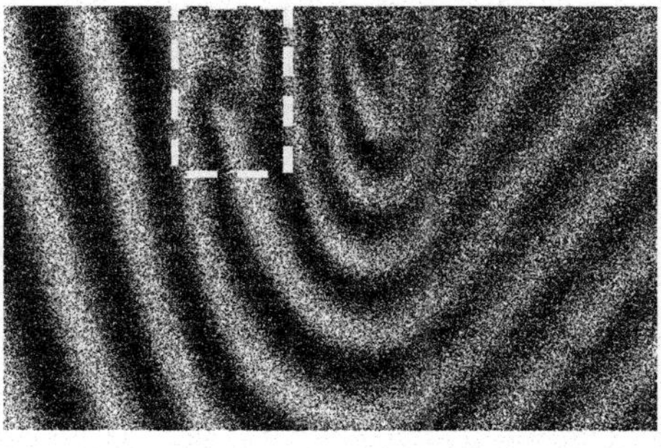

Fig. 4.27 Subtraction specklegram of the surface area of the steel beam specimen at load of 200 N containing the oblique narrow cut on the weld joint place from the specimen opposite side

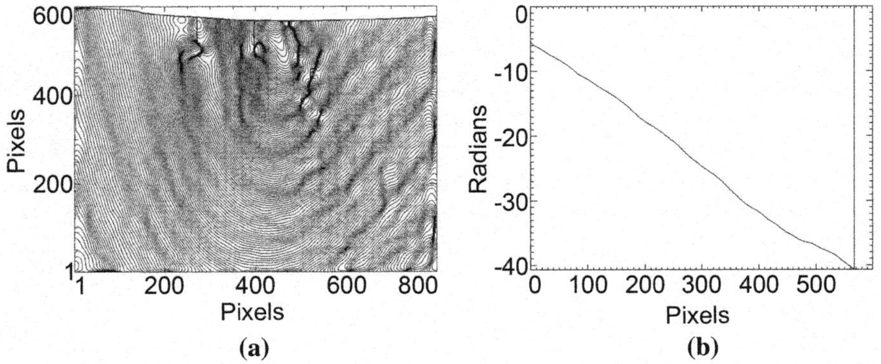

Fig. 4.28 Displacement phase fields of the steel beam specimen containing the hidden oblique narrow cat on the weld joint: displacement phase map of the surface area with isolines through 25 nm (**a**); the profile of the surface displacement phase map along the oblique narrow cut (**b**)

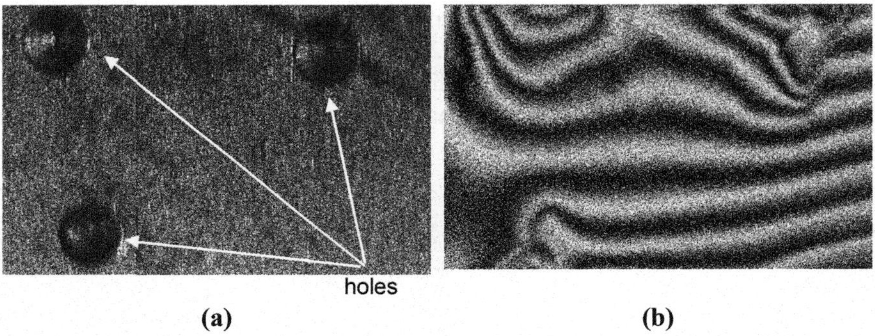

Fig. 4.29 Results of the study of the steel 20 beam specimen with three hidden holes having a diameter of 3.5 mm and a depth of 8 mm using the correlation DSPI: speckle pattern of the specimen surface area from the side of drilled holes, i.e. on the opposite side to the plane of observation (**a**); subtraction specklegram of the studied surface area (**b**)

corresponds $d_{\max}(z) = 2.67$ μm. As can be seen from Fig. 4.30, the location of the hidden holes can be determined quite accurately.

According to the performed experiments results, it was found that in most cases it is difficult to detect defects by correlation or phase shifting DSPI techniques at low specimen loads. Therefore, it is first necessary to apply a preload to the test specimen, and then observe its surface, gradually increasing the load. The level of preload depends both on the material of the specimen, the nature of the defect (size, shape, volume, voidness, etc.), the occurrence depth, and on the load circuit. Experiments have shown that preloads of the order of 100–3000 N and additional loads of the order of 100–200 N can be considered optimal for detecting artificial defects.

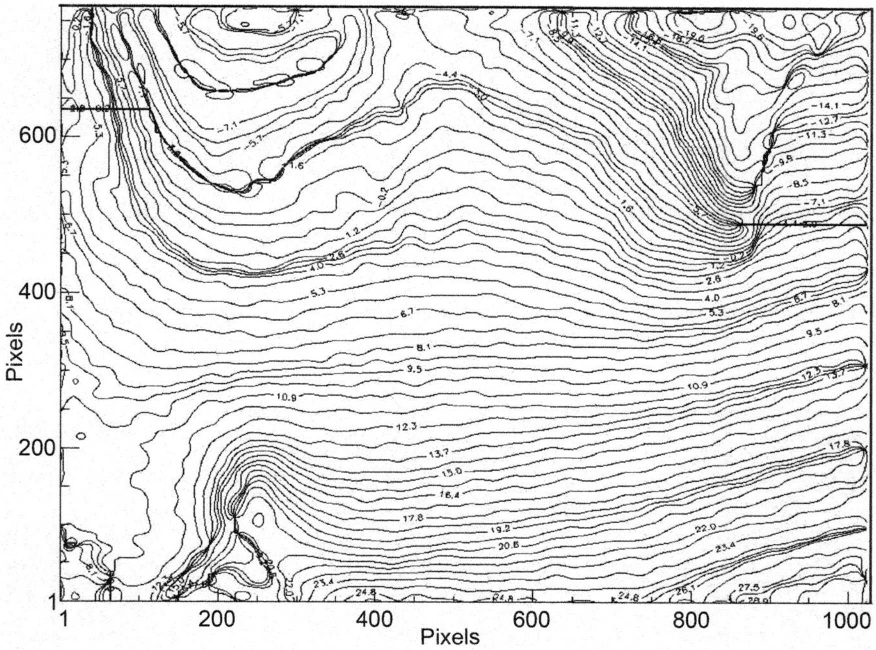

Fig. 4.30 Isolines of the displacement field of the steel 20 beam specimen surface area with three hidden holes drilled from the side opposite to the observation plane. Out-of-plane surface displacement phase map is divided into 76 levels at 35.25 nm intervals between adjacent isolines

Detection of surface and hidden defects by correlation fringes. Below are the results of detecting the operational and technological surface and hidden defects using correlation fringes obtained after applying the bending loads to the studied specimens.

During the experiments, a crack on the surface of an aluminum beam with a semicircular notch appeared at a load of less than 100 N (see Fig. 4.31). A break in the correlation fringes is recorded at the location of the crack, and a significant curvature is observed at some distance from its top. The length of the crack is about 5 mm, and the distance at which the curvature of the fringes is observed is about 7–10 mm.

The results of the study of another weld joint in a steel 20 beam are shown in Fig. 4.32. Without a preload, a hidden defect does not appear on the edge of the beam (see Fig. 4.32a), with a preload of 500 N it begins to appear (see Fig. 4.32b), and with a preload of 1500 N it clearly appears and its location can be determined (see Fig. 4.32c–d). The place of the weld joint is marked with dashed rectangles.

Figure 4.33 shows the correlation fringes of out-of-plane surface deformations characterizing the local welding zones of beam specimens made of different steel grades (steel 45 and steel 20), namely, in the zones of high-quality welding (Fig. 4.33a) and welding with a defect (Fig. 4.33b) [99]. The defect can be detected by fringe breaks. Fringe breaks, as well as fringes with a higher local density than

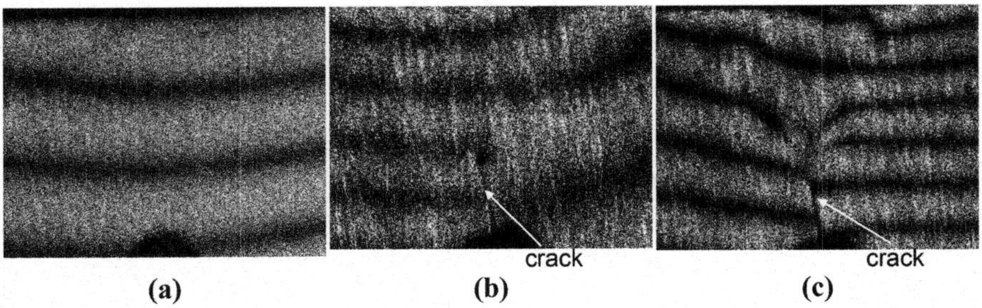

Fig. 4.31 Patterns of correlation fringes for aluminum beam with a semicircular notch: specimen without crack, load 50 N (**a**); specimen with a crack, load 50 N (**b**); specimen with a crack, load 100 N (**c**)

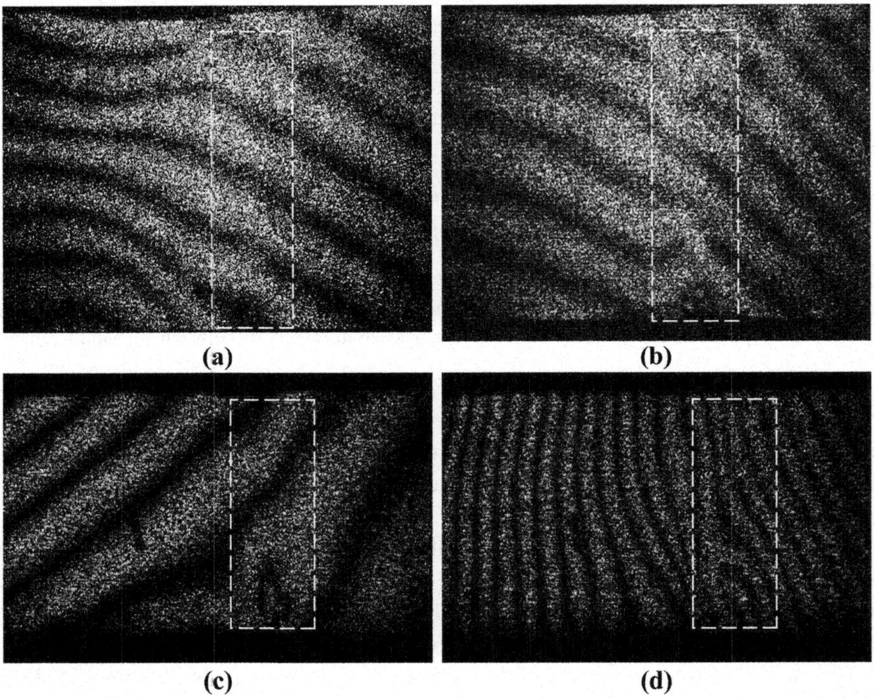

Fig. 4.32 Patterns of correlation fringes for the steel 20 beam welded specimen with a defect at the edge of the beam (defect location area highlighted): without preload, additional load 200 N (**a**); preload 500 N, additional load 200 N (**b**); preload 1500 N, additional load 100 N (**c**); preload 1500 N, additional load 200 N (**d**)

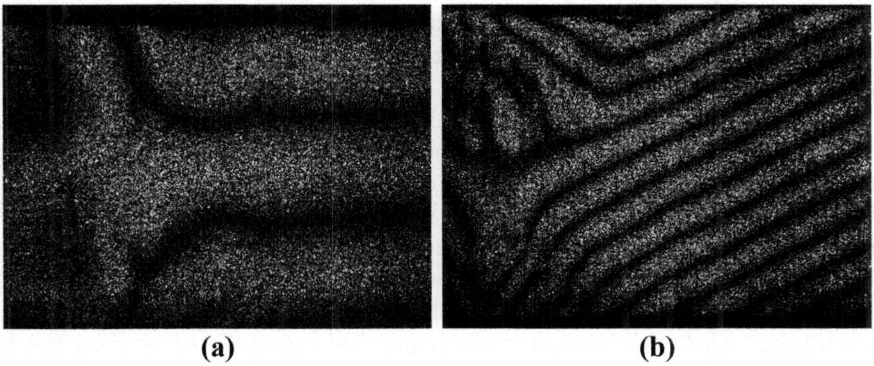

Fig. 4.33 Patterns of correlation fringes in the zones of high-quality welding (a) and welding with a defect (b)

the main ones, indicate sharp local changes in the surface geometry during loading of the specimens.

Examination of protective and restorative coatings. Similarly, it is possible to investigate the quality of protective and restorative coatings using the correlation DSPI, since the place of the coating delamination considered as a hidden defect [99, 100]. If it is problematic to load the specimens mechanically, thermal loading can be used. Unfortunately, it is difficult to apply the temporal PS DSPI technique in this case.

Studies were carried out with electrochemically applied protective coatings based on Zn, as well as with restorative coatings in the form of a composite material that is a mixture of 5% Cr, 2.5% B, 8% Al, 0.2% Si, 0.03% C and Fe, applied by cold spraying. Protective coatings were applied to the beam specimens, and during the research they were also loaded according to the four-point scheme. Since the greatest attention was paid to the delamination of coatings near stress concentrators, it was necessary to first study the displacement fields near these concentrators in the absence of delaminations. At low loads caused the out-of-plane surface displacements, the correlation fringe patterns (i.e. subtraction specklegrams) of surface areas of specimens made from different materials and containing stress concentrators, are approximately the same and differ insignificantly from the patterns corresponding to deformations of specimens without stress concentrators. Figure 4.34 shows the specklegrams corresponding to the out-of-plane surface displacements of a beam with a semicircular cutout without a coating preloaded to 500 N. Under such loads, a plastic zone begins to form in the vicinity of the notch (concentrator). The correlation fringe patterns for out-of-plane deformations for concentrators of various shapes have a characteristic form. Figures 4.34 and 4.35 show the characteristic subtraction specklegrams of surface areas of beam specimens containing stress concentrators of various shapes. They appear only under certain loads, the values of which depend on the material of the specimen and its geometry. Note that at low loads, the correlation fringe patterns at the out-of-plane deformations of beam specimens made of different materials and with different stress concentrators have the same character and differ slightly from the patterns corresponding to deformations of specimens without any stress concentrators.

In all specklegrams shown in Figs. 4.34 and 4.35, the correlation fringes in the vicinity of stress concentrators have a smooth curvature. Similar fringe patterns were observed for specimens with high-quality coatings, which are not delaminated under loading. For example, Fig. 4.36a shows a specklegram of the surface area of a high-quality protective zinc coating on a steel beam. When delamination, the fringes either break or sharply increase their curvature and density, and also change shape.

Figure 4.36b–c show examples of delaminations of protective zinc coatings from steel beams, where delamination zones are marked with dashed lines. Figure 4.37 shows the correlation fringes on the surface of a steel ring with a restorative coating, recorded at two different moments of the ring heating. The upper part of the ring has a high-quality coating. The coating at the bottom of the ring delaminates when heated.

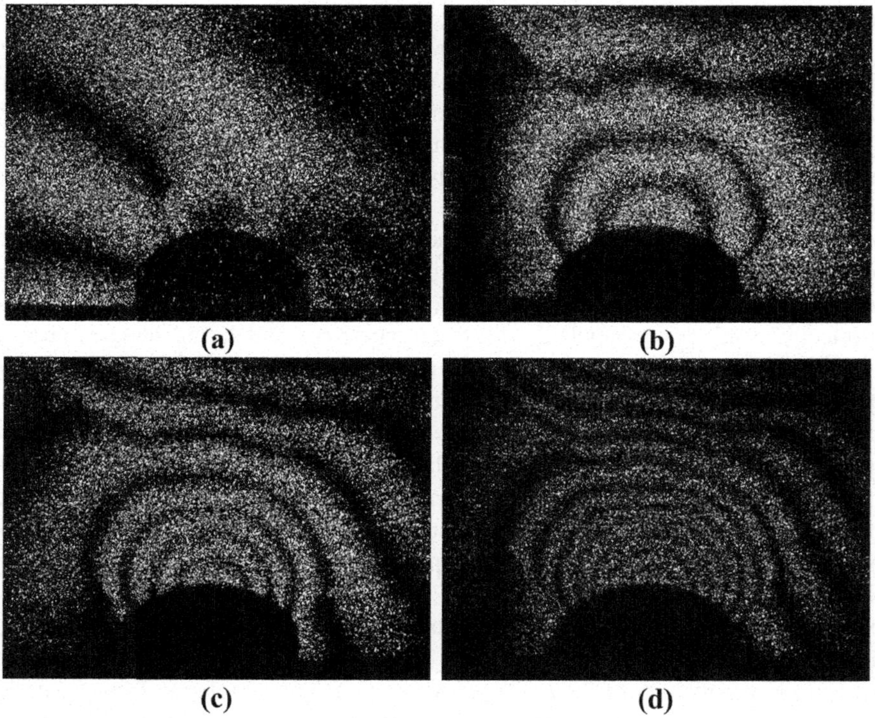

Fig. 4.34 Correlation fringe patterns on the surface of a steel beam specimen with a semicircular notch, preloaded up to 500 N, with additional loads: 100 N (**a**); 200 N (**b**); 300 N (**c**); 400 N (**d**)

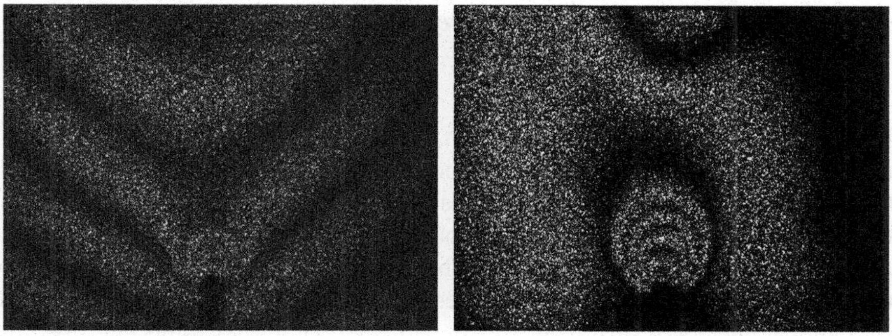

(a) (b)

Fig. 4.35 Patterns of correlation fringes on the surface of a duralumin beam with a stress concentrator in the form of a rounded notch

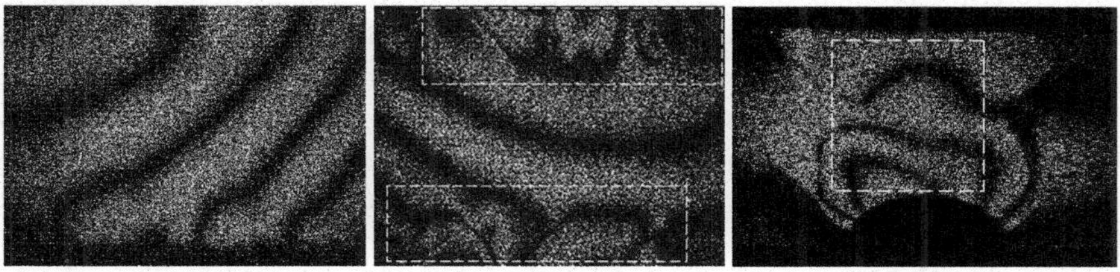

(a) (b) (c)

Fig. 4.36 Correlation fringes on surface areas with high-quality (**a**) and low-quality (**b, c**) protective zinc coatings

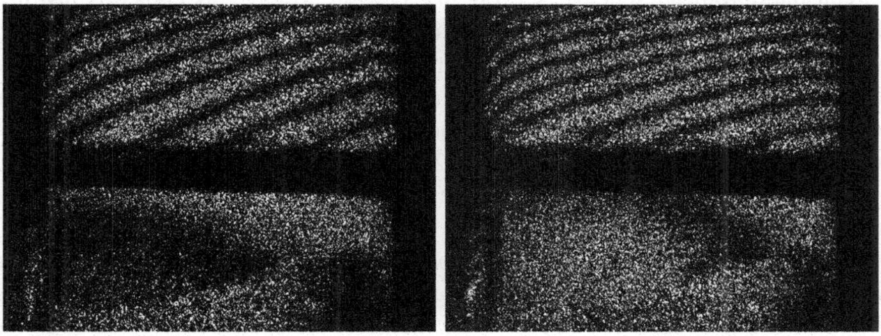

(a) (b)

Fig. 4.37 The subtraction specklegrams of the cut steel ring with the applied restorative coating (coating material is a mixture of Fe, 5% Cr, 2.5% B, 8% Al, 0.2% Si, 0.03% C; heat load) for two different heating moments (**a**) and (**b**): the upper and lower parts of the ring are high-quality and low-quality coatings, respectively

4.6.2 *Study of Regularities of Deformation and Damage of Composites*

New technologies that are used in modern industrial production require the development of high-performance experimental tools for TD and NDT of structural materials. Composite and metal-composite materials and structural elements play an important role in mechanical engineering and, in particular, in the aerospace industry. Their advantages are low specific gravity combined with high strength and rigidity, low thermal conductivity, resistance to aggressive environments, etc. Determining the strength parameters of such materials, in particular, laminated composites, is difficult due to the significant anisotropy associated with the inhomogeneity of the structure. Modern technologies for production of composites do not ensure their uniformity and defect-freeness. Defects in composites also appear during their operation. Therefore, the problem of detection and analysis of technological and operational defects to prevent the structure damage or predict residual life of composite structures using NDT methods is extremely relevant. The development of modern methods for retrieval of surface displacement and deformation fields make it possible to raise the issue of diagnostics of composite and metal-composite materials at a qualitatively new level. DSPI is one of the promising trends of Speckle Metrology capable to solve problems of this kind [16, 43, 70, 105].

Experimental setup and research methodology for composites. To study the processes of deformation and damage of structural composite materials specimens, an experimental setup was created that combines a digital speckle interferometer and a tensile testing machine. To construct a speckle interferometer, the Twyman–Green interferometer scheme (see Fig. 4.25) was chosen. This choice is justified by the results of preliminary experimental studies of beam and sheet specimens of structural materials. The fulfilled experiments have shown that with out-of-plane displacements it is better than with in-plane ones to detect and identify defects in deformed specimens, locate defects and their approximate sizes, as well as to monitor their occurrence and growth [99, 100, 102].

In a tensile testing machine, the specimens were loaded with tension, and the machine itself was attached to the optical table, on which the speckle interferometer was mounted. The schematic diagram of loading a specimen in tension in the direction of the y-axis by force F is shown in Fig. 4.38a. The diagram of the tensile testing machine is shown in Fig. 4.38b, and the image of the specimen 6 clamped in the machine with the help of two holders 5 and 7 is shown in Fig. 4.38c.

During the experiments, beam specimens of laminated CFRP (up to 12 layers) and CFRP with randomly distributed fibers were used (see Fig. 4.39). Observation sites with dimensions of 30×40 mm^2 were chosen in the vicinity of the stress concentrators in the form of round holes with a diameter of Ø3 mm in the middle of the laminated specimens and Ø5 mm in the middle of the specimens with randomly distributed fibers. A thin layer of finely dispersed aluminum paint was applied to the observation sites, which made it possible to form an optically rough surface,

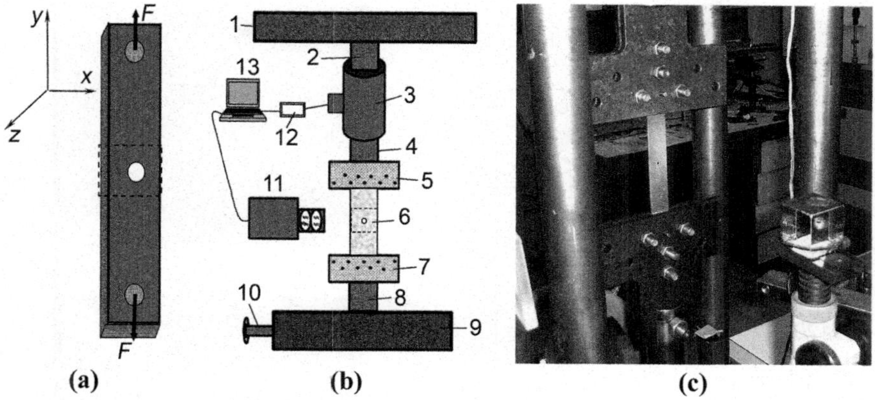

Fig. 4.38 Schemes of loading the specimen by tension (**a**) and the tensile testing machine (**b**); a general view of the tensile testing machine central part with a fixed composite specimen (**c**): fixed plate (1); tractions (2, 4, 8); dynamometer (3); holders (5, 7); specimen (6); gearbox with screw mechanism (9); hand drive flywheel (10); digital camera (11); indicator for load control (12); computer (13)

ensure uniform light scattering over the entire studied surface area, and obtain the subtraction specklegrams with contrasting correlation fringes.

Research results and discussion. Based on the results of a series of experiments with CFRP beam specimens, patterns of correlation fringes (subtraction specklegrams) were constructed for out-of-plane surface displacements. When mechanical loads were applied to the specimens, stress–strain diagrams were plotted simultaneously. One of such diagram for the laminated CFRP beam specimen, supplemented by the corresponding fringe patterns obtained at its marked points, is shown in Fig. 4.40, where point 1 corresponds to a preload $F_1 = 120$ kg, point 2 to a preload $F_2 = 500$ kg, point 3 to a preload $F_3 = 1000$ kg, point 4 to a preload $F_4 = 1200$ kg and point 5 to a preload $F_5 = 1300$ kg. Each of the subtraction specklegrams was obtained in the following sequence:

- applied a given preload;
- registered the first SI at this load;
- applied an additional load within 60 kg;

Fig. 4.39 Beam specimens of laminated CFRP (up to 12 layers) and CFRP with randomly distributed fibers

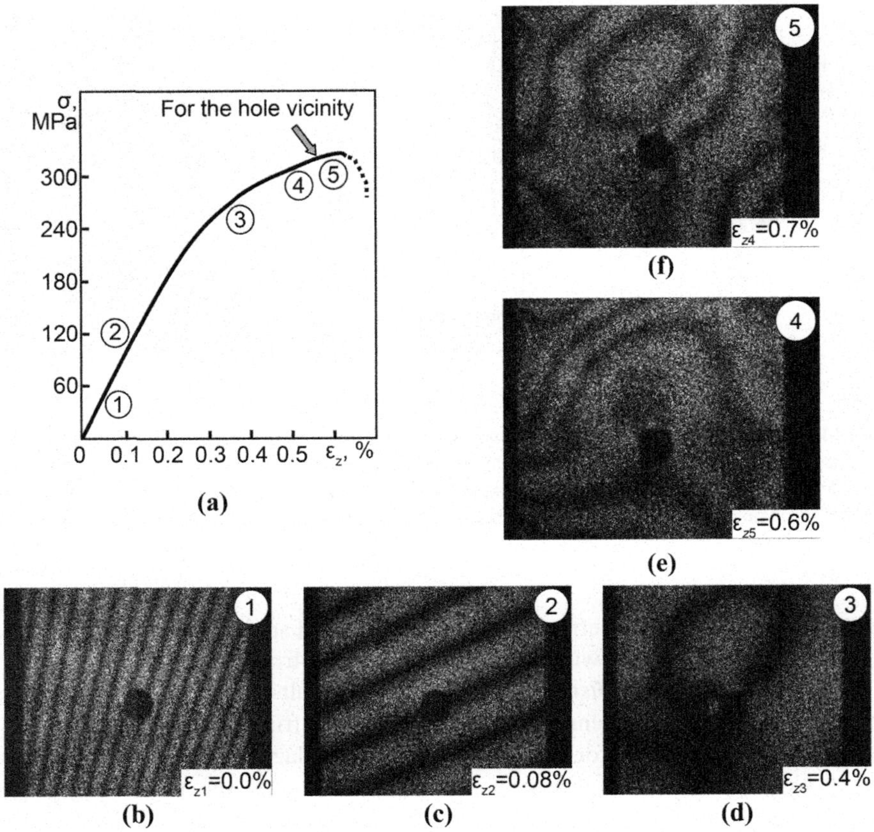

Fig. 4.40 Tensile diagram of the laminated CFRP beam specimen (**a**) and the correlation fringe pattern of the surface near the round hole, corresponding to certain values of the previous loads: $F_1 = 120$ kg (**b**), $F_2 = 500$ kg (**c**), $F_3 = 1000$ kg (**d**), $F_4 = 1200$ kg (**e**), $F_5 = 1300$ kg (**f**)

- registered the second SI;
- subtracted from the second SI the first one.

The obtained subtraction specklegrams were analyzed by linking them to the corresponding points of the tensile diagram and taking into account the fact that each fringe in the specklegram corresponds to the locus of points (pixels) of the same out-of-plane displacement, and the difference in surface displacements between adjacent fringes is $\lambda/2$. If a He–Ne laser is used in the speckle interferometer containing the Twyman–Green interferometer optical scheme (see Fig. 4.25), then the difference $\lambda/2 = 316.4$ nm.

As shown in Fig. 4.40b–f, the shape of the correlation fringes changes from straight to elliptical in the zone of subsurface defect localization with increasing preload. For example, if at $F_1 = 120$ kg and $F_2 = 500$ kg (see Fig. 4.40b, c) the fringes still retain a rectilinear shape, then at $F_3 = 1000$ kg, $F_4 = 1200$ kg and $F_5 = 1300$ kg

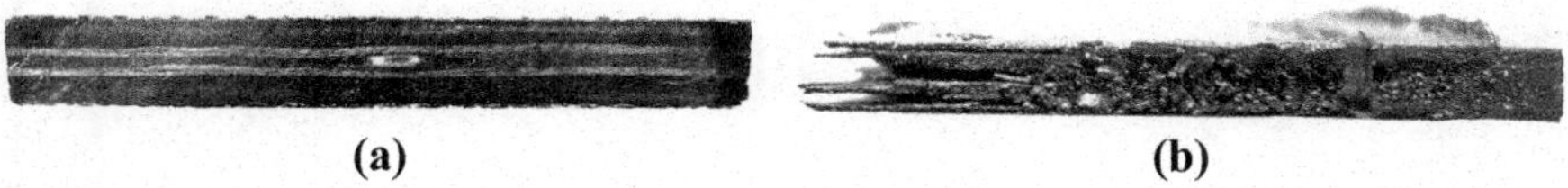

Fig. 4.41 Cross-sections of the laminated CFRP beam specimen: at the place of delamination (a), at the place of destruction (b)

(see Figs. 4.40d–f), i.e. with an increase in preliminary tensile loads, the fringes bend and close, concentrating around the round hole. Ring-shaped closed fringes indicate the appearance of convexity or concavity on the specimen surface, caused either by an internal technological or operational defect, or delamination of composite material during its exploitation. This assumption was confirmed after the specimen was cut before the complete destruction at the site of change in surface geometry and fringe density. Figure 4.41 shows cross-sectional images of the place of the specimen delamination (light elliptical spot) and the place of its destruction. Therefore, according to the results of experimental studies, the destruction of laminated CFRP beam specimens is preceded by their delamination, which can be established using correlation DSPI.

Beam specimen made of a CFRP with randomly distributed fibers was loaded with tension until its destruction. Figure 4.42 shows the correlation fringe patterns (subtraction specklegrams) for various loading forces corresponding to the points noted by the same numbers on the stress–strain diagram shown in Fig. 4.43. The incremental load for this specimen did not exceed 300 N. When the specimen was loaded up to 6000 N, the correlation fringes were linear and uniform (see Fig. 4.42a and point 1 in Fig. 4.43) and there was no effect of the stress concentrator. At higher loads, the correlation fringes were distorted in the vicinity of the stress concentrator, i.e. the round hole. This distortion increased with increasing preload, and the highest density of fringes corresponding to the specimen lateral deformation was observed in the vicinity of the stress concentrator (see Fig. 4.42b and point 2 in Fig. 4.43, as well as Fig. 4.42c and point 3 in Fig. 4.43). At a load of 13,600 N, the specimen was destructed (see Fig. 4.42d and point 4 in Fig. 4.43).

In this case, as well as for the laminated composite, the local increase in the density of the correlation fringes indicates the presence of a defect. For this specimen, the local fringe density at the defect site was 2 times or more high than the fringe density of the defect-free region. At the moment when the local density of the correlation fringes increased by more than 5 times (point 4 in Fig. 4.43), the specimen destruction begins in this site (see Fig. 4.42d). Therefore, the quantitative indicator of the normalized local density of these fringes may indicate the moment of the defect formation and the composite destruction. This indicator can be characterized as the coefficient of relative local density of fringes [56, 104]

$$k_f = \frac{n_{\text{pl}}}{n_{\text{el}}}, \tag{4.52}$$

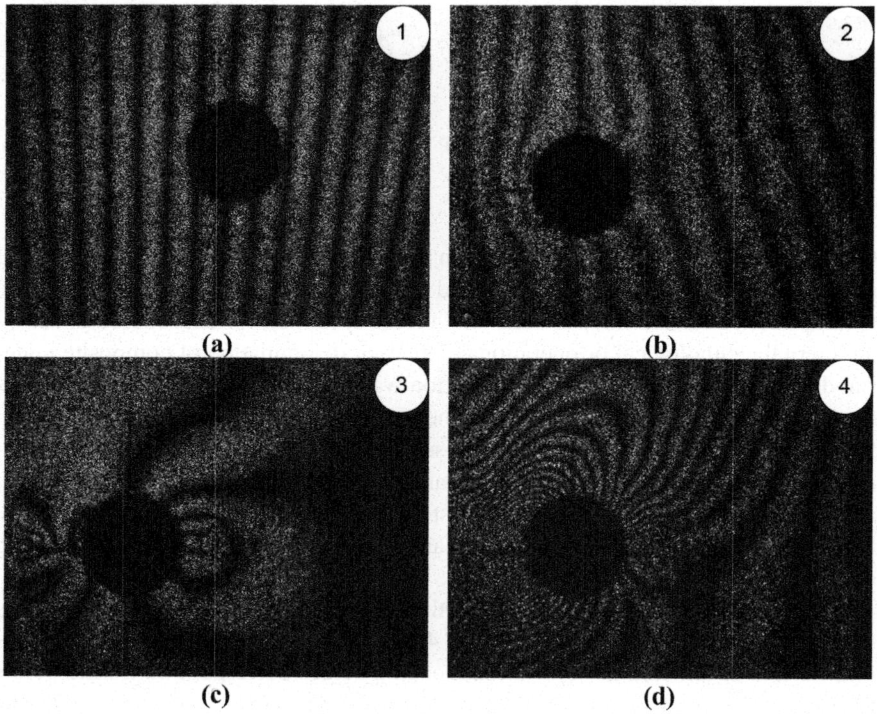

Fig. 4.42 Subtraction specklegrams at preloads of the beam specimen made of a CFRP with randomly distributed fibers: 6000 N (**a**), 10,000 N (**b**), 13,000 N (**c**), 13,600 N (**d**)

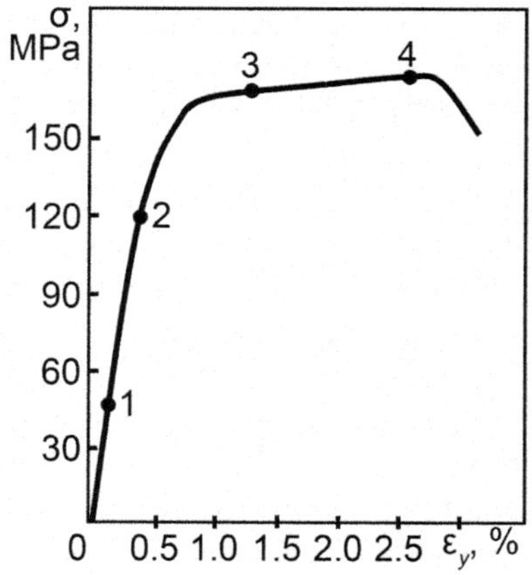

Fig. 4.43 Stress–strain diagram of the beam specimen made of a CFRP with randomly distributed fibers

where n_{pl} is the fringe density in the zone of plasticity, n_{el} is the fringe density of elastically deformed material.

Figure 4.44 shows the stress–strain diagram in the form of the dependence of k_f on the strain ε_y along the beam specimen made of a CFRP with randomly distributed fibers. In the elastic zone of material strain, where the strain ε_y does not exceed 0.2%, the relative density of the fringes does not change ($k_f = 1$), as evidenced by the horizontal line in the diagram (points 1, 1′, 1″ and 1‴). During the elastic plastic deformation of the material, when the strain ε_y is greater than 0.2%, a change in the k_f versus ε_y dependence is observed from a horizontal to an inclined line. At the beginning of the nonlinear section of the diagram (point 2), the strain magnitude in the vicinity of the stress concentrator increases 2 times compared to the elastically deformed section ($k_f = 2$). The directly proportional dependence k_f versus ε_y is observed over the entire period of elastic plastic deformation of the material (points 3, 3′, 3″ and 3‴) until it reaches the limit of its strength (points 4 and 4′). At points 4 and 4′ an internal defect is formed inside the specimen due to the triaxial stress state, leading to localized maximum normal stresses. In the elastic plastic region, the coefficient k_f increases 5–6 times, and before the destruction (points 4, 4′) more than an order of magnitude.

Considering the results of the research, it can be argued that the process of the occurrence of an internal defect in a composite material is associated with a change in the geometry of its surface. Therefore, under an external load, it is possible to establish the location of the occurrence of an internal defect with high accuracy by monitoring the change in the surface of the composite material. The results of experimental studies confirm the effectiveness of correlation DSPI for studying changes in the structure of surface displacement fields of composite specimens during their deformation and destruction. The proposed assessment to analyze the structure of surface displacement fields and its changes under the action of mechanical loads on the basis of the coefficient of relative fringe density makes it possible to identify both technological defects in composite materials and operational ones, as well as

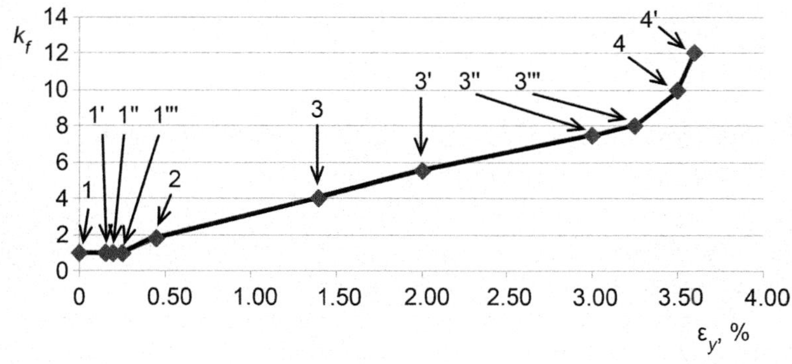

Fig. 4.44 Dependence k_f versus ε_y: points 1–4 correspond to points with the same numbers in Fig. 4.43

to determine the most probable damage places of composite specimens with stress concentrators.

4.7 Examination of Subsurface Defect in Composite Structures Using Subtractive Synchronized Digital Speckle Pattern Interferometry

Recently, the development of new techniques for the detection and localization of internal delaminations and other subsurface defects in laminated composite structures for purposes airspace, instrument making, automotive and building industries is one of the important problems of NDT and TD. The presence of such defects in composites as delaminations, cracks, disbonds, voids, inclusions, matrix cracking, etc., directly affects their strength and rigidity and contributes to their damage and destruction. Therefore, the development of effective methods and means for detecting and localizing subsurface defects in composite structures is an important and urgent problem [47]. Among the various methods, systems and instruments for detecting subsurface defects, the hybrid optical-acoustic techniques are widespread. They include both scanning and whole field approaches. Resonance spectroscopy technique that uses a laser vibrometer scan of each point of a studied specimen surface area is one of most a promising representative of the scanning approach [92–94]. On the other hand, the whole field approach is based on Speckle Metrology methods, which make it possible to perform the simultaneous control of large areas of the studied composite specimen or structural element. Additive–subtractive ESPI/shearography [22, 23, 78, 106] and synchronized reference updating ESPI [77] have common operational principles and use periodic US excitation of a studied composite specimen with variable frequencies from tens to hundreds of kilohertz. The lockin-interferometry technique that is also based on DSPI uses periodic thermal excitation of composite structures to detect the intrinsic defects [63]. It is slower than the techniques applying the US excitation, however the depth of defects detection in this case is larger.

4.7.1 Subtractive Synchronized Digital Speckle Pattern Interferometry Method

The subtractive synchronized DSPI method [56, 70, 71, 74] is similar to the mentioned above techniques, however it uses the sequence of N subtraction specklegrams $i_n(k, l)$ to increase the intensity of the output response from the searched subsurface defect and thereby raise the signal-to-noise ratio. The acoustic excitation of a studied composite panel is performed by a US transducer working in the frequency scanning mode in the range from 10 to 150 kHz. The searched subsurface defect can oscillate at its resonance frequencies, and if the frequency $f_{US} = 1/T_{US}$

of the acoustic wave is coincided with one of the resonance frequencies of the defect, it becomes oscillate. In this case, the surface area located directly above the defect, i.e. the region of interest (ROI), is induced by this defect and becomes oscillate with the same frequency. Extremes of the acoustic wave oscillations coincide with the maximum deviations of the ROI from the plane of the panel in both opposite directions. Surface oscillations at this frequency are very small or absent on the rest surface outside the ROI.

To obtain nth specklegram upon the composite panel US harmonic excitation, first the nth SI $i_{n,o}(k, l)$ is recorded at the maximum out-of-plane displacements of the ROI in certain direction. Then the nth SI $i_{n,e}(k, l)$ is recorded at the maximum out-of-plane displacements of the ROI in the opposite direction are generated ($n = 1, 2, ..., N$). Two nth SIs are accumulated by M elementary SIs $i_{m,n+}(k, l)$ recorded at the time gap τ at the maximum out-of-plane M displacements of the ROI in one direction during the frame time T, and by M elementary SIs $i_{m,n-}(k, l)$ recorded at the same time gap τ at the maximum out-of-plane M displacements of the ROI in the opposite direction during the same frame time T, that is

$$i_{n,o}(k, l) = \sum_{m=1}^{M} i_{m,+}(k, l), \; i_{n,e}(k, l) = \sum_{m=1}^{M} i_{m,-}(k, l), \tag{4.53}$$

where $i_{m,n+}(k, l)$ is one of the elementary SIs, which sequence generate the SI $i_{n,o}(k, l)$; $i_{m,n-}(k, l)$ is one of the elementary SIs, which sequence generate the SI $i_{n,e}(k, l)$; $m = 1, 2, ..., M$.

The registration of elementary SIs during time gap τ is synchronized respectively with maximum and minimum amplitudes of the US harmonic wave. As a rule, the time gap τ is very short and it can't be provided by the digital camera shutter. Therefore, a pulsed laser or an optical shutter such as an acousto-optical deflector or electro-optical modulator should be used to generate such short time gaps.

The accumulation of two sequences of SIs, namely, the odd sequence containing N SIs $i_{n,o}(k, l)$ and the even sequence containing N SIs $i_{n,e}(k, l)$ produce N pairs of SIs. The nth pattern of correlation fringes, i.e. the subtraction specklegram $i_n(k, l)$, is obtained by digital pixel-by-pixel subtraction modulo the nth even SI $i_{n,e}(k, l)$ from the nth odd SI $i_{n,o}(k, l)$ and is given by

$$i_n(k, l) = \left| i_{n,e}(k, l) - i_{n,o}(k, l) \right|. \tag{4.54}$$

The procedure of odd SI $i_{n,o}(k, l)$ and even SI $i_{n,e}(k, l)$ recording is shown in Fig. 4.45. Time gaps τ may have time delays δt concerning the maximums or minimums of the harmonic US wave with the time period T_{US}. Due to this operation, it is possible to adjust the contrast of interference fringes in obtained subtraction specklegrams if the excitation of the ROI by a defect occurs with some delay.

The resulted subtraction specklegram $i_n(k, l)$ allows to visualize the subsurface defect placed directly under the ROI. Frequently the contrast of the highlighted ROI

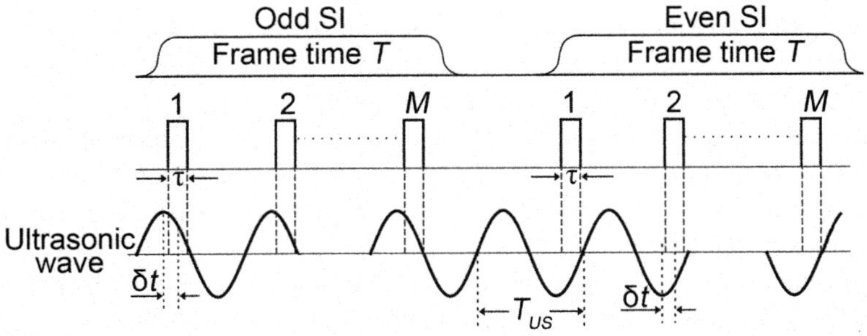

Fig. 4.45 Timing sequence of nth pair of SIs $i_{n,o}(k, l)$ and $i_{n,e}(k, l)$, produced by the accumulation of M elementary SIs $i_{m,+}(k, l)$ and $i_{m,-}(k, l)$ that are recorded with equal light exposure

is low. Therefore, the sequence of specklegrams $i_n(k, l)$ is summed and the resultant fringe pattern possessing higher contrast is obtained, i.e.

$$i_{\Sigma}(k, l) = \sum_{n=1}^{N} i_n(k, l). \tag{4.55}$$

4.7.2 Experimental Setup of Optical-Digital Speckle Interferometer

Functional diagram of the optical-digital speckle interferometer (ODSI) experimental setup performed the subtractive synchronized DSPI method is shown in Fig. 4.46. The ODSI contains an optoelectronic module (optical head) constructed according to the Twyman–Green interferometer scheme. The optical head generates SIs of the test specimen surface area with a size of 24 × 24 mm^2, which are recorded by the digital camera. Recorded digital SIs are stored in a computer. The optical head is attached to the movement and positioning device (MPD). If the ODSI setup is placed on the investigated structural element, the MPD moves the optical head in two mutually perpendicular directions. The area analyzed during the movement of the optical head using MPD without additional movement of the ODSI setup is 120 × 120 mm^2. The correlation fringe patterns are produced with speed equal about 2 frames/s. If to choose visual field of the ODS optical head that acquires an image of the studied specimen surface area equal to 24 × 24 mm^2 and to take into account the time necessary to displace the optical head into another position, the area equal to 24 × 120 mm^2 can be analyzed approximately for 30 s. The optomechanical part of the ODSI setup can be fixed on the studied surface inclined up to 45° to the horizontal plane. The ODSI setup also contains a unit of US excitation, synchronized with the start of SIs registration. The scheme of electronic modules and units of the ODSI setup is shown in Fig. 4.47. The ODSI setup also contains an electronic unit for

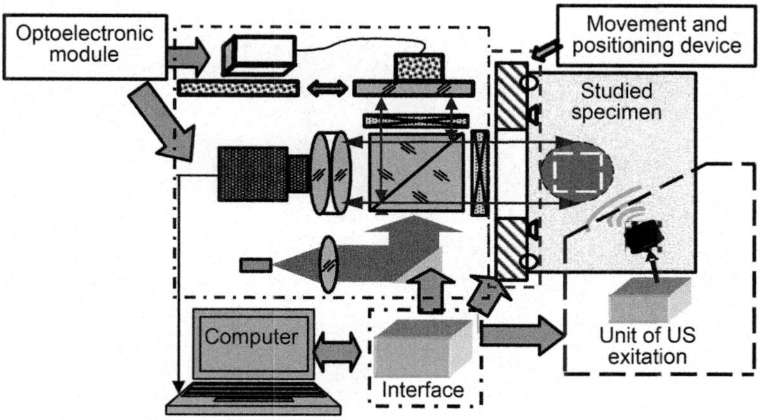

Fig. 4.46 Functional diagram of the optical-digital speckle interferometer experimental setup

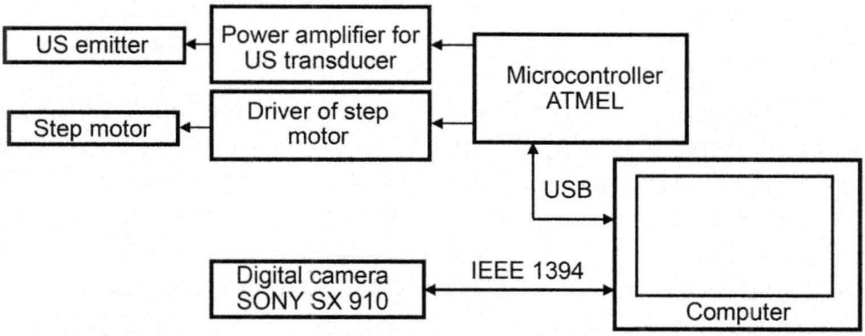

Fig. 4.47 Scheme of the ODSI electronic modules and units

generating US harmonic radiation in the range of 10–150 kHz. This unit provides continuous and pulsed US generation modes. Pulse generation is performed with a change of duration of 10–15 μs and a time interval between them of 0.5–1 s.

The appearance of the ODSI optical-mechanical part containing the optoelectronic module and the movement and positioning device is shown in Fig. 4.48.

4.7.3 Results of Experimental Research

Researches for detection and localization of subsurface defects in laminated composite structures, in particular, delaminations and debonds, were fulfilled with the help of the ODSI experimental setup using the developed subtractive synchronized DSPI method. The obtained experimental results demonstrated the ability of the ODSI to detect and localize subsurface delaminations and debonds. An example of correlation fringe patterns obtained under the action of US excitation to detect

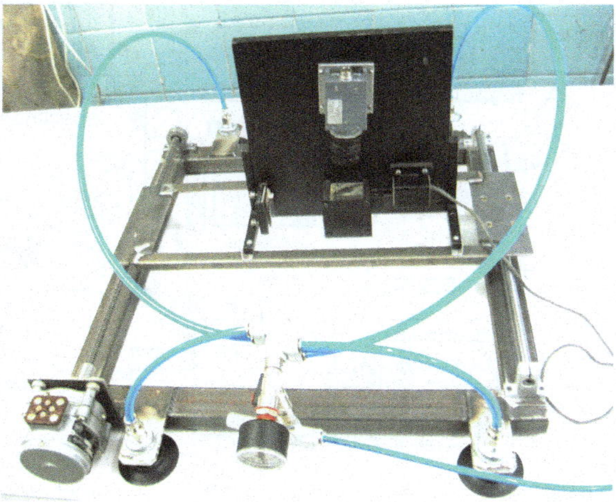

Fig. 4.48 Optical-mechanical part of the ODSI setup containing the optoelectronic module and the movement and positioning device

internal delaminations near mechanical damage in a 3 mm thick CFRP laminate specimen is shown in Fig. 4.49. The studied specimen was excited by US waves in the range of 27–60 kHz. It can be seen that the delaminations are visualized as light areas or light spots surrounded by a dark background. The squares of these areas and spots are approximately equal to the squares of the corresponding defects. As shown in Fig. 4.49, along the cross-section of the studied laminated composite specimen, cut after the experiments, the width of the delaminations recorded by this method varies in the range from several tens of microns to 1 mm.

An example of delaminations and defects detection caused by bearing strain in a fiberglass reinforced laminated composite is shown in Fig. 4.50. The places of delaminations in the composite were detected at the selected resonant frequencies of US excitation indicating different sizes of defects. The resulting light spots on a dark background characterize the square of the defect, and for most defects, surface nanodisplacements within the ROIs located directly above the defects correspond to one correlation fringe, i.e. about $\lambda/2$, where λ is the laser radiation wavelength.

Another example of the delaminations detection in the repaired laminated composite structural element of the AN aircraft wing flap is shown in Fig. 4.51. In this figure, the tested surface areas of the specimen are circled by white and grey solid closed curves. Results of delaminations detection are illustrated in Fig. 4.52a–d. These Figures indicate sharp bright correlation fringes within the ROIs located directly above the delaminations between 1st and 2nd layers on the specimen surface area 1 circled by a white solid closed curve (see Fig. 4.52a), 2nd and 3rd layers on the same specimen surface area (see Fig. 4.52b), 3rd and 4th layers on the specimen surface area 2 circled by a gray solid closed curve (see Fig. 4.52c), 4th and 5th layers on the specimen surface area 2 (see Fig. 4.52d).

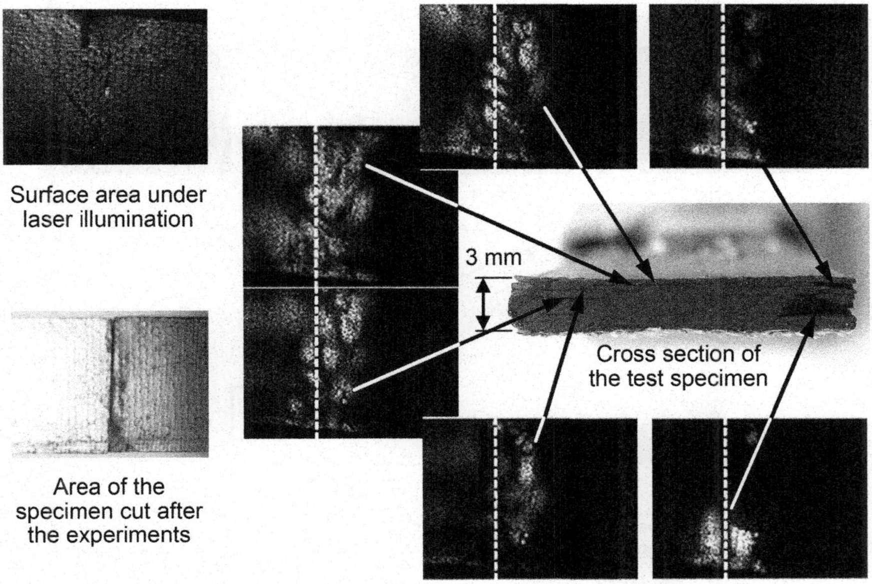

Fig. 4.49 Experimental results of detected subsurface defects in a laminated CFRP composite specimen during its US excitation. Composite layers represented by correlation fringe patterns are indicated by arrows. Vertical dashed lines show the location of the specimen cut after the experiment

The results of studies of the laminated composites of various thicknesses showed that delaminations and other subsurface defects can be detected at depth of up to 8 mm from the studied surface.

Thus, compact optoelectronic devices for NDT of laminated composite structures can be created on the basis of the proposed ODSI experimental setup. They can be used in the aerospace and engineering industries for detecting internal defects in composite and metal–composite structural elements without applying the mechanical loads.

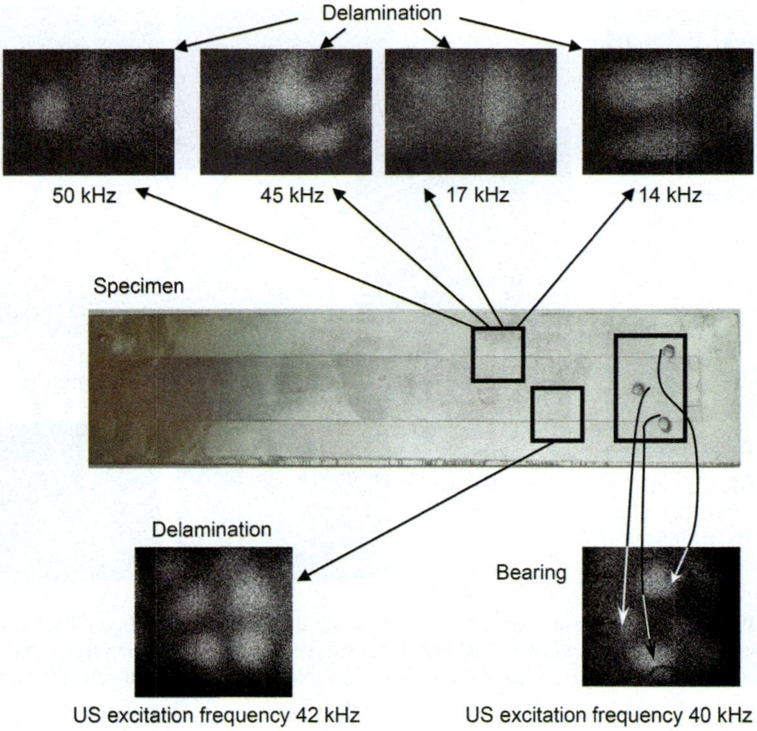

Fig. 4.50 Correlation fringe patterns in places of delaminations and bearing strains indicating the detection of subsurface defects in a fiberglass reinforced laminated composite

Fig. 4.51 Appearance of the repaired laminated composite structural element of the AN aircraft wing flap

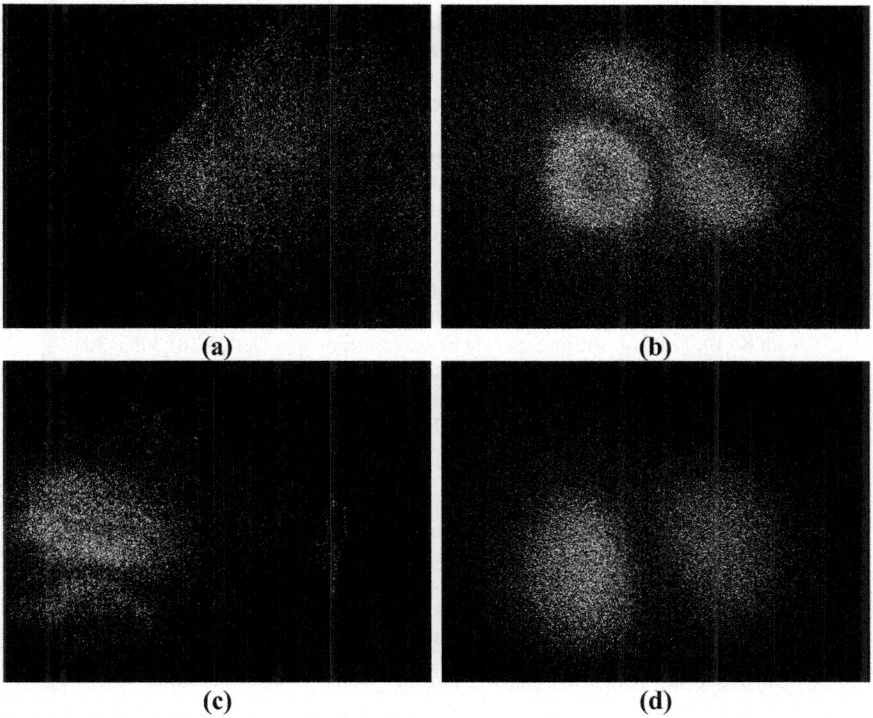

Fig. 4.52 Results of delaminations detection in the repaired laminated composite structural element of the AN aircraft wing flap: delamination between 1st and 2nd layers detected at the resonant frequency $f_{r1} = 15$ kHz on the specimen surface area 1 circled by a white solid closed curve (**a**); delamination between 2nd and 3rd layers detected at the resonant frequency $f_{r2} = 21$ kHz on the specimen surface area 1 circled by the same white curve (**b**); delamination between 3rd and 4th layers detected at the resonant frequency $f_{r3} = 19$ kHz on the specimen surface area 2 circled by a gray solid closed curve (**c**); delamination between 4 and 5th layers detected at the resonant frequency $f_{r4} = 17.5$ kHz on the specimen surface area 2 circled by the same gray curve (**d**)

References

1. Abbott JG, Thurstone FL (1979) Acoustic speckle: theory and experimental analysis. Ultrason Imag 1(4):303–324
2. Aebischer HA, Waldner S (1999) A simple and effective method for filtering speckle interferometric phase fringe patterns. Opt Commun 162:205–210
3. Almazan-Cuellar S, Malacara-Hernandez D (2003) Two-step phase-shifting algorithm. Opt Eng 42(12):3524–3531
4. Amjadi M, Kyung KU, Park I, Sitti M (2016) Stretchable, skin-mountable, and wearable strain sensors and their potential applications: a review. Adv Funct Mater 26(11):1678–1698
5. Baker RJ (2019) CMOS: circuit design, layout, and simulation, 4th edn. Wiley-IEEE Press, Piscataway, NJ
6. Baldi P, Brunak S, Chauvin Y, Andersen CA, Nielsen H (2000) Assessing the accuracy of prediction algorithms for classification: an overview. Bioinformatics 16(5):412–424
7. Bhaduri BM, Kothiyal P, Mohan NK (2008) A comparative study of phase-shifting algorithms in digital speckle pattern interferometry. Optik 119(3):147–152

8. Bobkov Yu N (1984) Korreliatsionnyye Metody i Ustrojstva Izmiereniya Sostavliayushchih Komponientov Signalov (Correlation Methods and Devices for Measuring Signal Components). Vyshcha Shkola, Lviv
9. Borza DN, Nistea IT (2012) High temporal and spatial resolution in time resolved speckle interferometry. Opt Lasers Eng 50(8):1075–1083
10. Butters JN, Leendertz JA (1971) Holographic and video techniques applied to engineering measurement. Meas Control 4(12):349–354
11. Carré P (1966) Installation et utilisation du comparateur photoélectrigue et Interferentiel du Bureau International des Poids ek Measures. Metrologia 1:13–23
12. Cirello A, Pasta A (2007) Determination of the stress intensity factor by means of the ESPI technique. In: Gdoutos EE (ed) Experimental analysis of nano and engineering materials and structures. Springer, Dordrecht, pp 395–396
13. Creath K (1985) Phase-shifting speckle interferometry. Appl Opt 24(18):3053–3058
14. Creath K (1988) V Phase-measurement interferometry techniques. In: Wolf E (ed) Progress in optics, vol 26. North-Holland, Amsterdam, pp 349–393
15. Dai X, Pu Q, Wang L, Yun H, Wang Y (2011) Measurement on fracture process and prediction of the load capacity of steel fiber reinforced concrete by electronic speckle pattern interferometry. Compos B Eng 42(5):1181–1188
16. De la Torre IM, Montes MDSH, Flores-Moreno JM, Santoyo FM (2016) Laser speckle based digital optical methods in structural mechanics: A review. Opt Lasers Eng 87:32–58
17. Dolinko AE, Kaufmann GH (2009) Measurement of the local displacement field generated by a microindentation using digital speckle pattern interferometry and its application to investigate coating adhesion. Opt Lasers Eng 47(5):527–531
18. Duncan DD, Kirkpatrick SJ (2008) Algorithms for simulation of speckle (laser and otherwise). In: Tuchin VV, Wang LV (eds) Complex dynamics and fluctuations in biomedical photonics V, vol 6855. SPIE, Bellingsham, WA, p 685505
19. Ettemeyer A (2000) Combination of 3-D deformation and shape measurement by electronic speckle-pattern interferometry for quantitative strain-stress analysis. Opt Eng 39(1):212–215
20. Ettemeyer A, Wang Z, Walz T (1997) Applications of 3D speckle interferometry to material and component testing. In: Gorecki C (ed) Optical Inspection and Micromeasurements II, vol 3098. SPIE. Bellingham, WA, pp 188–194
21. Flynn TJ (1997) Two-dimensional phase unwrapping with minimum weighted discontinuity. J Opt Soc Amer A 14(10):2692–2701
22. Fomitchov PA, Krishnaswamy S (1997) A compact dual-purpose camera for shearography and electronic speckle-pattern interferometry. Meas Sci Technol 8(5):581
23. Fomitchov P, Wang LS, Krishnaswamy S (1997) Advanced image-processing techniques for automatic nondestructive evaluation of adhesively-bonded structures using speckle interferometry. J Nondestr Eval 16(4):215–227
24. Françon M (1979) Laser speckle and applications in optics. Academic Press, New York
25. Galizzi GE, Federico A, Kaufmann GH (2017) Experimental evaluation of a 3D wavelet-based phase recovery method in temporal speckle pattern interferometry. Appl Opt 56(15):4412–4418
26. Ghiglia DC, Pritt MD (1998) Two-dimensional phase unwrapping: theory, algorithms, and software. Wiley-Interscience, New York
27. Gong XL, Toyooka S (1999) Investigation on mechanism of plastic deformation by digital speckle pattern interferometry. Exp Mech 39(1):25–29
28. Goodman JW (1975) Statistical properties of laser speckle patterns. In: Dainty JC (ed) Laser speckle and related phenomena. Springer-Verlag, Berlin, pp 9–75
29. Goodman JW (2007) Speckle phenomena in optics: theory and applications. Roberts and Company, Greenwood Village CO
30. Goodman JW (2015) Statistical optics. John Wiley & Sons
31. Greivenkamp JE (1984) Generalized data reduction for heterodyne interferometry. Opt Eng 23(4):350–352

32. Greivenkamp JE, Brunning JH (1992) Phase shifting interferometry. In: Malakara D (ed) Optical shop testing, 2nd edn. Wiley, New York, pp 501–598
33. Guelorget B, François M, Vial-Edwards C, Montay G, Daniel L, Lu J (2006) Strain rate measurement by electronic speckle pattern interferometry: a new look at the localization onset. Mater Sci Eng A 415(1):234–241
34. Hack E, Schumacher A (2007) DSPI strain measurement on an externally reinforced bending beam: a comparison of step-by-step addition and pixel shift correlation. Opt Lasers Eng 45(5):589–595
35. Hariharan P, Oreb BF, Eiju T (1987) Digital phase-shifting interferometry: a simple error-compensating phase calculation algorithm. Appl Opt 26(13):2504–2505
36. Höfling R, Osten W (1987) Displacement measurement by image-processed speckle patterns. J Mod Opt 34(5):607–617
37. Holst GC (1998) CCD arrays, cameras and displays. SPIE Press, Bellingham, WA
38. Hu B, Yoshida S, Gaffney J (2011) Stress and strain analysis of metal plates with holes. In: Proulx T (ed) Dynamic behavior of materials, vol 1. Springer, New York, pp 187–193
39. Huang YH, Hung SY, Janabi-Sharifi F, Wang W, Liu YS (2012) Quantitative phase retrieval in dynamic laser speckle interferometry. Opt Lasers Eng 50(4):534–539
40. Huang X, Liu Z, Xie H (2013) Recent progress in residual stress measurement techniques. Acta Mech Solida Sin 26(6):570–583
41. Huang X, Chen H, Sun B (2022) Experimental investigation on the concrete fracture process zone using electronic speckle pattern interferometry. Strain 58(1):e12402
42. Huntley JM (1989) Speckle photography fringe analysis: assessment of current algorithms. Appl Opt 28(20):4316–4322
43. Jacquot P (2008) Speckle interferometry: a review of the principal methods in use for experimental mechanics applications. Strain 44(1):57–69
44. Joenathan C, Haible P, Tiziani HJ (1999) Speckle interferometry with temporal-phase evaluation: influence of decorrelation, speckle size, and nonlinearity of the camera. Appl Opt 38(7):1169–1178
45. Jones R, Wykes C (1989) Holographic and speckle interferometry. Cambridge University Press, Cambridge
46. Kai L, Kemao Q (2011) Dynamic phase retrieval in temporal speckle pattern interferometry using least squares method and windowed Fourier filtering. Opt Express 19(19):18058–18066
47. Karbhari VM (ed) (2013) 1st edn. Woodhead Publishing, Cambridge, Philadelfia, New Delhi
48. Kerr D, Mendoza Santoyo F, Tyrer JR (1990) Extraction of phase data from electronic speckle pattern interferometric fringes using a single-phase-step method: a novel approach. J Opt Soc Am A 7(5):820–826
49. Kim K-S, Kang K-S, Kang Y-J, Cheong S-K (2003) Analysis of an internal crack of pressure pipeline using ESPI and shearography. Opt Las Technol 35(8):639–643
50. Kreis T (2005) Handbook of holographic interferometry: optical and digital methods. Wiley-VCH Verlag GmbH & Co., Weinheim
51. Leendertz JA (1970) Interferometric displacement measurement on scattering surfaces utilizing speckle effect. J Phys E: Sci Instrum 3(3):214–218
52. Liu K, Wu SJ, Gao XY, Yang LX (2015) Simultaneous measurement of in-plane and out-of-plane deformations using dual-beam spatial-carrier digital speckle pattern interferometry. In: Wang L, Wang C, Wang L, Zhang Q (eds) Applied mechanics and materials, vol 782. Trans Tech Publications Ltd., Baech, Switzerland, pp 316–325
53. Lobanov LM, Pivtorak VA, Savitsky VV, Tkachuk GI (2006) Procedure for determination of residual stresses in welded joints and structural elements using electron speckle-interferometry. Paton Weld J C/C Avtomaticheskaya Svarka 1:24–28
54. Lobanov L, Pivtorak V, Savitsky V, Tkachuk G (2013) Technology and equipment for determination of residual stresses in welded structures based on the application of electron speckle-interferometry. In: Kurz SJB, Mittemeijer EJ, Scholtes B (eds) Materials science forum, vol 768–769. Trans Tech Publications Ltd., Baech, Switzerland, pp 166–173

55. Lobanov LM, Pashchin NA, Mikhodui OL, Khokhlova JA (2016) Investigation of residual stresses in welded joints of heat-resistant magnesium alloy ML10 after electrodynamic treatment. J Magnes Alloys 4(2):77–82
56. Lobanov LM, Muravsky LI, Pivtorak VA, Voronyak TI (2017) Monitoryng napruzhenogo stanu elementiv konstruktsiy z vykorystanniam elektromagnitnyh hvyl' optychnogo diapazonu (Monitoring of structural elements stress state with the use of electromagnetic waves in the optical range). In: Nazarchuk ZT (ed) Tehnichna diagnostyka materialiv i konstruktsiy (Technical diagnostics of materials and structures), Reference manual in 8 volumes, vol 3. Publishing House "Prostir-M", Lviv
57. Løkberg OJ (1992) Speckles and speckle techniques for biomedical applications. Bioptics: optics in biomedicine and environmental sciences, vol 1524. SPIE, Bellingham, WA, pp 35–47
58. Lu M, Wang S, Bilgeri L, Song X, Jakobi M, Koch AW (2018) Online 3D displacement measurement using speckle interferometer with a single illumination-detection path. Sensors 18(6):1923
59. Macovski A, Ramsey D, Schaefer LF (1971) Time lapse interferometry and contouring using television systems. Appl Opt 10(12):2722–2727
60. Malacara D (2007) Twyman-Green interferometer. In: Malacara D (ed) Optical shop testing, 3rd edn. John Wiley & Sons, Hobocen, NJ, pp 46–96
61. Malacara D, Servín M, Malacara Z (2005) Interferogram analysis for optical testing, 2nd edn. Taylor & Francis, Boca Raton, FL
62. Mantravadi MV, Malacara D (2007) Newton, Fizeau, and Haidinger interferometers. In: Malacara D (ed) Optical shop testing. John Wiley & Sons, Hobocen, NJ, pp 1–45
63. Menner P, Gerhard H, Busse G (2011) Lockin-interferometry: principle and applications in NDE. Stroj Vestn/J Mech Eng 57(3):183–191
64. Mohan NK, Andersson A, Sjödahl M, Molin NE (2000) Optical configuration for TV holography measurement of in-plane and out-of-plane deformations. Appl Opt 39(4):573–577
65. Muravsky LI, Voronyak TI, Smolyak VI, Tkhoruk AL (2002) Doslidzhennia poverhnevyh deformatsiy metalichnyh balkovyh zrazkiv metodamy elektronnoyi spekl-interferometriyi (Study of surface strains of metal beam specimens by electronic speckle interferometry methods). Vidbir ta Obrobka Informatsiyi (Inf Extract Proc) 17:131–135
66. Muravsky LI, Ostash OP, Kmet' AB, Voronyak TI, Andreiko IM (2011) Two-frame phase shifting interferometry for retrieval of smooth surface and its displacements. Opt Lasers Eng 49(3):305–312
67. Muravsky LI, Kmet AB, Voronyak TI (2011) Retrieval of diffusing surface by two-frame interferometric method with blind phase shift of a reference wave. In: Kashyap R, Têtu M, Kleiman RN (eds) Photonics North 2011, vol 8007. SPIE, Bellingham, WA, pp 385–391
68. Muravsky LI, Kmet' AB, Voronyak TI (2012) Retrieving the relief of a low-roughness surface using a two-step interferometric method with blind phase shift of a reference wave. Opt Lasers Eng 50(11):1508–1516
69. Muravsky L, Kmet' A, Voronyak T (2013) Two approaches to the blind phase shift extraction for two-step electronic speckle pattern interferometry. Opt Eng 52(10):101909
70. Muravsky LI, Voronyak TI, Kmet' AB (2014) Lazerna interferometriya poverhni dlia potreb tehnichnoyi diagnostyky (Laser interferometry of surface for needs of technical diagnostics). Spolom, Lviv
71. Muravsky LI, Voronyak TI, Ivanytsky Ya L, Dzhala VR, Kuts OG (2017) Pryntsypy pobudovy aparatno-programnogo kompleksu dlia neruynivnogo kontrolyu kompozytnyh elementiv konstruktsiy aviakosmichnoyi tehniky (Principles for development of hardware-software complex for nondestructive testing of aircraft composite structural components). Tehnicheskaya Diagnostika i Nierazrushayushchiy Kontrol (Tech Diagn Non-Destructive Test) 1:37–42
72. Muravsky L, Kotsiuba Y, Kulynych Y (2020) Estimation of unknown phase shift between synthesized speckle interferograms using Pearson correlation coefficient. In: Proceedings of 2020 IEEE 15th international conference on computer sciences and information technologies (CSIT), vol 2. IEEE, Lviv, pp 58–61

73. Muravsky L, Kotsiuba Y, Kulynych Y (2020) Assessment of unknown phase shift for speckle interferometry using sample Pearson correlation coefficient. In: Conference on computer science and information technologies. Springer, Cham, pp 671–681
74. Nazarchuk Z, Muravsky L, Kuryliak D (2019) To the problem of the subsurface defects detection: theory and experiment. Procedia Struct Integrity 16:11–18
75. Parra-Michel J, Martinez A, Rayas JA (2010) Computation of crack tip elastic stress intensity factor in mode I by in-plane electronic speckle pattern interferometry. Rev Mex de Fís 56(5):394–400
76. Pfeifer T, Misho HK, Ettemeyer A, Wang Z, Wegner R (1998) Strain/stress measurements using electronic speckle pattern interferometry. In: Harding KG, Svetkoff DJ, Creath K, Harris JS (eds) Three-dimensional imaging, optical metrology, and inspection IV, vol 3520. SPIE, Bellingham WA, pp 262–271
77. Pouet B, Chatters T, Krishnaswamy S (1993) Synchronized reference updating technique for electronic speckle interferometry. J Nondestr Eval 12(2):133–138
78. Pouet BF, Krishnaswamy S (1994) Additive-subtractive phase-modulated electronic speckle interferometry: analysis of fringe visibility. Appl Opt 33(28):6609–6616
79. Povey MJW, Tong J, Nelson PV, Jones GM (1984) Method and apparatus for determining particle size distribution by acoustic speckle, US Patent No. 6,698,276
80. Rastogi PK (ed) (2000). John Wiley & Sons, Hoboken, NJ
81. Rathjen C (1995) Statistical properties of phase-shift algorithms. J Opt Soc Am A 12(9):1997–2008
82. Schnars U, Jueptner W (2005) Digital holography: digital hologram recording, numerical reconstruction, and related techniques. Springer, Dordrecht
83. Schreiber H, Brunning JH (2007) Phase shifting interferometry. In: Malacara D (ed) Optical shop testing, 3rd edn. John Wiley & Sons, Hoboken, NJ, pp 547–666
84. Servin M, Marroquin JL, Cuevas FJ (2001) Fringe-follower regularized phase tracker for demodulation of closed-fringe interferograms. J Opt Soc Am A 18(3):689–695
85. Sesselmann M, Gonçalves AA Jr (1998) Single phase-step algorithm for phase difference measurement using ESPI. In: Kujawinska M, Brown GM, Takeda M (eds) Laser interferometry IX: techniques and analysis, vol 3478. SPIE, Bellingham, WA, pp 153–158
86. Sesselmann M, Gonçalves AA Jr (2001) A robust spatial phase-stepping ESPI system. In: Albertazzi A (ed) Laser metrology for precision measurement and inspection in industry, vol 4420. SPIE, Bellingham, WA, pp 149–154
87. Shabadi R, Kumar S, Roven H, Dwarakadasa ES (2004) Characterisation of PLC band parameters using laser speckle technique. Mater Sci Eng A 364(1):140–150
88. Siebert T, El-Ratal W, Wegner R, Ettemeyer A (2002) Combine simulation and experiment in automotive testing with ESPI measurement. Exp Tech 26(3):42–47
89. Sirohi SR (2009) Optical methods of measurement: wholefield techniques, 2nd edn. Taylor & Francis, Boca Raton, FL
90. Sjödahl M, Benckert LR (1993) Electronic speckle photography: analysis of an algorithm giving the displacement with subpixel accuracy. Appl Opt 32(13):2278–2284
91. Sjödahl M, Benckert LR (1994) Systematic and random errors in electronic speckle photography. Appl Opt 33(31):7461–7471
92. Solodov I, Bai J, Busse G (2013) Resonant ultrasound spectroscopy of defects: case study of flat-bottomed holes. J Appl Phys 113(22):223512
93. Solodov I, Bai J, Bekgulyan S, Busse G (2011) A local defect resonance to enhance acoustic wave-defect interaction in ultrasonic nondestructive evaluation. Appl Phys Lett 99(21):211911
94. Solodov I, Rahammer M, Kreutzbruck M (2019) Analytical evaluation of resonance frequencies for planar defects: effect of a defect shape. NDT Int 102:274–280
95. Su RKL, Chen HHN, Kwan AKH (2012) Incremental displacement collocation method for the evaluation of tension softening curve of mortar. Eng Fract Mech 88:49–62
96. Sun P (2006) Three-dimensional displacement measurement by using reversed phase-shifting electronic speckle pattern interferometry. Opt Eng 45(9):093602

97. Svanbro A, Huntley JM, Davila A (2003) Optimal re-referencing rate for in-plane dynamic speckle interferometry. Appl Opt 42(2):251–258
98. Tendela LP, Federico A, Kaufmann GH (2011) Evaluation of the piezoelectric behaviour produced by a thick-film transducer using digital speckle pattern interferometry. Opt Lasers Eng 49(2):281–284
99. Voronyak TI, Muravsky LI (2006) Kontrol' poverhnevyh peremishchen' ta defektiv zrazkiv konstruktsijnyh materialiv metodamy elektronnoji spekl-interferometriji (Control of surface displacements and defects of constructive materials by electronic speckle interferometry methods). Metody ta Prylady Kontrolyu Yakosti (Meth Dev Q Control) 16:41–46
100. Voronyak TI, Muravsky LI, Gaskevych GI (2005) Vykorystannia riznytsevoyi spekl-interferometriji dlia neruynivnogo kontrolyu zahysnyh i vidnovliuvalnyh pokryviv (Use of differential electronic speckle interferometry for non-destructive testing of protective and restorative coatings). Fizychni metody ta zasoby kontrolyu seredovyshch, materialiv ta vyrobiv (Phys Meth Means Control Media, Mater Prod) 10:203–208
101. Voronyak TI, Kmet' AB, Lychak OV (2007) Single-step phase-shifting speckle interferometry. Mater Sci 43(4):554–567
102. Voronyak TI, Kmet' AB, Malanchuk TM (2007) Vyjavlennia pidpoverhnevyh defektiv v zrazkah konstruktsijnyh materialiv zasobamy elektronnoji spekl-interferometriji (Detection of subsurface defects in constructive materials specimens by means of electronic speckle interferometry). Fizychni metody ta zasoby kontrolyu seredovyshch, materialiv ta vyrobiv (Phy Meth Means Control Media, Mater Prod) 12:237–242
103. Voronyak TI, Kmet' AB, Muravs'kyi LI (2009) Determination of the 3D fields of displacements by the method of phase-shifting speckle interferometry. Mater Sci 45(3):372–377
104. Voronyak TI, Muravsky LI, Ivanytskyi YL (2012) Kontrol' protsesiv deformuvannia ta rujnuvannia kompozytnyh materialiv zasobamy elektronnoji spekl-interferometriji (Control of processes of composite materials deformation and destruction by means of electronic speckle interferometry). Metody ta Prylady Kontrolyu Yakosti (Meth Dev Q Control) 1:9–15
105. Voronyak TI, Ivanyts'kyi YL, Muravs'kyi LI, Semenets OI (2013) Investigation of the deformation and fracture of composites with the help of electronic speckle interferometry. Mater Sci 49(2):270–275
106. Wang LS, Krishnaswamy S (1996) Additive-subtractive speckle interferometry: extraction of phase data in noisy environments. Opt Eng 35(3):794–801
107. Wang L-S, Jambunathan K, Dobbins BN, He S-P (1996) Measurement of three-dimensional surface shape and deformations using phase stepping speckle interferometry. Opt Eng 35(8):2333–2340
108. Wang S, Lu M, Bilgeri LM, Jakobi M, Bloise FS, Koch AW (2018) Temporal electronic speckle pattern interferometry for real-time in-plane rotation analysis. Opt Express 26(7):8744–8755
109. Wegner R, Ettemeyer A (1999) Miniaturization of speckle interferometry for rapid strain analysis. In: Optical measurement systems for industrial inspection, vol 3824. SPIE, Bellingham, WA, pp 30–37
110. Wyant JC (1975) Use of an AC heterodyne lateral shear interferometer with real–time wavefront correction systems. Appl Opt 14(11):2622–2626
111. Yamaguchi I (2003) Fundamentals and applications of speckle. In: Gastinger K, Løkberg OJ, Winther S (eds) Speckle metrology 2003, vol 4933. SPIE, Bellingham, WA, pp 1–8
112. Yang L, Ettemeyer A (2003) Strain measurement by three-dimensional electronic speckle pattern interferometry: potentials, limitations, and applications. Opt Eng 42(5):1257–1266
113. Yang L, Xie X, Zhu L, Wu S, Wang Y (2014) Review of electronic speckle pattern interferometry (ESPI) for three-dimensional displacement measurement. Chin J Mech Eng 27(1):1–13
114. Yoshimura T, Zhou M, Yamahai K, Liyan Z (1995) Optimum determination of speckle size to be used in electronic speckle pattern interferometry. Appl Opt 34(1):87–91
115. Yu Q, Fu S, Yang X, Sun X, Liu X (2004) Extraction of phase field from a single contoured correlation fringe pattern of ESPI. Opt Exp 12(1):75–83

116. Yu Q, Fu S, Liu X, Yang X, Sun X (2004) Single-phase-step method with contoured correlation fringe patterns for ESPI. Opt Exp 12(20):4980–4985
117. Yu YJ, Wang LL, Zhang ZJ (2006) Design of phase-shifting ESPI system for 3D deformation measurement. J Phys Conf Ser 48(1):911–915
118. Yue K, Jia S, Tan Y (2000) DSPI technique for nanometer vibration mode measurement. In: Fourth international conference on vibration measurements by laser techniques: advances and applications, vol 4072. SPIE, Bellingham, WA, pp 341–345
119. Zhang J, Chang TC (1998) Fiber electronic speckle pattern interferometry and its applications in residual stress measurements. Appl Opt 37(28):6707–6715

Chapter 5
New Methods of Speckle Metrology in Analysis of Rough Surfaces

Abstract The method of three-frame digital speckle pattern interferometry (TF DSPI) with unknown phase shifts of a reference beam is presented in this chapter. In contrast to the two-step digital speckle pattern interferometry method discussed in Chap. 4, registration of the reference and object waves intensities is removed in this method. In this regard, speckle interferograms (SIs) are recorded using the integrating-bucket technique, which increases the rate of SIs registration before and after applying loads. In this method, the sample Pearson correlation coefficient can be used for extraction the unknown phase shift between SIs. The TF DSPI method was used in fringe projection interferometry for retrieval of the surface relief. During experiment several real surface areas were restored, for example, the 3D surface shape of the diamond drill, including its form and surface roughness.

Studies of an optically rough surface are an important trend in NDT and TD. Among the methods and means of such investigation based on various physical principles, the Speckle Metrology techniques are very important and relevant. High spatial resolution, the ability to form multidimensional data arrays and enter them into digital devices for further processing and generating the 3D surface height maps, as well as 3D displacement and deformation fields, allow these techniques to successfully compete with the others ones.

As it is shown in Chap. 4, to determine the nano- and microdisplacements fields in various problems of TD, NDT and experimental mechanics, the correlation and phase shifting digital speckle pattern interferometry techniques are widely used. Simultaneously, phase shifting DSPI can be used to estimate the parameters of the topography of the optically rough surface. To this end, interference fringes are projected onto the surface of the test specimen and, by performing the calibrated phase shifts of the interference fringes, a series of fringe patterns is recorded using a coherent optical system and a digital camera. Since the interference fringes are projected onto an optically rough surface, speckle patterns of these fringes are formed in the output plane of the coherent optical system, where the photosensor array of a digital camera is placed. These fringe patterns can be considered as speckle interferograms (SIs), as shown, for example, by Dhanasecar and Ramamoorthy [5].

Z. Nazarchuk et al., *Optical Metrology and Optoacoustics in Nondestructive Evaluation of Materials*, Springer Series in Optical Sciences 242,
https://doi.org/10.1007/978-981-99-1226-1_5

Fringe projection techniques that use optical projectors of incoherent monochromatic and polychromatic images, including multimedia projectors, are being developed quite intensively and have a wide practical application [10, 42]. On the other hand, the interferometric fringe projection methods are used to determine the surface shape [17], restore sharp jumps on the surface [35] and estimate the parameters of the surface topography [5].

To implement the phase shifting DSPI and fringe projection interferometry methods, the techniques of calibrated temporal or spatial phase shifts can be used. At least three SIs with calibrated phase shifts in the initial state of the surface and three SIs with the same calibrated phase shifts in its excited or deformed state during or after loading should be registered in the conventional temporal phase shifting DSPI. The phase shift calibration procedure is time consuming and requires complex and precise devices with appropriate drivers and software for accurate positioning of the phase shifting device in an optical speckle interferometer. To reduce the number of recorded SIs, several methods of two-step DSPI with calibrated and arbitrary phase shifts aforementioned in Chap. 4 have been proposed [1, 15, 29, 30, 40, 41]. The method of two-step digital speckle pattern interferometry (TS DSPI) with unknown phase shift of the reference beam in an optical speckle interferometer is described in Sect. 4.4. This method make it possible to extract the unknown phase shifts between two speckle interferograms (SIs) i_{11} and i_{12} in the initial state of the surface and two SIs i_{21} and i_{22} in its deformed state using the population Pearson correlation coefficient (PPCC) (see Eqs. (4.31)–(4.39)).

Despite the possibility of significant simplification of the equipment in the speckle interferometer due to the lack of need to produce the calibrated phase shifts using phase shifting element, the TS DSPI method has two disadvantages. First, it is necessary to record additionally the distributions of the intensities i_o and i_r of the reference and object waves except recording of all SIs, which takes a long time. Secondly, recording of the SIs i_{12} and i_{21} (see Sect. 4.4.5) should take place after the phase shift is performed with a delay of at least tens of seconds in order to stop the phase shifter oscillations.

To simplify the procedures for retrieving the surface nano- and microdisplacements, a new method of three-frame digital speckle pattern interferometry (TF DSPI) with arbitrary phase shifts of the reference wave is proposed [21]. Note that this method can also be used in projection fringe interferometry to restore the relief of an optically rough surface. To calculate the unknown phase shift, the so-called phase-step calibration (PSC) algorithm [36] discussed in Chaps. 2 and 4 can be used, although the same algorithm was first proposed by Bobkov and Molodov [2, 3] and modernized especially for the TS DSPI method by Muravsky et al. [19]. The choice of this algorithm to calculate the unknown phase shift in the TS DSPI method is based on the results of computer simulations, which showed that the absolute error $|\Delta\alpha_{21}|$ between the calculated and given phase shifts (see Eq. 4.42) decreases (see Table 4.1) with increasing surface roughness and already at $Sq = \lambda/2$ this error reaches a very small absolute value equal to about 10^{-3} rad. Note that Bobkov and Molodov [2, 3] also showed that as the level of random phase noise in a series of signals generated by phase meters increases, the systematic error of the phase shift between signals decreases. This error behavior indicates that the calculating the unknown phase shift can also be performed by the Pearson correlation coefficient for the sample (sample Pearson correlation coefficient (SPCC)) [24]. The use of the SPCC makes it possible to reduce the time required to calculate the phase shift angle and avoids the direct and inverse digital Fourier transforms of SIs recorded in the form of 2D data arrays, the execution of which leads to additional errors initiation in calculating the unknown phase shift.

Based on the foregoing, first the justification of the reliability of estimating the unknown phase shift between two SIs using the SPCC [22, 23] will be considered. Further, a new method of TF DSPI with unknown phase shifts of the reference wave, i.e. with unknown phase shifts between SIs, will be justified. And finally, a new method of three-frame fringe projection interferometry with unknown phase shifts of the reference wave will be proposed. The method is designed to restore the surface displacement field and the topography of an optically rough surface.

5.1 Method for Extraction the Unknown Phase Shift Between Speckle Interferograms Using Sample Pearson Correlation Coefficient

The PPCC was used to estimate the magnitude of unknown phase shift angle between two interferograms, which were considered as two variables having the form of two multidimensional centered vectors (see Sect. 2.4). The same approach was also used for two SIs i_{11} and i_{12} of the studied rough surface area in the initial state and two SIs i_{21} and i_{22} of the same surface area in the excited state, as well as for two specklegrams $(i_{11} - i_{21})$ and $(i_{12} - i_{22})$ (see Sect. 4.4 and Eqs. 4.31–4.39).

At the same time, it is known that the SPCC has a direct relationship with the angle α between variable vectors, if they are based on centered variables [25]. Assume that SIs i_{11} and i_{12}, as well as SIs i_{21} and i_{22} are such variable vectors. Then, SPCC for two SIs i_{11} and i_{12} is given by

$$r(i_{11}, i_{12}) = \frac{\sum_{i=0}^{N-1} \left(i_{11i} - \bar{i}_{11}\right)\left(i_{12j} - \bar{i}_{12}\right)}{\sqrt{\sum_{j=0}^{N-1} \left(i_{11j} - \bar{i}_{11}\right)^2} \sqrt{\sum_{j=0}^{N-1} \left(i_{12j} - \bar{i}_{12}\right)^2}}, \quad (5.1)$$

and the SPCC for two SIs i_{21} and i_{22} is given by

$$r(i_{21}, i_{22}) = \frac{\sum_{j=0}^{N-1} \left(i_{21j} - \bar{i}_{21}\right)\left(i_{22j} - \bar{i}_{22}\right)}{\sqrt{\sum_{j=0}^{N-1} \left(i_{21j} - \bar{i}_{21}\right)^2} \sqrt{\sum_{j=0}^{N-1} \left(i_{22j} - \bar{i}_{22}\right)^2}}, \quad (5.2)$$

where

$$\bar{i}_{11} = \frac{1}{N} \sum_{j=0}^{N-1} i_{11j}, \bar{i}_{12} = \frac{1}{N} \sum_{j=0}^{N-1} i_{12j} \quad (5.3)$$

are the sample means of the intensity distributions i_{11} and i_{12}, respectively;

$$\bar{i}_{21} = \frac{1}{N}\sum_{j=0}^{N-1} i_{21j}, \bar{i}_{22} = \frac{1}{N}\sum_{j=0}^{N-1} i_{22j} \tag{5.4}$$

are the sample means of the intensity distributions i_{21} and i_{22}, respectively; N is the sample size that can even reach the number of all pixels in each digital SI; j is the sample (pixel) number ($j = 0, 1, ..., N - 1$); i_{11j} and i_{12j}, as well as i_{21j} and i_{22j} are the intensities of pixels in SIs.

Since the dependence of the angle α between two centered multidimensional variable vectors on SPCC r is defined as [25]

$$\alpha = \arccos(r), \tag{5.5}$$

the unknown phase shift α_{21s} between i_{11} and i_{12} and the unknown phase shift $\tilde{\alpha}_{21s}$ between i_{21} and i_{22} are calculated, respectively, as

$$\begin{aligned}\alpha_{21s} &= \arccos[r(i_{11}, i_{12})] \\ &= \arccos\frac{\sum_{j=0}^{N-1}\left(i_{11j} - \bar{i}_{11}\right)\left(i_{12j} - \bar{i}_{12}\right)}{\sqrt{\sum_{j=0}^{N-1}\left(i_{11j} - \bar{i}_{11}\right)^2}\sqrt{\sum_{j=0}^{N-1}\left(i_{12j} - \bar{i}_{12}\right)^2}},\end{aligned} \tag{5.6}$$

and

$$\begin{aligned}\tilde{\alpha}_{21s} &= \arccos[r(i_{21}, i_{22})] \\ &= \arccos\frac{\sum_{j=0}^{N-1}\left(i_{21j} - \bar{i}_{21}\right)\left(i_{22j} - \bar{i}_{22}\right)}{\sqrt{\sum_{j=0}^{N-1}\left(i_{21j} - \bar{i}_{21}\right)^2}\sqrt{\sum_{j=0}^{N-1}\left(i_{22j} - \bar{i}_{22}\right)^2}}.\end{aligned} \tag{5.7}$$

Let us prove that the SPCC, like the PPCC, can be used to extract an unknown phase shift between two SIs. For this purpose, one can use the first and second order statistics [8, 9] and a simplified approach to describing each pixel as a sum of random phasors [8, 9] in the speckle patterns, namely the object speckle pattern $i_o(k, l)$ generated by the object wavefront and the reference one $i_r(k, l)$ generated by the reference wavefront in the speckle interferometer (see Eqs. 4.4–4.9). These speckle patterns can be considered as two sets of statistically independent random phasors. According to the random phasors properties, each pixel of a speckle pattern of the tested rough surface can be treated as a result of complex summation of large numbers of small independent phasors describing uncorrelated (independent) amplitude and phase of complex monochrome wave. Besides, in a developed speckle pattern, the intensity random distribution is described by exponential law and phases are uniformly distributed on the interval ($-\pi$,π) [8, 9]. In this case, the reference wavefront is also considered as an additive random noise embedded into the plane wave. Using properties of phasor sums and Eqs. (4.31) and (4.32) for intensities distributions $i_{11}(k,l)$ and $i_{12}(k,l)$ of two digital SIs before loading, the intensities of pixels i_{11j} and i_{12j} and sample means $\bar{i}_{11}$ and $\bar{i}_{12}$ of i_{11} and i_{12} can be expressed by formulae

$$\left.\begin{aligned} i_{11j} &= i_{Bj} + i_{Mj}\cos\varphi_j \\ i_{12j} &= i_{Bj} + i_{Mj}\cos(\varphi_j + \alpha_{21s}) \end{aligned}\right\}, \tag{5.8}$$

$$\begin{aligned} \bar{i}_{11} &= \frac{1}{N}\sum_{j=0}^{N-1} i_{11j} = \frac{1}{N}\sum_{j=0}^{N-1}\left(i_{Bj} + i_{Mj}\cos\varphi_j\right) \\ &= \frac{1}{N}\sum_{j=0}^{N-1} i_{Bj} + \frac{1}{N}\sum_{j=0}^{N-1} i_{Mj}\cos\varphi_j = \bar{i}_B + \bar{i}_M\frac{1}{N}\sum_{j=0}^{N-1}\cos\varphi_j, \end{aligned} \tag{5.9}$$

$$\begin{aligned} \bar{i}_{12} &= \frac{1}{N}\sum_{j=0}^{N-1} i_{12j} = \frac{1}{N}\sum_{j=0}^{N-1} i_{Bj} \\ &+ \frac{1}{N}\sum_{j=0}^{N-1} i_{Mj}\left(\cos\varphi_j\cos\alpha_{21s} - \sin\varphi_j\sin\alpha_{21s}\right) \\ &= \bar{i}_B + \bar{i}_M\cos\alpha_{21s}\frac{1}{N}\sum_{j=0}^{N-1}\cos\varphi_j - \bar{i}_M\sin\alpha_{21s}\frac{1}{N}\sum_{j=0}^{N-1}\sin\varphi_j, \end{aligned} \tag{5.10}$$

where i_{Bj} and i_{Mj} are the background and modulation intensities in each j th pixel, which are considered as random variables independent from random phases φ_j;

$$\bar{i}_B = \frac{1}{N}\sum_{j=0}^{N-1} i_{Bj}, \bar{i}_M = \frac{1}{N}\sum_{j=0}^{N-1} i_{Mj} \tag{5.11}$$

are the sample means of background and modulation intensity distributions, respectively.

During calculations, the well-known dependencies for random phase φ_j uniformly distributed in the interval (−π,π) are used [8, 9], that is

$$\overline{\cos\varphi_j\cos\varphi_n} = \overline{\sin\varphi_j\sin\varphi_n} = \begin{cases} 0 \text{ if} j \neq n \\ \frac{1}{2} \text{ if } j = n \end{cases}, \tag{5.12}$$

$$\overline{\cos\varphi_j\sin\varphi_n} = \begin{cases} \overline{\cos\varphi_j}\overline{\sin\varphi_n} = 0 \text{ if } j \neq n \\ \frac{1}{2}\overline{\sin 2\varphi_j} = 0 \quad \text{if } j = n \end{cases}, \tag{5.13}$$

where $n = 1, 2, ..., N$, and equality

$$\lim_{N\to\infty}\sum_{j=0}^{N-1}\left(i_{Bj}-\overline{i}_B\right)=0, \tag{5.14}$$

which are true for large N. Value of N is always large for speckle patterns.

In addition, the second moment of the intensity distribution i_M of the monochrome polarized light is also used. It is given by [9, 31]

$$\frac{1}{N}\sum_{j=0}^{N-1}i_{Mj}^2=\overline{i_M^2}=2\left(\overline{i}_M\right)^2. \tag{5.15}$$

Taking into account (5.12)–(5.15) and using (5.8)–(5.11), the numerator in Eq. (5.1) can be expressed as

$$\begin{aligned}\sum_{j=0}^{N-1}\left(i_{11j}-\overline{i}_{11}\right)\left(i_{12j}-\overline{i}_{12}\right)&=N^2\overline{i}_M^2\cos\alpha_{21s}\\&-N\overline{i}_M^2\cos\alpha_{21s}\sum_{j=0}^{N-1}\cos\varphi_j\cdot\frac{1}{N}\sum_{j=0}^{N-1}\cos\varphi_j\\&-N\overline{i}_M^2\cos\alpha_{21s}\sum_{j=0}^{N-1}\cos\varphi_j\cdot\frac{1}{N}\sum_{j=0}^{N-1}\cos\varphi_j\\&+\overline{i}_M^2\cos\alpha_{21s}\sum_{j=0}^{N-1}\frac{1}{N}\sum_{j=0}^{N-1}\cos\varphi_j\cdot\frac{1}{N}\sum_{j=0}^{N-1}\cos\varphi_j.\end{aligned} \tag{5.16}$$

To simplify Eq. (5.16), we use the trigonometric relation

$$\frac{1}{N^2}\left(\sum_{j=0}^{N-1}\cos\varphi_j\right)^2=\frac{1}{N}\frac{1}{N}\sum_{i=0}^{N-1}\cos^2\varphi_j=\frac{1}{2N} \tag{5.17}$$

and obtain

$$\begin{aligned}&\sum_{j=0}^{N-1}\left(i_{11j}-\overline{i}_{11}\right)\left(i_{12j}-\overline{i}_{12}\right)\\&=N^2\overline{i}_M^2\cos\alpha_{21s}\\&\quad-\frac{N}{2}\overline{i}_M^2\cos\alpha_{21s}-\frac{N}{2}\overline{i}_M^2\cos\alpha_{21s}+\frac{1}{2}\overline{i}_M^2\cos\alpha_{21s}\\&=\overline{i}_M^2\cos\alpha\left(N^2-N+\frac{1}{2}\right).\end{aligned} \tag{5.18}$$

The denominator in Eq. (5.1) is calculated in the same way. During calculations, we do not write down all the products in which the term $\sum_{j=0}^{N-1}\left(i_{Bj}-\bar{i}_B\right)$ presents, since they are equal to zero. Consequently,

$$\sum_{j=0}^{N-1}\left(i_{11j}-\bar{i}_{11}\right)^2=\sum_{j=0}^{N-1}\left[\left(i_{Bj}-\bar{i}_B\right)+i_{Mj}\cos\varphi_j-\bar{i}_M\frac{1}{N}\sum_{j=0}^{N-1}\cos\varphi_j\right]^2$$
$$=2N\bar{i}_M^2\sum_{j=0}^{N-1}\cos^2\varphi_j-2N\bar{i}_M^2\cdot\frac{N}{N}\sum_{j=0}^{N-1}\cos\varphi_j\cdot\frac{1}{N}\sum_{j=0}^{N-1}\cos\varphi_j$$
$$+\frac{\bar{i}_M^2}{N}\sum_{i=0}^{N-1}\frac{1}{N}\sum_{i=0}^{N-1}\cos^2\varphi_j$$
$$=2N^2\bar{i}_M^2\frac{1}{2}-2N\bar{i}_M^2\frac{N}{2N}+\frac{\bar{i}_M^2}{2}=\bar{i}_M^2\left(N^2-N+\frac{1}{2}\right);\tag{5.19}$$

$$\sum_{j=0}^{N-1}\left(i_{12j}-\bar{i}_{12}\right)^2=\sum_{j=0}^{N-1}\left(i_{Mj}\cos\varphi_j\cos\alpha_{21s}-i_{Mj}\sin\varphi_j\sin\alpha_{21s}\right.$$
$$\left.-\bar{i}_M\cos\alpha_{21s}\frac{1}{N}\sum_{j=0}^{N-1}\cos\varphi_j+\bar{i}_M\sin\alpha_{21s}\frac{1}{N}\sum_{j=0}^{N-1}\sin\varphi_j\right)^2$$
$$=2N\bar{i}_M^2\cos^2\alpha_{21s}N\frac{1}{2}+2N\bar{i}_M^2\sin^2\alpha_{21s}N\frac{1}{2}$$
$$+\bar{i}_M^2\cos^2\alpha_{21s}\frac{1}{2}+\bar{i}_M^2\sin^2\alpha_{21s}\frac{1}{2}$$
$$-2N\bar{i}_M^2\cos^2\alpha_{21s}\frac{N}{2N}-2N\bar{i}_M^2\sin^2\alpha_{21s}\frac{N}{2N}$$
$$=\bar{i}_M^2\left[N^2\left(\cos^2\alpha_{21s}+\sin^2\alpha_{21s}\right)\right.$$
$$\left.=+\frac{1}{2}\left(\cos^2\alpha_{21s}+\sin^2\alpha_{21s}\right)-N\left(\cos^2\alpha_{21s}+\sin^2\alpha_{21s}\right)\right]$$
$$=\bar{i}_M^2\left(N^2-N+\frac{1}{2}\right).\tag{5.20}$$

So, according to (5.19) and (5.20), the denominator is given by

$$\sqrt{\sum_{j=0}^{N-1}\left(i_{11j}-\bar{i}_{11}\right)^2}\cdot\sqrt{\sum_{j=0}^{N-1}\left(i_{12j}-\bar{i}_{12}\right)^2}$$
$$=\sqrt{\bar{i}_M^2\left(N^2-N+\frac{1}{2}\right)}\cdot\sqrt{\bar{i}_M^2\left(N^2-N+\frac{1}{2}\right)}$$

$$= \bar{i}_M^2 \left(N^2 - N + \frac{1}{2} \right). \tag{5.21}$$

Substituting (5.18) and (5.21) into (5.1), we obtain

$$r(i_{11}, i_{12}) = \frac{\left(\bar{i}_M\right)^2 \cos \alpha_{21s} \left(N^2 - N + \frac{1}{2}\right)}{\sqrt{\left(\bar{i}_M\right)^2 \left(N^2 - N + \frac{1}{2}\right)} \sqrt{\left(\bar{i}_M\right)^2 \left(N^2 - N + \frac{1}{2}\right)}} = \cos \alpha_{21s}. \tag{5.22}$$

The same result is also obtained for the SPCC $r(i_{21}, i_{22})$ (see Eq. 5.2). In this case, the intensities of pixels i_{21j} and i_{22j} and sample means $\bar{i}_{21}$ and $\bar{i}_{22}$ of i_{11} and i_{12} (see Eq. 5.4) can be expressed as

$$\left.\begin{aligned} i_{21j} &= i_{Bj} + i_{Mj} \cos\left(\varphi_j + \delta\varphi_j\right) \\ i_{22j} &= i_{Bj} + i_{Mj} \cos\left(\varphi_j + \delta\varphi_j + \tilde{\alpha}_{21s}\right) \end{aligned}\right\}; \tag{5.23}$$

$$\begin{aligned} \bar{i}_{21} &= \frac{1}{N} \sum_{j=0}^{N-1} i_{21j} = \frac{1}{N} \sum_{j=0}^{N-1} \left[I_{Bj} + I_{Mj} \cos\left(\varphi_j + \delta\varphi_j\right) \right] \\ &= \frac{1}{N} \sum_{j=0}^{N-1} i_{Bj} + \frac{1}{N} \sum_{j=0}^{N-1} i_{Mj} \cos\left(\varphi_j + \delta\varphi_j\right) \\ &= \bar{i}_B + \bar{i}_M \frac{1}{N} \sum_{j=0}^{N-1} \cos\left(\varphi_j + \delta\varphi_j\right); \end{aligned} \tag{5.24}$$

$$\begin{aligned} \bar{i}_{22} &= \frac{1}{N} \sum_{j=0}^{N-1} i_{22j} = \frac{1}{N} \sum_{j=0}^{N-1} i_{Bj} \\ &+ \frac{1}{N} \sum_{j=0}^{N-1} i_{Mj} \left[\cos\left(\varphi_j + \delta\varphi_j\right) \cos \tilde{\alpha}_{21j} - \sin\left(\varphi_j + \delta\varphi_j\right) \sin \tilde{\alpha}_{21s} \right] \\ &= \bar{i}_B + \bar{i}_M \cos \tilde{\alpha}_{21s} \frac{1}{N} \sum_{j=0}^{N-1} \cos\left(\varphi_j + \delta\varphi_j\right) \\ &- \bar{i}_M \sin \tilde{\alpha}_{21s} \frac{1}{N} \sum_{j=0}^{N-1} \sin\left(\varphi_j + \delta\varphi_j\right). \end{aligned} \tag{5.25}$$

In Eqs. (5.24) and (5.25), the background and modulation intensities i_{Bj} and i_{Mj} in each jth pixel determined by Eq. (5.11) are considered as random variables independent from random phases $\left(\varphi_j + \delta\varphi_j\right)$.

Calculating the unknown phase shifts α_{21s} and $\tilde{\alpha}_{21s}$ using SPCC (see Eqs. 5.6 and 5.7) is easier than calculating the unknown phase shifts α_{21} and $\tilde{\alpha}_{21}$ using PPCC (see Eqs. 4.35 and 4.36). The time spent on calculating the PPCC is primarily related to the need to calculate the covariance between two SIs. To speed up this procedure, especially if it concerns large information arrays, covariance is carried out in the frequency domain, using the direct and inverse Fourier transforms for two digital arrays of SIs [16]. However, the edge effects that occur in the resulting covariance matrix due to the use of the direct and inverse Fourier transforms introduce additional errors into the calculation of the unknown phase shift. Besides, in the numerator of the SPCC (see 5.1 and 5.2), only the pixel-by-pixel multiplication of two SIs is carried out. On the other hand, to calculate the covariance matrix, that is, the numerator of the PPCC (see 4.35 and 4.36), it is necessary, in addition to the pixel-by-pixel multiplication of two Fourier images, which are 2D Fourier transforms of two digital SIs, to perform also direct and inverse 2D Fourier transforms and complex conjugation of one of the Fourier images. Therefore, it is proposed to use the SPCC to calculate unknown phase shifts between two SIs.

5.2 Simulations to Determine Unknown Phase Shift Using Sample Pearson Correlation Coefficient

Since the number of pixels N in each SI is large and the law of large numbers can be applied to the calculation of the SPCC, this coefficient can be considered as a point estimator of the PPCC [24]. Such an estimator determines the presence of a certain confidence interval, which entails the initiation of systematic errors in the determination of the unknown phase shift by calculating it with the help of the SPCC. The computer simulation of the SPCC reliability testing to extract the unknown phase shift $\tilde{\alpha}_{21s}$ was performed according to the procedure given in Sect. 4.4.2. The sequence of SIs pairs of the deformed optically rough surface was synthesized. In this sequence, each SIs pair differed between itself only by the given phase shift of the reference wavefront. In all SIs, the same out-of-plane displacement was simulated by the standard 3D bipolar phase test depicted in Fig. 2.7. The dimensions of all SIs were $K \times L = 1024 \times 1024$ pixels, the ratio $Kp_x \big/ D = Lp_y \big/ D = 2$ (see Sect. 4.4.2) and the phase roughness root mean square (RMS) $Sq_\varphi = 4\pi$ that corresponds to the RMS of the roughness height $Sq = \lambda$, where λ is the laser wavelength in the Twyman – Green interferometer. An initial synthesized SI $i_{21}(k, l)$ of the deformed rough surface described by Eq. (4.33) and produced from the standard 3D bipolar phase test is shown in Fig. 5.1.

The absolute error of the unknown phase shift $\tilde{\alpha}_{21s}$ concerning the given phase shift α_0 was calculated in the same way as for the phase shift $\tilde{\alpha}_{21}$ using Eq. (4.43) in Sect. 4.4.3, that is

$$\Delta|\tilde{\alpha}_{21s}| = |\tilde{\alpha}_{21s} - \alpha_0| = |\arccos[r(i_{21}, i_{22})] - \alpha_0|. \tag{5.26}$$

Results of the absolute error calculations for simulated two surfaces with the phase roughness RMS $Sq_\varphi = 4\pi$, which generate the object and reference speckle patterns

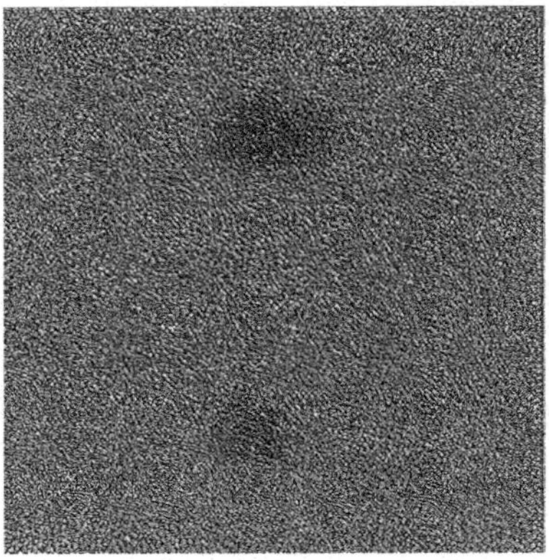

Fig. 5.1 Initial synthesized SI of the deformed rough surface generated from the standard 3D bipolar phase test: $K \times L = 1024 \times 1024$ pixels, $Kp_x / D = Lp_y / D = 2$, $Sq_\varphi = 4\pi$

in a Twyman–Green speckle interferometer, are presented in Table 4.1. Calculations were based on the choice the given phase shift α_0 and synthesis of SIs $i_{12}(k,l)$ and $i_{22}(k,l)$ using the procedure described in Sect. 4.4.2 and illustrated in Fig. 4.6. The synthesized spatial distributions of intensities $i_{12}(k,l)$ and $i_{22}(k,l)$ as well as their sample means $\overline{i}_{21}$ and $\overline{i}_{22}$ (see Eq. 5.4) were substituted into Eq. (5.7), and absolute error $\Delta\tilde{\alpha}_{21s}$ for each given phase shift α_0 was calculated using Eq. (5.26).

The results of the calculations of the absolute error $|\Delta\tilde{\alpha}_{21s}|$, given in Table 5.1, indicate its rather low level, which increase relatively slightly with phase shifts near 0 and 180 degrees. Therefore, the SPCC can be used to calculate the unknown phase shift in the TS DSPI method, as well as in the TF DSPI method with unknown phase shifts, which is proposed and given below.

5.3 Method of Three-Frame Digital Speckle Pattern Interferometry with Unknown Phase Shifts of Reference Wave

As it was shown in Sect. 4.4 and, in particular, in Sect. 4.4.5, the TS DSPI method, despite being simpler in implementation compared to the multistep phase shifting DSPI methods with given phase shifts, is time consuming and has some disadvantages, namely:

- it is necessary to record the intensity distributions of the object and reference wavefronts i_o and i_r;
- recording of the second SIs in the unloaded and loaded states should occur after the completion of the phase shifting element (PSE) arbitrary phase shift with a delay of at least several tens of seconds in order to stop the PSE oscillations.

Table 5.1 Absolute error $|\Delta\tilde{\alpha}_{21}|$ as a function of the given phase shift α_0 calculated with the help of Eq. (5.26) ($Sq_\varphi = 4\pi$, $Sq = \lambda$)

α_0 (deg)	$\lvert\Delta\tilde{\alpha}_{21s}\rvert$ (deg)	α_0 (deg)	$\lvert\Delta\tilde{\alpha}_{21s}\rvert$ (deg)
0	1.144	95	$3.07 \cdot 10^{-2}$
1	$5.18 \cdot 10^{-1}$	100	$2.98 \cdot 10^{-2}$
5	$1.253 \cdot 10^{-1}$	105	$2.73 \cdot 10^{-2}$
10	$5.84 \cdot 10^{-2}$	110	$2.52 \cdot 10^{-2}$
15	$3.36 \cdot 10^{-2}$	115	$2.29 \cdot 10^{-2}$
20	$1.942 \cdot 10^{-2}$	120	$2.07 \cdot 10^{-2}$
25	$9.41 \cdot 10^{-3}$	125	$1.868 \cdot 10^{-2}$
30	$1.546 \cdot 10^{-3}$	130	$1.693 \cdot 10^{-2}$
35	$5.03 \cdot 10^{-3}$	135	$1.567 \cdot 10^{-2}$
40	$1.071 \cdot 10^{-2}$	140	$1.509 \cdot 10^{-2}$
45	$1.565 \cdot 10^{-2}$	145	$1.544 \cdot 10^{-2}$
50	$1.993 \cdot 10^{-2}$	150	$1.705 \cdot 10^{-2}$
55	$2.29 \cdot 10^{-2}$	155	$2.05 \cdot 10^{-2}$
60	$2.66 \cdot 10^{-2}$	160	$2.66 \cdot 10^{-2}$
65	$2.90 \cdot 10^{-2}$	165	$3.78 \cdot 10^{-2}$
70	$3.08 \cdot 10^{-2}$	170	$6.03 \cdot 10^{-2}$
75	$3.20 \cdot 10^{-2}$	175	$1.258 \cdot 10^{-1}$
80	$3.25 \cdot 10^{-2}$	177	$2.08 \cdot 10^{-1}$
85	$3.24 \cdot 10^{-2}$	178	$3.02 \cdot 10^{-1}$
90	$3.18 \cdot 10^{-2}$	179	$5.18 \cdot 10^{-1}$

To eliminate these shortcomings, a new TF DSPI method was proposed [21]. Due to this method, the registration of i_o and i_r is absent, and the speed of the SIs registration is significantly increased.

In temporal phase shifting DSPI, several SIs of the test object surface area in an initial state are first recorded. The recorded SIs differ from each other only by the same calibrated phase shifts between the object and reference beams. At the next stage of the object loading or excitation, the same number of SIs of the given surface area is recorded with the same calibrated phase shifts at which the SIs were recorded in the initial state. In the proposed TF DSPI method, the SIs of the surface initial and deformed states are recorded at arbitrary phase shifts between the object and reference beams. Typically, these phase shifts are carried out using the reference beam. To formulate the theoretical basis of the method, let us first consider three SIs of the specimen surface area recorded before the specimen loading and three SIs after its loading. The Twyman–Green interferometer scheme is used to record SIs, although the proposed mathematical model can be used for any type of a two-beam interferometer. In each series containing three SIs, each SI differs from the previous one only by an arbitrary phase shift of the reference beam relative to the object one. The initial phase in each of the two SIs series is assumed to be zero. It is also assumed that during the recording of three SIs both in the first series (in the initial or intermediate state of the material surface) and in the second series (in the loaded state) there is a smooth and rapid linear phase shift in time. That is, to implement the 3F DSPI method, the high-speed integrating-bucket technique [11, 28, 37] is used as well as for the 3SI UPS method described in Sect. 2.6. Therefore, we can assume that during one phase saw-tooth shift from 0 to π, the change in the distributions of the background intensity $i_{B1}(k, l)$ and the modulation function $i_{M1}(k, l)$ for the first series of three SIs obtained in the unloaded state, as well as the background intensity $i_{B2}(k, l)$ and modulation function $i_{M2}(k, l)$ for the second series of three SIs obtained in the loaded state does not occur. However, such the change is occur between the first and second series, i.e. $i_{B1}(k, l) \neq i_{B2}(k, l)$ and $i_{M1}(k, l) \neq i_{M2}(k, l)$. Based on the foregoing, the intensities distributions of the first series of digital SIs can be expressed as

$$\left.\begin{aligned} i_{11} &= i_{B1} + i_{M1} \cos \varphi \\ i_{12} &= i_{B1} + i_{M1} \cos(\varphi + \alpha_{21s}) \\ i_{13} &= i_{B1} + i_{M1} \cos(\varphi + \alpha_{31s}) \end{aligned}\right\}, \tag{5.27}$$

where $i_{B1} = i_{o1} + i_{r1}$ is the background intensity distribution in the unloaded state, i_{o1} is the object wavefront intensity distribution in the unloaded state, i_{r1} is the reference wavefront intensity distribution in the unloaded state, $i_{M1} = 2\sqrt{i_{o1} \cdot i_{r1}}$ is the intensity modulation in the unloaded state, φ is the random phase distribution in the image plane caused by surface roughness of the studied surface area, α_{21s} and α_{31s} are the unknown (arbitrary) phase shifts of the reference wave relative to its initial position.

To find the unknown phase shifts α_{21s} and α_{31s}, one can use Eq. (4.35) based on the PPCC calculation [19]. Modifications of this equation for the TF DSPI method are given by

$$\alpha_{21s} = \arccos \rho(i_{11}, i_{12}) = \arccos \frac{\langle (i_{11} - \langle i_{11} \rangle)(i_{12} - \langle i_{12} \rangle) \rangle}{\sigma_{11}\sigma_{12}}, \tag{5.28}$$

$$\alpha_{31s} = \arccos \rho(i_{11}, i_{13}) = \arccos \frac{\langle (i_{11} - \langle i_{11} \rangle)(i_{13} - \langle i_{13} \rangle) \rangle}{\sigma_{11}\sigma_{13}}, \tag{5.29}$$

where σ_{11}, σ_{12} and σ_{13} are the RMS values of the SIs i_{11}, i_{12} and i_{13}, respectively.

One can also calculate the unknown phase shifts using the SPCC and Eq. (5.6). For the TF DSPI method, the formula for calculating α_{21s} is identical to (5.6) and the formula for calculating α_{31s} is given by

$$\begin{aligned} \alpha_{31s} &= \arccos[r(i_{11}, i_{13})] \\ &= \arccos \frac{\sum_{j=0}^{N-1} \left(i_{11j} - \bar{i}_{11}\right)\left(i_{13j} - \bar{i}_{13}\right)}{\sqrt{\sum_{j=0}^{N-1} \left(i_{11j} - \bar{i}_{11}\right)^2} \sqrt{\sum_{j=0}^{N-1} \left(i_{13j} - \bar{i}_{13}\right)^2}}. \end{aligned} \tag{5.30}$$

In the next loaded state of the studied specimen, the intensity distributions of the second series of digital SIs are given by

$$\left.\begin{aligned} i_{21} &= i_{B2} + i_{M2}\cos(\varphi + \delta\varphi) \\ i_{22} &= i_{B2} + i_{M2}\cos(\varphi + \delta\varphi + \tilde{\alpha}_{21s}) \\ i_{23} &= i_{B2} + i_{M2}\cos(\varphi + \delta\varphi + \tilde{\alpha}_{31s}) \end{aligned}\right\}, \tag{5.31}$$

where $i_{B2} = i_{o2} + i_{r2}$ is the background intensity distribution in the loaded state, i_{o2} is the object wavefront intensity distribution in the loaded state, i_{r2} is the reference wavefront intensity distribution in the loaded state, $i_{M2} = 2\sqrt{i_{o2} \cdot i_{r2}}$ is the intensity modulation in the loaded state.

In this system of equations, the unknown phase shifts $\tilde{\alpha}_{21s}$ and $\tilde{\alpha}_{31s}$ can be also calculated using both the PPCC and SPCC. Similarly to Eqs. (5.28) and (5.29) based on the PPCC, the unknown phase shifts $\tilde{\alpha}_{21s}$ and $\tilde{\alpha}_{31s}$ can be expressed as

$$\tilde{\alpha}_{21s} = \arccos\rho(i_{21}, i_{22}) = \arccos\frac{\langle(i_{21} - \langle i_{21}\rangle)(i_{22} - \langle i_{22}\rangle)\rangle}{\sigma_{21}\sigma_{22}}, \tag{5.32}$$

$$\tilde{\alpha}_{31s} = \arccos\rho(i_{21}, i_{23}) = \arccos\frac{\langle(i_{21} - \langle i_{21}\rangle)(i_{23} - \langle i_{23}\rangle)\rangle}{\sigma_{21}\sigma_{23}}, \tag{5.33}$$

where σ_{21}, σ_{22} and σ_{23} are the RMS values of the SIs i_{21}, i_{22} and i_{23}, respectively.

The calculation of $\tilde{\alpha}_{21s}$ using the SPCC is performed according to Eq. (5.7), and the formula for $\tilde{\alpha}_{31s}$ is given by looks like

$$\begin{aligned} \tilde{\alpha}_{31s} &= \arccos[r(i_{21}, i_{23})] \\ &= \arccos\frac{\sum_{j=0}^{N-1}\left(i_{21j} - \bar{i}_{21}\right)\left(i_{23j} - \bar{i}_{23}\right)}{\sqrt{\sum_{j=0}^{N-1}\left(i_{21j} - \bar{i}_{21}\right)^2}\sqrt{\sum_{j=0}^{N-1}\left(i_{23j} - \bar{i}_{23}\right)^2}}. \end{aligned} \tag{5.34}$$

To find the required surface displacement phase field $\delta\varphi(k, l)$, the phase surface $\varphi(k, l)$ before loading is separated from the system of Eq. (5.27), and the phase surface $[\varphi(k, l) + \delta\varphi(k, l)]$ after loading is separated from the system of Eq. (5.31). To perform this procedure, the systems of Eqs. (5.27) and (5.31) are reduced to the systems containing two equations. For this purpose, we first convert (5.27) and (5.31) to the following form:

$$\left.\begin{aligned} i_{11} - i_{B1} &= i_{M1}\cos\varphi \\ i_{12} - i_{B1} &= i_{M1}\cos(\varphi + \alpha_{21s}) \\ i_{13} - i_{B1} &= i_{M1}\cos(\varphi + \alpha_{31s}) \end{aligned}\right\}, \tag{5.35}$$

$$\left.\begin{aligned} i_{21} - i_{B2} &= i_{M2}\cos(\varphi + \delta\varphi) \\ i_{22} - i_{B2} &= i_{M2}\cos(\varphi + \delta\varphi + \tilde{\alpha}_{21s}) \\ i_{23} - i_{B2} &= i_{M2}\cos(\varphi + \delta\varphi + \tilde{\alpha}_{31s}) \end{aligned}\right\}, \tag{5.36}$$

and then divide every second and third equation in (5.35) and (5.36) by the first one. As a result, we get

$$\left.\begin{aligned}\frac{i_{12}-i_{B1}}{i_{11}-i_{B1}} &= \frac{\cos(\varphi+\alpha_{21s})}{\cos\varphi} = \cos\alpha_{21s} - \tan\varphi\sin\alpha_{21s} \\ \frac{i_{13}-i_{B1}}{i_{11}-i_{B1}} &= \frac{\cos(\varphi+\alpha_{31s})}{\cos\varphi} = \cos\alpha_{31s} - \tan\varphi\sin\alpha_{31s}\end{aligned}\right\}, \tag{5.37}$$

$$\left.\begin{aligned}\frac{i_{22}-i_{B2}}{i_{21}-i_{B2}} &= \frac{\cos(\varphi+\delta\varphi+\tilde{\alpha}_{21s})}{\cos(\varphi+\delta\varphi)} = \cos\tilde{\alpha}_{21s} - \tan(\varphi+\delta\varphi)\sin\tilde{\alpha}_{21s} \\ \frac{i_{23}-i_{B2}}{i_{21}-i_{B2}} &= \frac{\cos(\varphi+\delta\varphi+\tilde{\alpha}_{31s})}{\cos(\varphi+\delta\varphi)} = \cos\tilde{\alpha}_{31s} - \tan(\varphi+\delta\varphi)\sin\tilde{\alpha}_{31s}\end{aligned}\right\}. \tag{5.38}$$

Solution of the systems of Eqs. (5.37) and (5.38) concerning the phase φ is given by

$$\tan\varphi = -\frac{a_1}{b_1}, \tag{5.39}$$

where

$$a_1 = (i_{12} - i_{13}) + (i_{13} - i_{11})\cos\alpha_{21s} + (i_{11} - i_{12})\cos\alpha_{31s}, \tag{5.40}$$

$$b_1 = (i_{11} - i_{13})\sin\alpha_{21s} + (i_{12} - i_{11})\sin\alpha_{31s}; \tag{5.41}$$

and

$$\tan(\varphi + \delta\varphi) = -\frac{a_2}{b_2}, \tag{5.42}$$

where

$$a_2 = (i_{22} - i_{23}) + (i_{23} - i_{21})\cos\tilde{\alpha}_{21s} + (i_{21} - i_{22})\cos\tilde{\alpha}_{31s}, \tag{5.43}$$

$$b_2 = (i_{21} - i_{23})\sin\tilde{\alpha}_{21s} + (i_{22} - i_{21})\sin\tilde{\alpha}_{31s}. \tag{5.44}$$

The searched surface displacement wrapped phase map $\delta\varphi(k, l)$ is calculated using Eqs. (5.39) and (5.42) and is given by

$$\begin{aligned}\delta\varphi(k,l) &= [\varphi(k,l) + \delta\varphi(k,l)] - \varphi(k,l) \\ &= \arctan\left[-\frac{a_2(k,l)}{b_2(k,l)}\right] - \arctan\left[-\frac{a_1(k,l)}{b_1(k,l)}\right].\end{aligned} \tag{5.45}$$

As was mentioned above, all SIs of the first series, i.e. i_{11}, i_{12} and i_{13} and the second series, i.e. i_{21}, i_{22} and i_{23}, are recorded with the help of the integrating-bucket technique. The saw-tooth diagrams of one of possible version of these SIs recording by using the reference beam smooth linear phase shifts in the initial and loaded states of a studied surface area are shown in Fig. 5.2. In this figure, the upper

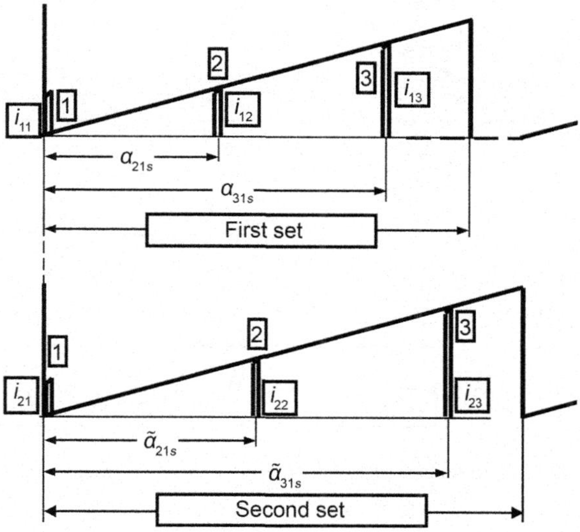

Fig. 5.2 Implementation of the integrating-bucket technique to record SIs in the TF DSPI method: recording of three SIs i_{11}, i_{12} and i_{13} in the unloaded state of the studied specimen (the upper diagram); recording of three SIs i_{21}, i_{22} and i_{23} in the loaded state of the studied specimen (the lower diagram)

diagram indicates the recording of three SIs i_{11}, i_{12} and i_{13} from the first series, that is, in the unloaded state of the studied specimen. The lower diagram shows the recording of three SIs i_{21}, i_{22} and i_{23} from the second series, that is, in the loaded state of the studied specimen.

The obtained wrapped phase map $\delta\varphi(k, l)$ completes only the first stage of the surface displacement field retrieval. This stage is finished by the double expanding of the wrapped phase map range. As a result, the wrapped phase map $\delta\varphi_w(k, l)$ is generated in the phase range adjustment $-\pi \leq \delta\varphi_w(k, l) \leq \pi$. To fulfill this adjustment, one can jump from the values (a_1/b_1) and (a_2/b_2), which are the input parameters in Eq. (5.45), to two pairs of input parameters a_1, b_1 and a_2, b_2, that are the functions of the sine and cosine values, respectively. This jump can be implemented by using the "atan2" function [4] (see Sect. 2.5.1). The flow-chart of the first stage of the TF DSPI method implementation is shown in Fig. 5.3. Note that both the SPCC and the PPCC can be used to determine the unknown phase shifts in this method. Retrieval of a final surface displacement phase map of the studied object from the wrapped phase map $\delta\varphi_w(k, l)$ is finished by the standard unwrapping procedure.

Thus, new TF DSPI method with unknown phase shifts of a reference wave is proposed that makes it possible to retrieve the displacement fields of the specimen surface during its dynamic changes. To perform this method, new correlation approach to extract the unknown phase shifts using the SPCC is developed [22, 23]. The method makes it possible to use the integrating-bucket technique [11, 28, 37], due to which the speed of SIs recording before and after applying the loads raises. Therefore, this method is more suitable for the fast reconstruction of in-plane and out-of-plane surface displacements compared to the TS DSPI method.

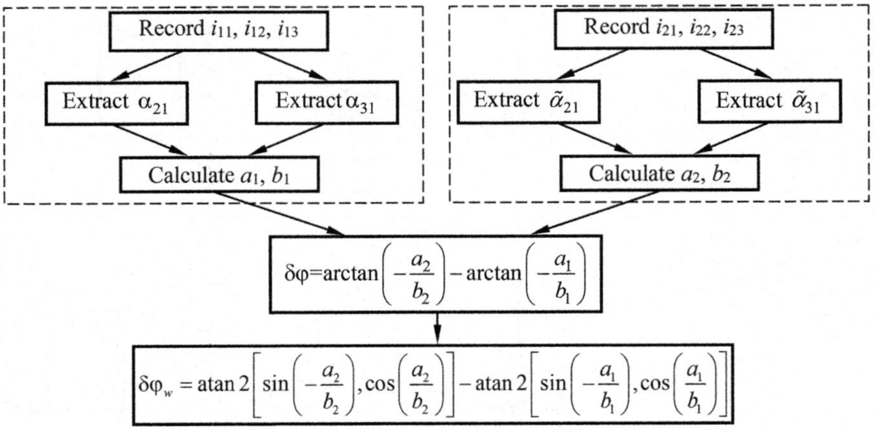

Fig. 5.3 Flow-chart of the first stage of the TF DSPI method implementation

5.4 Fringe Projection Interferometry for Surface Nondestructive Testing

Fringe projection techniques have become very popular in recent years due to application of relatively simple equipment and the ability to study the surfaces of large sizes with significant changes in shape and form [12, 34]. At the same time, the accuracy of these techniques depends on the available equipment, as well as the conditions and methods of the surface topography registration and restoration.

5.4.1 Main Principles of Fringe Projection Techniques

The fringe projection techniques use, as a rule, parallel linear fringes with sinusoidal or structured profiles across the fringes, which are projected on the investigated surface. The main direction of these techniques development uses commercial multimedia projectors, in particular, digital micromirror devices and liquid crystal display projectors for generating linear periodic sinusoidal fringes of variable spatial frequency or structured fringes with other types of profiles [14] and projecting them onto the studied surface [43]. Fringes can have various profiles, namely sinusoidal, U-shaped, triangular, etc. Different gratings also used to modulate the projected light [43]. Several other fringe projection methods are also used to produce fringe patterns on the studied surface [10]. They possess some advantages and lacks in comparison with the fringe projection using multimedia projectors. For example, the fringe projection method performed with the help of laser interferometric systems makes it possible to restore surfaces with big relief depth [5, 33]. Lasers and light-emitting diodes (LEDs) with low spatial coherence are used in these systems [5, 10, 26, 33].

Schemes of two different systems for designing periodic fringes are shown in Fig. 5.4. The first system is based on a commercial multimedia projector, and the second one uses an optical interferometer capable to generate a parallel wavefront and project interference fringe patterns onto the studied surface. Such types of interferometers can be used to evaluate the surface structure, because they provide the projection of interference fringe patterns on the studied surface with a high fringe density. This makes it possible to retrieve the waviness and form of the surface and even assess its roughness [5]. If a Twyman–Green interferometer is applied to produce fringe patterns (see Fig. 5.4b), the interference fringes with sinusoidal intensity distribution are formed due to a very small slope between two parallel laser beams, and the period p of fringes is given by

$$p = \frac{\lambda}{2\sin(\psi/2)}, \tag{5.46}$$

where λ is the laser radiation wavelength, ψ is the angle between two parallel beams.

As shows Eq. (5.46), one can use the angle ψ to adjust the fringe width and the period p. In particular, if the angle ψ decreases, the period p increases and vice versa. If the interference fringes are projected onto a flat surface at an incidence angle θ', then the period p_0 of fringes in its plane is given by

$$p_0 = \frac{p}{\cos\theta'} = \frac{\lambda_l}{2\sin(\psi/2)\cos\theta'}. \tag{5.47}$$

The fringe projection interferometry is based on the principle of triangulation [12]. If the sinusoidal fringe pattern falls on the studied diffuse 3D surface, then a fringe pattern of this surface is formed in the plane of the matrix photodetector of the digital camera. This pattern can be characterized as the surface interferogram or SI [5]. In the interferogram or in the SI, the linear fringes falling on the surface are

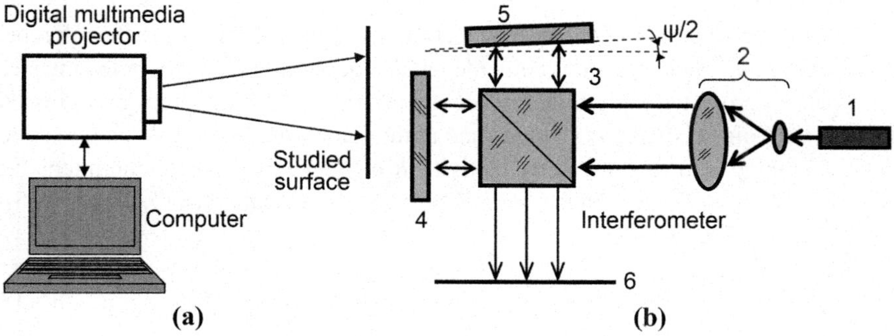

Fig. 5.4 Scheme of fringe projection system based on a multimedia projector (**a**); Twyman – Green interferometer optical scheme to generate linear equidistant fringe patterns (**b**): laser or LED (1); collimator (2); beam splitter (3); mirrors (4) and (5); studied surface (6)

distorted due to its irregularities and surface roughness. As in the previous chapters, the interferogram and SI intensity distributions are presented in the form of the fundamental equation for phase shifting interferometry, which differs from it only by the factor $R(k, l)$ describing the reflectivity of the studied surface [5, 12] and is given by

$$I(k,l) = R(k,l)[I_B(k,l) + I_M(k,l)\cos\varphi(k,l)] \tag{5.48}$$

for interferograms and

$$i(k,l) = R(k,l)[i_B(k,l) + i_M(k,l)\cos\varphi(k,l)] \tag{5.49}$$

for SIs.

Consider the one shown in Fig. 5.5 scheme of generating the interference fringes pattern on the studied object surface. Let a collimated parallel monochromatic beam falls on the surface at point D and virtually on the reference plane at point A. From the side of the observer, point A moves to a new location at point C. The distance $\Delta x = \mathrm{AC}$ between points A and C carries information about the object height $Z = h(x, y)$. A monochromatic pattern of interference fringes modulated by uneven surface topography is formed in an optical telecentric system and recorded by a digital camera. The sinusoidal pattern of straight parallel equidistant interference fringes projected onto the object reference plane has a period p_0. The intensity distribution of these fringes at an arbitrary point C is given by [12]

$$I_C = I_B + I_M\cos\left(2\pi \cdot \mathrm{OC}/\,p_0\right) \tag{5.50}$$

for interferograms or

$$i_C = i_B + i_M\cos\left(2\pi \cdot \mathrm{OC}/\,p_0\right) \tag{5.51}$$

for SIs.

It is assumed that point O is the intersection of the optical axis z with the reference plane xOy, and that this point is coincided with a maximum intensity of the interference pattern projected onto the reference plane. A matrix photodetector records the intensity at point C of the xOy plane and at the same time at point D on the object surface. Taking into account the reflectivity R of the studied object, the intensities observed at points D and A in the xOy plane are the same, and Eqs. (5.50) and (5.51) can be represented as [12]

$$I_D = R\left[I_B + I_M\cos\left(2\pi \cdot \mathrm{OA}/\,p_0\right)\right] \tag{5.52}$$

for interferograms or

$$i_D = R\left[i_B + i_M\cos\left(2\pi \cdot \mathrm{OA}/\,p_0\right)\right] \tag{5.53}$$

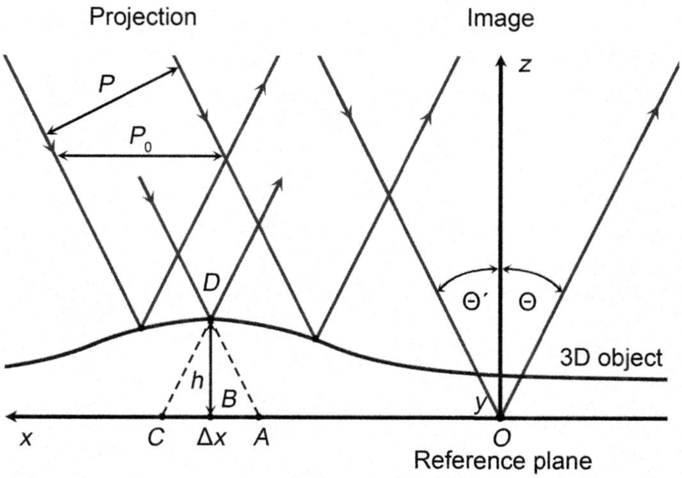

Fig. 5.5 Triangulation scheme for projecting a pattern of strait interference fringes onto the specimen surface: θ' is the angle of incidence, θ is the angle of observation

for SIs.

To relate the phase difference φ_{CD} in phase values between points C and D, which is observed by the same matrix photodetector, with the metric distance AC, it is sufficient to consider the following proportion between phase and metric values:

$$\frac{p_0}{\mathrm{AC}} = \frac{2\pi}{\varphi_{\mathrm{CD}}}. \tag{5.54}$$

This proportion allows determining the distance AC, that is

$$\mathrm{AC} = \frac{p_0 \varphi_{\mathrm{CD}}}{2\pi}. \tag{5.55}$$

On the other hand, the distance AC is related with the object high $h = \mathrm{BD}$ that can be expressed as [12]

$$h_\Lambda = \frac{\mathrm{AC}}{\tan\theta' + \tan\theta} = \frac{p_0 \varphi_{\mathrm{CD}}}{2\pi(\tan\theta' + \tan\theta)}, \tag{5.56}$$

where θ is the angle of observation.

If the direction of observation is perpendicular to the reference xOy plane, then $\tan\theta = 0$ and Eq. (5.56) is given by

$$h_{\lambda'} = \frac{p_0 \varphi_{\mathrm{CD}}}{2\pi \cdot \tan\theta'}, \tag{5.57}$$

and vice versa—if the direction of the light beam incidence is perpendicular to the reference xOy plane, then $\tan\theta' = 0$ and Eq. (5.56) is given by

$$h_\lambda = \frac{p_0 \varphi_{CD}}{2\pi \cdot \tan\theta}. \quad (5.58)$$

Depending on how the collimated light beam is incident and reflected, one can determine the equivalent or effective wavelength [12] for the interferometric fringe projection system.

So, if $\tan\theta = 0$ and $\tan\theta' \neq 0$, the equivalent wavelength λ'_e is given by

$$\lambda'_e = p_0 \big/ \tan\theta', \quad (5.59)$$

and Eq. (5.57) is rewritten as

$$h_{\lambda'} = \frac{\lambda'_e \cdot \varphi_{CD}}{2\pi}. \quad (5.60)$$

If $\tan\theta' = 0$ and $\tan\theta \neq 0$, the equivalent wavelength λ_e is given by [12]

$$\lambda_e = p_0 \big/ \tan\theta, \quad (5.61)$$

and Eq. (5.58) is rewritten as

$$h_\lambda = \frac{\lambda_e \cdot \varphi_{CD}}{2\pi}. \quad (5.62)$$

And finally, if $\tan\theta' \neq 0$ and $\tan\theta \neq 0$, the equivalent wavelength Λ_e is given by

$$\Lambda_e = p_0 \big/ \left(\tan\theta' + \tan\theta\right), \quad (5.63)$$

and Eq. (5.56) is rewritten as

$$h_\Lambda = \frac{\Lambda_e \varphi_{CD}}{2\pi}. \quad (5.64)$$

5.4.2 *Application of Method of Three-Frame Fringe Projection Interferometry to Retrieve Surface Relief Using Interference Fringe Patterns*

The new method of three-frame fringe projection interferometry (TF FPI) with unknown phase shifts of the reference wave is proposed to retrieve the phase map of the studied optically rough surface area by projecting equidistant straight sinusoidal interference fringes. It is based on both the TF DSPI method and the 3SI UPS method considered in Sect. 2.6. In the TF FPI method, it is enough to register only three SIs at the fixed state of the studied surface area. These SIs can be described by the system of fundamental Eq. (5.27), which for convenience are rewritten as

$$\left.\begin{array}{l} i_{11} = i_{B1} + i_{M1}\cos(\varphi_{\mathrm{CD}})_w \\ i_{12} = i_{B1} + i_{M1}\cos\left[(\varphi_{\mathrm{CD}})_w + \alpha_{21s}\right] \\ i_{13} = i_{B1} + i_{M1}\cos\left[(\varphi_{\mathrm{CD}})_w + \alpha_{31s}\right] \end{array}\right\}, \tag{5.65}$$

where $[\varphi_{\mathrm{CD}}(k,l)]_w$ is the wrapped coarse phase map of phase differences between points C and D in each its k, lth pixel (see Fig. 5.5 and Eq. 5.54), α_{21s} and α_{31s} are the unknown phase shifts of the reference wave relative to its initial position calculated using the PPCC or the SPCC and Eqs. (5.6), (5.28)–(5.30).

Algorithm implementing the TF FPI method consists of two stages, and its flowchart is shown in Fig. 5.6. At the algorithm first stage marked by the dashed-dotted contour in Fig. 5.6, the wrapped coarse phase map $[\varphi_{\mathrm{CD}}(k,l)]_w$ of the test surface area is defined using Eq. (5.37), which can be rewritten as

$$\left.\begin{array}{l} \frac{i_{12}-i_{B1}}{i_{11}-i_{B1}} = \frac{\cos\left[(\varphi_{\mathrm{CD}})_w+\alpha_{21s}\right]}{\cos(\varphi_{\mathrm{CD}})_w} = \cos\alpha_{21s} - \tan\varphi\sin\alpha_{21s} \\ \frac{i_{13}-i_{B1}}{i_{11}-i_{B1}} = \frac{\cos\left[(\varphi_{\mathrm{CD}})_w+\alpha_{31s}\right]}{\cos\varphi} = \cos\alpha_{31s} - \tan\varphi\sin\alpha_{31s} \end{array}\right\}. \tag{5.66}$$

The solution of these equations with respect to the phase $[\varphi_{\mathrm{CD}}(k,l)]_w$ is given by

$$\tan(\varphi_{\mathrm{CD}})_w = -\frac{a_1}{b_1}, \tag{5.67}$$

where a_1 and b_1 are calculated from Eqs. (5.40) and (5.41).

As in previous algorithms implementing the 3SI UPS and TF DSPI methods, the first stage is finished by the double expanding of the wrapped phase map $(\varphi_{\mathrm{CD}})_w$ range. As a result, this map is generated in the phase range adjustment $-\pi \le \varphi_{\mathrm{CD}w} \le \pi$.

As a rule, the studied surface area, on which the interference fringes are projected, is optically rough. Therefore, the second stage of this algorithm makes it possible to retrieve only the macrorelief phase map $\varphi_M(k,l)$ containing waviness and form, while removing the surface microrelief containing speckle noise. Compared to the algorithm that implements the 3SI UPS method (see Sect. 2.6.1), the second stage of which is marked by a dashed contour in Fig. 2.20 (see Sect. 2.5.5), the second stage of the algorithm implementing the TF FPI method has some differences. First iteration in its second stage is the same as first iterations of the algorithms performing the ITSI UPS (see Sect. 2.5.5) and 3SI UPS (see Sect. 2.6.1) methods. After the first iteration, the coarse wrapped phase map $\tilde{\varphi}_{\mathrm{mc}}(k,l)$ of the surface macrorelief is extracted using the cut-off spatial filtering of the expanded coarse phase map $(\varphi_w)_{\exp 1}$ with cut-off frequency f_{c0}. As mentioned in Sect. 2.5.5, this frequency is determined by calculation the maximum frequency of a saw-tooth fringe structure

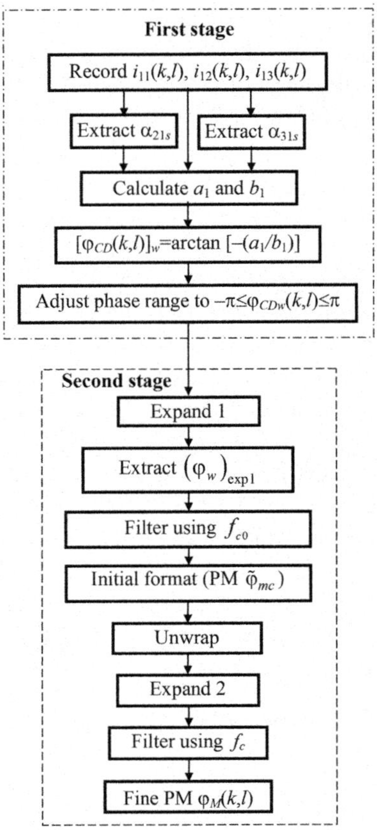

Fig. 5.6 Flow-chart of the modified two-stage algorithm performing the method of three-frame fringe projection interferometry with unknown phase shifts of the reference wave

of the initial wrapped phase map $\varphi_{\text{CD}w}(k, l)$ using the algorithm for estimating the fringe orientation and fringe density [39]. The level of speckle noise in the phase map $\varphi_{\text{CD}w}(k, l)$ depends on the coherence length of the light source in the interferometer. The longer the coherence length, the higher the speckle noise level. If a single mode laser is used, the level of the speckle noise is comparable with the microrelief surface roughness and often exceeds them. Since surface microrelief cannot be separated from speckle noise, the second iteration is used to extract only the fine phase map of the macrorelief.

Since the main task of the second iteration at the second stage of the algorithm performance is to cut-off the speckle noise and extract the macrorelief phase map $\varphi_M(k, l)$, the choice of the optimal filter is not limited to the choice of a low-frequency linear *L*-filter [13] (see Sect. 2.5.5), which separates the roughness from the waviness and form. Therefore, the choice of frequency f_c at the second filtering of the coarse phase map $(\varphi_w)_{\exp 1}$ depends of several factors, including the type of the light source in the interferometer and the level of surface roughness of the studied object.

To avoid the edge effects arising due to application of the ideal cut-off filter, the Gaussian and Butterworth filters with cut-off frequencies f_{c0} and f_c were used [20, 32]. During performing this algorithm, phase unwrapping procedures were implemented using the Flynn [6], Goldstain [7] and CPULSI [38] algorithms.

The ability to restore the height map can be estimated using Eqs. (5.60), (5.62) and (5.64). As it was shown in Sects. 2.5.6 and 2.7.3, the 2π ambiguity problem will not occur if the height difference δh_{adj} between adjacent samples of the surface relief does not exceed $\lambda/4$ [18, 27] (see also Eqs. 2.2, 2.14). In fringe projection interferometry, the permissible relief height increment between two adjacent pixels also should not exceed one-fourth of the equivalent wavelength, that is

$$\max(h_{\lambda'}) = \frac{\lambda'_e}{4},\ \max(h_{\lambda}) = \frac{\lambda_e}{4},\ \max(h_{\Lambda}) = \frac{\Lambda}{4}. \tag{5.68}$$

Taking into account (5.68) and using (5.59), (5.61) and (5.63), it is possible to determine the permissible height increments, which allows restoring the surface topography with sufficient accuracy by fringe projection interferometry. For example, if in the fringe projection interferometer $p_0 = 30\ \mu\text{m}$, $\theta = 15°$ and $\theta' = 0°$, the maximum permissible height increment between two adjacent pixels according Eqs. (5.61) and (5.68) is 28.0 μm.

Thus, the fringe projection interferometry makes it possible to significantly expand the range of the surface relief roughness retrieving, which opens wide prospects for its practical use.

5.4.3 *Experimental Results for Surface Relief Retrieval Using Interference Fringe Patterns*

To generate linear interference fringes on the specimen surface, an optical scheme of the Twyman–Green interferometer was constructed in an optical-digital experimental setup dedicated for recording the interference fringe patterns and retrieving the studied surface relief. The optical scheme in the interferometer can be considered as the telecentric one due to application of a "Jupiter-37A" lens with a specially designed adapter that connects the lens to a digital camera. Figure 5.7 shows the experimental setup scheme containing a laser source 1 (a laser diode PL 450B with wavelength of $\lambda = 450$ nm and a small coherence length of about 1.0 mm), an optical collimator 2 that expands the laser beam and forms a parallel laser beam with a diameter of 20 mm and a Gaussian intensity distribution in the cross-section of ~ 90% at the edge of the beam from its intensity in the center. The parallel beam inputs to the Twyman–Green interferometer scheme, due to which the linear parallel equidistant interference fringes with spatial period p falls on the studied surface. To adjust the optical scheme and precisely align the optical patches in the two arms of the interferometer, the mirror 5 was fixed in a micropositioning device. If the fringes are projected at an angle θ' relative to the reference plane (see Fig. 5.7), then their spatial period becomes equal to p_0 (see Eq. 5.47 and Fig. 5.5). A digital camera 9 records the interferograms I_1, I_2 and I_3, if the specimen surface area is smooth, and SIs i_{11}, i_{12} and i_{13}, if the surface area is optically rough. The SONY XCD-SX910 CCD camera with a 1/2 "IT-CCD-SXGA" CCD matrix was used to record SIs. The unit 10 was used to control the PSE 4, which provided a mode of smooth linear phase shift from 0 to 180° according to the integrating-bucket technique by applying a saw-tooth voltage to the PSE.

In this setup, three interferograms I_1, I_2 and I_3 or three SIs i_{11}, i_{12} and i_{13} are recorded during the time of applying one voltage saw to the PSE. The operation of the experimental setup is driven by a controller 10. The obtained interferograms or SIs are processed in computer 11 with the help of the software product realizing the modified two-stage algorithm for the surface relief retrieval, which flow-chart is shown in Fig. 5.6. The unwrapped phase map can be easily converted into a height map using Eqs. (5.60), (5.62) and (5.64), which were applied depending on the configuration of the optical system generating and recording the interference fringes.

For a clear example of demonstrating the reliability of the surface relief retrieval using the fringe projection interferometry, the medallion shown in the Fig. 5.8 was chosen. Interference fringe patterns of the medallion surface area marked with a rectangular contour were recorded using the integrating-bucket technique. To generate these patterns, periodic interference fringes with period $p = 176.43\ \mu\text{m}$ were formed by the interferometer optical scheme and projected on the chosen surface area with a linear magnification of $M = 0.52$. The sum of the incidence and observation angles

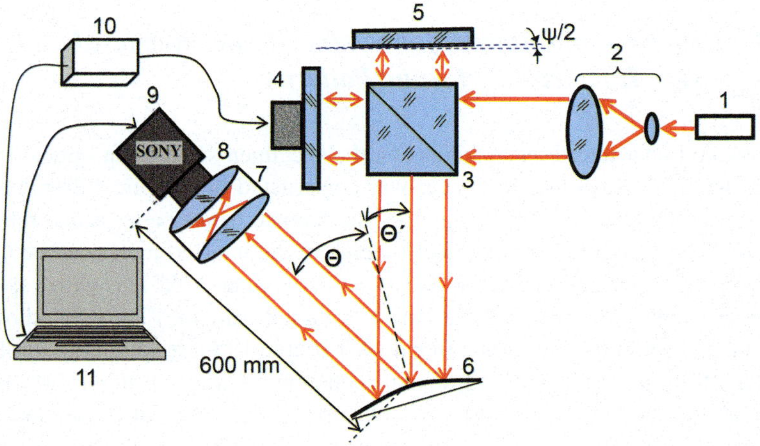

Fig. 5.7 Optical-digital experimental setup for recording the interference fringe patterns and retrieving the studied surface relief: laser source containing a laser diode PL 450B ($\lambda = 450$ nm) (1); optical collimator (2); beam splitter (3); phase shifting element (4); deflecting mirror (5); studied surface area (6); lens "Jupiter-37A" (7); lens adapter (8); SONY XCD-SX910 CCD camera (9); controller (10); computer (11)

Fig. 5.8 Studied object. The medallion surface area marked with a rectangular contour

was $\theta' + \theta = 21.2°$. The recorded series of three interference fringes is shown in Fig. 5.9.

The studied surface area retrieval was performed using the modified two-stage algorithm, which flow-chart is shown in Fig. 5.6. An ideal cut-off filter f_{c0} and a Gaussian filter f_c were chosen taking into account the saw-tooth fringe structure of the initial wrapped phase map $\varphi_w(k, l)$ and the level of speckle noise generated by the laser source 1 shown in Fig. 5.7. The form (general shape) of the studied surface area, which was filtered from the roughness of this area and speckle noise, is shown in Fig. 5.10.

The correct restoration of the selected surface area using the recorded interference fringes patterns made it possible to verify the reliability of the proposed method and to conduct experiments for retrieving the surface relief of the structural material. In particular, the fracture surface relief of a diamond drill 8 mm in diameter shown in Fig. 5.11a was retrieved. To reduce speckle noise, the laser source containing a laser diode PL 450B was used to produce interference fringes projected on the diamond drill fracture surface in the optical-digital experimental setup shown in Fig. 5.7. Figure 5.11b shows the retrieved full relief of the fracture surface, which is filtered from the nanorelief and speckle noise using Gaussian filter with cut-off frequency f_c. Unfortunately, we have not yet developed an approach to choosing the optimal f_c-filter, since its choice depends on several factors, including both the level of

(a) (b) (c)

Fig. 5.9 Three interference fringe patterns (**a**), (**b**) and (**c**) of the surface area marked on the medallion, which were recorded using the integrated-bucket technique

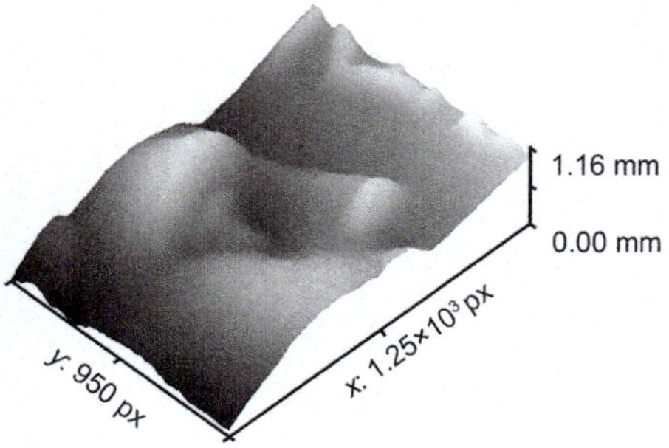

Fig. 5.10 Form (general shape) of the surface area marked on the medallion, which was obtained after digital processing of the recorded interference fringe patterns shown in Fig. 5.9, using the modified two-stage algorithm (see Fig. 5.6) performing the TF FPI method. The surface area size is $11.5 \times 8.6\ \text{mm}^2$

speckle noise and the level of surface roughness, waviness and form. Therefore, by trial and error, the 3D surface shape was retrieved including the form and microrelief, using the modified two-stage algorithm (see Fig. 5.6) performing the TF FPI method.

Figure 5.12a shows the fracture surface of the diamond drill and the retrieved surface relief profile (black line) located across the slip bands of the fracture surface. The distribution of form and microroughness along this profile is shown in Fig. 5.12b.

Thus, the new method of fringe projection interferometry for retrieving the complex 3D surface relief containing sharp drops in peaks and valleys is developed. In contrast to the similar method developed by Dhanasekar and Ramamoorthy [5], in which the conventional five frame algorithm with given phase shifts is used, our method uses only two arbitrary phase shifts to record three SIs during continuous phase scanning of the surface using the integrated-bucket technique. The possibilities of the three-frame projection interferometry method have not yet been fully studied. Therefore, its further development will be a theoretical and experimental analysis of the relationship between the limit of retrieving the maximum drops in surface

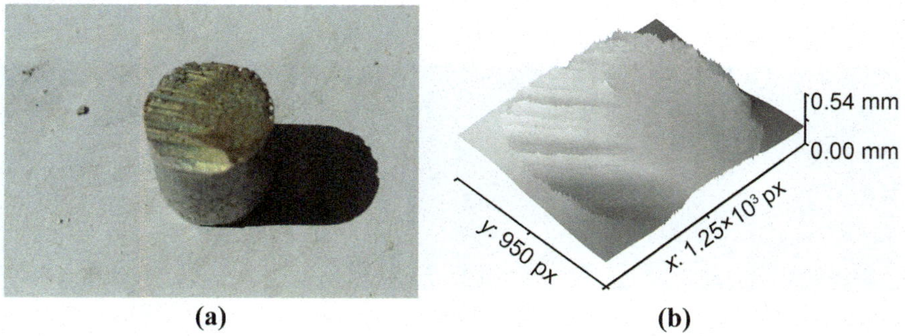

Fig. 5.11 Appearance of the diamond drill Ø8 mm (**a**) and its retrieved 3D surface shape, including the form and microrelief (**b**)

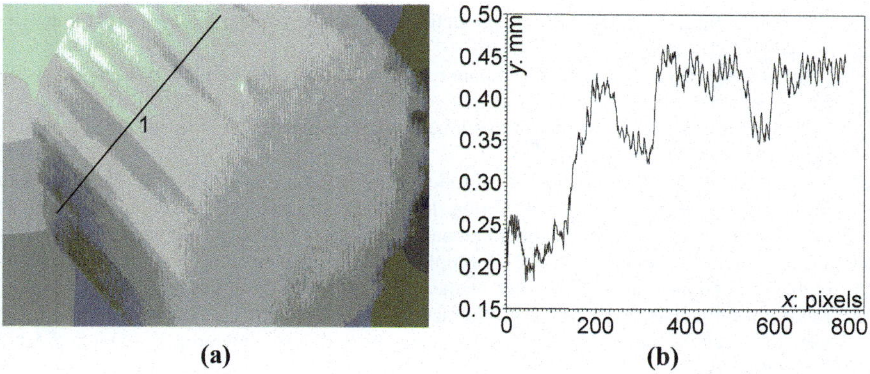

Fig. 5.12 Retrieved fracture surface of the diamond drill, on which the profile located across the slip bands is marked by a black line (**a**); distribution of the shape and microroughness along the indicated profile of the fracture surface (**b**)

peaks and valleys and the limit of the optical-digital system resolution when a 3D surface topography is restored. Such an analysis will make it possible to predict the parameters of the optical system of the interferometer, at which it would be possible to retrieve the roughness, waviness and form of the structural material surface. Since fringe projection interferometry is based on the triangulation approach, it can be used to retrieve surfaces of materials with high roughness, as well as to retrieve sharp surface differences, which is especially important when analyzing surfaces of drilling tools and other surface fractures, as well as products with an odd-shaped surface.

References

1. Almazan-Cuellar S, Malacara-Hernandez D (2003) Two-step phase-shifting algorithm. Opt Eng 42(12):3524–3531
2. Bobkov YN, Molodov VD (1969) Sposob izmierieniya fazovogo sdviga (Method for phase shift measurement), USSR Patent No. 256,867. Bulletin of Inventions 35, Moscow
3. Bobkov YN (1984) Korreliatsionnyye mietody i ustroystva izmierieniya sostavliayushchih komponientov signalov (Correlation methods and devices for measuring signal components). Vyshcha Shkola, Lviv
4. Burger W, Burge MJ (2013) Principles of digital image processing—advanced techniques, vol 3. Springer-Verlag, London
5. Dhanasekar B, Ramamoorthy B (2008) Digital speckle interferometry for assessment of surface roughness. Opt Lasers Eng 46(3):272–280
6. Flynn TJ (1997) Two-dimensional phase unwrapping with minimum weighted discontinuity. J Opt Soc Am A 14(10):2692–2701
7. Goldstein RM, Zebker HA, Werner CL (1988) Satellite radar interferometry: two-dimensional phase unwrapping. Radio Sci 23(4):713–720
8. Goodman JW (2000) Statistical optics. John Wiley and Sons, New York
9. Goodman JW (2007) Speckle phenomena in optics: theory and applications. Roberts and Company Publishers, Greenwood Village CO

10. Gorthi SS, Rastogi P (2010) Fringe projection techniques: whither we are? Opt Lasers Eng 48(ARTICLE):133–140
11. Greivenkamp JE (1984) Generalized data reduction for heterodyne interferometry. Opt Eng 23(4):350–352
12. Halioua M, Liu HC (1989) Optical three-dimensional sensing by phase measuring profilometry. Opt Lasers Eng 11(3):185–215
13. ISO 25178–2012 (2012) Geometrical Product Specification (GPS)—Surface Texture: areal. Surface texture indications (Part 1); Terms, definitions and surface texture parameters (Part 2); Specification operators (Part 3). International Organization for Standardization, Geneva
14. Jia P, Kofman J, English C (2007) Multiple-step triangular-pattern phase shifting and the influence of number of steps and pitch on measurement accuracy. Appl Opt 46(16):3253–3262
15. Kerr D, Santoyo FM, Tyrer JR (1990) Extraction of phase data from electronic speckle pattern interferometric fringes using a single-phase-step method: a novel approach. J Opt Soc Am A 7(5):820–826
16. Lyon DA (2010) The discrete fourier transform, part 6: cross-correlation. J Object Technol 9:17–22
17. Maas AA (1992) Shape measurement using phase-shifting speckle interferometry. In: Pryputniewicz R (ed) Laser interferometry IV: computer-aided interferometry, vol 1553. SPIE, Bellingham, WA, pp 558–568
18. Malacara D, Servín M, Malacara Z (2005) Interferogram analysis for optical testing, 2nd edn. Taylor & Francis, Boca Raton, FL
19. Muravsky L, Kmet' A, Voronyak T (2013) Two approaches to the blind phase shift extraction for two-step electronic speckle pattern interferometry. Opt Eng 52(10):101909
20. Muravsky LI, Kmet' AB, Stasyshyn IV, Voronyak TI, Bobitski YV (2018) Three-step interferometric method with blind phase shifts by use of interframe correlation between interferograms. Opt Lasers Eng 105:27–34
21. Muravsky LI (2019) Three-step electronic speckle pattern interferometry method with arbitrary phase shifts of reference wave. Vidbir ta Obrobka Informatsiyi (Inf Extract Proc) 47:54–58
22. Muravsky L, Kotsiuba Y, Kulynych Y (2020) Estimation of unknown phase shift between synthesized speckle interferograms using pearson correlation coefficient. In: Proceedings of 2020 IEEE 15th international conference on computer sciences and information technologies (CSIT), vol 2. IEEE, Lviv, pp 58–61
23. Muravsky L, Kotsiuba Y, Kulynych Y (2020) Assessment of unknown phase shift for speckle interferometry using sample Pearson correlation coefficient. In: Conference on computer science and information technologies. Springer, Cham, pp 671–681
24. Pearson correlation coefficient. https://en.wikipedia.org/wiki/Pearson_correlation_coefficient. Last edited on 20 September 2022
25. Rodgers JL, Nicewander WA (1988) Thirteen ways to look at the correlation coefficient. Am Stat 42:59–66
26. Sánchez JR, Martínez-García A, Rayas JA, León-Rodríguez M (2022) LED source interferometer for microscopic fringe projection profilometry using a Gates' interferometer configuration. Opt Lasers Eng 149:106822
27. Schmit J, Creath K, Wyant JC (2007) Surface profilers, multiple wavelength, and white light intereferometry. In: Malacara D (ed) Optical shop testing, 3rd edn. John Wiley & Sons, Hoboken, NJ, pp 667–755
28. Schreiber H, Brunning JH (2007) Phase shifting interferometry. In: Malacara D (ed) Optical shop testing, 3rd edn. John Wiley & Sons, Hoboken, NJ, pp 547–666
29. Sesselmann M, Gonçalves AA Jr (1998) Single phase-step algorithm for phase difference measurement using ESPI. In: Kujawinska M, Brown GM, Takeda M (eds) Laser interferometry IX: techniques and analysis, vol 3478. SPIE, Bellingham, WA, pp 153–158
30. Sesselmann M, Gonçalves AA Jr (2001) A robust spatial phase-stepping ESPI system. In: Albertazzi A (ed) Laser metrology for precision measurement and inspection in industry, vol 4420. SPIE, Bellingham,WA, pp 149–154

31. Sirohi SR (2009) Optical methods of measurement: wholefield techniques, 2nd edn. Taylor & Francis, Boca Raton, FL
32. Stasyshyn IV, Kotsiuba YM, Muravsky LI, Voronyak TI (2018) Retrieving the surface relief components using phase-shifting interferometry and Gaussian filter. Vidbir ta Obrobka Informatsiyi (Inf Extract Proc) 46:50–53
33. Stoykova E, Harizanova J, Sainov V (2008) Pattern projection profilometry for 3D coordinates measurement of dynamic scenes. In: Ozaktas HM, Onural L (eds) Three-dimensional television: capture, transmission, display. Springer, Berlin, Heidelberg, pp 85–164
34. Su X, Zhang Q (2010) Dynamic 3-D shape measurement method: a review. Opt Lasers Eng 48(2):191–204
35. Takeda M, Yamamoto H (1994) Fourier-transform speckle profilometry: three-dimensional shape measurements of diffuse objects with large height steps and/or spatially isolated surfaces. Appl Opt 33(34):7829–7837
36. van Brug H (1999) Phase-step calibration for phase-stepped interferometry. Appl Opt 38(16):3549–3555
37. Wyant JC (1975) Use of an ac heterodyne lateral shear interferometer with real–time wavefront correction systems. Appl Opt 14(11):2622–2626
38. Xia H, Montresor S, Guo R, Li J, Yan F, Cheng H, Picart P (2016) Phase calibration unwrapping algorithm for phase data corrupted by strong decorrelation speckle noise. Opt Express 24(25):28713–28730
39. Yang X, Yu Q, Fu S (2007) An algorithm for estimating both fringe orientation and fringe density. Opt Commun 274(2):286–292
40. Yu Q, Fu S, Yang X, Sun X, Liu X (2004) Extraction of phase field from a single contoured correlation fringe pattern of ESPI. Opt Express 12(1):75–83
41. Yu Q, Fu S, Liu X, Yang X, Sun X (2004) Single-phase-step method with contoured correlation fringe patterns for ESPI. Opt Express 12(20):4980–4985
42. Zhang S (2010) Recent progresses on real-time 3D shape measurement using digital fringe projection techniques. Opt Lasers Eng 48(2):149–158
43. Zhang ZH (2012) Review of single-shot 3D shape measurement by phase calculation-based fringe projection techniques. Opt Lasers Eng 50(8):1097–1106

Chapter 6
Methods for Processing and Analyzing the Speckle Patterns of Materials Surface

Abstract Study of the stress–strain state of structural materials using an optical-digital speckle correlation (ODSC) method and digital image correlation (DIC) technique are discussed in this chapter. Here, we also consider a method for detecting subsurface defects in laminated composites by analyzing dynamic speckle patterns. The developed ODSC method makes it possible to narrow the peak width and increase the signal-to-noise ratio, thereby providing a more reliable restoration of displacement and deformation fields of the studied specimens near the edges and crack tips. This method uses various architectures of the joint transform correlator (JTC) to implement a cross-correlation operation between respective subsets of two speckle patterns to generate in-plane displacement and deformation vector fields. Computer simulations and experimental studies of the displacement fields of real specimens using six JTC architectures have shown that the optical-digital implementations of the JTC with adaptive median thresholding and the JTC with ring median thresholding have the smallest surface displacements errors and demonstrate the highest robustness to output noise. On the other hand, to improve the accuracy of evaluating the stress–strain state of quasi-brittle materials near the crack tip, a method for determining the components of the stress field based on the DIC technique was developed. A new method for subsurface defects detection based on generation of dynamic speckle patterns series and the use of speckle decorrelation and speckle blurring phenomenon is described. This method allows generating light responses from subsurface defects containing in laminate composite panels. The hybrid optical-digital system with a frequency swept ultrasonic excitation implementing this method makes it possible to detect real subsurface defects in composite and metal–composite elements of aircraft structures.

Recently, the techniques of speckle pattern processing and analyzing have been rapidly and intensively developed. Unlike DSPI and DH techniques, they cannot capture and restore phase data from the recorded optical wavefronts. However, the intensity distributions recorded in the form of digital speckle patterns also contain very valuable data about the properties of the surfaces and inhomogeneous media

Z. Nazarchuk et al., *Optical Metrology and Optoacoustics in Nondestructive Evaluation of Materials*, Springer Series in Optical Sciences 242,
https://doi.org/10.1007/978-981-99-1226-1_6

of various objects under study, including constructive materials, structural elements, biotissues, flows, fluids, particles, etc. [9, 29, 31, 91, 102, 111, 127, 129]. Among such techniques, we single out the optical-digital speckle correlation (ODSC) [65, 66, 67, 69, 70, 75] and dynamic laser speckle analysis [73, 81, 82, 134] methods. ODSC method can be considered as the one of the numerous branches of Digital Image Correlation (DIC). DIC, along with DSPI and DH, is currently the leading direction in Optical Speckle Metrology that makes it possible to process and analyze not only real speckle patterns of rough surfaces, but also any artificial images or speckle patterns created using surface spray painting or two-dimensional random distributions of painted spots [85, 111]. The basic principles of DIC and its application to the problems of experimental mechanics were developed by pioneering work of a research group at the University of South Carolina [15, 87, 111, 114]. Such techniques as digital particle image velocimetry (DPIV) [44, 127], electronic speckle photography [101–104], digital speckle correlation [26, 135], as well as ODSC method [65, 66, 67, 69, 70, 75] were developed using basics of DIC.

Dynamic laser speckle analysis (DLSA) is a technique that allows one to study the dynamics of speckle patterns generated by moving surfaces and various other time-varying phenomena associated, for example, with the biological material activity or mechanical oscillations of the specimen surfaces. The temporal activity of studied objects is performed by means of statistical processing in order to determine the correlation dependencies [46, 83, 90], local speckle contrast [9, 47, 119] or different estimators [19, 90] characterizing dynamic phenomena recorded by sets of speckle patterns. Methods for detecting the subsurface defects in composite structures using laser speckle contrast imaging (LSCI) and speckle decorrelation [73, 81, 82], as well as assessing fruit shelf life [134] and cellular changes in muscle tissue [57] can be considered as perspective directions of DLSA, developed by researchers of the Karpenko Physico-Mechanical Institute of the National Academy of Sciences of Ukraine (PMI).

6.1 Main Approaches for Determination the Surface Displacements and Strains Using Speckle Patterns

DIC is considered as a general scientific direction for measuring the displacement and deformation fields of various objects or their surfaces. At least two digital images, including both laser speckle patterns and artificial white light random gray intensity patterns of the object surface are used to measure the surface displacement and deformation fields in all the DIC methods [85]. Such digital images can be obtained at the next conditions: (i) Surface area is illuminated by coherent or incoherent light, intensity distribution of which is practically the same over all the area; (ii) surface area is optically rough; (iii) if the surface area is smooth, it is treated in order to create an artificial pattern by spray painting the surface or by applying the painted spots randomly distributed over the entire area. Further calculation of in-plane 2D displacement and deformation fields is based on the correlation comparison of each m, nth ($m = 1, 2 \ldots, M; n = 1, 2, \ldots, N$) square subset from the digital image of the deformed surface area with the corresponding reference m, nth square subset from the digital image of the same surface area in the initial (undeformed) state. For this purpose, two digital images of the studied surface area are recorded, entered into a computer, and then divided into $M \times N$ square subsets. The subset sizes contain an odd or even number of pixels in two orthogonal directions, such as 15×15; 16×16; 31×31; 32×32. In this chapter, we will consider only the speckle patterns generated by artificial and real rough surfaces.

Among all the developed DIC methods, one can distinguish two approaches to determining the surface displacement and deformation fields. The first approach involves measuring only the displacements of the studied surface points with the subsequent determination of the strain components as derivatives of their displacements [12, 44, 65, 69, 102, 127]. The second approach consists in the simultaneous full-field determination of the surface displacements and strains [14, 111, 112]. The methods in which the first approach is implemented are usually referred to DIC, although some of them have their own names, for example, DPIV [44, 127], electronic speckle photography [102, 103], or digital speckle-displacement measurement technique [12]. On the other hand, the second approach is directly related to DIC, and the technique developed by the scientific group at the University of South Carolina had just such a name.

6.1.1 First Approach to Determine the Surface Displacement and Deformation Fields

The first approach to measure the displacements and deformations is simpler to implement, but requires precise determination of displacement fields. To calculate the surface deformation fields, the numerical differentiation operators are applied to the obtained displacement fields. If small bases are used for calculation of strains, the errors can be significant. This approach is used in numerous methods of surface displacement and deformation fields.

The essence of this approach is to determine the spatial displacement of the maximum signal of cross-correlation between the reference and deformed subsets (correlation peak) relative to the maximum autocorrelation signal of the reference subset (autocorrelation peak), which is the zero of the measuring system. If a reference $r_{mn}(x, y)$ and deformed $g_{mn}(x, y)$ subsets extracted from a reference speckle pattern spatial intensity distribution $r(x, y)$ and deformed speckle pattern intensity distribution $g(x, y)$, respectively, are entered into the speckle correlator input, then the cross-correlation function described the output correlation signal can be expressed as

$$c_{mn}(x', y') = \int_{-\infty}^{\infty}\int r_{mn}(x, y) g^*_{mn}(x + x', y + y') \mathrm{d}x\mathrm{d}y, \tag{6.1}$$

where index * denotes the complex conjugate function.

As a result, the correlation peak with maximum in point (x', y') is produced. To reduce the computations, this function can be calculated in a frequency domain [104], that is

$$c_{mn} = F_d^{-1}(R_{mn} G^*_{mn}), \tag{6.2}$$

where $R_{mn} = F_d(r_{mn})$, $G_{mn} = F_d(g_{mn})$, F_d and F_d^{-1} are the operators of direct and inverse digital Fourier transforms.

To increase the accuracy of determination the surface displacement, a method of digital speckle-displacement measurement (DSDM) was proposed [12]. In this method, the Kumar–Hasselbrook filter was used to narrow the width of the correlation peak and increase the signal-to-noise ratio [49]. In this case, the output correlation signal is described by the normalized cross-correlation function, which is given by

$$\tilde{c}_{mn} = F_d^{-1}\left\{\frac{R_{mn} G^*_{mn}}{\left|R_{mn} G^*_{mn}\right|^{0.5\chi}}\right\}, \tag{6.3}$$

where χ is the filter parameter that acquires only the real values.

If $\chi = 0$, the Kumar–Hasselbrook filter becomes a classical matched one and Eq. (6.3) becomes a standard correlation function. If $\chi > 0$, the high-frequency components of the spatial frequency spectrum are amplified, and the correlation peak narrows. Simultaneously, the filter amplifies the high-frequency noise, and the peak-to-noise ratio, as a rule, increases insignificantly. Choosing $\chi = 1$, we get the so-called phase filter, which can be considered as optimal for many cases. Its use amplifies the high-frequency noise, which makes it impossible to determine the position of the peak with the desired accuracy. In this regard, after filtering, a biparabolic interpolation algorithm was used to improve the accuracy of the peak coordinates [12]. When using the normalized cross-correlation function (6.2) instead of the standard one described by Eq. (6.1), the influence of the nonuniformity of the studied surface illumination on the result of calculation the correlation peak position is weakened. An appropriate optical system with sufficient depth of field (of the order of a few millimeters) allows almost completely eliminating the out-of-plane strain influence on the parameters of cross-correlation function, and thus reducing the errors in determining in-plane displacements and strains of a studied specimen.

The displacement of each *m*, *n*th subset in the specimen surface plane is determined by the matrix of displacement vectors

$$\vec{d}_{mn} = d_{mnx}\vec{n}_x + d_{mny}\vec{n}_y, \tag{6.4}$$

where d_{mnx} and d_{mny} are the displacements of the m, nth subset in two orthogonal directions along the axes x and y.

The displacement vector $\vec{d}_{mn}$ of each m, nth subset determines the displacement of the correlation peak maximum from the initial subset position, i.e. the position of its autocorrelation peak maximum, to the point $(x', y') = (x + d_{mnx}, y + d_{mny})$ of the deformed subset along the coordinate axes x and y. All these vectors generate a displacement field in the form of a matrix of local displacements of deformed subsets. Thus, the shift of the maximum of each m, nth correlation peak from the center corresponds to the displacement of the m, nth subset by an integer number of pixels. It is also possible to determine the displacement of the peak with subpixel accuracy, which requires the use of optimization and interpolation algorithms. To find the position of the correlation peak top with such an error, bilinear [131], biparabolic and bicubic interpolation algorithms [12, 13], as well as the Sjödahl–Benckert algorithm of trigonometric interpolation [103] are used. The bilinear, biparabolic and bicubic interpolation algorithms make it possible to find the peak displacement with an error of 3–5% of the pixel size, and the Sjödahl–Benckert algorithm determines the position of the maximum value of the correlation peak with an error of 1% of the pixel.

The strain of each mnth subset in the specimen surface plane is determined by the matrix of the strain vectors

$$\vec{\varepsilon}_{mn} = \frac{d_{mnx}}{b_x}\vec{n}_x + \frac{d_{mny}}{b_y}\vec{n}_y. \tag{6.5}$$

Here, b_x, b_y are the minimum bases for determining the surface strains along the x and y directions, which are equal to the distances between the centers of adjacent subsets along these axes. As a rule, choose $b_x = b_y$. One can also choose bases that are multiples of $b_x = b_y$, i.e. $2b_x = 2b_y$, $3b_x = 3b_y$, etc.

The size of the subsets significantly affects the accuracy and resolution of certain displacement fields. As a rule, a certain optimal value of the subset size is set in order to provide a given error/resolution ratio. Strains are determined by obtaining derivatives in the direction of displacement fields. This is achieved using the processing of displacement fields by window algorithms with numerical differentiation operators (see, for example, [8]). The obtained matrices of displacement and strain vectors display the two-dimensional surface displacements or deformation fields, taking into account the linear increase of the optoelectronic registration system.

The considered DSDM method does not possess a sufficient flexibility in the study of various specimens with different physical and mechanical properties, since in this method the parameter χ is always constant, i.e. $\chi = 1$. For this reason, a modified digital speckle-displacement measurement (MDSDM) method was proposed. In this method, the choice of the parameter χ was based on the results of experimental studies of the given specimen surface and comparison of signal-to-noise ratios for the same surface at different values of this parameter [58, 65, 67–69]. In addition, in contrast to the procedure of finding the subpixel position of the correlation peak by the DSDM method using the biparabolic interpolation algorithm [12], the MDSDM method used either the Sjödahl–Benckert interpolation algorithm [103, 104] or the algorithm for determining the center of gravity of the correlation peak [67, 69].

Typical displacement fields of the surface area of the duralumin plate D16T similar to alloy 2024-T4 are shown in Fig. 6.1a–d, which were obtained by the MDSDM method when applying uniaxial tensile loads to the plate [58, 68]. In the central part of the plate with dimensions of $250 \times 200 \times 1.5$ mm^3, a slit stress concentrator 0.2 mm wide and $2l_0 = 23.5$ mm long was made. A fatigue crack was grown before the tests. Plate was loaded on a universal tensile-testing machine EU-40. The maximum fracture force for specimen series of this type was approximately 40,000 N. Changes in the spatial distribution of in-plane field displacements were investigated at the uniaxial successively increasing loads applied to the specimen perpendicular to the line of the slit stress concentrator. Speckle patterns of the deformed plate surface area were recorded with a digital camera, and each speckle pattern was compared with the initial speckle pattern of the same undeformed surface area. The size of the studied surface area was 50×37.5 mm^2. All the speckle patterns were divided into $M \times N$ subsets of 16×16 pixels. Simultaneously, the fracture diagram shown in Fig. 6.2 was plotted using a strain gauge installed near the fatigue crack edges. The diagram shows the points of the applied load values, which correspond to the displacement fields in Fig. 6.1, namely points ***a***, ***b***, ***c***, and ***d*** correspond to the displacement fields in Fig. 6.1a–d , respectively. In these figures, two black small squares to the left of the stress concentrator indicate the location of the strain gauge.

The obtained results indicate that the MDSDM method can be successfully used for experimental studies of the deflected mode of the structural material surfaces during their elastic–plastic deformation and fracture.

6.1.2 Second Approach to Determine the Surface Displacement and Deformation Fields

The second approach, which is directly related to DIC, makes it possible to determine simultaneously the displacements and strains of the test specimen surface with the same accuracy [15, 87, 111, 113, 114]. Through DIC, the maximization of the normalized correlation coefficient (or minimization of the decorrelation coefficient) of speckle patterns of undeformed and deformed surfaces is achieved by optimization of nonlinear algorithm and simultaneously correcting the results of calculation both surface displacements and strains using interpolation algorithms [111, 113]. The accuracy of determining in-plane strains of the object surface is sufficient to establish the deformation fields near the stress concentrators.

One of the effective trends for the DIC technique is its combination with the finite element method [93], which makes it possible to determine the starting values of displacements and strains for the optimization procedure. However, such a combination has some disadvantages, namely (i) the need for a preliminary high-precision

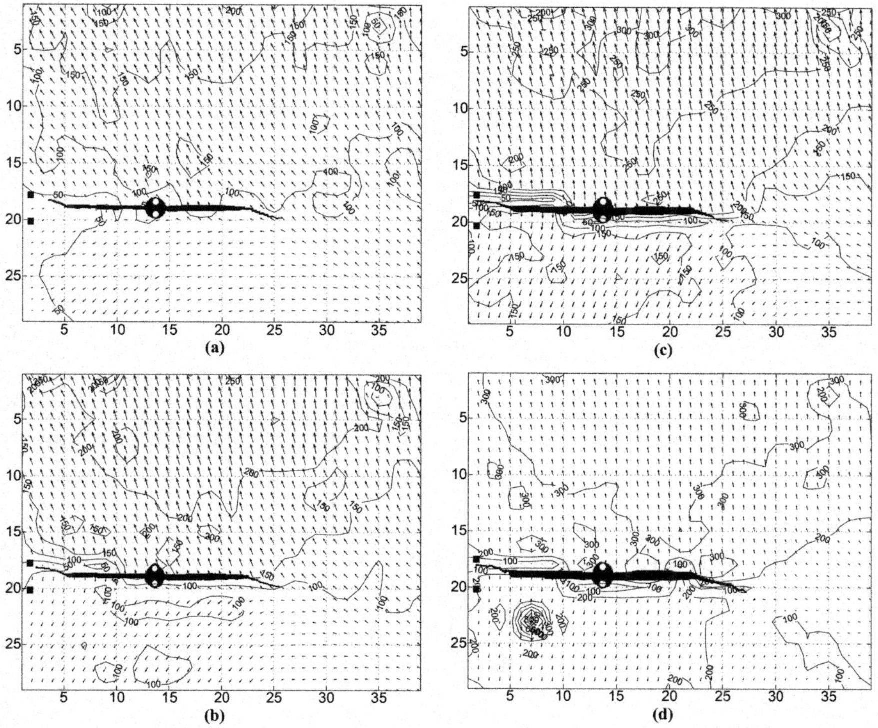

Fig. 6.1 Distribution of in-plane surface displacement fields of the duralumin D16T plate with a slit stress concentrator, on one of the edges of which a fatigue crack is grown: load change from 0 to 24,500 N (**a**); load change from 0 to 34,300 N (**b**); load change from 0 to 39,200 N (**c**); load change from 0 to 39,690 N (**d**). Samples along the *x* and *y* axes correspond to the *mn*th number of the speckle pattern subset surface displacement fields are given in μm

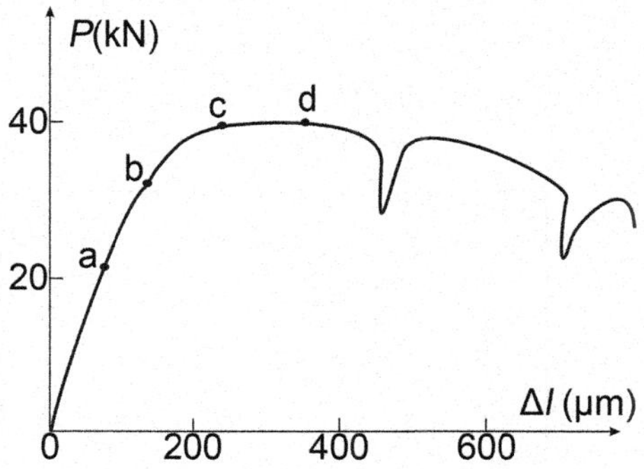

Fig. 6.2 Diagram of the sheet duralumin D16T plate fracture when an uniaxial tensile load is applied: *P* is the tensile load, Δ*l* is the change in the distance between the strain gauge legs

determination of the coordinates of the anchor point in the finite element method (for example, the crack tip) for effective operation of the corresponding algorithms; (ii) the low accuracy of determining the displacement and strain field at the stress concentrator and edges of crack-like defects in the material of the test specimen, which are an area of special interest for fracture mechanics. The disadvantage of DIC is also the small radius of convergence of optimization procedures, which leads to limitation of the range of measured values.

Consider the procedure for implementing the DIC technique in its simplified version. To this end, we select subsets of two real or artificial speckle patterns, namely the subset of an undeformed specimen with the spatial intensity distribution $r_{mn}(x, y)$, and the subset of a deformed one with the spatial intensity distribution $g_{mn}(x^*, y^*)$, as shown in Fig. 6.3. The method provides for the simultaneous determination of the displacement and deformation fields of the studied surface by solving a system of nonlinear equations using the decorrelation coefficient as an optimization parameter, that is

$$S(x, y, u, v, \frac{du}{dx}, \frac{du}{dy}, \frac{dv}{dx}, \frac{dv}{dy}) = 1 - \frac{\sum [r_{mn}(x, y) \times g_{mn}(x^*, y^*)]}{\left[\sum r_{mn}^2(x, y) \times \sum g_{mn}^2(x^*, y^*)\right]^{0.5}}. \quad (6.6)$$

If in the telecentric optical system the in-plane deformations of the object are parallel to the image plane, then the relations between the coordinates (x, y) and (x^*, y^*) are given as [10]

$$\left.\begin{aligned} x^* &= x + d_{mn,x} + \left[\tfrac{d}{dx}(d_{mn,x})\right]\Delta x + \left[\tfrac{d}{dy}(d_{mn,x})\right]\Delta y \\ y^* &= y + d_{mn,y} + \left[\tfrac{d}{dx}(d_{mn,y})\right]\Delta x + \left[\tfrac{d}{dy}(d_{mn,y})\right]\Delta y \end{aligned}\right\}, \quad (6.7)$$

where $d_{mn,x}$ and $d_{mn,y}$ are the displacements of the point in the directions of the x and y axes, respectively; Δx and Δy are normal projections of the distance from the center of the speckle pattern subset to the point (x, y); the partial derivatives determine the magnitudes of the corresponding strains.

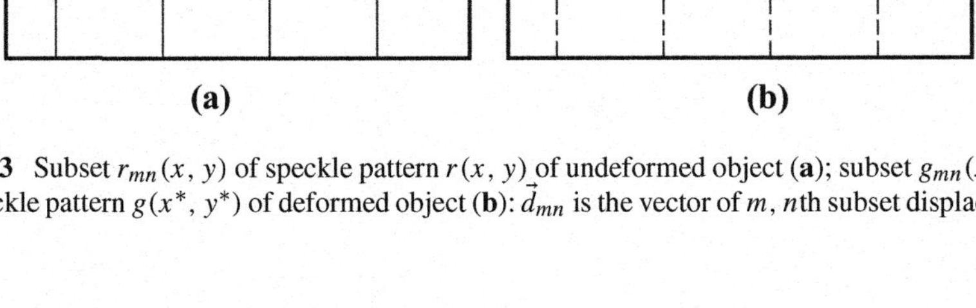

Fig. 6.3 Subset $r_{mn}(x, y)$ of speckle pattern $r(x, y)$ of undeformed object (**a**); subset $g_{mn}(x^*, y^*)$ of speckle pattern $g(x^*, y^*)$ of deformed object (**b**): $\vec{d}_{mn}$ is the vector of m, nth subset displacement

In particular, in the case of a plane problem, the components of the deformation field are given by [10]

$$\left.\begin{aligned} \varepsilon_{xx} &\cong \frac{\partial}{\partial x}\left(d_{mn,x}\right) + \frac{1}{2}\left\{\left[\frac{\partial^2}{\partial x^2}\left(d_{mn,x}\right)\right] + \left[\frac{\partial^2}{\partial x^2}\left(d_{mn,y}\right)\right]\right\} \\ \varepsilon_{yy} &\cong \frac{\partial}{\partial y}\left(d_{mn,y}\right) + \frac{1}{2}\left\{\left[\frac{\partial^2}{\partial y^2}\left(d_{mn,x}\right)\right] + \left[\frac{\partial^2}{\partial y^2}\left(d_{mn,y}\right)\right]\right\} \\ \varepsilon_{xy} &\cong \frac{1}{2}\left[\frac{\partial}{\partial y}\left(d_{mn,x}\right) + \frac{\partial}{\partial x}\left(d_{mn,y}\right)\right] \\ &+ \frac{1}{2}\left[\frac{\partial\left(d_{mn,x}\right)\partial\left(d_{mn,y}\right)}{\partial x \partial y} + \frac{\partial\left(d_{mn,y}\right)\partial\left(d_{mn,x}\right)}{\partial x \partial y}\right] \end{aligned}\right\}, \tag{6.8}$$

where ε_{xx}, ε_{yy} and ε_{xy} are the components of the deformation field for point on the speckle pattern.

As a rule, in applied calculations the partial derivatives of higher orders are neglected. In this case, the system of Eq. (6.8) is given by

$$\left.\begin{aligned} \varepsilon_{xx} &\cong \tfrac{\partial}{\partial x}\left(d_{mn,x}\right) \\ \varepsilon_{yy} &\cong \tfrac{\partial}{\partial y}\left(d_{mn,y}\right) \\ \varepsilon_{xy} &\cong \tfrac{1}{2}\left[\tfrac{\partial}{\partial y}\left(d_{mn,x}\right) + \tfrac{\partial}{\partial x}\left(d_{mn,y}\right)\right] \end{aligned}\right\}. \tag{6.9}$$

The optimization algorithm for determining displacements and deformations is based on the Newton–Raphson method [10]. At each ith step of the iterative procedure, the following matrix equation is solved:

$$\Delta P_i = -H^{-1}(P_{i-1}) \times \nabla(P_{i-1}), \tag{6.10}$$

where

$$P_i^T = \left\{u \; v \; \tfrac{\partial u}{\partial x} \; \tfrac{\partial u}{\partial y} \; \tfrac{\partial v}{\partial x} \; \tfrac{\partial v}{\partial y}\right\}, \tag{6.11}$$

P_i is the column vector containing the searched components of displacements and strains, H is the Hessian matrix of the decorrelation coefficient S second order partial derivatives and ∇ is the Jacobian symbol.

Simultaneous determination of displacements and strains makes it possible to obtain their values with the same relative error. The standard deviations in determining the strains does not exceed 0.1% if the instrumental noise of a digital camera with an encoding depth of 8 bits was averaged over 20 consecutive frames and a subset size is 20 × 20 pixels [10].

The measurement system implementing this approach can use an uncoherent narrow bandwidth light source and a digital camera containing matrix photodetector at least of 1000 × 1000 pixels and dynamic range of 256 grayscale. An important requirement is also the need to ensure the telecentricity of the speckle pattern registration, i.e. in-plane displacements and strains of the studied object should be parallel to the plane of the speckle pattern formation. Thus, the research is formally performed within the plane problem of mechanics in order to ensure the fulfillment of Eqs. (6.7) and (6.8).

6.2 Optical-Digital Speckle Correlation Method

DIC techniques are widely used to solve many problems of experimental mechanics, materials science, TD and NDT of materials and structural elements. Powerful interpolation algorithms are used to restore the surface displacement and deformation fields with absolute errors commensurate with hundredths of the pixel size. However, there are two main factors that reduce the effectiveness of most DIC methods for satisfactorily determining these fields in the immediate vicinity of stress concentrators, in particular, near crack tips and edges. First factor is associated with the occurrence of significant speckle decorrelation near the stress concentrators, which leads to a decrease in the intensity of correlation peaks and an increase in the level of noise and interference, resulting in a significant increase in errors when determining the positions of peak maxima. The second factor is caused by the relatively large width of the correlation peaks, which is determined by the size of the selected subsets of the recorded speckle patterns. This circumstance makes it possible to form a vector of sufficiently accurate displacement of the subset in the specimen surface plane only at a certain distance from the crack tip and crack edges, which does not always allow one to correctly solve specific problems of fracture mechanics. One of the possible solutions to these problems is the development of such speckle correlation methods that would narrow the peak width and increase the signal-to-noise ratio, thereby providing a more precise definition of displacement and deformation fields near the crack tips and edges. The considered ODSC method was developed to solve the mentioned problems [65, 66, 69, 74, 75].

Known methods of correlation of speckle patterns using the optical correlators make it possible either to compare two speckle patterns with each other, or to determine only the rigid body motion of the entire surface, i.e. to determine only one displacement vector. The first methods of optical speckle correlation were associated with the assessment of the degree of identity of two speckle patterns based on the determination of the cross-correlation signal and its comparison with the autocorrelation one. To compare two speckle patterns, Marom [61] suggested using the Vander Lugt correlator architecture [53]. The correlation peak formed at the output of the correlator was a measure of the similarity or difference between the surfaces of test objects. Using the same correlator architecture, the quality of solder joints was investigated under cyclic changes in temperature [42], the process of fatigue failure of the aluminum specimen surface was controlled [61], and the quality of ceramic materials was assessed [62]. Optical correlation methods were also used to study in-plane and out-of-plane rigid body displacements, rigid body elastic or plastic surface strains [60, 98, 121, 122], as well as to measure the rigid body inclination relative to the optical axis of the optical correlator [121, 122].

In contrast to the considered approaches of optical correlation, the ODSC method provides the generation of 2D fields of surface displacements and deformations. In addition, it makes it possible to achieve a significant increase in the intensity of correlation peaks, narrow their width, increase the peak-to-noise ratio (PNR) and reduce the level of random impulse noises in the determined field of surface displacements. The ODSC method is based on the use of a cross-correlation operation between the corresponding subsets of two speckle patterns that are compared with each other to form in-plane displacement and deformation fields. To implement this operation, it is proposed to use the architecture of the joint transform correlator (JTC) [30, 92, 126]. It is more expedient to use such a correlator to implement the ODSC method, because it simultaneously modulates two subsets of the compared speckle patterns, without first making a Fourier hologram for one of the subsets, as it should be done in the Vander Lugt correlator [30, 53]. In addition, the JTC architecture can provide processing of speckle pattern subsets in real time. The JTC also has greater software and algorithmic flexibility, which is manifested in the possibility of its relatively simple technical implementation in optical, hybrid optical-digital and digital versions. In this method, all the operations and procedures performed on speckle patterns are similar to those performed using the first approach to determining the surface displacement and deformation fields (see Sect. 6.1.1).

In the ODSC method, linear and nonlinear spatial filters can be synthesized to transform a joint power spectrum (JPS) generated by two corresponding subsets r_{mn} and g_{mn}, which are extracted from the reference and deformed speckle patterns $r(x, y)$ and $g(x, y)$, respectively. These filters are used in optical pattern recognition to improve the signal characteristics of correlation peaks and reduce the noise and interference [2, 4, 5, 33, 41]. As already mentioned in Sect. 6.1.1, pure phase filters, i.e. linear Kumar–Hasselbrook filters [49], can also be used in the DSDM [12] and MDSDM [58, 65, 67, 69] methods to narrow the correlation peak and increase the accuracy of determining the position of its maximum. However, a sharp increase in the noise level when using these filters does not significantly increase the PNR for the output correlation signals. In this regard, special attention should be paid to the use of known and synthesis of new nonlinear spatial filters in order to achieve both peak narrowing and an increase in PNR by increasing the intensity of peaks and reducing the level of the ambient noise and interference. In [65, 69, 74, 75], it was proposed to use the well-known nonlinear filters synthesized for optical pattern recognition (binary filters with median and subset median thresholds [41, 124], linear fringe-adjusted filters (FAFs) [2, 5] and adaptive binary filters [33]), and a new synthesized binary filter with ring median threshold taking into account the structure of the speckle pattern power spectral density (PSD).

Transformations of speckle pattern subsets in the joint transform correlator. Consider the procedure for transforming the speckle patterns and their subsets in the

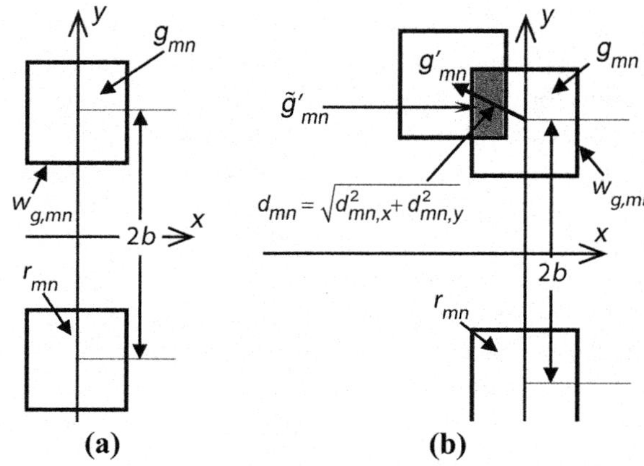

Fig. 6.4 Location of subsets r_{mn} and g_{mn} in the input plane of the JTC first stage (**a**); location of the shifted subset g'_{mn} concerning the input square window $w_{g,mn}$ (**b**)

conventional optical JTC. The rectangular speckle pattern of undeformed surface area S_1 with intensity distribution $i_1(x, y)$ and the speckle pattern of the same deformed one S_2 with intensity distribution $i_2(x, y)$ are recorded by a digital camera. For clarity and convenience, we will label these speckle patterns with functions $r(x, y)$ (Pattern 1) and $g(x, y)$ (Pattern 2), respectively, that is, $i_1(x, y) = r(x, y)$ and $i_2(x, y) = = g(x, y)$. These functions are divided on two rectangular matrices of square subsets $r_{mn} = r_{mn}(x - ma, y - na)$ and $g_{mn} = g_{mn}(x - ma, y - na)$ of the same size $a \times a$, and the corresponding mnth subset pair is entered into input plane (x, y) of the first JTC stage, as it is shown in Fig. 6.4a. The complex amplitude transmission of this pair in the input plane (x, y) is given by

$$t_{mn}(x, y) = r_{mn}(x - ma, y - na + b) + g_{mn}(x - ma, y - na - b), \quad (6.12)$$

where $2b$ is the distance between centers of subsets.

Assume that each shifted subset g'_{mn} can be considered as the subset r_{mn} of the initial speckle pattern displaced in an arbitrary direction at a distance $d_{mn} = \sqrt{d^2_{mn,x} + d^2_{mn,y}}$ from the subset g_{mn} and corrupted by uncorrelated noise $\eta_{mn}(x, y)$ [12, 70, 75], that is

$$\begin{aligned} g'_{mn}(x, y) &= g_{mn}\left(x - d_{mn,x}, y - d_{mn,y}\right) \\ &= r_{mn}\left(x - d_{mn,x}, y - d_{mn,y} - b\right) + \eta_{mn}(x, y). \end{aligned} \quad (6.13)$$

However, the searched displacement $\vec{d}_{mn} = d_{mn,x}\vec{n}_x + d_{mn,y}\vec{n}_y$ of the subset g'_{mn} is unknown, and this subset is only the part of the subset g_{mn} displaced in a square window

$$w_{g,mn}(x, y) = \text{rect}\left(\frac{x - ma}{a}, \frac{y - na - b}{a}\right). \quad (6.14)$$

Taking into account (6.13), the part of the subset g'_{mn} displaced within the window $w_{g,mn}$ can be expressed as

$$\tilde{g}'_{mn} = \tilde{r}_{mn}\left(x - ma - d_{mn,x}, y - na - b - d_{mn,y}\right) + \tilde{\eta}_{mn}(x, y), \tag{6.15}$$

and Eq. (6.12) is rewritten as

$$\begin{aligned} t_{mn}(x, y) = r_{mn}(x - ma, y - na + b) \\ + \tilde{r}_{mn}\left(x - ma - d_{mn,x}, y - na - b - d_{mn,y}\right) \\ + s_{mn}(x - ma, y - na - b), \end{aligned} \tag{6.16}$$

where $\tilde{\eta}_{mn}(x, y)$ is the noise term of the subset part $\tilde{g}'_{mn}$, $s_{mn}(x, y) = \tilde{\eta}'_{mn}(x, y) + h_{mn}(x, y)$ is the noise term of the subset g_{mn}, and $h_{mn}(x, y)$ is a sum of noise parts of adjacent subsets (for example, subsets $g_{m-1,n-1}$, $g_{m,n-1}$ and $g_{m-1,n}$) within the window $w_{g,mn}$ (see Fig. 6.4b).

The JPS of the two subsets r_{mn} and g_{mn} that are described by Eq. (6.16) is obtained at the JTC spatial frequency plane (u, v) as a result of 2D Fourier transform of the input signal $t_{mn}(x, y)$ and is given by

$$\begin{aligned} |T_{mn}(u, v)|^2 = |R_{mn}(u, v)|^2 + \left|\tilde{R}_{mn}(u, v)\right|^2 + |S_{mn}(u, v)|^2 \\ + \tilde{R}_{mn}(u, v)S^*_{mn}(u, v)\exp\left[j2\pi\left(d_{mn,x}u + d_{mn,y}v\right)\right] \\ + \tilde{R}^*_{mn}(u, v)S_{mn}(u, v)\exp\left[-j2\pi\left(d_{mn,x}u + d_{mn,y}v\right)\right] \\ + R_{mn}(u, v)\tilde{R}^*_{mn}(u, v)\exp\left[-j2\pi\left(d_{mn,x}u + d_{mn,y}v\right)\right]\exp(-j4\pi bv) \\ + R_{mn}(u, v)S^*_{mn}(u, v)\exp(-j4\pi bv) \\ + R^*_{mn}(u, v)\tilde{R}_{mn}(u, v)\exp\left[j2\pi\left(d_{mn,x}u + d_{mn,y}v\right)\right]\exp(j4\pi bv) \\ + R^*_{mn}(u, v)S_{mn}(u, v)\exp(j4\pi bv), \end{aligned} \tag{6.17}$$

where $T_{mn}(u, v) = F[t_{mn}(x, y)]$, $R_{mn}(u, v) = F[r_{mn}(x, y)]$, $\tilde{R}_{mn}(u, v) = F\left[\tilde{r}_{mn}(x, y)\right]$, $S_{mn}(u, v) = F[s_{mn}(x, y)]$, F is the 2D Fourier transform operator, $*$ denotes complex conjugate.

In an output plane $\left(x', y'\right)$ of the scale invariant JTC, the output signal spatial distribution is obtained by the inverse Fourier transform of the JPS expressed by Eq. (6.17) and consists of seven spatially separated terms. Since it is always possible to choose a distance $2b$ such that $2b \gg d_{mn,x}$ and $2b \gg d_{mn,y}$, these terms can be grouped into a term $C^0_{mn}\left(x', y'\right)$ placed at the diffraction center (zero diffraction order), and two terms $C^+_{mn}\left(x', y'\right)$ and $C^-_{mn}\left(x', y'\right)$ in +first and −first diffraction orders.

According to Fig. 6.4a, the positional relationship of each subset pair extracted from Patterns 1 and 2 in the scale invariant linear JTC input plane is given by

$$\tilde{t}_{mn}(x,y) = r_{mn}(x, y+b) + g_{mn}[x - (0.5M - m)a, y - (0.5N - n)a - b]. \quad (6.18)$$

Taking in account Eqs. (6.17) and (6.18), the grating of correlation peaks produced at the +first diffraction order of JTC output plane (x', y') as a result of cross-correlation of all the corresponding subset pairs can be expressed as

$$C^{+}(x', y') = \sum_{m=1}^{M}\sum_{n=1}^{N} c_{mn}\left[x' - (0.5M - m)a - d_{mn,x}, y' - (0.5N - n)a - 2b - d_{mn,y}\right], \quad (6.19)$$

where c_{mn} is an output spatial signal of a mnth correlation field.

The coordinates of intensity maximums $I_{mn}^{\max}$ of correlation peaks p_{mn} in points $(0.5Ma - ma + d_{mn,x}, 0.5Na - na + 2b + d_{mn,y})$ of the output plane (x', y') correspond to the coordinates of central points $(0.5Ma - ma + d_{mn,x}, 0.5Na - na + b + d_{mn,y})$ of displaced subsets g'_{pq} in the input plane (x, y). So, the coordinates of these peaks determine the displacement vectors $\vec{d}_{mn}$ of subsets g'_{mn} of Pattern 2.

The procedure for transformation of Pattern 1 and Pattern 2 and generation a grating of correlation peaks in the +1st diffraction order of the conventional JTC is shown in Fig. 6.5. As shown in Fig. 6.5 and Eq. (6.19), the shift of each peak is performed by setting the reference subset r_{mn} to the same place in the input plane (x, y).

6.3 Spatial Filtering of the Speckle Pattern Subsets in Joint Transform Correlator

The proposed ODSC method is practically indistinguishable in efficiency from the known methods of electronic speckle photography [102], as well as the DSDM and MDSDM methods [12, 58, 68], if the conventional JTC architecture is used. This is evidenced by approximately the same parameters of correlation peaks, namely the width, intensity and PNR, which are formed with the help of these methods. However, if some spatial linear and nonlinear filters are used in the ODSC method, these parameters of correlation peaks can be significantly improved.

In optical pattern recognition, linear and nonlinear spatial filtering of the JPS is used to improve the parameters of correlation signals at the output of the JTC [2, 5, 30, 41]. Since the implementation of the ODSC method is carried out with the help of the JTC architecture, it is quite reasonable to use these operations in this method. Through the use the spatial linear and nonlinear filters, it is possible to increase the intensity, sharpness and steepness of the correlation peaks, reduce the level of the ambient noise and thereby increase the PNR.

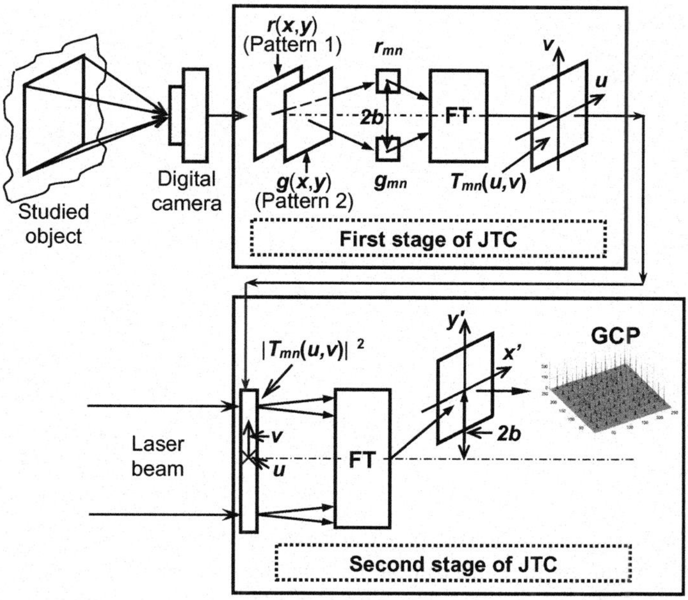

Fig. 6.5 Generation of the grating of correlation peaks in the conventional JTC: (x, y), input plane; r_{mn}, reference subset of Pattern 1; g_{mn}, input subset of Pattern 2; $2b$, distance between subsets; FT, Fourier transform; u and v, axes of frequency plane (u, v); $T_{mn}(u, v)$, Fourier spectrum of input spatial signal $t_{mn}(x, y)$; $|T_{mn}(u, v)|^2$, JPS; x' and y', axes of output plane $\left(x', y'\right)$; GCP, grating of correlation peaks

6.3.1 Binarization of Joint Power Spectrum by Median and Subset Median Thresholds

Nonlinear filtering methods of median and subset median thresholding [41, 124] make it possible to increase the correlation peak intensity, narrow its width and reduce the noise and pulse interference. These methods are based on binarization of the JPS $|T_{mn}(u, v)|^2$ by using the global or subset median thresholds. The median threshold is determined by the median value of the JPS described by Eq. (6.17), and the subset median threshold corresponding to the median value for the JPS rows or columns is given by

$$V_{T_m n} = \text{med}\left\{\text{hist}\left[|T_{mn}(u, v)|^2\right]\right\}, \tag{6.20}$$

where “hist” is the histogram of the row or column and “med” is the median. Those values of the JPS, which exceed the threshold value $V_{T_m n}$, equal to 1, and other values of the JPS are equal 0 or -1. The binarization procedure is shown in Fig. 6.6.

Thresholding methods allow to improve significantly the parameters of correlation signals at the output of the JTC by taking into account the noise of the input images during nonlinear transformations of the JPS. However, some disadvantages of these

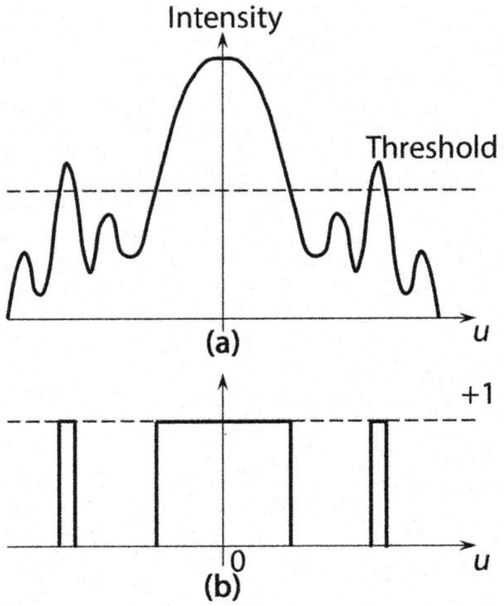

Fig. 6.6 JPS binarization: JPS row before binarization (**a**); the same JPS row after binarization (**b**)

methods limit the possibilities of binary correlators. Among them are a large amount of calculations during the binarization of JPS in the frequency domain, the presence of intense zero diffraction order, which leads to additional noise in the first diffraction orders, and a high probability of pulsed interference.

6.3.2 Comparative Analysis of Three JTC Computer Models

Consider briefly the procedure for implementing the ODSC method using the computer models of three types of the JTC, namely conventional JTC, JTC with median thresholding and JTC with subset median thresholding. The developed procedure was dedicated to compare the correlation peaks from two corresponding *mn*th subsets of Pattern 1 and Pattern 2 obtained with these three models of correlators. So, Patterns 1 and 2 were divided on square subsets of 32 × 32 pixels. Next, the corresponding *mn*th subsets of Patterns 1 and 2 were introduced into the input plane of each correlator, and their JPS was calculated. In the conventional JTC, the correlation peak was produced after inverse Fourier transform of the calculated JPS. In the JTC with median thresholding, the median threshold for each row of the JPS was calculated and binarized, and the correlation peak was generated after the inverse Fourier transform of the binarized JPS. The same procedure was performed for the computer model of the JTC with subset median thresholding. For each type of the JTC, the above procedures were repeated for all the other pairs of corresponding subsets of Patterns 1 and 2, and three rectangular gratings of correlation peaks were obtained.

The single correlation signals obtained as a result of joint correlation of the corresponding subsets of Pattern 1 and Pattern 2 using the computer models of the conventional JTC, JTC with median thresholding and JTC with subset median

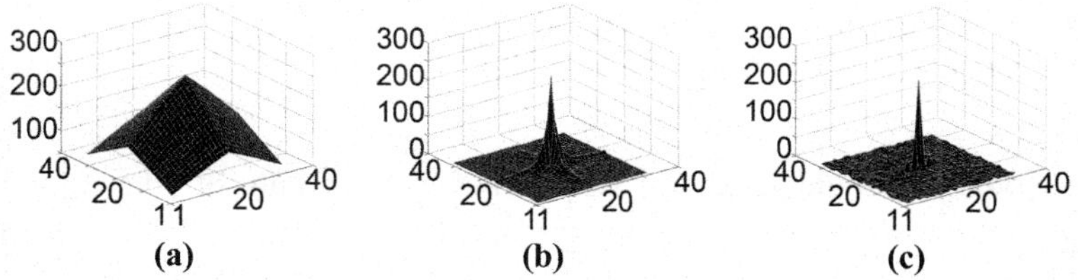

Fig. 6.7 Output correlation peaks generated in computer models of the joint transform correlators: peak at the conventional JTC output (**a**); peak at the JTC with median thresholding output (**b**); peak at the JTC with subset median thresholding output (**c**)

thresholding are shown in Fig. 6.7a–c. These figures clearly show a significant influence of nonlinear transformations of the JPS on the peak sharpness and on the PNR. Fulfilled calculations also show that the PNR is highest for the JTC with subset median thresholding and lowest for the conventional JTC.

6.3.3 Fringe-Adjusted Filters

Through the use of FAFs developed and improved by Alam et al. [1–5], the aforementioned lacks of thresholding filters can be eliminated or attenuated. The FAF belongs to the class of apodization filters, and its amplitude transfer function is defined by

$$H_F(u,v) = B(u,v)\left[D(u,v) + |R(u,v)|^{\gamma}\right]^{-1}, \tag{6.21}$$

where $R(u,v)$ is the Fourier transform of the reference function $r(x,y)$ in the JTC input plane, γ is the constant value, $B(u,v)$ and $D(u,v)$ are either constants or functions depending on u and v.

The JPS described by Eq. (6.17) is multiplied by the FAF before performing the inverse Fourier transform operation. As a result, the filtered JPS is given by

$$|T_{mn}(u,v)|_F^2 = |T_{mn}(u,v)|^2 H_F(u,v). \tag{6.22}$$

If $\gamma = 0$, $D(u,v) = 0$ and $B(u,v) = 1$, then the JTC containing this filter is the conventional; if $\gamma = 2$, $D(u,v) \ll |R(u,v)|^2$ and $B(u,v) = 1$, then it is the fringe-adjusted JTC; if $\gamma = 1$, $D(u,v) \ll |R(u,v)|^2$ and $B(u,v) = 1$, then it is the amplitude-modulated phase-only JTC [1, 3]. To eliminate the zero diffraction order at the output, an image subtraction method was proposed [3]. However, the FAF in combination with the image subtraction method does not take into account the influence of the input image on the correlation response. This drawback has insignificant effect on the response if Pattern 2 is shifted by a short distance relative to Pattern 1. In this case, the subsets r_{mn} and g_{mn} can be assumed to be the same. If the distance between Patterns 1 and 2 becomes large, then the lack of input image data may affect the PNR and the peak may be significantly attenuated.

6.3.4 *Binarization of Joint Power Spectrum by Adaptive Median Threshold*

The constant threshold used in the JPS binarization cuts off the high spatial frequencies containing more information about the correlation signal. A changeable threshold, adapted to the local contours of the JPS intensity distribution, makes it possible to amplify the high spatial frequencies to the same values as the lower ones. The adaptive median thresholding method is based on determining the local average value for almost each pixel of the JPS and binarization of the pixel intensity, using the average value as a threshold [33]. A one-dimensional example of the implementation of this method for the JPS selected line is shown in Fig. 6.8. This figure shows both the adaptive threshold and the fixed threshold. Use of the adaptive threshold for the JPS binarization allows improving the correlation signal characteristics in comparison with the methods of median and subset median thresholding.

Fig. 6.8 Joint power spectrum cross-section before binarization (**a**), binarized JPS cross-sections after using subset median threshold (**b**) and adaptive median threshold (**c**)

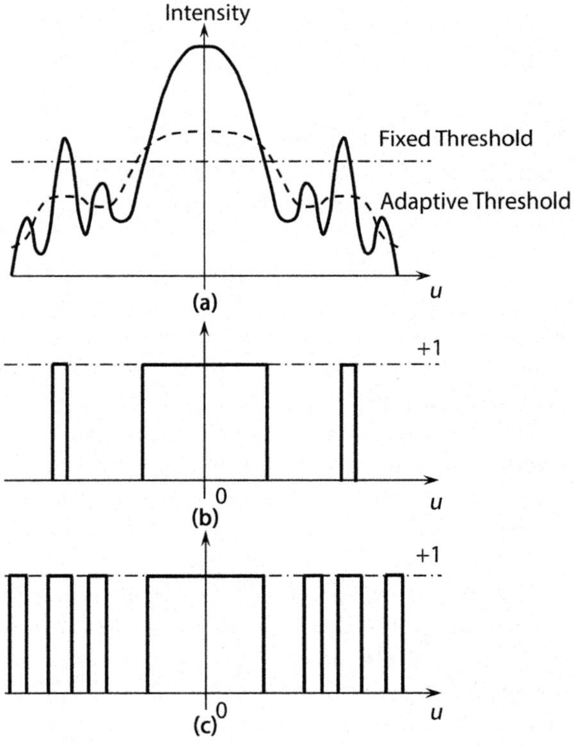

6.3.5 *Binarization of Joint Power Spectrum by Ring Median Threshold*

The global and subset median thresholding methods, as well as the method for JPS filtering using FAFs can improve the parameters of correlation peaks, which is very important to increase the reliability of restored surface displacement and deformation fields near the stress concentrators. However, these methods, which are used for optical image recognition, do not take into account the specific spatial structure of the Fourier spectra and the PSDs of the speckle patterns. In order to further improve the parameters of the correlation peaks, a new method of the JPS binarization by a ring median threshold (or the method of the JPS ring median binarization) was proposed. This method takes into account the specific structure of the speckle pattern power spectrum [74, 75].

In the conventional JTC, the JPS from the two subsets $r_{mn} = r_{mn}(x, y + b)$ and $g_{mn} = g_{mn}(x, y - b)$ of Patterns 1 and 2 is given by [126]

$$\begin{aligned} |T_{mn}(u, v)|^2 &= R^*_{mn}(u, v)R_{mn}(u, v) + G^*_{mn}(u, v)G_{mn}(u, v) \\ &+ R^*_{mn}(u, v)G_{mn}(u, v)\exp(-j4\pi bv) \\ &+ R_{mn}(u, v)G^*_{mn}(u, v)\exp(j4\pi bv), \end{aligned} \tag{6.23}$$

where $G_{mn}(u, v) = F[g_{mn}(x, y)]$.

Equation (6.23) can be considered as the simplified version of Eq. (6.17). Let's note that in Eq. (6.23), as in Eq. (6.17), the transfer characteristics of an optical system and digital camera, as well as the electronic noise of the camera, are not taken into account.

The first two components of Eq. (6.23) are the PSDs of the subsets r_{mn} and g_{mn}, respectively. According to the well-known formula for the speckle pattern PSD, or the speckle halo, the PSD in the frequency domain can be written as [28]

$$\begin{aligned} W_i(u, v) &= \langle i_s\rangle^2\left(\delta(u, v) + \left(\frac{\lambda f}{D}\right)^2\frac{4}{\pi}\left\{\arccos\left(\frac{\lambda f}{D}\rho\right) - \frac{\lambda f}{D}\rho\left[1 - \left(\frac{\lambda f}{D}\rho\right)^2\right]^{0.5}\right\}\right) \\ &= \langle i_s\rangle^2\left[\delta(u, v) + \frac{4}{\pi\rho^2_{\max}}\left(\arccos\tilde{\rho} - \tilde{\rho}\sqrt{1 - \tilde{\rho}^2}\right)\right], \end{aligned} \tag{6.24}$$

where $\langle i_s\rangle$ is the mean value of the speckle pattern intensity distribution; f is the focal length of the Fourier lens in the correlator first stage; $\rho = \sqrt{u^2 + v^2}$ is the radial spatial frequency; D is the lens aperture producing the speckle pattern; λ is the laser wavelength; $\rho_{\max} = D/\lambda f$ is the maximum spatial frequency of the PSD; $\tilde{\rho} = \rho/\rho_{\max}$ is the normalized (relative) frequency. Therefore, the speckle pattern PSD has the limited spectrum with the maximum spatial frequency $\rho_{\max}$, and its envelop function has conical shape and circular symmetry.

The next two terms in Eq. (6.23) are the cross-PSDs of the subsets r_{mn} and g_{mn}. In contrast to the PSD of a single speckle pattern, the high frequencies in the cross-PSD are suppressed by aberrations of imaging system and by decorrelation of the shifted subset g_{mn}. Suppression of the cross-PSD high-frequency components restricts its bandwidth and leads to broadening of the output correlation peak. Moreover, the interference fringe pattern from subsets r_{mn} and g_{mn} generates the spurious modulation of the cross-PSD.

The cross-PSD terms $R^*_{mn}(u, v)G_{mn}(u, v)$ and $R_{mn}(u, v)G^*_{mn}(u, v)$ lose circular symmetry, but they possess central symmetry and their maximum spatial frequency does not exceed $\rho_{\max}$. On the other hand, to eliminate the aliasing, one should take into account the spatial sampling of the speckle patterns by a matrix photodetector and follow the Nyquist criterion when choosing the sizes of speckle patterns. Power spectrum of an elementary square photosensor of size $a \times a$ has central symmetry and is given by [101]

$$W_P(u, v) = W_i(u, v) \cdot \sin c^2(au) \sin c^2(av). \tag{6.25}$$

If the photosensor pitch and speckle size satisfy conditions (4.2) and (4.3), the PSD replicas of the speckle pattern do not overlap, i.e. there is no aliasing. Therefore, the influence of the power spectrum $W_P(u, v)$ is insignificant and can be neglected.

Thus, it can be argued that the JPS $|T_{mn}(u, v)|^2$ average value within any annular zone concentric with respect to the spectrum center can be a threshold value for this zone during its binarization. Based on this inference, a series of narrow concentric ring zones was formed in the frequency domain (see Fig. 6.9a), and the JPS intensity distributions within each ring were binarized using the median threshold for each selected ring. The widths of the ring zones can be selected from hundreds to tenths of the entire frequency band $0 \leq \rho \leq \rho_{\max}$ depending on the speckle size, the size of the windows limiting the speckle pattern subsets, the subsets contrast, as well as the noise that causes decorrelation of speckle patterns. The efficiency of the ring median threshold binarization of the JPS is significantly affected by the central lobe and the sidelobes of higher orders produced by the input square windows $w_{r,mn}$ and $w_{g,mn}$ restricting the subsets r_{mn} and g_{mn}, respectively. The sidelobes and the central lob create a cross-shaped spatial structure. To remove the sidelobes, first, a homogeneous background with intensity equal to the average intensity of Pattern 1 and Pattern 2 was introduced into the windows $w_{r,mn}$ and $w_{g,mn}$ located in the input plane. The Fourier transform of these two images was then performed to obtain a JPS in the form of sidelobes producing a cruciform spatial structure. The influence of the sidelobes on the JPS of the subsets r_{mn} and g_{mn} was minimized by subtracting the obtained cruciform spatial structure from the JPS of the subsets r_{mn} and g_{mn}. To eliminate the effect of the central lobe, a low-frequency rejecting spatial filter was used.

Procedure of the JPS binarization by the ring median threshold is shown in Fig. 6.9. It is similar to the adaptive threshold binarization procedure, however, in this case, a constant median threshold is used for each selected ring of the JPS.

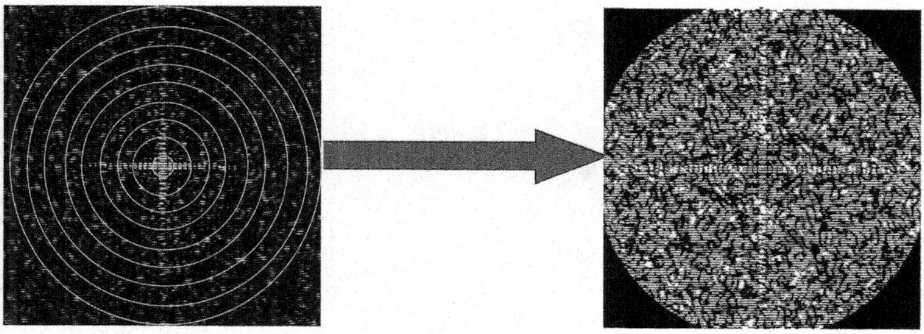

(a) (b)

Fig. 6.9 Binarization of the JPS of two subsets r_{mn} and g_{mn} by the ring median threshold: splitting the JPS into rings (**a**); JPS after binarization and subtraction of the sidelobes cruciform structure (**b**)

6.4 Digital Implementation of Optical-Digital Speckle Correlation Method

There are three main configurations of the JTC, namely a coherent optical, hybrid optical-digital, and digital JTC. All these versions are considered as two-stage image processors. In the optical JTC, two stages are the coherent optical Fourier processors [30]. However, all the operations with spatial filtering of the JPS should be performed with computer or a specialized digital processor. Besides, the input of the transformed JPS into the second stage of the optical correlator, as a rule, is performed with the help of corresponding digital hardware. The hybrid optical-digital JTC may contain the first digital stage and the second optical one, or vice versa—the first optical stage, and the second digital one. All these types of JTCs can be used to implement the ODSC method. To check its effectiveness using the above linear filters and nonlinear JPS binarization procedures, it is better to choose the digital version of the JTC, since thanks to this version it is possible to work with both synthesized and real speckle patterns.

For digital implementation of the JTC, it is enough to use a conventional computer, in which the speckle patterns are introduced using an optical system projected these patterns on a digital camera. In addition, computer can generate the synthesized speckle patterns, which are convenient to use for comparative analysis of the efficiency of linear and nonlinear filters.

The scheme of the optical-digital system for input and correlation analysis of the speckle patterns with a digital implementation of the JTC in a computer is shown in Fig. 6.10. The system contains a light source illuminating an optically rough surface of a structural material specimen, a recording system including a lens (L) and a digital camera, as well as a computer with special software. Halogen lamps, light-emitting diodes (LEDs) or semi-conductor lasers can be used as the light source. The speckle pattern of the specimen surface is recorded with the digital camera and entered into the computer. Speckle patterns can be processed using the digital version of the JTC, which implemented all the operations of linear filtering and threshold binarization of

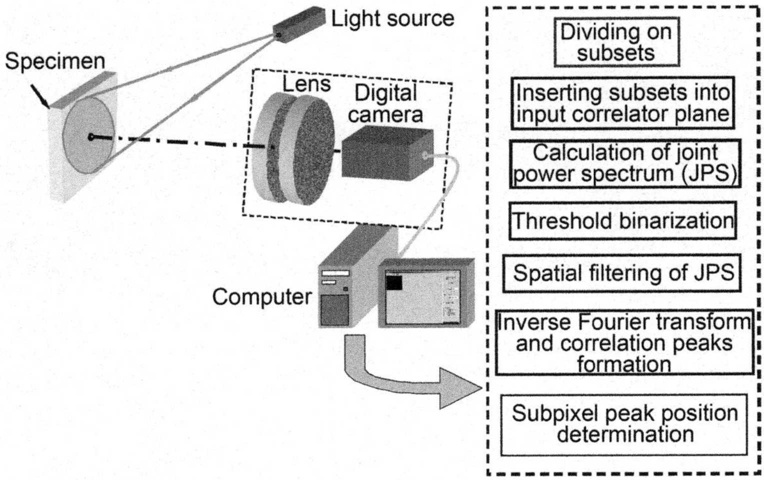

Fig. 6.10 Optical-digital system for correlation analysis of the speckle patterns with digital implementation of the JTC

the JPS considered above. Using the created system, synthesized speckle patterns and real speckle patterns obtained as a result of experiments with specimens of structural materials were studied.

The developed digital model of the JTC provides for the implementation of a number of operations and procedures, namely

- processing of graphic files in .bmp and .tif format;
- non-linear transformation and filtering of the JPS from two corresponding subsets r_{mn} and g_{mn} of real or test speckle patterns r and g in order to improve such information characteristics of the output signal as the correlation peak intensity, PNR, peak steepness and width;
- the possibility of the JPS monitoring;
- saving the results in graphical form and as a data file;
- use of algorithms for determining the subpixel position of the correlation peak, in particular, the Sjödahl–Benckert algorithm (Algorithm 1) [103, 104] and the algorithm for determining the center of gravity (Algorithm 2) [67, 69].

The operation algorithm of the JTC digital model includes the following operations:

1. Reading graphic files whose names are specified by the user.
2. Selection of the image area for analysis of the chosen subset size.
3. Dividing the speckle patterns r and g into subsets r_{mn} and g_{mn}.
4. Pairwise input of subsets into the input plane of the JTC.
5. Calculation of the measurement range.
6. Choice of the correlator type: conventional JTC (JTC1); JTC with median thresholding (JTC2); JTC with subset median thresholding (JTC3); JTC with the FAF

(JTC4); JTC with adaptive median thresholding (JTC5); JTC with ring median thresholding (JTC6).

6.1. The median threshold calculation (see Sect. 6.3.1) and the JPS binarization (JTC2).
6.2. The subset median threshold calculation (see Sect. 6.3.1) and the JPS binarization (JTC3).
6.3. The FAF calculation (see Sect. 6.3.3) and the JPS filtering (JTC4).
6.4. The adaptive median threshold calculation (see Sect. 6.3.4) and the JPS binarization (JTC5).
6.5. The ring median threshold calculation (see Sect. 6.3.5) and the JPS binarization (JTC6).

7. Calculation the inverse Fourier transform of the binarized or filtered JPS.
8. Determining the displacements of subsets using Algorithm 1 and Algorithm 2.
9. Saving data as a graphic file and as a data file.

The flow-chart of the algorithm for implementing the ODSC method using the JTC digital models is shown in Fig. 6.11.

The developed computer model allows obtaining all the data necessary for a detailed study of the informative characteristics of the correlation signals. In addition, the program makes it possible to enter the speckle patterns or images of real surfaces and obtain the surface displacement and deformation fields of structural materials during their experimental studies.

6.5 Comparative Evaluation of Two Methods for Speckle Patterns Correlation

As noted above, six types of JTC computer models can implement the ODSC method using linear and nonlinear transformations of the JPS. The algorithm implementing the MDSDM method was developed by Maksymenko et al. [58] and can be considered as the modified version of the algorithm that implements the DSDM method [12]. The developed digital models of the ODSC and MDSDM methods were used to compare the informative parameters of output correlation signals generated by subsets of synthesized speckle patterns.

6.5.1 Synthesized Speckle Patterns in Analysis of Two Methods for Speckle Patterns Correlation

Test digital speckle patterns were synthesized to provide an accurate shift of the speckle pattern (Pattern 2) and to analyze systematic and random errors during the in-plane rigid body motion (RBM) of the studied optically rough flat surface. The

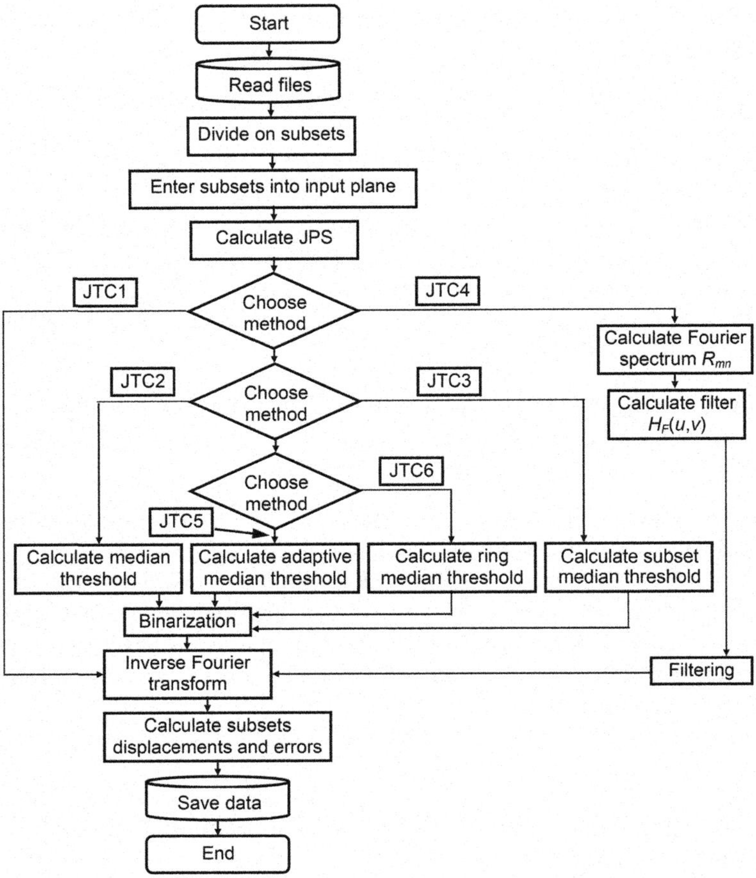

Fig. 6.11 Flow-chart of the algorithm of the JTC models that implement the ODSC method

procedure for their synthesis was similar to the procedure for generating the simulated speckle patterns described in Sect. 4.4.2. The average speckle size S_p in these patterns was chosen taking into account the Nyquist criterion and the pixel pitches p_x and p_y in a digital camera photosensor matrix. According to Eq. (4.3), we have $S_p \geq 2.44(\max|p_x, p_y|)$. To simplify the calculations, it was assumed that the pixels of the subsets coincide with the pixels of the photosensor matrix. In order to suppress the influence of rectangular spatial signals generated by the input windows $w_{r,mn}$ and $w_{g,mn}$ on the location of correlation peaks, the normalized average intensity value of each subset was subtracted from the intensity value of each pixel in this subset. Decorrelation parameter Δ_d was used to take into account the influence of out-of-plane and large in-plane surface displacements of a rigid body and additional noise caused by local random changes in the microstructure of the speckle patterns. This parameter was simulated by adding a new synthesized speckle pattern g_u not correlated with patterns r (Pattern 1) and g (Pattern 2) to the displaced shifted speckle pattern g in accordance with [103, 104]. Then, the resulting speckle pattern g_d (Pattern 3) can be expressed as [103]

$$g_d = (1 - \Delta_d)g + g_u, \tag{6.26}$$

where the decorrelation parameter $0 \le \Delta_d \le 1$ can be considered as a measure of decorrelation between Pattern 2 and Pattern 3. Measurement and analysis of errors in subset displacements was carried out at the decorrelation parameters $\Delta_d = 0;\ 0.25;\ 0.5$.

Synthesized speckle patterns with a dimension of 256×256 pixels were chosen for research, and subsets r_{mn} and g_{mn} with size of 32×32 pixels were selected. To calculate the average displacement $\langle d \rangle = (MN)^{-1} \sum_{m,n} d_{mn}$ of all the subsets in Pattern 2 or Pattern 3, in-plane surface displacements of the test speckle patterns simulating the RBM were considered. In this case, only diagonal displacements $d = \sqrt{d_x^2 + d_y^2}$ of the speckle patterns were performed with the same surface displacements d_x and d_y along the x and y axes. The mean bias error $\langle \theta \rangle$ of the displacement $\langle d \rangle$ measurement was calculated as the mean value of the measurement errors θ_{mn} of all relevant displacements d_{mn}, and the random error $\langle s \rangle$ of the displacement $\langle d \rangle$ measurement was calculated as the mean value of the root mean square (RMS) s_{mn} of all the relevant displacements d_{mn} measurement. The pixel size was chosen per unit of displacement.

To compare the efficiency of the ODSC and MDSDM methods, the metrics proposed by Horner [37] and Kumar and Hassebrook [49] for optical correlators were used. The objective measure of the correlator efficiency is the peak-to-correlation energy metric, i.e. the peak-to-output noise ratio (PONR) metric [37]. If the output noise is the square array of the discrete correlation field $c_{mn}(k', l')$ surrounding the correlation peak, the PONR can be given by

$$\mu_{mn} = \left[|p_{mn}|^2 - \left(\overline{|n_{k'l'}|^2_{mn}} \right) \right] / \sigma \left(|n_{k'l'}|^2_{mn} \right), \tag{6.27}$$

where p_{mn} is the complex amplitude of the mnth peak; $n_{k'l'}$ is the complex amplitude of $k'l'$th count of the output noise; σ is the RMS symbol; $k' = 1, \ldots, K_m$; $l' = 1, \ldots, L_m$ are the noise counts in the output correlation field $c_{mn}(k', l')$; $K_m = K/M$; $L_n = L/N$; $K_m = L_n$, where

$$\overline{|n_{k'l'}|^2_{mn}} = \sum_{k'=1}^{K_m} \sum_{l'=1}^{L_n} \left| n_{k'l'}(x', y') \right|^2_{mn} / (K_m L_n - 1). \tag{6.28}$$

Moreover, in this correlation field, one count was assigned to the correlation peak, and $K_m L_n - 1$ counts were assigned to the ambient noise. In this case, the output noise RMS is given by

$$s\left(|n_{k'l'}|_{mn}^2\right) = \sqrt{\sum_{k'=1}^{K_m}\sum_{l'=1}^{L_n}\left[|n_{k'l'}(x', y')|_{mn}^2 - \overline{|n_{k'l'}|_{mn}^2}\right]^2 / (K_m L_n - 1)}. \tag{6.29}$$

When measuring the RBM of the real specimen or simulating the RBM using synthesized speckle patterns, for corresponding displacement $\langle d \rangle$ of Pattern 2 or Pattern 3 it is necessary to determine the PONR mean value, which is given by

$$\langle \mu \rangle = \sum_{m=1}^{M}\sum_{n=1}^{N} \mu_{mn}/MN. \tag{6.30}$$

The objective measure of variability of the correlation peaks in the output correlation fields $c_{mn}(k', l')$ is the peak-to-input noise ratio (PINR) [49], which for this case is defined as

$$\eta_{mn} = \overline{|p_{mn}|^2} / \sqrt{\sum_{m=1}^{M}\sum_{n=1}^{N}\left(|p_{mn}|^2 - \overline{|p_{mn}|^2}\right)^2 / MN}, \tag{6.31}$$

where $\overline{|p_{mn}|^2} = \sum_{m=1}^{M}\sum_{n=1}^{N}|p_{mn}|^2/MN$.

Using Eq. (6.31), it was assumed that all subsets have approximately the same statistical characteristics and are recorded by the digital camera in a linear area of the light characteristic.

The PINR mean value for the displacement $\langle d \rangle$ of Pattern 2 or Pattern 3 is given by

$$\langle \eta \rangle = \sum_{m=1}^{M}\sum_{n=1}^{N} \eta_{mn}/MN. \tag{6.32}$$

6.5.2 *Estimation of Errors in Shifts of Synthesized Speckle Patterns Subsets Using Modified Digital Speckle-Displacement Measurement Method*

According to the results of experimental studies using the MDSDM method, it was found that the optimal value of the parameter χ in Kumar–Hasselbrook filter for the surfaces of the test steel specimens is achieved at the lens "Helios-44" diaphragm number $f_{\#} = 28$ and is in the range from 1.4 to 1.8 [58]. Specimens were made from steel 20, which is similar to steels 1020, 1023, G10200 and G10230, and steel 45, which is similar to steels 1044, 1045, 1045H and G10450. A protective coating was applied to the surface of steel 45 specimens. Therefore, to conduct research with synthesized speckle patterns by this method, the parameter $\chi = 1.6$ was chosen.

Evaluation of mean bias and random displacement errors of synthesized speckle patterns with decorrelation parameters $\Delta_d = 0; 0.25; 0.5$ using the MDSDM method and determining the correlation peak subpixel position was fulfilled using the Sjödahl–Benckert algorithm (Algorithm 1) [103, 104] and the algorithm for determining the center of gravity (Algorithm 2) [64, 67, 69]. To this end, after the input displacement d of Pattern 3 in the diagonal direction $d = \sqrt{d_x^2 + d_y^2}$, where $d_x = d_y$, the average values of displacements $\langle d_x \rangle = (MN)^{-1} \sum_{m=1}^{M} \sum_{n=1}^{N} d_{mn,x}$ along x axis and $\langle d_y \rangle = (MN)^{-1} \sum_{m=1}^{M} \sum_{n=1}^{N} d_{mn,y}$ along y axis were calculated. The mean bias errors $\langle \theta_x \rangle$ and $\langle \theta_y \rangle$ along x and y axes, as well as the RMS errors $\langle s_x \rangle$ and $\langle s_y \rangle$ of displacements along x and y axes were also calculated. All numeric data was given in pixels. Pattern 2 and Pattern 3 were shifted from 0.5 to 12 pixels simultaneously in two orthogonal directions, that is, in the diagonal direction.

Analysis of the obtained results shows that Algorithm 1 gives better results than Algorithm 2. However, if $\Delta_d = 0$, then the mean bias error and RMS error calculated by Algorithm 1 increase sharply for displacements of the Pattern 2 by a noninteger number of pixels. The results obtained using Algorithm 2 also indicate large fluctuations of the random error, and these fluctuations are commensurate with the fluctuations obtained using Algorithm 1. Using Algorithm 1, the mean bias error of the subpixel position of the peak when the subset g_{mn} is displaced by an integer number of pixels is about 0.01 pixels, while when the subset is displaced by an odd number of half pixels, the error is about 0.1 pixels. If Algorithm 2 is used, then the mean bias error of the subpixel position of the peak when the subset g_{mn} is displaced by an integer number of pixels is about 0.06 pixels, while when displaced by an odd number of half pixels, the error is about 0.25 pixels.

The results obtained by displacing Pattern 3 with decorrelation parameters equal to $\Delta_d = 0.25$ and $\Delta_d = 0.5$ indicate the effect of decorrelation on increasing the mean bias error and random error. In this case, Algorithm 1 also gives better results than Algorithm 2. These errors increase when Pattern 3 is displaces both by an integer and a noninteger number of pixels. For example, if $\Delta_d = 0.25$ and Algorithm 1 is used, the mnth peak subpixel position RMS error is about 0.015 pixel when the subset $(g_d)_{mn}$ is displaced by an integer number of pixels, and about 0.09 pixel when the subset $(g_d)_{mn}$ is displaced by an odd number of half pixels. If $\Delta_d = 0.25$ and Algorithm 2 is used, the corresponding RMS errors are 0.06 and 0.27 pixels. With an increase in the decorrelation parameter Δ_d, there is a tendency to increase the mean bias error and the RMS error. For example, if the decorrelation parameter $\Delta_d = 0.5$ and Algorithm 1 is used, the mnth peak subpixel position RMS error is about 0.03 pixel when the subset $(g_d)_{mn}$ is displaced by an integer number of pixels, and about 0.23 pixel when the subset $(g_d)_{mn}$ is displaced by an odd number of half pixels. If $\Delta_d = 0.5$ and Algorithm 2 is used, the corresponding RMS errors are 0.16 and 0.33 pixels.

A significant increase in errors when displacing by a noninteger number of pixels is primarily due to the expanding of the frequency band of the JPS $\tilde{T}_{mn}$ for any two subsets r_{mn} and $(g_d)_{mn}$ selected from Pattern 1 and Pattern 3. This increase in error is caused mainly due to the fact that the number of samples (number of pixels) of the JPS $\tilde{T}_{mn}$ frequency band is the same as the number of samples (number of pixels) of the frequency band of the unexpanded JPS T_{mn} [see Eq. (6.3)] at Pattern 3 displacement by an integer number of pixels. In general, if Pattern 2 or Pattern 3 are displaced by a noninteger number of pixels, then the JPS $\tilde{T}_{mn}$ is broadened compared to the JPS T_{mn}. For example, let's Pattern 3 containing square pixels of size $a \times a$ is displaced along the axis x by a noninteger number of pixels equal to $k' + \frac{\Delta a}{a}$. Then, for $\Delta a \leq 0.5a$, the JPS $\tilde{T}_{mn}$ expands $\frac{a}{\Delta a}$ times, and for $0.5a \leq \Delta a < a$, this spectrum expands $\frac{a}{a-\Delta a}$ times in comparison with the JPS T_{mn}. However, the number of counts for the JPS $\tilde{T}_{mn}$ is the same as for the JPS T_{mn}.

6.5.3 Estimation of Errors in Shifts of Synthesized Speckle Patterns Subsets Using Optical-Digital Speckle Correlation Method

Similar results of evaluation of mean bias error and random displacement error of the synthesized speckle patterns with decorrelation parameters $\Delta_d = 0; 0.25; 0.5$ were obtained using the ODSC method and determining the subpixel position of the cross-correlation peak with Algorithm 1 and Algorithm 2. Digital models of the JTC1–JTC6 were considered.

Analysis of mean bias errors $\langle \theta_x \rangle$ and $\langle \theta_y \rangle$ and random errors $\langle s_x \rangle$ and $\langle s_y \rangle$ calculated by the ODSC method for the JTC1–JTC6 digital models using Algorithm 1 and Algorithm 2 at $\Delta_d = 0$ shows that these errors and their fluctuations calculated for an integer and noninteger number of pixels become especially large for the JTC1 model. The fluctuations of the errors calculated both Algorithm 1 and Algorithm 2 for this model are approximately the same (up to 0.9 pixel for the mean bias error and up to 2.4 pixel for the RMS error). However, the RMS error calculated by Algorithm 1 at displacements of Pattern 2 by distances that are multiples of half a pixel (up to 1.5 pixels inclusive) is about 0.1 pixel, and the same error calculated by Algorithm 2 is about 0.38 pixel. Besides, calculation of displacement errors for the JTC1 digital model shows that the noise level increases with increasing displacement, and at $d/\sqrt{2} > 4$, it is practically impossible to correctly measure the displacement of the artificial speckle pattern due to the high noise level. Such fluctuations are also observed in JTC2–JTCP6, but to a much lesser extent. For all JTC models, one can also note the tendency for errors to increase with increasing the displacement $d = \sqrt{d_x^2 + d_y^2}$. Note that as d increases, the errors increase most significantly in JTC4. Calculations of errors in this model show that with the displacement of 8 pixels, the RMS errors obtained by Algorithm 1 is 0.03 pixel, and by Algorithm 2 it is 0.11 pixel. This behavior of the errors is due to the influence of Pattern 2 noise, since this pattern is not taken into account during the FAF synthesis. The smallest values of random errors were observed in JTC6 for both Algorithm 1 and Algorithm 2. For the displacement of $d/\sqrt{2} = 8$ pixels, performed in JTC6 model, the RMS errors are 0.01 pixel using Algorithm 1 and 0.02 pixel using Algorithm 2.

The results of the mean bias errors calculations show that the smallest error values were obtained in JTC2 and JTC3 and are equal to 0. Only when $d/\sqrt{2} = 8$, the values of the mean bias errors are different from 0. If Algorithm 2 is used, then the smallest values of mean bias errors are obtained in JTC6 and are equal to 0. In addition, there are fluctuations in the errors calculated for integer and noninteger number of pixels. For noninteger number of pixels, these fluctuations are larger. They are especially large for JTC1. The smallest values of mean bias errors are obtained in JTC2, JTC3 and JTC6.

Thus, Algorithm 1 provides greater accuracy when calculating displacements. The only advantage of Algorithm 2 is its simplicity and shorter calculation time. Therefore, only Algorithm 1 was used to calculate the displacement errors of the JTC correlator digital models for $\Delta_d = 0.25;\ 0.5$.

The results of calculating the displacement errors of Pattern 3 for six digital models of the JTC correlator (JTC1–JTC6) implementing the ODSC method for $\Delta_d = 0.25$ and $\Delta_d = 0.5$ using Algorithm 1 were also analyzed. These results indicate the growth of mean bias errors and RMS error during the displacements of Pattern 3 from 0.5 to 8 pixels. The random error increases most significantly in the JTC1, and this model can be excluded from the further comparative analysis of JTC2–JTC6 models. An increase in error fluctuations when Pattern 3 is displaced by an integer and non-integer number of pixels is observed with an increase in the decorrelation parameter Δ_d. For example, if $\Delta_d = 0.25$, then the RMS error of displacing Pattern 3 by an integer number of pixels (up to 2 pixels inclusive) is 0.02 pixels for JTC2 and JTC3, 0.04 pixels for JTC4, and 0.01 pixels for JTC5 and JTC6. On the other hand, if $\Delta_d = 0.25$, then the RMS error of Pattern 3 displacement at distances that are multiples of half a pixel (up to 1.5 pixels inclusive) is 0.12 pixels for JTC2, 0.13 pixels for JTC3, 0.17 pixels for JTC4, JTC5 and JTC6. If $\Delta_d = 0.5$, then the RMS errors of displacing Pattern 3 by an integer number of pixels (up to 2 pixels inclusive) is 0.04 pixels for JTC2 and JTC3, 0.72 pixels for JTC4, and 0.014 pixels for JTC5 and JTC6. On the other hand, if $\Delta_d = 0.5$, then the RMS error of Pattern 3 displacement at distances that are multiples of half a pixel (up to 1.5 pixels inclusive) is 1.22 pixels for JTC2, 1.66 pixels for JTC3, 3.35 pixels for JTC4, and 0.68 pixels for JTC5 and JTC6. The greater Δ_d, the greater the ratio between the RMS error of Patter 3 displacements by an integer and non-integer number of pixels for all JTC2–JTC6 models.

The largest fluctuations of random errors were observed for the JTC4. Note that for this correlator model, if $\Delta_d = 0.5$, it was impossible to obtain the data on Pattern 3 displacements by 1.5 and 8 pixels. This is primarily due to the influence of Pattern 3 noise, since it is not taken into account when synthesizing the FAF. The smallest values of random errors were observed for the JTC5 and JTC6.

6.5.4 Comparative Analysis of Optical-Digital Speckle Correlation and Modified Digital Speckle-Displacement Measurement Methods Using Peak-to-Output Noise Ratio and Peak-to-Input Noise Ratio Mean Values

The results of calculating the PONR mean values $\langle\mu\rangle$ by using Eqs. (6.27)–(6.30) and the PINR mean values $\langle\eta\rangle$ using Eqs. (6.31) and (6.32) for six digital models of the JTC correlator (JTC1–JTC6) implementing the ODSC method at $\Delta_d = 0; 0.25; 0.5$ showed that these values are much larger for JTC5 and JTC6 digital models than for JTC1–JTC4 models. Therefore, the values $\langle\mu\rangle$ and $\langle\eta\rangle$ for correlation peaks resulting from displacements of Patterns 2 and 3 are given in Tables 6.1 and 6.2 only for JTC5 and JTC6.

To compare the efficiency of the MDSDM and ODSC methods, the PONR mean values $\langle\mu\rangle$ and PINR mean values $\langle\eta\rangle$ were calculated for the computer model of the MDSDM method with the Kumar–Hasselbrook filter ($\chi = 1.6$) for displaced

Table 6.1 PONR mean values $\langle\mu\rangle$ for correlation peaks resulting from Pattern 2 and Pattern 3 displacements in JTC5 and JTC6 digital models at $\Delta_d = 0;\ 0.25;\ 0.5$

#	$\frac{d}{\sqrt{2}}$	JTC5			JTC6		
		Δ_d			Δ_d		
		0	0.25	0.5	0	0.25	0.5
1	0.5	22	25	11	23	24	12
2	1.0	1244	1086	584	1234	933	558
3	1.5	32	28	15	37	35	15
4	2.0	1399	809	225	1061	828	195
5	4.0	984	498	60	739	479	58
6	8.0	320	191	21	338	294	27

Table 6.2 PINR mean values $\langle\eta\rangle$ for correlation peaks resulting from Pattern 2 and Pattern 3 displacements in JTC5 and JTC6 digital models at $\Delta_d = 0;\ 0.25;\ 0.5$

#	$\frac{d}{\sqrt{2}}$	JTC5			JTC6		
		Δ_d			Δ_d		
		0	0.25	0.5	0	0.25	0.5
1	0.5	8	7.1	5.0	7.8	6.7	5.3
2	1.0	38	25.8	14.9	38.5	25.4	13.8
3	1.5	8.2	6.5	6.1	7.2	6.8	6.4
4	2.0	30	22.5	18	32.1	22.3	17.6
5	4.0	20	12.5	8.7	20.2	13.1	9.3
6	8.0	11	7.0	5.5	10.4	5.6	6.4

Table 6.3 PONR mean values $\langle\mu\rangle$ and PINR mean values $\langle\eta\rangle$ calculated by the MDSDM method computer model with the Kumar–Hasselbrook filter ($\chi = 1.6$) for displaced Pattern 2 and Pattern 3 at $\Delta_d = 0;\ 0.25;\ 0.5$

#	$\frac{d}{\sqrt{2}}$	$\langle\mu\rangle$			$\langle\eta\rangle$		
		Δ_d			Δ_d		
		0	0.25	0.5	0	0.25	0.5
1	1.0	126.0	62.0	21.4	30.0	26.7	14.2
2	1.5	13.3	9.2	5.1	8.8	7.5	4.1

Pattern 2 and Pattern 3 at $\Delta_d = 0; 0.25; 0.5$. The maximal values of $\langle\eta\rangle$ and $\langle\mu\rangle$ for displacement $d/\sqrt{2} = 1.0$ pixel and the minimal values of $\langle\eta\rangle$ and $\langle\mu\rangle$ for $d/\sqrt{2} = 1.5$ pixel calculated by the MDSDM in the displacement range from 0.5 to 8 pixels are given in Table 6.3.

According to the results presented in Tables 6.1, 6.2 and 6.3, the PONR mean values are much larger in the JTC5 and JTC6 digital models in comparison with same values in the MDSDM method digital model. Therefore, the JTC5 and JTC6 have much better robustness to output noise. The PINR mean values in the JTC5 and JTC6 digital models, as well as in the MDSDM method digital model, are approximately the same. Therefore, these models have approximately the same robustness to input noise.

6.5.5 *Assessment of Correspondence the Subset Position to the Correlation Peak Position*

When implementing the methods of linear filtering and nonlinear transformation of JPS of subsets r_{mn} and g_{mn}, it is important to establish the correspondence between the coordinates of the correlation peak and the coordinates of the corresponding subset g_{mn} of Pattern 2. Such an assessment makes it possible to determine the methodical error of one or another method and choose among them the most suitable ones for the implementation of speckle correlation procedures and algorithms.

Computer models JTC1–JTC6 were used to calculate the speckle correlation between all subsets r_{mn} and g_{mn} in two Patterns 1 and 2. Pattern 2 and all its subsets did not move, that is, the calculations were performed at $d_x = d_y = 0$. To determine the mean absolute errors $\langle\theta_x^a\rangle$ and $\langle\theta_y^a\rangle$ and random errors $\langle s_x\rangle$ and $\langle s_y\rangle$ in measuring the positions of correlation peaks relative to the position of Pattern 2 subsets, the positions of all the peaks formed in a rectangular grating in the first diffraction order for all digital models of JTCs were calculated [96]. Table 6.4 shows the result of calculations of random and mean absolute errors of peaks positions measurement in digital models of JTC1–JTC6 using Algorithm 1 and Algorithm 2 for determining the subpixel position of the peak.

Table 6.4 Mean absolute and random errors of correlation peak position coincidence with the centers of Pattern 2 subsets for digital models of JTC1–JTC6 using Algorithm 1 and Algorithm 2 at $d_x = d_y = 0$

Type of correlator	$\langle d_x \rangle$	$\langle d_y \rangle$	$\langle \theta_x^a \rangle$	$\langle \theta_y^a \rangle$	$\langle s_x \rangle$	$\langle s_y \rangle$
Algorithm 1						
JTC1	0.002	0.003	0.002	0.003	0.01	0.01
JTC2	0.002	0.002	0.002	0.002	0.01	0.01
JTC3	0.001	0.003	0.001	0.003	0.01	0.01
JTC4	0.002	0.002	0.002	0.002	0.01	0.01
JTC5	0.002	0.002	0.001	0.001	0.01	0.01
JTC6	0.002	0.001	0.002	0.001	0.01	0.01
Algorithm 2						
JTC1	0	0	0	0	0	0
JTC2	0	0	0	0	0.002	0.002
JTC3	0	0	0	0	0.001	0.002
JTC4	0	0	0	0	0.002	0.002
JTC5	0	0	0	0	0.002	0.003
JTC6	0	0	0	0	0.003	0.003

The obtained results indicate a high coincidence of the positions of the correlation peaks with the positions of the corresponding subsets in all JTC digital models. All the models are characterized by approximately the same values of random and mean absolute errors. Table 6.4 also shows that the random errors calculated using Algorithm 2 are about 5–10 times lower than the errors calculated with Algorithm 1. Mean absolute errors in determining the subset positions when using Algorithm 2 for all the digital models of correlators are equal to 0. As can be seen from the Table 6.4, Algorithm 2 gives greater accuracy when calculating peak positions. Based on the obtained data, it can be asserted that there are no random overshoots and false signals on the outputs of all JTC1–JTC6 correlators. This testifies to the high reliability of the proposed methods of the JPS linear and nonlinear filtering, as they prevent the formation of false signals in the output correlation plane.

The measurement results represented in Table 6.4 show that the proposed JPS filtering and binarization methods are correct and the implementation of additional filtering procedures and nonlinear spectrum transformations does not lead to an increase in systematic and random errors in determining the displacements of the speckle pattern subsets.

6.5.6 Experimental Comparison of Modified Digital Speckle-Displacement Measurement and Optical-Digital Speckle Correlation Methods

Experimental implementation of the MDSDM and ODSC methods was performed using the developed corresponding digital models. Real speckle patterns of specimen surfaces were entered into computer using an experimental setup shown in Fig. 6.12. The setup was developed for experimental studies of the specimen surface in-plane displacements without its loading in order to simulate the RBM. The setup contained an adjusting table 1 having two micrometer screws 7 and a mount for fixing a specimen 2, a light source 3 (a LED "Kingbright L-1513SURC", peak wavelength $\lambda_p = 640$ nm, half-width of the radiation spectral line $\Delta\lambda_{0.5} = 27$ nm), a CCD camera (640×480 pixels) with a lens 4, and a computer 6. The setup was mounted on an interference table 5. The specimen was illuminated at an angle ϕ to the normal to its surface. The object of study was the surface of a steel 45 beam specimen with a protective coating deposited on the surface by gas-thermal treatment. The chosen surface area speckle pattern of this specimen is shown in Fig. 6.13.

As was mentioned in Sect. 6.5.2, the optimum value of the parameter in Kumar–Hasselbrook filter was determined equal to $\chi = 1.6$ for steel 20 and steel 45 specimen surfaces. The same value of this parameter was chosen for experimental implementation of the MDSDM method. All calculations to determine the correlation peaks parameters in digital models of correlators were fulfilled for pairs of subsets with a size of 32×32 pixels, which corresponds to the specimen surface area from 0.6×0.6 mm^2 to 1.5×1.5 mm^2, depending on the lens magnification.

The results of the experimental evaluation of random errors of the in-plane displacements along the x axis using the digital implementation of the MDSDM method at $\chi = 1.6$ and JTC1–JTC6 digital models are given in Table 6.5. The calculation of the subpixel position of the peaks was carried out using the Algorithm 1. Note that only random errors were calculated, since it was impossible to calculate

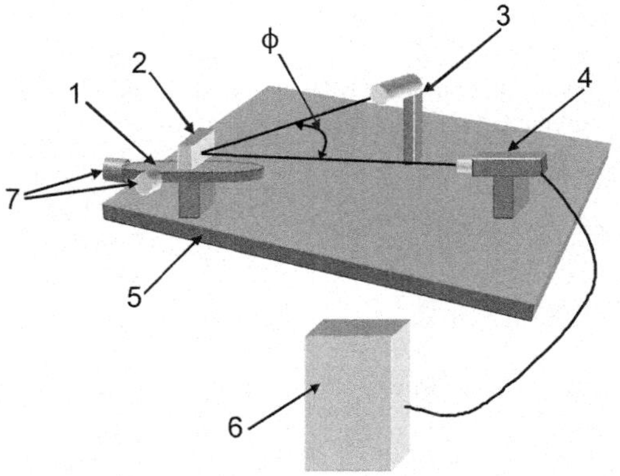

Fig. 6.12 Scheme of the experimental setup for studying in-plane RBMs of the specimen without applying loads to it: adjusting table (1), mount for fixing specimen (2), light source (3), CCD camera with lens "Helios-44" (4), interference table (5), computer (6), micrometer screws (7)

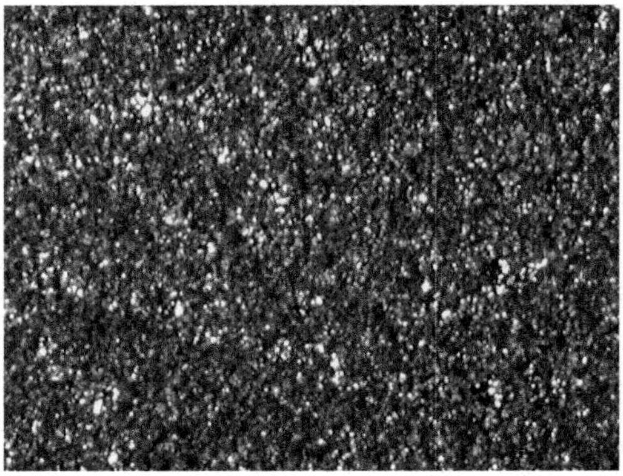

Fig. 6.13 Speckle pattern of the steel 45 beam specimen surface area. Surface was formed after the gas-thermal deposition of a protective coating

the mean bias errors due to the lack of technical means for precision movement of the specimen surface.

The values of RMS errors given in Table 6.5 show that the lowest level of random errors was obtained in the JTC5 and JTC6 digital models, that is as for the synthesized speckle patterns. The highest level of errors is in the JTC1 and in the digital implementation of the MDSDM method.

The PONR mean values $\langle \mu \rangle$ for the MDSDM method and the JTC1–JTC6 digital models were also calculated. The calculation results are represented in Table 6.6, where the specimen displacements d_x are given in μm.

Experiments with the real specimen given in Table 6.6 show that the highest PONR mean values are achieved for the JTC5. For the JTC6, these values are approximately two times smaller, and for the JTC2 and JTC3, they are approximately four times smaller. Higher PONR mean values were obtained for the JTC4 than for the JTC2 and JTC3, which is not entirely consistent with the results of computer simulations discussed above. The smallest values of $\langle \mu \rangle$ were obtained in the JTC1, which is quite understandable, since only joint correlation of subsets is carried out in this model without additional filtering operations or nonlinear transformations of the JPS.

Table 6.5 RMS errors $\langle s_x \rangle$ of the specimen displacements d_x without applying loads for the digital implementation of the MDSDM method and JTC1–JTC6 digital models (all numerical values are presented in microns)

d_x (μm)	$\langle s_x \rangle$ (μm)						
	MDSDM	JTC1	JTC2	JTC3	JTC4	JTC5	JTC6
10	2.1	1.6	1.3	2.5	1.6	0.6	0.8
20	4.7	9.0	2.0	9.4	3.1	1.8	1.9
30	8.0	6.0	7.1	13.0	9.2	3.1	3.5
40	10.5	11.5	4.8	6.5	5.5	4.2	5.2
60	12.5	16.0	10.5	9.0	10.5	6.1	7.3

Table 6.6 Results of PONR mean values $\langle \mu \rangle$ calculation of the specimen displacements d_x without applying loads using the digital models of the MDSDM method and JTC1–JTC6

d_x (μm)	$\langle \mu \rangle$						
	MDSDM	JTC1	JTC2	JTC3	JTC4	JTC5	JTC6
10	27.6	11.4	105	80	167	345	192
20	25.8	11.1	66	48	86	321	168
30	22.4	10.8	56	44	66	207	138
40	20.7	10.5	65	54	74	285	141
60	19.5	9.9	55	49	63	330	159

Thus, the experimental verification of the ODSC method showed the high reliability of such digital correlator models as the JTC5 and JTC6, and their resistance to noise and interference. In comparison with the digital model of the MDSDM method, the JTC5 and JTC6 models have significantly better information parameters. The digital models JTC2 and JTC3 in terms of such indicators as the level of random errors of displacements and the JTC4 in terms of mean values $\langle \mu \rangle$ also have an advantage over the MDSDM method.

The developed experimental setup for studying in-plane RBMs of the specimen was modified and combined with the EU-40 universal tensile-testing machine to determine the surface displacement fields of structural materials under uniaxial tensile loads. To this end, sheet specimens made of steel 08kp were studied. These specimens contained stress concentrators in the form of narrow slits, in which fatigue cracks were grown according to the methodical recommendations given in GOST [32]. The studied surface areas in each specimen were chosen at the location of the fatigue crack and the crack tip and were equal to 7×6 mm^2. The size of recorded speckle patterns was 512×480 pixels, and the subset size was 32×32 pixels. Experiments were fulfilled using the JTC1, JTC2, JTC5 and JTC6 digital models, as well as the digital model of the MDSDM method. The output peak gratings were used to calculate the PONR mean values $\langle \mu \rangle$ for each digital model at different tensile loads [65].

The experimental results have shown that the largest values of $\langle \mu \rangle$ are obtained in the JTC5 and JTC6 digital models, and values of $\langle \mu \rangle$ obtained in the JTC5 are approximately 1.5 times larger than these values in the JTC6. In turn, the values of $\langle \mu \rangle$ obtained in the JTC6 digital model are about 7–8 times larger than those in the digital model of the MDSDM method. For example, the PONR values in the JTC5 model are $\langle \mu \rangle = 280$–$300$, in the JTC6 model are $\langle \mu \rangle = 170$–$190$, and in the model of the MDSDM method are $\langle \mu \rangle = 22$–26, if the uniaxial tensile loads of 39–40 kN were applied to specimens. The smallest values of $\langle \mu \rangle$, as expected, were obtained in the JTC1 model.

Thus, the developed digital models, especially the JTC5 and JTC6 models, can be used to study the surface displacement and deformation fields in structural materials near the stress concentrators, including cracks.

6.6 Creation of Hybrid Optical-Digital Speckle Correlator Experimental Setup

The digital models of JTCs, considered in Sects. 6.4 and 6.5, showed high efficiency to generate the surface displacement fields of test synthesized objects and structural material specimens. Compared to them, the optical versions of JTCs with first and second optical stages are deprived of a number of advantages the correlator digital models, in particular, the simplicity of dividing speckle patterns into subsets, as well as the accurate and fast input and positioning of the two corresponding subsets in the input plane of the correlator. Moreover, producing of the JPS from two subsets r_{mn} and g_{mn} in the JTC needs to use complex and precise optical-mechanical devices for positioning the input speckle patterns, collimate the laser beam at the JTC first stage, suppress the influence of diffraction patterns from the square input windows in the JPS and in the first diffraction order at the JTC output and eliminate the influence of intense illumination of the diffraction center. However, in the digital correlator implementing the ODSC method, it is necessary to perform direct and inverse digital Fourier transforms. These operations lead to additional errors caused by edge effects, the influence of which on the reconstruction of the roughness, waviness and total fine phase maps was considered in Sect. 2.5.5. To suppress these errors, additional digital image processing operations in the spatial frequency domain are required [76]. On the other hand, the hybrid version of the JTC with the first digital stage and the second optical-digital one makes it possible to reduce the edge errors caused by the inverse Fourier transform of the JPS and improve the accuracy of determining the position of the correlation peaks, provided that a high-quality Fourier objective is used in the JTC second stage. At the same time, the specified optical-digital speckle correlator performing the joint Fourier transform preserves all the advantages of the first stage of the digital correlator, since all time-consuming operations on subsets of speckle patterns are performed in this stage. However, in order to ensure the necessary accuracy to determine the correlation peak position in this correlator, the optimal dimensions of all optical and optoelectronic elements of its second stage should be selected and the geometric parameters should be calculated.

The experimental setup of the hybrid optical-digital speckle correlator (HODSC) was created on the basis of the JTC correlator with median and circular median thresholds, i.e. the JTC2 and JTC6 were used to build the first digital stage of this setup. Scheme of the experimental setup is shown in Fig. 6.14.

A conventional computer was used to create the first stage of the setup. All the speckle patterns were recorded by a digital CCD camera SONY XCD-SX910 (1/2" IT CCD-SXGA, 1280 (H) × 960 (V), pixel size 4.65 μm × 4.65 μm and pixel pitch $p_x = p_y = 4.65$ μm). Recorded Pattern 1 and Pattern 2 were transformed into digital images with dimensions $K \times L = 512 \times 512$ pixels, which were divided on subsets r_{mn} and g_{mn} with dimensions $K_m \times L_n = 64 \times 64$ pixels. Each JPS of the subsets r_{mn} and g_{mn} was calculated and binarized using the digital models of JTCs with median thresholding (JTC2) and circular median thresholding (JTC6). To provide the sampling theorem at the binarized JPS calculation, the same dimensions

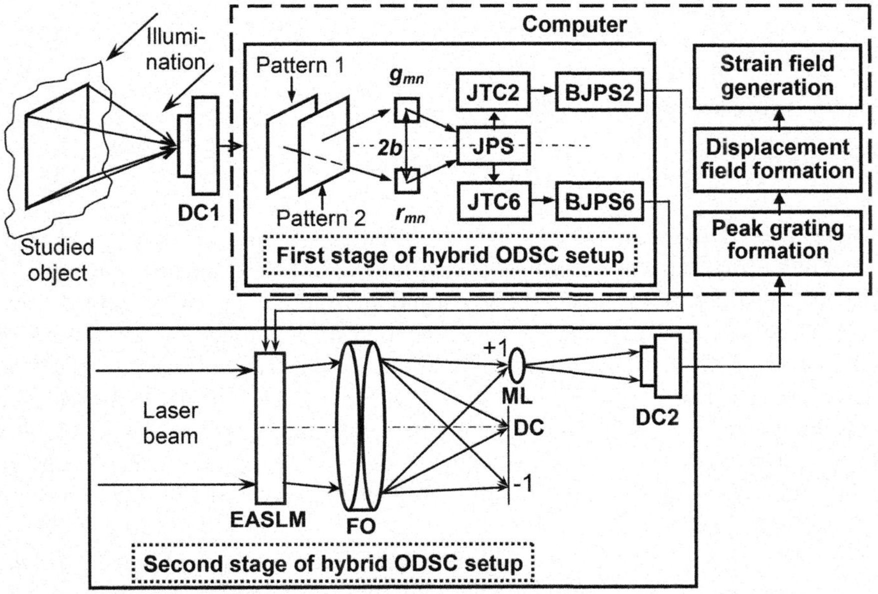

Fig. 6.14 Experimental setup of the HODSC containing first digital stage and second optical-digital stage. First stage includes digital camera (DC1), speckle patterns of initial surface area (Pattern 1) and deformed surface area (Pattern 2), input subsets r_{mn} and g_{mn}, joint power spectrum of input subsets (JPS), first stages of JTC2 and JTC6 digital models (JTC2, JTC6), binarized joint power spectrum (BJPS); the distance $2b$ between subsets r_{mn} and g_{mn}. Second stage include: electrically addressed spatial light modulator "Transducent SVGA SLM" (EASLM), Fourier objective (FO); diffraction center (DC) and $+1$ and -1 diffraction orders (+1, -1), microlens (ML), digital camera (DC2)

$K \times L = 512 \times 512$ pixels were chosen for input and frequency planes in the setup, and the next conditions between these planes were determined

$$\Delta_x = (2u_{\max})^{-1},\ \Delta_y = (2v_{\max})^{-1}, \tag{6.33}$$

where Δ_x and Δ_y are the sample pitches at the input JTC plane (x, y), $u_{\max}$ and $v_{\max}$ are the maximum spatial frequencies in the frequency domain (u, v).

Therefore, if the distance $2b$ between the subsets r_{mn} and g_{mn} with sizes

$$a_x \times a_y = (K_m \Delta_x) \times \left(L_n \Delta_y\right) = \left(\frac{K}{8}\Delta_x\right) \times \left(\frac{L}{8}\Delta_y\right) \tag{6.34}$$

is equal to

$$2b = 2L_m \Delta_y = \frac{L}{4}, \tag{6.35}$$

the carrier frequency is defined by

$$\nu_{car} = \frac{(L/4)}{L}\nu_{max} = \frac{1}{4}\frac{1}{2\Delta_y}, \tag{6.36}$$

and the fringe spacing is given by

$$D_s = (\nu_{car})^{-1} = 8\Delta_y, \tag{6.37}$$

where $a_x = a_y = a$ are the size of subsets r_{mn} and g_{mn}, $K_m = L_n$, $\Delta_x = \Delta_y$, $K = L$.

In the second stage of the setup, an electrically addressed spatial light modulator (EASLM) "Transducent SVGA SLM" of the transmission type with a matrix 800 × 600 pixels and 32 μm pixel pitches was used for input the binarized JPS with dimensions 512 × 512 pixels from the first stage output. The EASLM was placed in the front focal plane of a Fourier objective (FO) with focal length $f_{FO} = 210$ mm and was illuminated by a collimated parallel laser beam with wavelength $\lambda = 441$ nm. For lateral magnification of the generated correlation grating in the +1st diffraction order, the microlens 10MO-M-42 with focal length $f_{MO} = 18.14$ and numerical aperture 0.20 was used. The enlarged +1st diffraction order had sizes $2a'_x \times 2a'_y = 12.9 \times 12.9$ mm^2, and the distance between the FO rear focal plane P_F and the plane P_M of the enlarged order formation was 360 mm. A digital camera (DC2) placed behind the plane P_M and recorded only part of the enlarged diffraction order. Due to this configuration of the optical system in the second stage of the setup, a lateral magnification equal to $M_l = 21.8$ was achieved.

Both synthesized speckle patterns and the speckle patterns of the surfaces of real specimens were used for research. Figures 6.15 and 6.16 show the correlation peaks obtained by comparing the corresponding subset pairs of two synthesized test speckle patterns (Pattern 1 and Pattern 2).

Figure 6.15 shows two correlation peaks in the 0th and +1st diffraction orders, obtained in the HODSC experimental setup using the digital model of the JTC with ring median thresholding (JTC6) in its first stage. Figure 6.16 shows the correlation peaks in the 0th and +1st diffraction orders, which confirm the priority of choosing the JTC with ring median thresholding compared to the JTC with median thresholding. Correlation peaks from synthesized subsets for the JTC2 digital model (JTC

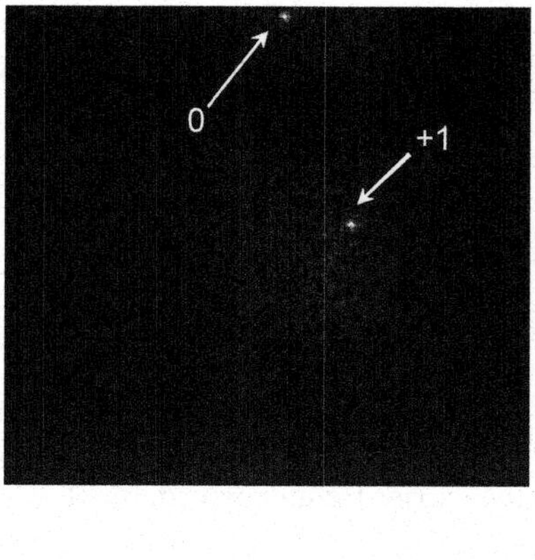

Fig. 6.15 Correlation peak from the subsets r_{mn} and g_{mn} in the +1st order of the output plane P_M of the experimental setup second stage based on the JTC6 digital model

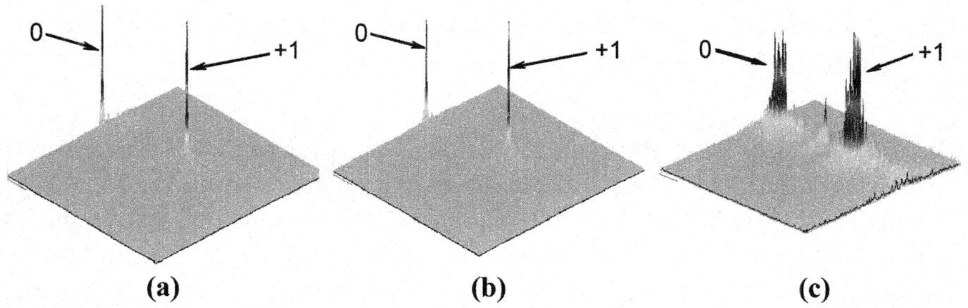

Fig. 6.16 Correlation peaks in the 0th and +1st diffraction orders at the output of the hybrid optical-digital speckle correlator obtained with the JTC2 digital model (**a**), JTC6 digital model (**b**), JTC1 digital model (**c**)

with median thresholding) are shown in Fig. 6.16a and for the JTC6 digital model (JTC with ring median thresholding) are shown in Fig. 6.16b. In addition, Fig. 6.16c shows the correlation peaks obtained in the conventional JTC (JTC1 digital model). They are surrounded by intense noise components, from which it is quite difficult to extract the desired peak.

To calculate the surface displacement errors during the rigid body motion of the steel 45 beam specimen with a protective coating (see Sect. 6.5.6), the HODSC experimental setup based on JTC6 architecture was used. The procedure for estimating these errors can be conditionally divided into several stages. First, the autocorrelation peak from the subset r_{mn} was formed in the +1st diffraction order for determination of a reference point corresponding to the initial position of this subset. The displacement of the correlation peak as a result of cross-correlation of the subsets r_{mn} and g_{mn} was calculated by subtracting the coordinates of this shifted peak from the coordinates of the reference point. Next, a correlation peaks grating was generated in the +1st diffraction order and displacements d_{mn} of all correlation peaks in this grating were determined. Finally, the displacement mean value $\langle d \rangle$ and the random error $\langle s \rangle$ of displacements were calculated. For example, if the speckle pattern was displaced by 60.0 mm in the HODSC experimental setup, then the RMS error of this speckle pattern displacement was $\langle s \rangle = 0.23\ \mu\text{m}$, which corresponds to approximately 0.02 pixel. On the other hand, in the experimental setup implementing the MDSDM method (see Sect. 6.5.6), the RMS error was $\langle s \rangle = 0.68\ \mu\text{m}$ with the same displacement of this speckle pattern. This error corresponds to approximately 0.06 pixel.

Thus, experimental studies have shown that the developed HODSC experimental setup can generate narrow and sharp correlation peaks and provide the necessary lateral magnification of the peaks grating to find the peak position with high accuracy. This position in the optical scheme is equivalent to a subpixel position, the random error of which is 0.02 pixel. The use of more precise optics will, in our opinion, increase the accuracy of determining the position of the peak. The disadvantage of the proposed ODSC method is the presence of complex optical elements and expensive optoelectronic equipment, in particular, the EASLM.

6.7 Using Digital Image Correlation to Assess Stress–Strain State of Material Near Crack

DIC technique [85, 111] briefly considered in Sect. 6.1.2 has some advantages in comparison with DPIV [44, 127], electronic speckle photography [102, 103], DSDM technique [12], MDSDM [58, 65, 67–69] and ODSC methods [63, 65, 67, 69–71, 75, 79]. Using the DIC, it is quite accurate to determine the surface strains and displacements in a wide measurement range. The resolution limit of DIC is very high and depends on the dimensions of the subsets. DIC provides smaller errors in determining the surface displacements and strains, which can significantly improve the accuracy of diagnosing the stress–strain state of materials and structural elements containing cracks and other mechanical damage [34, 94, 115]. On the other hand, the technical implementation of surface displacements in 2D DIC is more complicated, since it requires exclusively in-plane movements and strains, and hence the absence of normal strains.

DIC technique allows solving different problems of fracture mechanics. Among them, the study of features of cracking and crack resistance of materials is of particular interest. Irwin [40] postulated three types of independent kinematic displacements of the upper and lower surfaces of the crack relative to each other. These three types of displacements (deformations), namely normal opening (mode I), transverse shear (mode II) and longitudinal shear (mode III) describe all possible types of crack behavior. Each of the three basic types of crack surfaces displacements is associated with components of the stress field in the crack tip vicinity. These stresses, in turn, are expressed in terms of the stress intensity factors (SIFs) K_{I}, K_{II} and K_{III} [40], which are considered as a measure of stress singularity near the crack tip.

Standard measurement techniques of the fracture toughness of materials [32] provide for the determination of the SIFs critical values $K_{\mathrm{I}C}$, $K_{\mathrm{II}C}$ and $K_{\mathrm{III}C}$, which are the force criteria for crack resistance, as well as the critical values of crack opening (deformation criterion δ_c) and J-integral (energy criterion J_c). These techniques are destructive and not universal enough; therefore, they are not appropriate for the diagnosis of specific constructions, as it involves significant restrictions on a specimen size and shape. In addition, they are unsuitable for mixed types of loads, which are most often used in practice.

Thus, a reliable determination of the SIF and the geometric characteristics of a crack in material (the position of the crack tip, its increment, the direction of crack propagation, its characteristic size) is a key factor for diagnosing the stress state of the material during quasi-brittle cracking. To improve the reliability and accuracy of diagnosing the stress state near a crack, taking into account the SIF and the crack geometric parameters, it is necessary to use approaches that will better describe the fields of mechanical stresses in a loaded material specimen with a crack-like defect for certain types of loads. Development of such approaches is possible by using the DIC methodology.

6.7.1 Development of Stress Field Components Determination Technique Using Digital Image Correlation

Due to the inaccurate determination of the crack tip location, the errors in the SIF calculation increase significantly [132]. To decrease these errors, DIC techniques were used. Lopez–Crespo et al. [52] determined the position of the crack tip to calculate the SIF under a mixed load. To this end, the Sobel algorithm was applied to displacement fields measured by the DIC means. Du et al. [20] assessed the speed of crack growth at mixed load and calculated the corresponding SIF values. For this purpose, the displacements were calculated by the DIC technique and the position of the crack tip was determined using the developed software and the Sobel algorithm. The frequent use of this algorithm is due to the possibility of its applying to data containing significant noise. However, for precise determination of the stress state parameters in a material with a crack, the coordinates of its tip, obtained from the displacement fields using the Sobel algorithm, are usually used only as a certain starting approximation. This circumstance is due both to the errors of this algorithm, and to the fact that real cracks have a complex curvilinear front in the bulk of the material.

Conventional DIC and similar techniques use only square or rectangular subsets, whereas it is necessary to record surface displacement fields in close proximity to uneven crack edges, uneven fractures and irregular edges of specimens. To overcome this problem, image adaptive segmentation approaches for producing of subsets with irregular shape were developed by Tong [120] and Cofaru et al. [18]. The modified technique of adaptive segmentation of specimen surface speckle patterns which allows producing irregular subsets and takes into account a subset speckle structure and dimension was proposed by Sakharuk et al. [97]. In order to increase the accuracy of displacement definition, the correlation algorithm with fractional power filter was used. The adaptive segmentation procedure contains several steps of speckle pattern digital processing, namely image binarization by Otsu technique, morphological erosion, morphological thinning, connected components labeling and Voronoi tessellation. After the image segmentation, correlation of each corresponding subset pair is performed, and the displacement vector is defined. Obtained displacement fields for duralumin beam specimen D16T similar to alloy 2024-T4 under a load of 980 N are presented in Fig. 6.17.

The proposed technique makes it possible to form subsets with taking into account nonlinear edges, cracks and other stress concentrators. Meanwhile, it generates more accurate displacement fields near stress concentrators and near specimen nonlinear edges and does not require surface painting to produce speckles in comparison with other techniques with adaptive segmentation.

To increase the accuracy of diagnosing the stress–strain state of quasi-brittle materials near the crack tip, the stress field components determination (SFCD) technique using the DIC was developed [51, 54–56]. This technique makes it possible to calculate the stress fields in the vicinity of the crack tip on basis of the deformation field

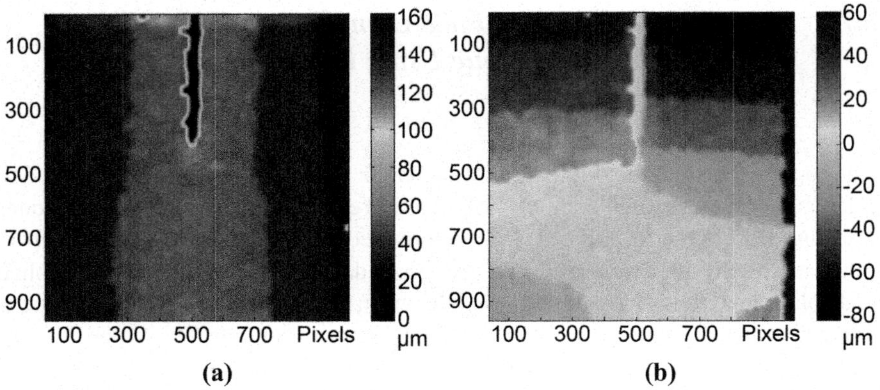

Fig. 6.17 Displacement fields along x (**a**) and y (**b**) axes for duralumin beam specimen D16T under a load of 980 N

components computed from the speckle patterns of the material surface using the DIC.

Testing of static crack resistance of materials and structural elements involves ensuring the conditions for in-plane deformation of the studied specimens [32]. The implementation of the selected DIC technique for determining the deformation fields also implies the need to consider displacements and strains within the frameworks of the plane elasticity problem. For such the problem, the well-known formulas for searched components of the stress field can be used that are given as follows [100]:

$$\begin{aligned} &E\varepsilon_{xx} = \sigma_{xx} - \mu(\sigma_{yy} + \sigma_{zz}); \quad E\varepsilon_{yy} = \sigma_{yy} - \mu(\sigma_{xx} + \sigma_{zz}); \\ &G\varepsilon_{xy} = \sigma_{xy}; \quad \sigma_{xx} = \mu(\sigma_{xx} + \sigma_{yy}), \end{aligned} \tag{6.38}$$

where ε_{xx}, ε_{yy} and ε_{zz} are the deformation field components; σ_{xx}, σ_{yy}, σ_{xy} and σ_{zz} are the stress field components; μ is the Poisson ratio; E is the elastic modulus; G is the shear modulus.

When solving (6.38), the following relations are obtained for the desired components of the stress field:

$$\left.\begin{aligned} \sigma_{xx} &= \tfrac{E}{(1-2\mu)(1-\mu)}\left[(1-\mu)\varepsilon_{xx} + \mu\varepsilon_{yy}\right] \\ \sigma_{yy} &= \tfrac{E}{(1-2\mu)(1-\mu)}\left[(1-\mu)\varepsilon_{yy} + \mu\varepsilon_{xx}\right] \\ \sigma_{xy} &= G\varepsilon_{xy} \end{aligned}\right\}. \tag{6.39}$$

The majority of modern structural materials can be characterized as quasi-brittle during cracking. In such a material, when it is loaded, a certain plastic zone appears around the crack tip (as a rule, of small dimensions compared to the characteristic crack length). The zone is surrounded by a surface area of disproportionate deformations, which has a much larger size. The conditions of stresses and strains proportionality within this area are not fulfilled, that is, the linear elastic model cannot be used. This imposes the first limitation on the selection of data for determining the parameters of the stress–strain state of the material from the components of the stress field around the crack tip. The second limitation concerns the provision of a given accuracy in determining the stress field.

Thus, the selection and processing of data on the components of the stress field involves the following steps:

- establishing the boundaries of the surface area of disproportionality;
- setting the boundaries of the surface area with a critical signal-to-noise ratio;
- sampling of data of a given dimension between these areas.

The initial data in the SFCD technique are the surface displacement and deformation fields. The proportionality limit during tensile deformation of the material is set by its standard parameter σ_T (tensile yield stress). The boundary of the surface area of disproportionality is set on the basis of certain total normal components of the stress field in material $(\sigma_{xx} + \sigma_{yy})$, where the criterion for the surface area boundary is the proportionality limit σ_T of the material.

The starting approximation for the position of the crack tip in material is determined by applying the Sobel algorithm to the d_y-component of the displacement field. This is achieved by moving the sliding window of one of the typical Sobel algorithms [22]:

$$\begin{bmatrix} 1 & 0 & -1 \\ 2 & 0 & -2 \\ 1 & 0 & -1 \end{bmatrix}. \tag{6.40}$$

Speckle patterns of the test object surface are recorded before the load and at each step of the load, taking into account the size of the averaging sample. The deformed speckle patterns are obtained as a result of the static load application with the provision of plane deformation conditions. The recorded speckle patterns are then averaged within each sample. Averaged speckle patterns of the deformed and undeformed surfaces of the test specimen with a crack are fed to the input of the software-implemented digital speckle correlator. Next, the components of the displacement field (d_x, d_y) and the deformation field $(\varepsilon_{xx}, \varepsilon_{yy}, \varepsilon_{xy})$ of the specimen surface are determined by the procedure described in Sect. 6.1.2. Starting approximations of the crack tip coordinates (X_0, Y_0) are found from the d_y-component (perpendicular to the edges of the crack) of the displacement field using the Sobel algorithm.

The components of the stress field $(\sigma_{xx}, \sigma_{yy}, \sigma_{xy})$ are calculated from the components of the deformation field using Eq. (6.39) and the known mechanical characteristics of the material. After that, the boundaries of the data selection area are determined based on the known proportionality limit σ_T of the studied material, and the boundaries of the area of disproportionality and the boundaries of the area with small values of the stress field components are also set. Taking into account the boundaries of the data selection area and the required number of points of the stress field components, the density of the data acquisition grid is calculated and the data of the grid nodes with the nodes coordinates are selected.

6.7.2 Using DIC to Determine the Crack Propagation Angle

There are several approaches to predicting or determining the direction of the crack start and propagation in metal that use the force and energy parameters of the stress–strain state of such material [23, 36, 39, 48, 51, 86, 99]. Erdogan and Sih [23] proposed the approach of the highest breaking stress, which is based on the postulation of the following conditions: The crack propagates from the top in the direction perpendicular to the highest breaking stress $\sigma_{\theta\theta}$. That is, for the crack propagation direction angle θ_c, the following conditions must be satisfied [23]:

$$\left.\frac{\partial\sigma_{\theta\theta}}{\partial\theta}\right|_{\theta=\theta c}=0;\quad \left.\frac{\partial^2\sigma_{\theta\theta}}{\partial\theta^2}\right|_{\theta=\theta c}<0. \tag{6.41}$$

The value θ_c is found by substituting in Eq. (6.41) the formulas describing the stresses around the crack tip through the values K_{I} and K_{II}, that is [48]

$$\theta_c=2\mathrm{arctg}\left[\frac{1}{4}\left(\frac{K_{\mathrm{I}}}{K_{\mathrm{II}}}-\mathrm{sign}(K_{\mathrm{II}})\sqrt{8+\frac{K_{\mathrm{I}}^2}{K_{\mathrm{II}}^2}}\right)\right], \tag{6.42}$$

or by using the stress field components $\sigma_{r\theta}$ and $\sigma_{\theta\theta}$ in polar coordinate system. The predicted direction of crack propagation coincides with the direction determined by the points of intersection of the figure contours $\sigma_{r\theta}=0$ and $\sigma_{\theta\theta}=\sigma_T$.

This approach is quite widespread, however, it has a significant drawback, since the components of the stress field in the immediate vicinity of the crack tip cannot be reliably described using elastic strains.

There are also known energy approaches to determining the angle θ_c based on the maximum strain energy release rate [39] and the minimum strain energy density factor [99]. Hussain et al. [39] prove that the crack grows along the line with the maximum rate of release of strain energy G, which is given by

$$G(\theta)=\frac{4(1+\mu)}{E}\left(\frac{1}{3+\cos^2\theta}\right)^2\left(\frac{1-\pi/\theta}{1+\pi/\theta}\right)^{\frac{\theta}{\pi}}\left[K_{\mathrm{I}}^2(1+3\cos^2\theta)\right.$$
$$\left.+8K_{\mathrm{I}}K_{\mathrm{II}}\sin\theta\cos\theta+K_{\mathrm{II}}^2(9-5\cos^2\theta)\right]. \tag{6.43}$$

In this case, the conditions for the direction of crack propagation are as follows:

$$\left.\frac{\partial G(\theta)}{\partial\theta}\right|_{\theta=\theta c}=0;\quad \left.\frac{\partial^2 G(\theta)}{\partial\theta^2}\right|_{\theta=\theta c}<0. \tag{6.44}$$

Therefore, in order to determine the angle θ_c, one should find the site of the supremum for the function (6.43) according to the conditions (6.44).

The energy approach proposed by Sih [99] is based on the axiomatic assumptions that the crack grows along the direction $\theta_c \in [-\pi, \pi]$ of the minimum value of the strain energy density factor $S(\theta)$, which is defined as

$$S(\theta) = \frac{1+\mu}{32\pi E}\big[K_{\mathrm{I}}^2(1-\cos\theta)(k-\cos\theta) + 2K_{\mathrm{I}}K_{\mathrm{II}}\sin\theta(2\cos\theta - k + 1) \\ + K_{\mathrm{II}}^2((k+1)(1-\cos\theta) + (1+\cos\theta)(3\cos\theta - 1))\big], \quad (6.45)$$

where $k = 4 - 3\mu$ for plane-strain conditions. The value of the angle θ_c can be found from the conditions that determine the infimum of the function $S(\theta)$ and are given by

$$\left.\frac{\partial S(\theta)}{\partial\theta}\right|_{\theta=\theta c} = 0; \quad \left.\frac{\partial^2 S(\theta)}{\partial\theta^2}\right|_{\theta=\theta c} > 0. \quad (6.46)$$

Papadopoulos [86] proposed an approach that is based on the full components of the stress field (the so-called third stress invariant) and involves using the critical value of the stress tensor determinant as a crack growth criterion. This value is defined as

$$\mathrm{Det}(\sigma_{ij}) = \begin{vmatrix} \sigma_{xx} & \sigma_{xy} \\ \sigma_{xy} & \sigma_{yy} \end{vmatrix}. \quad (6.47)$$

Upon reaching a certain critical value $\mathrm{Det}(\sigma_{ij})_{cr}$ in Eq. (6.47), which is a mechanical characteristic of the studied material and takes into account all the mechanisms of its deformation and destruction, the direction of crack start θ_c can be determined from the maximum of $\mathrm{Det}(\sigma_{ij})$ using the next dependencies:

$$\left.\frac{\partial\big[\mathrm{Det}(\sigma_{ij})\big]}{\partial\theta}\right|_{\theta=\theta c} = 0; \quad \left.\frac{\partial^2\big[\mathrm{Det}(\sigma_{ij})\big]}{\partial\theta^2}\right|_{\theta=\theta c} < 0. \quad (6.48)$$

The critical value of stress tensor determinant for the studied material D16T, which was calculated on the basis of the material ultimate tensile and shear strength, is $\mathrm{Det}(\sigma_{ij})_{cr} \approx 115 \cdot 10^{15}$ Pa^2. However, the proposed approach during real measurements is inaccurate, since the differentiation of the data for finding the global maximum in the vicinity of the crack tip in the presence of significant noise and nonlinearities is approximate.

Therefore, a new method for determining θ_c from the calculated stress tensor determinant critical value $\mathrm{Det}(\sigma_{ij})_{cr}$ was proposed and implemented [36, 51]. The method requires finding the direction from the crack tip to the mass center of the determinant $\mathrm{Det}(\sigma_{ij})$ distribution within a closed figure limited by the values of $\mathrm{Det}(\sigma_{ij})$. In Fig. 6.18, the distribution of $\mathrm{Det}(\sigma_{ij})$ is reconstructed with the help of determined components of the stress field. The physical size of the image is 6 × 6 mm^2.

The results of numerical calculations for the above methods for predicting the direction of the crack propagation for loads of 686 N and 980 N and deviations from the calculated values are given in Table 6.7. The data presented in this table show that different approaches give approximately the same (±1.5°) values of the crack propagation angle. The smallest deviation from the crack propagation angle determined for different loads was obtained by the proposed method [36, 51] based on the use of the critical third stress invariant [86]. The proposed method differs from the well-known approach [86] in that it uses a data set extracted within the figure bounded by the value of the third invariant, while the well-known approach extracts a much smaller data set lying only on the contour of this figure.

The obtained results show that the smallest deviation in determining the crack propagation angle σ_c is achieved using the developed method [36, 51]. Example of

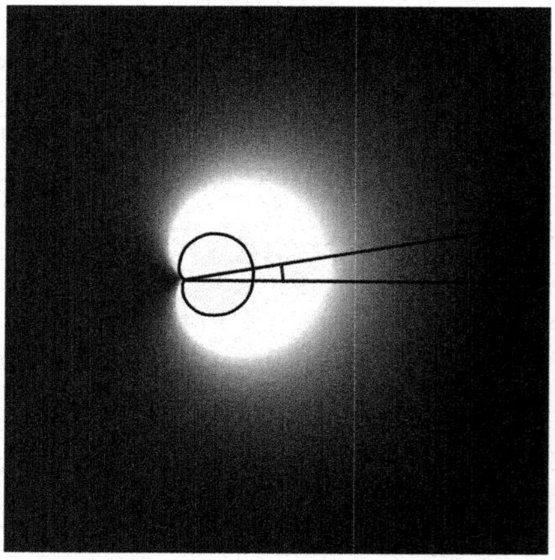

Fig. 6.18 Direction of crack propagation, determined by the guide to the mass center of the critical third stress invariant

Table 6.7 Values of the crack propagation angle θ_c in the studied D16T specimen with a size 200 × 10 × 20 mm^3 under three-point bending, determined by different approaches and their relative deviations

According to Kim and Paulino [48]	According to Hussain and Pu [39]	According to Sih [99]	According to the developed method [36, 51]
Load 686 N			
9.41°	9.47°	9.37°	9.89°
Load 980 N			
10.53°	11.65°	11.42°	9.10°
Relative deviations			
11.23%	20.6%	19.7%	8.32%

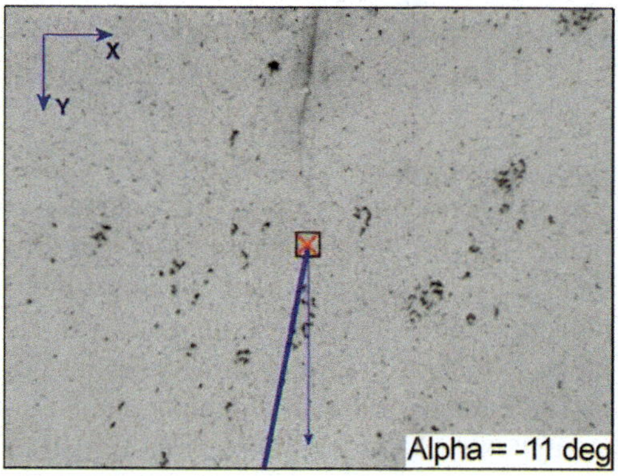

Fig. 6.19 Determination of the crack propagation direction in a gantry crane structural element

technical implementation of the method for determination of the crack propagation angle in a metal structural element is shown in Fig. 6.19.

6.7.3 Conclusion

In last years, DIC, which also includes digital particle image velocimetry, electronic speckle photography, as well as DSDM, MDSDM, ODSC methods and various DIC applications to assess the stress–strain state of material near crack, has become the powerful tool for NDT and TD of materials and structural elements. They are used in different industrial branches and scientific researches due to high reliability and possibility to measuring the surface displacement and deformation fields of studied specimens. The wide application of these techniques in such prominent branches as aerospace, automotive and building industries testifies their importance, availability and competitiveness.

Several devices and systems implementing 2D and 3D DIC and similar techniques were created. In particular, the Laser Speckle Extensometer (Dewetron Inc.), ARAMIS Digital Image Correlation System (Trilion Quality Systems Company), Digital Image Correlation System Q-400 (Dantec Dynamics A/S) and other high-performable devices are widely used in scientific researches and industrial applications. The Hand Held Optical Digital Speckle Correlator (HHODSC) has been created in PMI [59, 65] and modified in last years [51]. This device and its modifications are used for NDT and TD of constructional materials. In contrast to the mentioned above devices and systems, the HHODSC possesses higher reliability of displacement field reconstruction near stress concentrators due to reaching the larger peak-to-noise ratio and sharper and narrower correlation peaks. The device and its modification can work both in autonomous regime and in combination with measuring complexes containing test machines and corresponding cells. Testing of specimens can be fulfilled, if they are mechanically loaded under tension, compression or fatigue.

6.8 Detecting Subsurface Defects in Composite Structures Using Speckle Decorrelation

The subtractive synchronized DSPI method [51, 72, 77, 81] considered in Sect. 4.7.1, as well as the additive–subtractive electronic speckle pattern interferometry (ESPI)/Shearography [24, 25, 88, 125] and synchronized reference updating ESPI [89] techniques, which are close to it, make it possible to detect subsurface defects in laminated composite structures. However, the techniques based on ESPI are very sensitive to the external vibrations and air or heat flows, which practically makes it impossible to use in natural conditions. Shearographic systems are robust enough to such external influences. But all of the above methods are characterized by a high level of decorrelation of the recorded fringe patterns, i.e. speckle interferograms and shearograms, which are used to detect defects. In these methods, it is not possible to reliably identify the subsurface defects of relatively large sizes due to the high level of decorrelation between two adjacent fringe patterns, which are recorded at the maximum out-of-plane displacements of the region of interest (ROI) in two opposite directions and mutually subtracted (see Sect. 4.7). In addition, all the systems performing these techniques contain quite complex and precise equipment, which also complicates their use in field conditions.

In connection with the foregoing, the development of new methods for NDT of composite structures to detect subsurface defects and corresponding systems, which would have relatively simple equipment and be able to work under natural conditions, is an important scientific and technical problem. One of the possible ways to solve it consists in the study of dynamic speckle patterns for their analysis and processing. Such speckle patterns are formed in the optical system when coherent or quasi-coherent light is reflected from the optically rough surface of the laminated composite, which is excited by harmonic ultrasonic (US) radiation. Sequences of speckle patterns are recorded by a digital camera and, after their additional processing, defects are highlighted. The studies carried out in this direction made it possible to develop a new method for detecting the subsurface defects in composite panels using a sequence of dynamic speckle patterns of the surface [73, 78, 80–82]. The experimental setup created for the technical implementation of this method confirmed its reliability to detect both artificial and real defects. This method has analogs among other methods of dynamic speckle patterns analysis [11, 43, 45, 109, 128], which use the surface tilt to analyze vibrations and thermal stresses in beams, membranes and other types of materials and structural elements.

6.8.1 Method for Detection the Subsurface Defects in Laminated Composites Using Dynamic Speckle Pattern Sequences

Studying of speckle fluctuations on the surfaces and in the subsurface layers of various objects is the urgent direction of Speckle Metrology. Analysis of speckle fluctuations successfully uses to research blood flows [9, 91], freshness and aging of food products (fruits, vegetables, muscle tissues of animals) [59, 91, 134], kinetics of surface corrosion of metals and alloys [27], paint drying [7], in-plane rotation of diffuse objects [133]. They use the temporal evolution of laser speckles, which are generated by the stationary or moving surfaces and subsurface layers of biological or technical objects. The proposed method for subsurface defects detecting in laminated composites uses out-of-plane displacements of thin layers of local surface areas placed directly above the defects, i.e. ROIs that oscillate at resonant frequencies.

Due to US excitation of the laminated thin composite panel, acoustic flexural waves propagating in thin layers activate the bottom defects at their resonant frequencies [16, 17, 106–108]. When these waves excite the composite structure with a variable frequency, the ROI begins to vibrate at one of its own resonant frequencies in the direction transverse to the direction of the US wave propagation, that is, to the plane of the surface. If the harmonic US excitation is generated in the frequency scanning mode, then all defects in the given frequency range are excited on their own resonant frequencies. The ROI located directly above the excited defect can be considered as a thin edge-clamped membrane. The excitation of the defect leads to vibration of this membrane on one of its own resonant frequency. That is, the excitation of the defect under resonant frequency is inextricably linked to the vibrations of the ROI due to the flexural waves propagating in the thin layer above the defect, i.e. in the ROI [82].

Suppose that the optically rough surface area of the laminated composite, within which a subsurface defect is located, is illuminated by an expanded laser beam, and the US harmonic excitation has a frequency that coinciding with one of the resonant frequencies of the defect. Then, a series of speckle patterns of the excited surface is formed and registered during the vibrations of the ROI with the help of an optical system. If the ROI vibrates at one of the resonant frequencies, this frequency is not resonant for the area of the surface outside the ROI, and this area does not vibrate in contrast to the ROI. The spatial structure and contrast of the local section of the speckle pattern, i.e. the local speckle pattern (LSP), generated by the ROI changes synchronously with the ROI vibrations. At the same time, the rest of the speckle pattern outside the LSP, which is generated by the area of the surface around the ROI, remains unchanged. This difference between the LSP and the rest of the speckle pattern makes it possible to highlight the LSP or its spatial elements and thereby detect a subsurface defect.

Analysis of publications devoted to temporal changes of dynamic laser speckles during rough surface vibration shows that two main factors affect the structure and contrast of speckles. The first factor (Factor 1) is caused by the spatial frequency shift in the aperture plane of the lens of the optical scheme forming a speckle pattern, which leads to the transformation of this pattern spatial structure [11, 26, 129, 130]. The second factor (Factor 2) is caused by the out-of-plane movement of the diffuse surface, which leads to a decrease in the speckle contrast and the speckle blurring [84].

Influence of Factor 1 on transformation of dynamic speckle patterns structure. The transformation of the LSP generated by the ROI during its oscillations occurs due to the tilt of the ROI spatial elements, which leads to a shift of the spatial frequency spectrum in the plane of the lens aperture forming speckle patterns in the image plane of a coherent or quasi-coherent imaging system [26, 129, 130]. The decorrelation between the transformed LSP and the initial one leads to a decrease in the local speckle contrast in this pattern. Therefore, it is possible to extract the LSP from the whole speckle pattern, if a pair of such speckle patterns or their entire sequence is processed appropriately.

Let us analyze the effect of out-of-plane displacements of the ROI placed in the input plane (x, y) of the imaging system on the LSP transformation in the image plane (x', y') of the lens. The optical scheme for recording speckle patterns generated by the surface area, including the ROI and the tilted spatial elements of the ROI, is shown in Fig. 6.20.

Let the ROI spatial element that tilts when the ROI vibrates has a flat surface before vibration. Its surface coincides with the flat surface of the composite plate, and its roughness does not differ from the composite plate surface roughness. The complex amplitudes of the LSP part formed by this ROI element before tilting and during tilting in the image plane can be respectively defined as

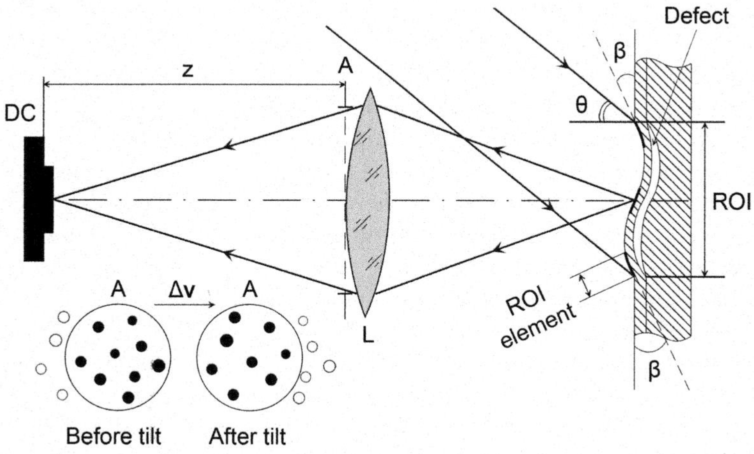

Fig. 6.20 Optical scheme for recording speckle patterns generated by the ROI and the ROI tilted spatial elements; θ is the incidence angle of coherent or quasi-coherent light beam; z is the distance between the lens aperture and the digital camera (DC) matrix photosensor; β is the tilt angle between the tilted ROI spatial element and the unexcited surface plane; L is the lens; $\Delta\nu$ is the spatial frequency shift in the lens aperture plane; A is the lens aperture

$$\tilde{a}(r) = \tilde{g}(r) \otimes h(r), \tag{6.49}$$

$$a(r) = g(r)\exp(-j2\pi \Delta\nu r) \otimes h(r), \tag{6.50}$$

where $r = \left(x', y'\right)$ is the point in the LSP part; $\tilde{g}$ and g are the complex amplitudes of the LSP part before and during the ROI spatial element tilting within the limits of the geometrical optics approximation; h is the point spread function of the optical imaging system; $\Delta\nu = (\Delta u, \Delta v)$ is the spatial frequency shift in the lens aperture plane; $\otimes$ is the convolution symbol.

Fourier spectra of the amplitude distributions $\tilde{a}$ and a expressed by Eqs. (6.49) and (6.50) are given by

$$\tilde{A}(\nu) = \tilde{G}(\nu)H(\nu), \tag{6.51}$$

$$A(\nu) = G(\nu + \Delta\nu)H(\nu), \tag{6.52}$$

where H is the pupil function of the lens aperture, $\tilde{A}(\nu) = F\left[\tilde{a}(r)\right]$, $A(\nu) = F[a(r)]$, $G(\nu) = F[g(r)]$, $\tilde{G}(\nu) = F\left[\tilde{g}(r)\right]$, F is the Fourier transform symbol, $\nu = (u, v)$ is the spatial frequency.

Formula (6.52) shows that the spatial frequency shift in the lens aperture plane is caused by the tilt of the ROI spatial element. The linear dependence of the spatial frequency shift $\Delta\nu$ versus tilt angle β of the ROI spatial element relative to the surface plane (see Fig. 6.20) can be expressed as [26]

$$\Delta\nu = \frac{1 + \cos\theta}{\lambda} \cdot \frac{1}{M_l}\beta, \tag{6.53}$$

where λ is the wavelength, θ is the incidence angle of coherent or quasi-coherent light, M_l is the linear magnification of the optical system forming speckle patterns.

This equation shows that the maximum spatial frequency shift corresponds to the maximum tilt angle $\beta_{\max}$ between the tilted ROI spatial element and the unexcited surface plane. Therefore, the largest shift in the spatial frequency of the amplitude spectrum in the lens aperture leads to the largest changes in the spatial structure of the LSP.

Ciampa et al. [16, 17] and Solodov [105] developed mathematical models described the nonlinear elastic effects of the local defect resonance, which can characterize both damage sites and all points on the composite surface area. Damage in the form of a blind flat-bottomed hole can be interpreted as a subsurface defect with a thin layer of a composite panel between the defect and ROI. This thin layer we consider as a thin membrane with clamped edges [82]. If the subsurface defect have the rectangular shape and its ROI is a rough surface of the same size as the defect, then this ROI can be given by a shape function of a thin rectangle membrane of the same size. In a simplified linear approximation, without taking into account nonlinear transverse loads applied to the composite plate and other physical parameters, the equation of out-of-plane sinusoidal vibrations of a thin rectangle membrane, i.e. its shape function, according to [6, 45] is expressed as

$$\begin{aligned} U_{k,n}(x, y, t) &= u_{k,n}(t)u_{k,n}(x, y) \\ &= A_{k,n}\sin\bigl(2\pi f_{k,n}t + \varphi_{k,n}\bigr)\sin(\lambda_k x)\sin(\mu_n y), \end{aligned} \tag{6.54}$$

where k and n are the modes of eigenfrequencies $f_{k,n}$ of the membrane vibrations; $\varphi_{k,n}$ is the phase shift of vibrations; $\lambda_k = k\pi/l$, $\mu_n = n\pi/\mathrm{m}$; l and m are the rectangle membrane sides; $A_{k,n}$ is the amplitude of the membrane vibrations $u_{k,n}(x, y) = \sin(\lambda_k x)\sin(\mu_n y)$.

During one vibration period T_{US} of the membrane excited by the US wave, the maximum tilt angles $|\beta_{\max}|$ in two opposite directions reach twice and cause maximum shifts of spatial frequency $|\Delta\nu_{\max}|$. These shifts, in its turn, generate the largest transformations of the LSP. It is quite obvious that the LSP parts, which reach maximum transformations, correspond to the ROI spatial elements with maximum tilt angles and, accordingly, maximum gradients within the ROI. In order to find the locations of these maximum gradients, one can use Eq. (6.54) to evaluate the spatial distribution of the gradient vector value $\left\|\nabla u_{k,n}(x, y)\right\|$ similar to how it was done by Keene and Chiang [45] for an edge-clamped rectangular membrane. To this end, Eq. (6.54) for a rectangle membrane mode (2,3) is given by

$$\begin{aligned} U_{2,3}(x, y, t) &= u_{2,3}(t)u_{2,3}(x, y) \\ &= A_{2,3}\sin\bigl(2\pi f_{2,3}t + \varphi_{2,3}\bigr)\sin\frac{2\pi x}{l}\sin\frac{3\pi y}{m}, \end{aligned} \tag{6.55}$$

and after differentiation by coordinates x and y of Eq. (6.55) we get

$$\begin{aligned} \bigl[U_{2,3}(x, y, t)\bigr]_{\mathrm{grad}} &= u_{2,3}(t)\left\|\nabla u_{2,3}(x, y)\right\| = A_{2,3}\sin\bigl(2\pi f_{2,3}t + \varphi_{2,3}\bigr) \\ &\times \sqrt{\frac{4\pi^2}{l^2}\cos^2\left(\frac{2\pi x}{l}\right)\sin^2\left(\frac{3\pi y}{m}\right) + \frac{9\pi^2}{m^2}\sin^2\left(\frac{2\pi x}{l}\right)\cos^2\left(\frac{3\pi y}{m}\right)}. \end{aligned} \tag{6.56}$$

If the defect and, accordingly, the ROI are square and $l = m = 1$, $f_{2,3} = 1/T_{\mathrm{US}}$, $t = t_1 = T_{\mathrm{US}}/4$, $\varphi_{2,3} = 0$, then Eq. (6.56) takes the form

$$\left\|\nabla U_{2,3}\right\| = A_{2,3}\sqrt{4\pi^2\cos^2(2\pi x)\sin^2(3\pi y) + 9\pi^2\sin^2(2\pi x)\cos^2(3\pi y)}. \tag{6.57}$$

The full-field spatial distribution of the antinodes and nodes of the rectangle membrane for the mode (2,3) calculated using Eq. (6.57) is shown in Fig. 6.21. Here, the nodes are marked with bright spots, and the antinodes are dark. If the membrane vibrating in the mode (2,3) is considered as the ROI, then the ROI elements are tilted at its nodes at an angle β during oscillations and reach the maximum tilt angle β_{max}.

The ROI spatial elements tilt by the angle β and the corresponding shift of the spatial frequency $\Delta\nu$ according to Eq. (6.53) lead to the transformation of the LSP structure generated by the ROI and to decorrelation between the initial LSP formed by the ROI before US excitation of the composite panel and the same transformed LSP recorded during US excitation. The relationship between the decorrelation induced by the tilt of the ROI spatial element and the spatial frequency shift $\Delta\nu$ can be determined using the Yamaguchi correlation factor (YCF) for a circular aperture [84, 129, 130], which is given by

$$C_Y = \frac{4}{\pi^2}\left[\arccos\left(\frac{|\Delta\boldsymbol{\nu}|}{D_\nu}\right) - \frac{|\Delta\boldsymbol{\nu}|}{D_\nu}\sqrt{1-\left(\frac{|\Delta\boldsymbol{\nu}|}{D_\nu}\right)^2}\right]^2, \tag{6.58}$$

where

$$D_\nu = D/\lambda z \tag{6.59}$$

is the lens aperture diameter in the frequency domain; D is the lens aperture diameter; z is the distance between the lens aperture and the matrix photosensor of the digital camera (DC).

Using Eq. (6.53), one can calculate the YCF if the relevant parameters of the optical system forming the speckle patterns (see Fig. 6.20) and the spatial frequency shift $\Delta\nu$ are known. It should be noted that the transformed parts of the LSP generated by the tilted elements of the ROI are completely decorrelated with the same LSP parts generated by the same flat nontilted ROI elements, if $\Delta\nu_t = D_\nu$, where $\Delta\nu_t$ is the minimum spatial frequency shift, for which the YCF reaches zero value. In this case, Eq. (6.53) takes the form

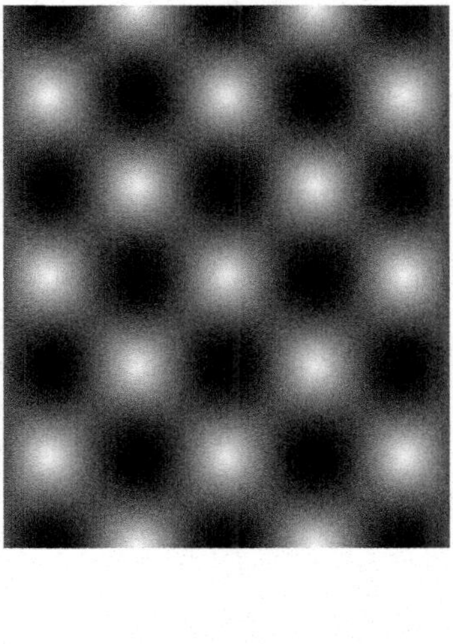

Fig. 6.21 Spatial distribution of the mode (2,3) gradient $\|\nabla u_{2,3}\|$ for the square edge-clamped membrane (i.e. ROI). Membrane nodes are depicted by light spots, and antinodes by dark spots

$$D_\nu = \frac{1 + \cos\theta}{\lambda} \cdot \frac{1}{M_l} \beta_t, \tag{6.60}$$

where β_t is the threshold tilt angle, i.e. the minimum tilt angle of the ROI spatial element, at which the YCF for the frequency $\Delta\nu_t = D_\nu$ reaches zero.

The threshold tilt angle β_t can be calculated in the imaging system schematically shown in Fig. 6.20 using formula, which obtains by inserting (6.59) into (6.60), that is

$$\beta_t = \frac{M_l D}{z(1 + \cos\theta)}. \tag{6.61}$$

According to (6.61), the threshold tilt angle β_t can be calculated if the parameters D, θ, M_l and z of the optical imaging system shown in Fig. 6.20 are known. In particular, if θ, M_l and z have given values, the angle β_t depends linearly on the lens aperture diameter D and tends to zero as D decreases.

The ROI elements that reach the maximum tilt during vibrations in the nodes of the excited ROI, which is considered as a thin fixed membrane, generate the maximum decorrelation in the corresponding LSP parts. The locations of these LSP parts with a certain error correspond to the location of the specified ROI elements in the membrane nodes. Correlative comparison of two LSPs registering in two opposite phases of the vibrating ROI by their subtraction leads to changes in the contrast of these parts. As a result, these parts are highlighted with bright spots, and the subsurface defect is detected in the resulting defect map. Accumulation of sequences of such subtractive LSPs makes it possible to increase the intensity of the spatial response from the desired defect relative to its surrounding background.

Influence of Factor 2 on transformation of dynamic speckle patterns structure. Factor 2 influencing changes in the LSP is associated with the ROI out-of-plane displacements relative to the composite surface plane, which directly leads to a decrease in the speckle contrast and speckle blurring [21, 84]. If the surface does not move, the spatial coherence between the optical waves reflected from the surface does not change and has a constant level. If the surface begins to vibrate, the optical waves acquire random phase differences during the vibration cycle, caused by phase distortions introduced by the moving rough surface [21]. As a result, the spatial coherence decreases when the LSP is generated by the moving ROI, and the LSP loses its contrast. In addition, the registration of the speckle patterns generated by a rough surface containing the vibrating ROI always has some finite exposure time; and temporal averaging of dynamic speckles during their registration entails additional blurring of speckles in the obtained defect map.

The considered approach for detecting rectangular test subsurface defects can be extended to the similar circular ones. The surface layer above such a defect can be considered as a ROI in the form of edge-clamped circular membrane. Using the vibration equations for a thin clamped circular membrane, given, for example, in [6], one can find the location of nodes where the largest tilts are reached in the circular ROI for resonant US frequencies.

6.8.2 Technical Implementation of the Method for Detection the Subsurface Defects

To check the reliability of subsurface defects detection using the developed method, an experimental setup of a hybrid optical-digital system (HODS) with acoustic harmonic excitation of laminated composite panels was created [73, 78, 81, 82]. Propagation of the transverse flexural acoustic waves occurs by their radiation with the help of a broadband piezoelectric transducer (PZT), which is in acoustic contact with the composite panel. Acoustic waves are introduced to the PZT from an ultrasound generator operating in the scanning mode of acoustic waves in the range from 4 to 150 kHz. If a subsurface defect is present in the composite panel, the acoustic flexural waves excite it when the frequency of the scanning wave coincides with the fundamental or one of the multiple resonant frequencies of the defect. The ROI located above this defect begins to vibrate at the resonant frequency as an edge-clamped membrane. Although the method considers ROIs located over regular defects extending along the composite panel layers directions or between adjacent layers, this setup can also detect irregular subsurface defects, including internal delaminations, debonds and fiber breaks.

The procedure of speckle patterns recording in the HODS setup has some features in common with the procedure of speckle interferograms recording in the ODSI experimental setup given in Sect. 4.7.2. The studied composite panel, which is installed in the HODS setup, is excited by an US wave in a given frequency range. The frequencies are increased sequentially in steps with given equal increments. At each acoustic frequency, the sequences of speckle patterns are recorded by opening the digital camera shutter at Q odd and Q even exposures, which alternate one after another and have the same frame time T. At odd exposures, the sequence of P initial speckle patterns (ISPs) $I_{p,q1}(k,l)$ is recorded at the same time gaps τ within each vibration cycle equal to the period T_{US} of the US wave, where $p = 1, 2, \ldots, P$ is the number of the ISP; $q = 1, 2, \ldots, Q$ is the number of odd or even exposure; $k = 1, 2, \ldots, K$ and $l = 1, 2, \ldots, L$ are the pixel numbers in the recorded speckle patterns. The time gap τ for each ISP $I_{p,q1}(k,l)$ ends when the amplitude of the vibrating ROI, corresponding to the amplitude of the US wave, passes its maximum. The time gap τ is provided by an acousto-optic deflector (AOD), which acts as an optical shutter in the optical scheme of the HODS setup. All ISPs during odd exposures are accumulated by the digital camera, resulting in the formation of total odd speckle patterns $I_{q,o}(k,l) = \sum_q I_{p,q1}(k,l)$. The same operation with registration of speckle images is repeated during even exposures of a digital camera, recording the sequence of P ISPs $I_{p,q2}(k,l)$ at the same time gaps τ within each cycle of vibrations equal to the period T_{US} of the US wave. The time gap τ for each $I_{p,q2}(k,l)$ ends when the amplitude of the vibrating ROI, corresponding to the amplitude of the US wave, passes its minimum. As a result of the accumulation of initial even speckle pattern by the digital camera, total even speckle patterns $I_{q,e}(k,l) = \sum_q I_{p,q2}(k,l)$ are formed. ISPs $I_{p,q1}(k,l)$ are recorded during the maximum tilts of the ROI spatial elements in one direction, normal to the plane of the composite panel flat surface, and ISPs $I_{p,q2}(k,l)$ are recorded during the maximum tilts of the ROI spatial elements in the opposite direction.

The procedure of odd total speckle pattern $I_{q,o}(k,l)$ and even one $I_{q,e}(k,l)$ recording is shown in Fig. 6.22. Time gaps τ may have time delays δt concerning the maximums or minimums of the harmonic US wave with the time period T_{US}. This operation makes it possible to adjust the intensity of the spatial response from a subsurface defect and suppress the surrounding background.

To highlight the spatial response from a subsurface defect or several spatial responses from several defects that are in the field of view of the optical imaging system, it is proposed to use a difference data descriptor for the accumulated total speckle patterns $I_{q,o}(k,l)$ and $I_{q,e}(k,l)$. As a result, the accumulated total difference speckle pattern (DSP) is obtained, which is given by

$$I_q(k,l) = \left|I_{q,o}(k,l) - I_{q,e}(k,l)\right|. \quad (6.62)$$

Such the data descriptor is similar to the descriptors used to evaluate the statistical characteristics of dynamic speckles [19]. The implementation of the data descriptor (6.62) and the formation of the total DSP makes it possible to achieve decorrelation of speckles and blurring of the LSP, which is generated by the ROI due to its opposite normal movements during the registration of qth even and odd total speckle patterns $I_{q,o}(k,l)$ and $I_{q,e}(k,l)$. The total DSP $I_q(k,l)$ can be considered as a defect map, containing the spatial response from the subsurface defect, as well as the surrounding background, which is generated by the composite panel surface area placed outside the ROI.

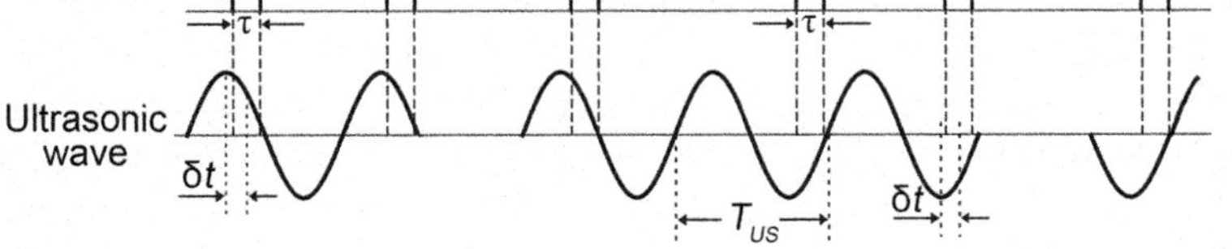

Fig. 6.22 Timing sequence of qth pair of total speckle patterns $I_{q,o}(k,l)$ and $I_{q,e}(k,l)$, produced by the accumulation of P ISPs $I_{p,q1}(k,l)$ and $I_{p,q2}(k,l)$ that are recorded with equal light exposure

In order to obtain a more intense spatial response within the LSP, it is proposed to register the total speckle patterns $I_{q,o}(k,l)$ and $I_{q,e}(k,l)$ Q times and sum the received total DSPs $I_1(k,l)$, …, $I_q(k,l)$, …, $I_Q(k,l)$. The resulting defect map $I_\Sigma(k,l)$ can be obtained using a cumulative data descriptor [73, 78, 81] given by the formula

$$I_\Sigma(k,l) = (\gamma Q)^{-1} \sum_q \left| I_{q,o}(k,l) - I_{q,e}(k,l) \right| = (\gamma Q)^{-1} \sum_q I_q(k,l), \tag{6.63}$$

where γ $\left(Q^{-1} \leq \gamma \leq 1\right)$ is the coefficient for choosing the optimal intensity level of the resulting defect map. This descriptor is implemented using the Difference Averaging Algorithm (DAA) [78].

The scheme of the HODS experimental setup, which implements the DAA completely using the cumulative data descriptor evaluated by Eq. (6.63) or partially using the formula (6.62) for obtaining the accumulated total DSP, is presented in Fig. 6.23. The experimental setup includes a Nd:YAG laser (1) ($\lambda = 532$ nm, $P = 120$ mW), an AOD (2) that performs the functions of an optical shutter, a divergent lens (3) that illuminates a selected surface area of a laminated composite panel (4), a lens "Carl Zeiss Flektogon 20 mm F2.8" (5), a lens adapter (6), a DC "Sony-XCD-V60" (7) with dimensions of 640 × 480 pixels, an ultrasound generator (8), a broadband PZT (9), a controller (10) and computer (11). An axial laser beam propagating along the optical axis of the illumination system containing laser (1) and lens (3) is incident on the composite panel surface at an angle θ. Acoustic radiation with a sweep frequency is introduced into the studied specimen from the ultrasound generator (8) and through the PZT (9), due to which the ROIs, if they are in the field of view of the optical system, begin to vibrate at the resonant frequencies of subsurface defects. The formation of ISPs $I_{p,q1}(k,l)$ and $I_{p,q2}(k,l)$ is carried out using AOD (2), which opens the laser beam for time gaps $\tau \ll T$. The ISPs are accumulated by the DC and the total speckle patterns $I_{q,o}(k,l)$ and $I_{q,e}(k,l)$ are entered into computer (11). Controller (10) synchronizes elements and units of the HODS setup. In computer (11), DSPs $I_q(k,l)$ are obtained [see Eq. (6.62)], which can be used as the defect maps. The generated DSPs can also be summed to obtain the defect map according to the DAA [see Eq. (6.63)]. A more detailed description of the HODS setup operation is given in [73, 78, 80].

In the created HODS setup, the optical system for speckle patterns formation was synthesized taking into account similar optical schemes for generation of speckle patterns from tilted rough surfaces [11, 26]. In this optical system, the threshold angle β_t is reached between the tilted ROI spatial elements if they are in antiphase with each other and reach maximum gradients in opposite directions of the out-of-plane ROI displacements. In this case, the tilt angle β, shown in Fig. 6.20, is $\beta = \frac{\beta_{\max}}{2} \geq \frac{\beta_t}{2} = \beta_{\text{thr}}$. If the parameters of the optical system D, M_l, z and θ are known, then it is possible to calculate the threshold tilt angle β_{thr} by using Eq. (6.61) obtained for β_t, that is

$$\beta_{\text{thr}} = \frac{M_l D}{2z(1+\cos\theta)} = \frac{M_l F}{2z f_\#(1+\cos\theta)}, \tag{6.64}$$

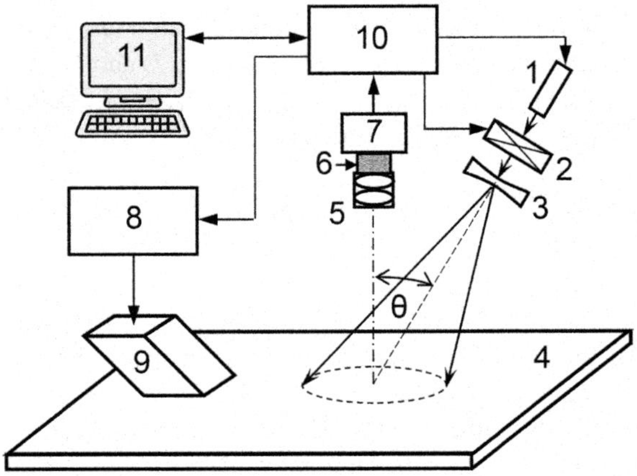

Fig. 6.23 Scheme of the HODS experimental setup with acoustic harmonic excitation of laminated composite panels

where $D = \frac{F}{f_{\#}}$, F is the lens focal length, $f_{\#}$ is the f-number of the lens.

In order to create a compact device for detecting the subsurface defects on the basis of the developed HODS experimental setup, one should know its limiting overall parameters, at which the efficiency of defect detection will not decrease. One of the main factors affecting the overall parameters of such devices is the incidence angle θ of the laser beam, since the dimensions of the device grow with its increase. On the other hand, it is undesirable to reduce this angle, bringing it closer to zero, so as not to interfere with the optical scheme for the formation and registration of speckle patterns, since they are best recorded by the DC in a direction orthogonal to the surface of the composite panel. It is also important to know the values of β_{thr}, which also depend on the parameters of the optical system, since the smaller they are, the lower the acoustic radiation power required to excite ROIs on the surface of the composite panel. For this purpose, the dependencies of the β_{thr} on the angle θ for different f-numbers of the lens were determined using Eq. (6.64) and the known parameters of the optical system of the setup, namely $M_l = 0.09$, $z = 24.1$ mm, the focal length of the lens 2 (see Fig. 6.23) is $F = 20$ mm and $f_{\#} = 2.8, 4, 5.6, 8, 11,$ 16 and 22. Dependencies of β_{thr} on angle θ for given $f_{\#}$ are shown in Fig. 6.24.

The dependencies shown in Fig. 6.24, grow very weakly with increasing angle of incidence θ. Therefore, it is enough to choose the angle θ in the range from 5° to 30° for large relative apertures of the lens, while for small relative apertures, that is, for f-numbers $f_{\#} = 16$ and $f_{\#} = 22$, it practically does not matter at what angle θ to illuminate the surface, that is, the threshold tilt angle β_{thr} does not actually increase. So, for example, for $f_{\#} = 22$, the increment of the β_{thr} in the range $0° \leq \theta \leq 45°$ is $\Delta\beta_{\text{thr}}(0° - 45°) = 0.0083°$, i.e. 17.2% of $\beta_{\text{thr}}(0°) = 0.0486°$ at $\theta = 0°$, and in the range $0° \leq \theta \leq 30°$ the increment is $\Delta\beta_{\text{thr}}(0° - 30°) = 0.0035°$, i.e. 7.1% of $\beta_{\text{thr}}(0°)$. If we take into account the overall parameters of the HODS setup, especially when the incident beam can be partially overlapped by the lens 5 and DC 7 (see Fig. 6.23), then the angle of incidence θ should not be less than 5°. But dependencies shown in Fig. 6.24 indicate that it makes no sense to reduce it, since at $\theta \leq 5°$ the threshold tilt angle β_{thr} practically remains unchanged. So, for example, if $f_{\#} = 2.8$, then in

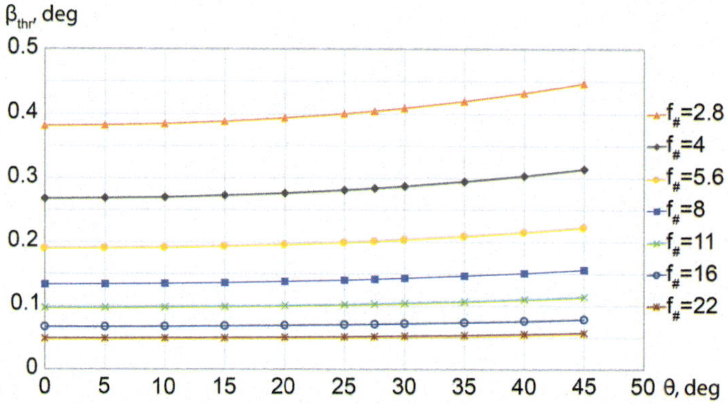

Fig. 6.24 Dependencies "β_{thr} versus θ" for f-numbers 2.8, 4, 5.6, 8, 11, 16 and 22 of the lens "Carl Zeiss Flektogon 20 mm F2.8"

the range $0° \leq \theta \leq 5°$ the increment is $\Delta\beta_{thr}(0°) = 0.00073°$, i.e. 0.0019% of $\beta_{thr}(0°) = 0.382°$, and if $f_{\#} = 22$, then the increment in the same range $0° \leq \theta \leq 5°$ is $\Delta\beta_{thr}(0°) = 0.000093°$ and also 0.0019% of $\beta_{thr}(0°) = 0.0486°$.

Also it should be noted that the twofold reduction of the threshold angle β_{thr} compared to the angle β_t, as it is shown by Eqs. (6.61) and (6.64), makes it possible to expand the capabilities of the HODS setup when detecting subsurface defects, in particular, defects of larger sizes and located at a greater depth from the composite panel surface. Moreover, owing to a twofold decrease in the threshold angle, one can reduce significantly the power of the US waves exciting a specimen and, accordingly, the power of the ultrasound generator used in the HODS setup.

6.8.3 Algorithms for Processing Total Difference Speckle Patterns

The developed software was used to record the initial and accumulated total DSPs and process them in order to increase the signal-to-noise ratio in defect maps. In addition to the DAA [73, 78, 80, 81] implementing the cumulative data descriptor [see Eq. (6.63)], additional algorithms for the DSPs processing were used. The pairwise sum of differences algorithm (PSDA) [80, 110] and the spatial speckle contrast algorithm (SSCA) [9, 35, 45] were selected among a variety of known algorithms for processing the dynamic speckle patterns [9, 19, 90, 91].

Pairwise Sum of Differences Algorithm (PSDA). The sequence of total speckle patterns $I_{q,o}(k, l)$ and $I_{q,e}(k, l)$ is used to implement this algorithm. The absolute values of difference speckle patterns $S_q^-(k, l)$ and $S_q^+(k, l)$ of each pair of adjacent patterns $I_{q,o}(k, l)$ and $I_{q,e}(k, l)$ are binarized with respect to a given threshold value z, that is

$$T_q^-(k,l) = \begin{cases} 1, \left|S_q^-(k,l)\right| > z \\ 0, \left|S_q^-(k,l)\right| < z \end{cases}, \quad (6.65)$$

$$T_q^+(k,l) = \begin{cases} 1, \left|S_q^+(k,l)\right| > z \\ 0, \left|S_q^+(k,l)\right| < z \end{cases}, \quad (6.66)$$

where

$$\left|S_q^-(k,l)\right| = \left|I_{q,e}(k,l) - I_{q,o}(k,l)\right|, \quad (6.67)$$

$$\left|S_q^+(k,l)\right| = \left|I_{q+1,o}(k,l) - I_{q,e}(k,l)\right|, \quad (6.68)$$

and $1 \le q, o \le Q - 1, 1 \le q, e \le Q$.

Obtained sequence is summed up and the resulting defect map (DM) is given by

$$\text{DM} = \frac{255}{(2Q-1)} \sum_{q=1}^{2Q-1} \left[T_q^-(k,l) + T_q^+(k,l)\right]. \quad (6.69)$$

Spatial Speckle Contrast Algorithm (SSCA). This algorithm is well-known and described in detail by He and Briers [35] and further by Keene and Chiang [45]. If each subset on the defect maps with dimensions of $K \times L$ have dimensions $K_m \times L_n$ ($K_m = L_n$), where K_m and L_n are the odd numbers, this algorithm for the accumulated total DSP $I_q(k,l)$ [see Eq. (6.62)] and for the resulting defect map $I_\Sigma(k,l)$ [see Eq. (6.63)] can be given by

$$c\left(\tilde{k},\tilde{l}\right) = \frac{\sigma\left(\tilde{k},\tilde{l}\right)}{\overline{I}\left(\tilde{k},\tilde{l}\right)}, \quad (6.70)$$

where

$$\overline{I}\left(\tilde{k},\tilde{l}\right) = \frac{\sum_{\tilde{k}=1}^{K-\frac{K_m+1}{2}} \sum_{\tilde{l}=1}^{L-\frac{L_n+1}{2}} I\left(\tilde{k} + \frac{K_m+1}{2}, \tilde{l} + \frac{L_n+1}{2}\right)}{K_m L_n} \quad (6.71)$$

is the mean intensity within each subset; I is the intensity in each $\tilde{k}, \tilde{l}$th pixel of the defect map,

$$\sigma\left(\tilde{k},\tilde{l}\right) = \sqrt{\frac{\left[\sum_{\tilde{k}=1}^{K-\frac{K_m+1}{2}} \sum_{\tilde{l}=1}^{L-\frac{L_n+1}{2}} I\left(\tilde{k} + \frac{K_m+1}{2}, \tilde{l} + \frac{L_n+1}{2}\right) - \overline{I}\left(\tilde{k},\tilde{l}\right)\right]^2}{K_m L_n}} \quad (6.72)$$

is the standard deviation of the mean intensity $\overline{I}\left(\tilde{k},\tilde{l}\right)$. The subset dimensions were $K_m \times L_n = 9 \times 9$ pixels.

6.8.4 Experiments to Detect the Artificial Subsurface Defects

Experimental verification of the developed method for detecting the artificial subsurface defects in thin-layered composite structures was carried out using the developed HODS experimental setup with acoustic excitation of studied composite panels in the acoustic frequency sweep mode. The field of view of the Sony-XCD-V60 CCD camera with dimensions of 640 × 480 pixels, which captured the image of a section of the panel surface, was 52.6 × 39.5 mm^2, and the angle of view along the diagonal of the captured image was 15.6°. The f-number of the lens was chosen equal to $f_{\#} = 22$, since with such a lens aperture the average speckle size satisfies the Nyquist sampling criterion [104]. When the camera sensitivity was not enough to register speckle patterns, the lens aperture was increased to $f_{\#} = 16$. The wideband PZT was connected to the panel using a special gel.

Production of test specimens of fiberglass composite panels. To make the first composite panel with artificial subsurface defects (Panel 1), a single-layer isotropic fiberglass panel with dimensions of 248 × 215 mm^2 and a thickness of 2 mm (Panel 1a) was used. Square and round through holes were cut into Panel 1a. The geometry of the holes in Panel 1a and their dimensions are shown in Fig. 6.25. Another fiberglass panel with dimensions of 207 × 178 mm^2 and a thickness of 0.5 mm (Panel 1b) was glued to Panel 1a with E-5 epoxy glue so that it covered all the through holes. Panel 1b was spray coated with white to achieve the surface roughness needed to generate high-quality speckle patterns in the DC 7 containing in the HODS setup [see Fig. 6.23]. The image of the produced two-layer composite panel (Panel 1) with flat blind holes from the side of Panel 1b is shown in Fig. 6.26.

To produce the second composite panel with artificial subsurface defects (Panel 2), a single-layer isotropic fiberglass panel with dimensions of 250 × 200 mm^2 and a thickness of 4 mm (Panel 2a) was chosen. Through holes of the correct shape were cut in Panel 2a. The geometry of the holes locations and their dimensions are shown in Fig. 6.27. Another fiberglass panel with dimensions of 180 × 100 mm^2 and 1 mm thick (Panel 2b) was glued to Panel 2a with E-5 epoxy glue so that it covered all through holes. Panel 2b was also white spray coated to achieve the surface roughness necessary for the formation of high-quality speckle patterns.

Experimental results. The results of artificial subsurface defects detection in Panel 1 and Panel 2 showed that LSPs stands out against the surrounding background on defect maps obtained using the aforementioned algorithms for processing total DSPs. In this case, the LSP generated by the ROI is approximately located within the limits determined by the ROI contour. A number of obtained LSPs can be unambiguously

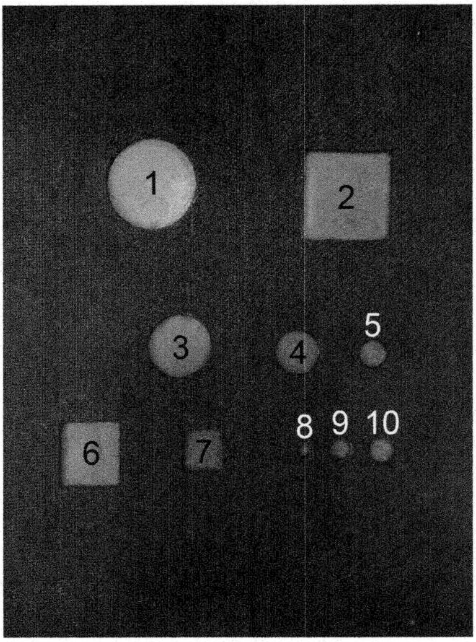

Fig. 6.25 Geometry of the through holes locations in Panel 1a: round hole ⌀33 mm (1); square hole 33.5 × 33.5 mm^2 (2); round hole ⌀24 mm (3); round hole ⌀17 mm (4); round hole ⌀11 mm (5); square hole 23 × 24 mm^2 (6); square hole 15.5 × 15.5 mm^2 (7); round hole ⌀6 mm (8); round hole ⌀7.5 mm (9); round hole ⌀10 mm (10)

Fig. 6.26 Image of two-layer Panel 1 with blind flat holes from the side of glued Panel 1b

interpreted as vibrating nodes of the edge-clamped membrane, which, due to their tilt, form bright spots in the LSP.

Let's first consider and analyze the detection results of vibrating square blind holes on Panel 1 and Panel 2. To excite the rectangular hole 6 with dimensions of 23 × 24 mm^2 on Panel 1, the PZT was placed in such a way that the direction of the exciting US waves was strictly perpendicular to the horizontal sides of the hole equal to 23 mm (see Fig. 6.25). During the US excitation of the composite Panel 1,

Fig. 6.27 Geometry of the through holes locations in Panel 2a: square hole 34 × 34 mm^2 (1); round hole ⌀ 19.5 mm (2); round hole ⌀9.5 mm (3); round hole ⌀7.5 mm (4); round hole ⌀6 mm (5); round hole ⌀26.5 mm (6); round hole ⌀18 mm (7); round hole ⌀11 mm (8); square hole 20 × 20 mm^2 (9); triangular hole (10): the hole vertical side is 22 mm, the hole horizontal height is 18 mm; round hole ⌀10 mm (11)

a flexural US wave formed in it acts on the ROI in the direction perpendicular to the wave propagation and practically does not act in the longitudinal direction. Such a property of the flexural waves during the excitation of flat blind holes at resonant frequencies is described, for example, by Ciampa et al. [16]. As a result, the flexural waves excite the ROI above the subsurface hole, and if the hole has the shape of a square, then during resonance the ROI can be considered as a square edge-clamped membrane that vibrates in a certain mode depending on the multiple of the resonance frequency. Figure 6.28a shows a model of a square membrane vibrating in mode (2,3) at the resonance frequency $f_{2,3}$. If a flexural US wave falls on the membrane in a direction perpendicular to its horizontal side, then this wave acts on the membrane only in orthogonal direction. The only modes transverse to the direction of wave propagation, i.e. along the drawn dashed lines, are excited in this case. Three rows of nodes are formed in the membrane. In these nodes, the membrane surface acquires maximum opposite tilts during resonant vibrations. Then, the light responses inside the LSP are formed as a result of speckle decorrelation in the LSP parts corresponding to the membrane transverse nodes, which have the form of a 3 × 3 matrix. Nodes on the membrane model in Fig. 6.28a almost coincide with the nodes on the defect maps shown in Fig. 6.28b, c and obtained from the blind hole 6 at the resonant frequency $f_{2,3} = 29$ kHz.

Note that the nodes of longitudinal modes of the membrane are practically invisible on these maps. The antinodes of the membrane are also not visible, since the speed of their movement is not yet sufficient to blur the speckles in places where light responses from membrane antinodes should form on the LSP. At the same time, the received light responses inside the LSP from the nodes of the transverse modes can be not only due to the passage of the flexural waves through the Panel 1, but also as a result of the reflection of these waves from the opposite end of Panel 1 and the formation of standing flexural waves within the defect. The nonuniformity of the intensities of the light responses matrix obtained within the LSP may indicate

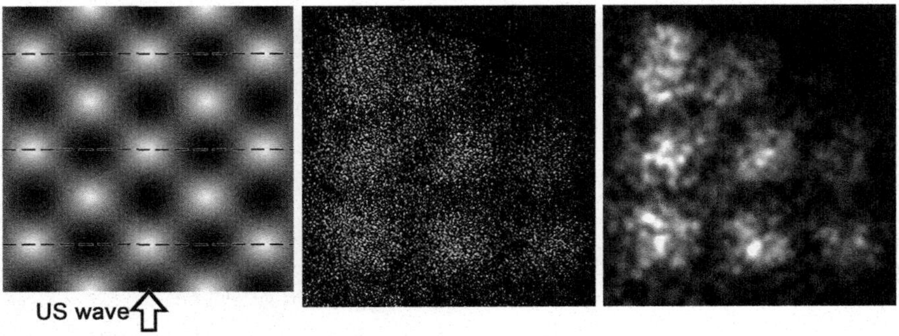

(a) (b) (c)

Fig. 6.28 Matrixes of light responses from the transverse nodes of the square blind hole 6 on Panel 1, which vibrates on the mode (2,3) of the square membrane with a resonant frequency of $f_{2,3} = 29$ kHz: the model of the square membrane and the direction of propagation of the flexural wave (**a**); defect map obtained using the PSDA (**b**); (**c**) DSP in the form of the defect map [see Eq. (6.62)] after low-pass filtering

inaccuracies in the fabrication of the test defect, as well as the deviation of direct and reflected flexural waves from the direction of propagation orthogonal to the horizontal sides of the defect 6 (see Fig. 6.25). The size of the LSP on the defect map shown in Fig. 6.28b is 20.4×22.4 mm^2. This size is slightly smaller than the actual size of the defect 6.

If the resonant frequency in the (2,3) mode is $f_{2,3} = 29$ kHz, then the resonant frequency in the (1,2) mode of the clamped rectangular membrane is 18 kHz [95]. During the experiment, the mode (1,2) of the investigated hole 6 at a frequency of $f_{1,2} = 18$ kHz was detected, while the PZT was located at an angle different from the right angle to the horizontal side of this hole. After the obtained DSP processing, light responses from all four transverse nodes of the blind hole appeared here. Change in the direction of the acoustic wave propagation, the deviation of the hole from a rectangular shape, the influence of reflected flexural waves, and the approximation of the ROI by the model of an ideal thin membrane without taking into account the influence of nonlinear components on the oscillations of this ROI lead to such intensity distributions of light responses shown in Fig. 6.29. Note that the correspondence between the model membrane transverse nodes and nodes of the subsurface defect 6 on this frequency still exists.

At the same time, at higher frequencies, the order of modes is disturbed due to the influence of nonlinear effects and uncontrolled reflections of flexural US waves from the edges of Panel 1. The results of the formation of such defect maps at high frequencies of US waves are shown in Fig. 6.30.

Figure 6.31 shows the LSP generated by the ROI located directly above the subsurface round hole 11 (⌀10 mm) when the Panel 1 is excited with a resonant US frequency $f_{11} = 40$ kHz. Since the blind hole is round, the surface above it can be considered as the ROI in the form of the edge-clamped circular membrane. The shape of the generated LSP before digital processing (Fig. 6.31a) and after it (Fig. 6.31b)

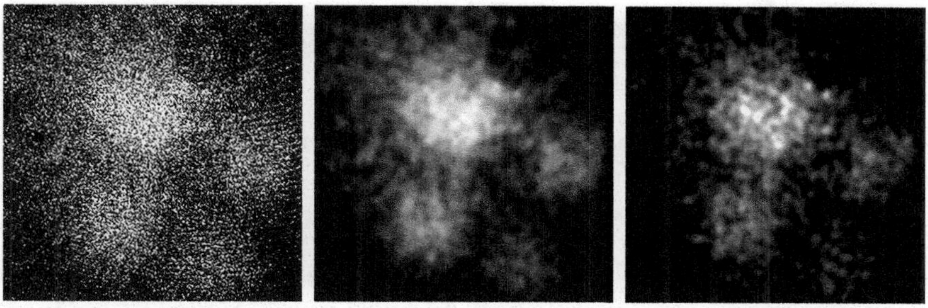

(a) (b) (c)

Fig. 6.29 Matrix of light responses from the transverse nodes of the square blind hole 6 on Panel 1, which vibrates on the mode (1,2) of a thin square membrane with a frequency of $f_{1,2} = 18$ kHz: defect map obtained using the PSDA (**a**); defect map obtained using the PSDA after low-pass filtering (**b**); DSP in the form of the defect map [see Eq. (6.62)] after low-pass filtering (**c**)

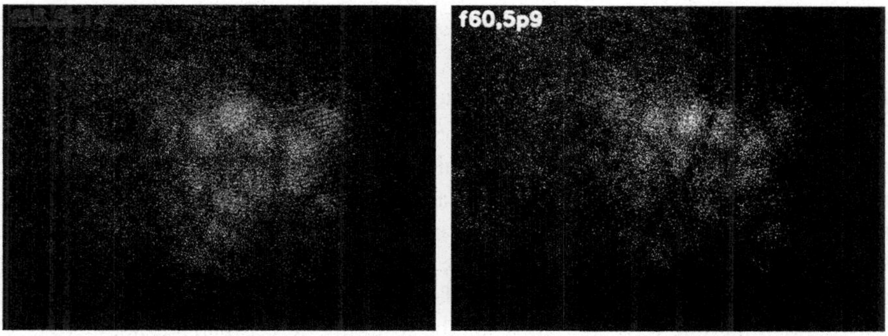

(a) (b)

Fig. 6.30 Matrix of light responses from modes of the ROI of the square blind hole 6 vibrating on frequencies 55.5 kHz (**a**) and 60.5 kHz (**b**)

corresponds to the mode (1,1) of the circular membrane. As can be seen from the LSP containing two symmetrical light spots, Factor 1 and Factor 2 play an equal role in their formation. The peripheral zones of these two spots are formed due to Factor 1, and the central zones are formed due to Factor 2. The previous experiments have shown that the fundamental mode (0,1) of the round hole 11 was formed at a resonant US frequency equal to $f_{0,1} = 25.6$ kHz [82]. Since the ratio between the frequencies in the (1,1) and (0.1) modes in thin edge-clamped circular membranes is $\tilde{f}_{11}/f_{01} = 1.59$ [95], the resonant frequency in the (1,1) mode should be equal to $\tilde{f}_{11} = 40.7$ kHz. Hence, the frequency $f_{11} = 40$ kHz was found almost correctly. The size of the LSP from the blind hole 11 (⌀10 mm) is 10.1 mm.

Figure 6.32 shows two LSPs generated by the ROIs of two blind round holes of Panel 2, namely round hole 6 (⌀26.5 mm) and round hole 7 (⌀18 mm). The LSPs formed at a frequency of 8 kHz are shown in Fig 6.32a. As can be seen, this frequency is resonant for hole 7, whose LSP is placed at the lower right side of this

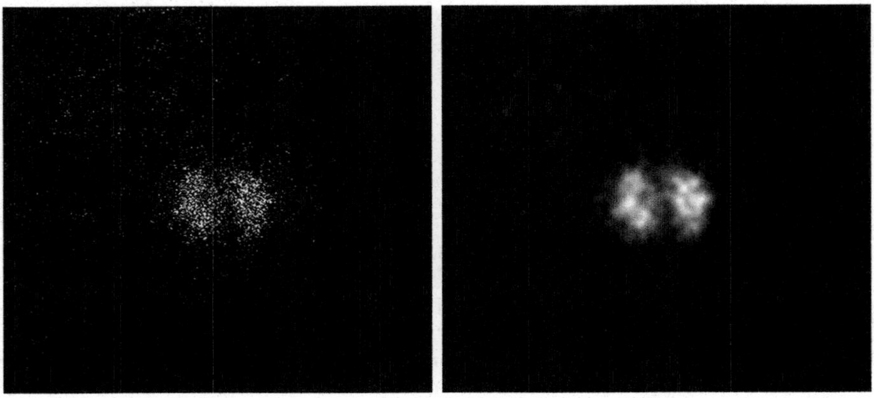

Fig. 6.31 Defect maps containing the LSP of the subsurface round hole 11 (⌀10 mm) when Panel 1 is excited by resonant US frequency $f_{11} = 40$ kHz: defect map obtained using the PSDA (**a**); the same defect map after low-pass filtering (**b**)

figure, and intermediate for hole 6, whose LSP is placed at the top of the figure. The ring-like structure is evident on the LSP generated by the vibrating ROI located directly above the hole 7 (⌀18 mm), which indicates the influence of Factor 1 on this LSP formation. At the same time, Factor 2 does not affect it, probably due to the relatively small speed of oscillation of the ROI in its central zone, i.e. due to such a speed that is not enough to blur the speckles in the central zone. Figure 6.32b shows one LSP obtained at a frequency of 4.2 kHz and generated by the ROI located above the blind hole 6 (⌀26.5 mm). As the resulting defect map shows, there is no light response from the hole 7 due to the fact that the main resonant frequency of its ROI is approximately 2 times higher than this frequency. Consequently, the LSP from the hole 7 is absent, and for the hole 6, the frequency of 4.2 kHz is close to the fundamental resonant one. The results of measurements of the hole sizes according to the sizes of their LSPs, i.e. their spatial responses, showed that the size of the response from the hole 6 is 26.5–27 mm, which basically coincides with its real size, and the size of the response from the hole 7 is 14.5 mm, which is 3.5 mm smaller than its real size.

6.8.5 Experiments to Detect Real Subsurface Defects

Real subsurface defects were also analyzed in composite and metal–composite elements of aircraft structures using the HODS experimental setup with a frequency swept US excitation. In particular, internal defects were found in composite honeycomb panels attached to the lower part of the aircraft fuselage. These panels prevent the impact of small stones on the fuselage during the landing and takeoff of the aircraft.

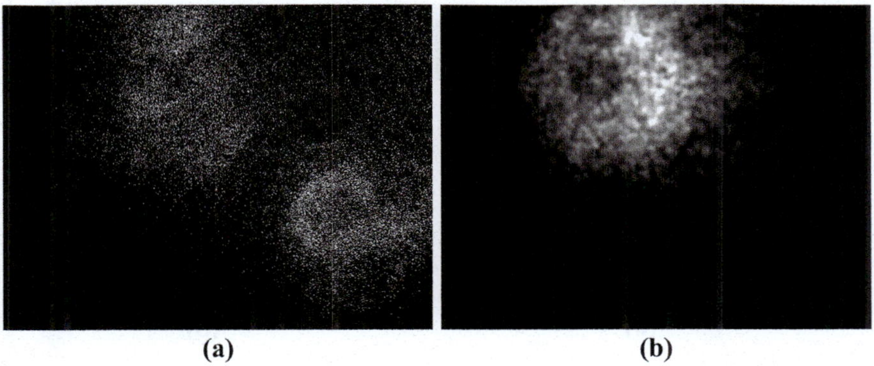

Fig. 6.32 Defect maps in which light responses are generated by two ROIs located directly above two round blind holes on Panel 2, namely round hole 6 of ⌀26.5 mm (top in Fig. 6.32a, b) and round blind hole 7 of ⌀18 mm (lower right in Fig. 6.32a): defect map obtained at the frequency of 8 kHz after processing using the PSDA (**a**); DSP as a defect map [see Eq. (6.62)] obtained at 4.2 kHz after low-pass filtering (**b**)

Figure 6.33 shows an image of the honeycomb composite panel on both sides. Figure 6.34 shows the results of the internal defect detection in this panel using the HODS setup. The defect was detected at an US excitation frequency equal to 21 kHz. Figure 6.34a shows the detected defect after processing the series of DSPs using the DAA. Figure 6.34b shows the defect map with detected defect after processing the series of DSPs using the PSDA.

Internal defects were also detected in metal-composite joints. Using the HODS setup, a light response from an operational internal defect was detected in a 4-mm-thick carbon fiber composite panel, which is part of the metal–composite joint used in aircrafts [82]. The defect arose in the composite panel after applying a static load of 4,500 Kg to the metal–composite joint using a tensile testing machine. Such a load led to the fracture of the panel inner layers.

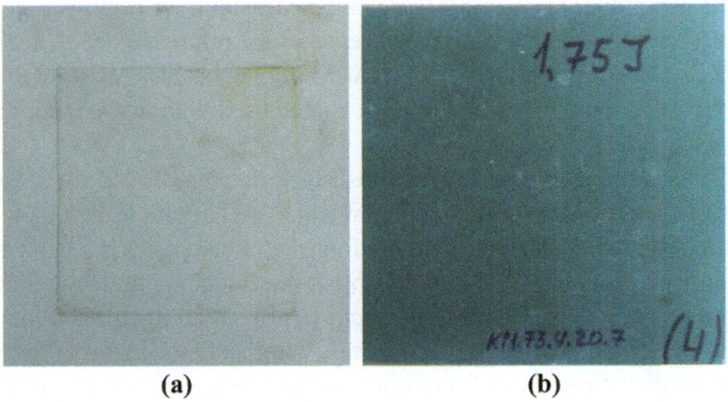

Fig. 6.33 Images of the honeycomb panel on both sides

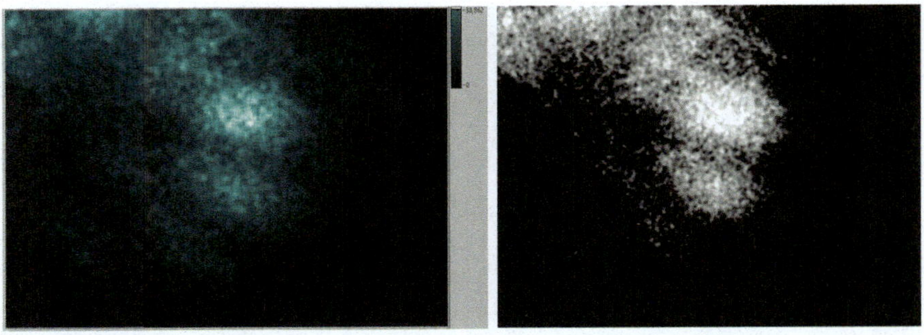

(a) (b)

Fig. 6.34 Detection of internal defects in the honeycomb panel with the help of the HODS setup on a panel area of 52.6×39.5 mm^2: defect maps obtained using the DAA (**a**) and PSDA (**b**)

6.8.6 Conclusion

The new method for detection of subsurface defects in laminated composite panels has been developed using speckle decorrelation and speckle blurring. Due to Factor 1 and Factor 2, the LSP in the form of light spots indicates the presence of a subsurface defect within the studied composite surface area. The light responses are generated in the coherent or quasi-coherent optical system due to vibrations of the ROI placed directly above the defect. Previous studies have shown that the method can provide detection of defects with sizes from 5 to 35 mm. However, it has potentially greater capabilities and is able to detect defects of much larger sizes, since with an increase in the amplitude of ROI vibrations, the level of speckle decorrelation increases, due to which it will be possible to detect large defects. The size of defects that can be detected using this method will be limited mainly by the level of laser radiation intensity.

The HODS experimental setup created using the proposed method is much simpler than similar systems based on ESPI [25, 88, 89, 125], shearography [24, 38, 50, 116, 123] or DH techniques [117, 118], since only an object laser beam is used here, and there is no a reference one. Due to the absence of the reference beam, devices based on the proposed method will not only be much simpler, but also resistant to vibrations and other external factors. Therefore, they can be used both in the laboratory and in natural conditions.

The results of studies of artificial subsurface defects proved the possibility of experimental determination of the fundamental and multiple resonant vibration frequencies of these defects. However, finding such frequencies is a laborious procedure. In addition, it is impossible to establish the fundamental resonant frequency with sufficient accuracy, since the LSP structure corresponding to this frequency practically does not change in the frequency range within a few kilohertz. Therefore, it is difficult to predict the values of multiple resonant frequencies during experiments, which can lead to false conclusions. It is even more difficult to find the fundamental frequency for real defects, in particular for crack-type ones. In connection with this situation, it is extremely important to further develop the theory of elastic waves propagation in layered composite structures containing internal defects. In order to establish the values of resonance frequencies for such defects, it is necessary to develop a mathematical model of elastic waves interaction with interface crack-type defects and, based on this model, to determine the fundamental and multiple resonance frequencies for each defect. One possible solution to this problem is suggested in Chap. 7.

References

1. Alam MS (1995) Fractional power fringe-adjusted joint transform correlation. Opt Eng 34(11):3208–3216
2. Alam MS (2000) Phase-encoded fringe-adjusted joint transform correlation. Opt Eng 39(5):1169–1176
3. Alam MS, Karim MA (1993) Fringe-adjusted joint transform correlation. Appl Opt 32(23):4344–4350
4. Alam MS, Ochilov S (2010) Spectral fringe-adjusted joint transform correlation. Appl Opt 49(10):B18–B25
5. Alam MS, Chen XW, Karim MA (1997) Distortion-invariant fringe-adjusted joint transform correlation. Appl Opt 36(29):7422–7427
6. Aramanovich IG, Levin II (1969) Uravnieniya matematicheskoy fiziki, vtoroye izd (Equations of mathematical physics), 2nd edn. Nauka, Moscow
7. Arizaga R, Grumel EE, Cap N, Trivi M, Amalvy JI, Yepes B, Ricaurte G (2006) Following the drying of spray paints using space and time contrast of dynamic speckle. J Coat Technol Res 3(4):295–299
8. Bing P, Hui-Min X, Bo-Qin X, Fu-Long D (2006) Performance of sub-pixel registration algorithms in digital image correlation. Meas Sci Technol 17(6):1615–1621
9. Briers D, Duncan DD, Hirst ER, Kirkpatrick SJ, Larsson M, Steenbergen W, Stromberg T, Thompson OB (2013) Laser speckle contrast imaging: theoretical and practical limitations. J Biomed Opt 18(6):066018
10. Bruck HA, McNeill SR, Sutton MA, Peters WH (1989) Digital image correlation using Newton-Raphson method of partial differential correction. Exp Mech 29(3):261–267
11. Bruno L, Pagnotta L, Poggialini A (2000) Laser speckle decorrelation in NDT. Opt Lasers Eng 34(1):55–65
12. Chen DJ, Chiang FP, Tan YS, Don HS (1993) Digital speckle-displacement measurement using a complex spectrum method. Appl Opt 32(11):1839–1849
13. Chen J, Xia G, Zhou K, Xia G, Qin Y (2005) Two-step digital image correlation for micro-region measurement. Opt Lasers Eng 43(8):836–846
14. Cheng P, Sutton MA, Schreier HW, McNeill SR (2002) Full-field speckle pattern image correlation with B-spline deformation function. Exp Mech 42(3):344–352
15. Chu TC, Ranson WF, Sutton MA (1985) Applications of digital-image-correlation techniques to experimental mechanics. Exp Mech 25(3):232–244
16. Ciampa F, Scarselli G, Meo M (2017) On the generation of nonlinear damage resonance intermodulation for elastic wave spectroscopy. J Acoust Soc Am 141(4):2364–2374
17. Ciampa F, Pickering SG, Scarselli G, Meo M (2017) Nonlinear imaging of damage in composite structures using sparse ultrasonic sensor arrays. Struct Control Health Monit 24(5):e1911
18. Cofaru C, Philips W, Van Paepegem W (2012) A novel speckle pattern–adaptive digital image correlation approach with robust strain calculation. Opt Lasers Eng 50(2):187–198

19. Dai Pra AL, Meschino GJ, Guzmán MN, Scandurra AG, González MA, Weber C, Trivi M, Rabal H, Passoni LI (2016) Dynamic speckle image segmentation using self-organizing maps. J Opt 18(8):085606
20. Du Y, Diaz-Garrido FA, Burguete RL, Patterson EA (2009) Stress intensity factor evaluation using DIC on an aerospace component. In: Proceedings of the 2009 SEM annual conference and exposition on experimental and applied mechanics. Society for Experimental Mechanics
21. Eliasson B, Mottier FM (1971) Determination of the granular radiance distribution of a diffuser and its use for vibration analysis. J Opt Soc Am 61(5):559–565
22. ERDAS (Firm) (1997) ERDAS field guide. Erdas
23. Erdogan F, Sih GC (1963) On the crack extension in plates under plane loading and transverse shear. ASME J Basic Eng 85(4):519–525
24. Fomitchov PA, Krishnaswamy S (1997) A compact dual-purpose camera for shearography and electronic speckle-pattern interferometry. Meas Sci Technol 8(5):581
25. Fomitchov P, Wang LS, Krishnaswamy S (1997) Advanced image-processing techniques for automatic nondestructive evaluation of adhesively-bonded structures using speckle interferometry. J Nondestr Eval 16(4):215–227
26. Fricke-Begemann T (2003) Three-dimensional deformation field measurement with digital speckle correlation. Appl Opt 42(34):6783–6796
27. Fricke-Begemann T, Gülker G, Hinsch KD, Wolff K (1999) Corrosion monitoring with speckle correlation. Appl Opt 38(28):5948–5955
28. Goodman JW (1975) Statistical properties of laser speckle patterns. In: Dainty JC (ed) Laser speckle and related phenomena. Springer, Berlin, Heidelberg, pp 9–75
29. Goodman JW (2000) Statistical optics. Wiley, New York
30. Goodman JW (2005) Introduction to Fourier optics, 3rd edn. Roberts and Company Publishers, Greenwood Village, CO
31. Goodman JW (2007) Speckle phenomena in optics: theory and applications. Roberts and Company Publishers, Greenwood Village, CO
32. GOST 25.506-85 (1985) Raschety i ispytaniya na prochnost'. Metody mehanicheskih ispytaniy metallov. Opredeleniye kharakteristik treshchinostoykosti (vyazkosti razrusheniya) pri staticheskom nagruzhenii (Strength analyses and tests. methods for mechanical testing of metals. Determination of the crack resistance characteristics (Fracture toughness) under static Loading). Izd. Standartov, Moscow
33. Hahn WB Jr, Flannery DL (1992) Design elements of binary joint transform correlation and selected optimization techniques. Opt Eng 31(5):896–905
34. Han G, Sutton MA, Chao YJ (1994) A study of stationary crack-tip deformation fields in thin sheets by computer vision. Exp Mech 34(2):125–140
35. He XW, Briers JD (1998, July) Laser speckle contrast analysis (LASCA): a real-time solution for monitoring capillary blood flow and velocity. In: Hoffman EA (ed) Medical imaging 1998: physiology and function from multidimensional images, vol 3337. SPIE, Bellingham, WA, pp 98–107
36. Holynskyy IS, Muravsky LI, Polovynko TI (2010) Otsinyuvannia dynamiky zminy poliv deformatsiy poverhni sharuvatogo kompozytu metodom prostorovo-chasovoyi spekl-koreliatsiyi (Assessment of change of deformation fields dynamics of the laminated composite surface by the spatial-temporal correlation method). Elektromagnitni ta akustychni metody neruynivnogo kontrolyu materialiv ta vyrobiv LEOTEST–2010 (Electromagnetic and Acoustical Methods for Nondestructive Testing of Materials and Products LEOTEST–2010) 15:91–92
37. Horner JL (1992) Metrics for assessing pattern-recognition performance. Appl Opt 31(2):165–166
38. Hung YY, Chen YS, Ng SP, Shepard SM, Hou Y, Lhota JR (2007) Review and comparison of shearography and pulsed thermography for adhesive bond evaluation. Opt Eng 46(5):051007
39. Hussain MA, Pu SL, Underwood J (1974) Strain energy release rate for a crack under combined mode I and mode II. In: Fracture analysis, proceedings of the 1973 national symposium on fracture mechanics, part II, University of Marylend, College Park, MD, August 1073. ASTM STP 560, Philadelphia, PA, pp 2–28

40. Irwin GR (1960) Fracture testing of high-strength sheet materials under conditions appropriate for stress analysis. United States Naval Research Laboratory
41. Javidi B, Wang J (1991) Binary nonlinear joint transform correlation with median and subset median thresholding. Appl Opt 30(8):967–976
42. Jenkins RW, McIlwain MC (1971) Holographic analysis of printed circuit boards. Mater Eval 29:199–204
43. Jones R, Wykes C (1989) Holographic and speckle interferometry (No 6). Cambridge University Press
44. Keane RD, Adrian RJ (1992) Theory of cross-correlation analysis of PIV images. Appl Sci Res 49(3):191–215
45. Keene L, Chiang FP (2009) Real-time anti-node visualization of vibrating distributed systems in noisy environments using defocused laser speckle contrast analysis. J Sound Vib 320(3):472–481
46. Kelly DP, Ward JE, Gopinathan U, Hennelly BM, O'Neill FT, Sheridan JT (2006) Generalized Yamaguchi correlation factor for coherent quadratic phase speckle metrology systems with an aperture. Opt Lett 31(23):3444–3446
47. Khaksari K, Kirkpatrick SJ (2016) Combined effects of scattering and absorption on laser speckle contrast imaging. J Biomed Opt 21(7):076002
48. Kim JH, Paulino GH (2007) On fracture criteria for mixed-mode crack propagation in functionally graded materials. Mech Adv Mater Struct 14(4):227–244
49. Kumar BV, Hassebrook L (1990) Performance measures for correlation filters. Appl Opt 29(20):2997–3006
50. Lai WL, Kou SC, Poon CS, Tsang WF, Ng SP, Hung YY (2009) Characterization of flaws embedded in externally bonded CFRP on concrete beams by infrared thermography and shearography. J Nondestruct Eval 28(1):27–35
51. Lobanov LM, Muravsky LI, Pivtorak VA, Voronyak TI (2017) Monitoryng napruzhenogo stanu elementiv konstruktsiy z vykorystanniam elektromagnitnyh hvyl' optychnogo diapazonu (Monitoring of structural elements stress state with the use of electromagnetic waves in the optical range), vol 3. In: Nazarchuk ZT (ed) Tehnichna diagnostyka materialiv i konstruktsiy (Technical diagnostics of materials and structures), Reference manual in 8 volumes. Publishing House "Prostir-M", Lviv
52. Lopez-Crespo P, Burguete RL, Patterson EA, Shterenlikht A, Withers PJ, Yates JR (2009) Study of a crack at a fastener hole by digital image correlation. Exp Mech 49(4):551–559
53. Lugt AV (1964) Signal detection by complex spatial filtering. IEEE Trans Inf Theor 10(2):139–145
54. Lychak OV, Holyns'kyi IS (2014) Evaluation of stress field reconstruction errors near the crack tip of body under plain strain conditions. Vidbir i obrobka informatsiyi (Information Extraction and Processing) 41:63–69
55. Lychak OV, Holyns'kiy IS (2016) Evaluation of random errors in Williams' series coefficients obtained with digital image correlation. Meas Sci Technol 27(3):035203
56. Lychak OV, Holyns'kiy IS (2016) Improving the accuracy of derivation of the Williams' series parameters under mixed (I+II) mode loading by compensation of measurement bias in the stress field components data. Meas Sci Technol 27(12):125203
57. Maksymenko OP, Muravsky LI, Berezyuk MI (2015) Application of biospeckles for assessment of structural and cellular changes in muscle tissue. J Biomed Opt 20(9):095006
58. Maksymenko OP, Muravsky LI, Lytvyn MO (2003) Vybir parametra inversnogo filtra dlia tsyfrovogo metodu vymiryuvannia zmishchen' spekliv (Choice of inverse filter parameter for digital method of measuring speckle displacements). Fizychni metody ta zasoby kontroyu seredovyshch, materialiv ta vyrobiv (Physical Methods and Means for Control of Media, Materials and Products) 8:151–156
59. Maksymenko OP, Muravsky LI, Sakharuk OM (2008, September) Portable optical-digital speckle correlator for non-destructive control of constructional elements. In: Proceedings of the topical meeting on optoinformatics. St. Petersburg State University for Information Technologies, Mechanics and Optics, St. Petersburg, pp 105–108

60. Marom E (1970) Real–time strain measurements by optical correlation. Appl Opt 9(6):1385–1391
61. Marom E (1974) Holografic correlation. In: Erf RK (ed) Holographic nondestructive testing. Academic Press, New York, pp 149–181
62. McLysaght F, Slevin JA (1991) Holographic evaluation of ceramic materials. Appl Opt 30(7):780–787
63. Muravskii LI, Fit'o NV (2005) Estimating the displacements of the surfaces of deformable objects and of solid bodies by an optical speckle-correlation technique. J Opt Technol 72(5):410–414
64. Muravskii LI, Sakharuk AN, Maksimenko AP (2004) Using optical speckle correlation to study breakdown processes in structural materials. J Opt Technol 71(7):461–465
65. Muravsky LI (2010) Metody spekl-koreliatsiyi dlia doslidzhennia mehanichnyh vlastyvostey konstruktsiynyh materialiv (Methods of speckle correlation for studying the mechanical properties of structural materials). Naukova Dumka, Kyiv
66. Muravsky L, Fityo N, Sakharuk OM (2005, May) Estimation of surface displacement field by optical speckle correlation technique. In: Proceedings of III international conference on materials testing, Nurnberg. AMA Service GmBH, Wunstdorf, Germany, pp 163–168
67. Muravsky LI, Maksymenko OP, Sakharuk OM (2003) Comparative analysis of digital and optical speckle displacement techniques for study of in-plane surface strains. In: Kastinger K, Løkberg OJ, Winther S (eds) Speckle metrology 2003, vol 4933. SPIE, Bellingham, WA, p 149–154
68. Muravsky LI, Maksymenko OP, Sakharuk OM (2003) Otsinka poperechnyh zsuviv poverhni material metodamy spekl-koreliatsiyi (Estimation of in-plane displacements of the material surface by speckle correlation methods). Vidbir ta Obrobka Informatsiyi (Information Extraction and Processing) 18:95–99
69. Muravsky LI, Maksymenko OP, Sakharuk OM (2004) Use of a joint transform correlator architecture for study of speckle displacements. Opt Commun 240(4–6):275–291
70. Muravsky LI, Sakharuk OM, Byczkowska-Lipińska L (2009) Accurate definition of correlation peak position by optical speckle-displacement correlation technique. Opt Lasers Eng 47(6):712–720
71. Muravsky LI, Sakharuk OM, Yezhov PV (2009, June) Evaluation of aliasing influence on the performance of hybrid optical-digital speckle correlator. In: Lehmann PH (ed) Optical measurement systems for industrial inspection VI, vol 7389. SPIE, Bellingham, WA, pp 785–796
72. Muravsky LI, Voronyak TI, Kmet' AB (2014) Lazerna interferometriya poverhni dlia potreb tehnichnoyi diagnostyky (Laser interferometry of surface for needs of technical diagnostics). Spolom, Lviv
73. Muravsky L, Kuts O, Gaskevych G, Suriadova O (2019, September) Detection of subsurface defects in composite panels using dynamic speckle patterns. In: 2019 XIth international scientific and practical conference on electronics and information technologies (ELIT). IEEE, Lviv, pp 7–10
74. Muravsky LI, Sakharuk AN, Fitio NV, Pavlov AV (2006) Ring median binarization of joint power spectrum at optical correlation of speckle patterns. Bull Russ Acad Sci: Phys 70(1):155–160
75. Muravsky LI, Sakharuk OM, Fityo NV, Yezhov PV (2007) Increase of reliability of surface displacement field recovery by optical speckle-displacement correlation technique. Opt Lasers Eng 45(10):993–1000
76. Muravsky LI, Kmet' AB, Stasyshyn IV, Voronyak TI, Bobitski YV (2018) Three-step interferometric method with blind phase shifts by use of interframe correlation between interferograms. Opt Lasers Eng 105:27–34
77. Muravsky LI, Voronyak TI, Ivanytsky YaL, Dzhala VR, Kuts OG (2017) Pryntsypy pobudovy aparatno-programnogo kompleksu dlia neruynivnogo kontrolyu kompozytnyh elementiv konstruktsiy aviakosmichnoyi tehniky (Principles for development of hardware-software complex for nondestructive testing of aircraft composite structural components). Tehnichna

Diagnostyka ta Neruynivnyy Kontrol (Technical Diagnostics and Non-Destructive Testing) 1:37–42

78. Muravsky LI, Kuts OG, Voronyak TI, Gaskevych GI, Kotsiuba YM, Stasyshyn IV (2019) Sposib vyyavlennia vnutrishnih defektiv u zrazkah konstruktsiynyh materialiv ta elementiv konstruktsiy (Method for detecting internal defects in specimens of structural materials and structural elements). Patent Ukrayiny na Korysnu Model (Ukrainian patent for a utility model #136987). Ministry of Economy of Ukraine
79. Muravsky LI, Sakharuk OM, Kostyukevych SO, Maksymenko OP, Kostyukevych KV (2007, June) Performance of optical speckle displacement technique near stress concentrators. In: Osten W, Gorecki C, Novak EL (eds) Optical measurement systems for industrial inspection V, vol 6616. SPIE, Bellingham, WA, pp 980–988
80. Muravsky LI, Voronyak TI, Ivanytskyy YL, Hvozdyuk MM, Maksymenko OP Kuts OG, Suriadova OD (2020) Novi metodyky ta tehnichni zasoby kontrolyu i pidvyshchennia resursu roboty metalo-kompozytnyh zyednan' aviatsiynoyi tehniky (New methods and technical means of monitoring and increasing the service life of metal-composite joints of aviation equipment). Tehnichna Diagnostyka ta Neruynivnyy Kontrol (Technical Diagnostics and Non-Destructive Testing) 3:26–31
81. Nazarchuk Z, Muravsky L, Kuryliak D (2019) To the problem of the subsurface defects detection: theory and experiment. Procedia Struct Integr 16:11–18
82. Nazarchuk ZT, Muravsky LI, Kuts OG (2022) Nondestructive testing of thin composite structures for subsurface defects detection using dynamic laser speckles. Res Nondest Eval 33(2):59–77
83. Nemati M, Presura CN, Urbach HP, Bhattacharya N (2014) Dynamic light scattering from pulsatile flow in the presence of induced motion artifacts. Biomed Opt Exp 5(7):2145–2156
84. Owner-Petersen M (1991) Decorrelation and fringe visibility: on the limiting behavior of various electronic speckle-pattern correlation interferometers. J Opt Soc Am A 8(7):1082–1089
85. Pan B, Qian K, Xie H, Asundi A (2009) Two-dimensional digital image correlation for in-plane displacement and strain measurement: a review. Meas Sci Technol 20(6):062001
86. Papadopoulos GA (1987) The stationary value of the third stress invariant as a local fracture parameter (Det.-criterion). Eng Fract Mech 27(6):643–652
87. Peters WH, Ranson WF (1982) Digital imaging techniques in experimental stress analysis. Opt Eng 21(3):427–431
88. Pouet BF, Krishnaswamy S (1994) Additive-subtractive phase-modulated electronic speckle interferometry: analysis of fringe visibility. Appl Opt 33:6609–6616
89. Pouet B, Chatters T, Krishnaswamy S (1993) Synchronized reference updating technique for electronic speckle interferometry. J Nondestr Eval 12(2):133–138
90. Pra ALD, Passoni IL, Sendra GH, Trivi M, Rabal H (2014) Biospeckle signal descriptors: a performance comparison. arXiv preprint arXiv:1411.7522
91. Rabal HJ, Braga Jr RA (eds) (2018) Dynamic laser speckle and applications. CRC Press, Boca Raton, eBook
92. Refregier P, Laude V, Javidi B (1994) Nonlinear joint-transform correlation: an optimal solution for adaptive image discrimination and input noise robustness. Opt Lett 19(6):405–407
93. Réthoré J, Hild F, Roux S (2008) Extended digital image correlation with crack shape optimization. Int J Numer Meth Eng 73(2):248–272
94. Riddell WT, Piascik RS, Sutton MA, Zhao W, McNeill SR, Helm JD (1999) Determining fatigue crack opening loads from near-crack tip displacement measurements. ASTM Spec Tech Publ 1343:157–174
95. Russell D (2018) Acoustic and vibration animations. Graduate Program in Acoustic. The Pennsilvania State Univ, https://www.acs.psu.edu/drussell/Demos/rect-membrane/rect-mem.html
96. Sakharuk OM, Muravsky LI (2007) Otsinka vidpovidnosti polozhen' fragmenta spekl-zobrazhennia ta koreliatsiynogo pika (Assessment of correspondence between the positions of the speckle pattern subset and the correlation peak). Fizychni metody ta zasoby kontrolyu

seredovyshch, materialiv ta vyrobiv (Physical Methods and Means for Control of Media, Materials and Products) 12:229–232

97. Sakharuk OM, Muravs'kyi LI, Holyns'kyi IS, Lychak OV (2014). Determination of the field of local displacements by the digital speckle correlation method with adaptive segmentation of the images. Mater Sci 49(5):660–666
98. Sigler D, Haworth WL (1981) Strain measurement by optical correlation. J Nondestr Eval 2(2):125–132
99. Sih GC (1974) Strain-energy-density factor applied to mixed mode crack problems. Int J Fracture 10(3):305–321
100. Sih GC, Liebowitz H (1968) Mathematical theories of brittle fracture. In: Liebowitz H (ed) Mathematical fundamentals of fracture, vol 2. Academic Press, New York, pp 67–190
101. Sjödahl M (1994) Electronic speckle photography: increased accuracy by nonintegral pixel shifting. Appl Opt 33(28):6667–6673
102. Sjödahl M (1998) Some recent advances in electronic speckle photography. Opt Lasers Eng 29(2–3):125–144
103. Sjödahl M, Benckert LR (1993) Electronic speckle photography: analysis of an algorithm giving the displacement with subpixel accuracy. Appl Opt 32(13):2278–2284
104. Sjödahl M, Benckert LR (1994) Systematic and random errors in electronic speckle photography. Appl Opt 33(31):7461–7471
105. Solodov I (2014) Resonant acoustic nonlinearity of defects for highly-efficient nonlinear NDE. J Nondestr Eval 33(2):252–262
106. Solodov I, Bai J, Busse G (2013) Resonant ultrasound spectroscopy of defects: case study of flat-bottomed holes. J Appl Phys 113(22):223512
107. Solodov I, Rahammer M, Kreutzbruck M (2019) Analytical evaluation of resonance frequencies for planar defects: effect of a defect shape. NDTE Int 102:274–280
108. Solodov I, Bai J, Bekgulyan S, Busse G (2011) A local defect resonance to enhance acoustic wave-defect interaction in ultrasonic nondestructive evaluation. Appl Phys Lett 99(21):211911
109. Spagnolo GS, Paoletty D, Zanetta P (1996) Local speckle correlation for vibration analysis. Opt Commun 123(1–3):41–48
110. Suriadova OD (2019) Analiz dynamiky lazernyh spekliv dlia vyyavlennia pidpoverhnevyh defektiv u kompozytnyh paneliah (Analysis of laser speckle dynamics for detection of subsurface defects in composite panels). In: Problems of materials science and surface engineering: conference abstract of MSSE2019, Lviv, pp 192–195
111. Sutton MA, Orteu JJ, Schreier H (2009) Image correlation for shape, motion and deformation measurements: basic concepts, theory and applications. Springer Science & Business Media
112. Sutton MA, Turner JL, Bruck HA, Chae TA (1991) Full-field representation of discretely sampled surface deformation for displacement and strain analysis. Exp Mech 31(2):168–177
113. Sutton MA, Mingqi C, Peters WH, Chao YJ, McNeill SR (1986) Application of an optimized digital correlation method to planar deformation analysis. Image Vis Comput 4(3):143–150
114. Sutton MA, Wolters WJ, Peters WH, Ranson WF, McNeill SR (1983) Determination of displacements using an improved digital correlation method. Image Vis Comput 1(3):133–139
115. Sutton MA, Zhao W, McNeill SR, Helm JD, Piascik RS, Riddell WT (1999) Local crack closure measurements: development of a measurement system using computer vision and a far-field microscope. ASTM Spec Tech Publ 1343:145–156
116. Tao N, Anisimov AG, Groves RM (2022) Shearography non-destructive testing of thick GFRP laminates: numerical and experimental study on defect detection with thermal loading. Compos Struct 282:115008
117. Thomas BP, Pillai SA, Narayanamurthy CS (2017) Investigation on vibration excitation of debonded sandwich structures using time-average digital holography. Appl Opt 56(13):F7–F13
118. Thomas BP, Pillai SA, Narayanamurthy CS (2021) Computed time average digital holographic fringe pattern under random excitation. Appl Opt 60(4):A188–A194
119. Thompson O, Andrews M, Hirst E (2011) Correction for spatial averaging in laser speckle contrast analysis. Biomed Opt Exp 2(4):1021–1029

120. Tong W (2004) Technical application series: an adaptive backward image correlation technique for deformation mapping of a growing crack in thin sheets. Exp Tech 28(3):63–67
121. Tripathi R, Pati GS, Kumar A, Singh K (1998) Three-dimensional displacement measurement using chirp modulation in a photorefractive correlator. Opt Eng 37(11):2979–2987
122. Tripathi R, Pati GS, Kumar A, Singh K (1998) Object tilt measurement using a photorefractive speckle correlator: theoretical and experimental analysis. Opt Eng 37(11):2988–2997
123. Vandenrijt JF, Xiong H, Lequesne C, Blain P, Georges M (2019, June) Shearography inspection of monolithic CFRP composites: finite element modeling approach for assessing an adequate strategy of artificial defects representing delamination. In: Lehmann P, Osten W, Gonsalves Jr AA (eds) Optical measurement systems for industrial inspection XI, vol 11056. SPIE, Bellingham, WA, pp 107–113
124. Wang J, Javidi B (1994) Multiobject detection using the binary joint transform correlator with different types of thresholding methods. Opt Eng 33(6):1793–1805
125. Wang LS, Krishnaswamy S (1996) Additive-subtractive speckle interferometry: extraction of phase data in noisy environments. Opt Eng 35(3):794–801
126. Weaver CS, Goodman JW (1966) A technique for optically convolving two functions. Appl Opt 5(7):1248–1249
127. Willert CE, Gharib M (1991) Digital particle image velocimetry. Exp Fluids 10(4):181–193
128. Wong WO (1997) Vibration analysis by laser speckle correlation. Opt Lasers Eng 28(4):277–286
129. Yamaguchi I (1981) Speckle displacement and decorrelation in the diffraction and image fields for small object deformation. Opt Acta Int J Opt 28(10):1359–1376
130. Yamaguchi I (1993) Theory and applications of speckle displacement and decorrelation. In: Sirohi RC (ed) Speckle metrology. CRC Press, Boca Raton, FL, pp 1–40
131. Yoneyama S, Kitagawa A, Kitamura K, Kikuta H (2006) In-plane displacement measurement using digital image correlation with lens distortion correction. JSME Int J Ser A 49(3):458–467
132. Yoneyama S, Ogawa T, Kobayashi Y (2007) Evaluating mixed-mode stress intensity factors from full-field displacement fields obtained by optical methods. Eng Fract Mech 74(9):1399–1412
133. Yura HT, Rose B, Hanson SG (1998) Speckle dynamics from in-plane rotating diffuse objects in complex ABCD optical systems. J Opt Soc Am A 15(5):1167–1173
134. Zdunek A, Muravsky LI, Frankevych L, Konstankiewicz K (2007) New nondestructive method based on spatial-temporal speckle correlation technique for evaluation of apples quality during shelf-life. Int Agrophys 21(3):305–310
135. Zhou P, Goodson KE (2001) Subpixel displacement and deformation gradient measurement using digital image/speckle correlation. Opt Eng 40(8):1613–1620

Chapter 7
Mathematical Modeling of Elastic Waves Interaction with Interface Crack-Type Defects

Abstract The simple model of the elastic SH-plane wave scattering from the interface crack in a joint of two half-spaces is considered using the Wiener–Hopf technique. Its approximate solution is obtained for the wide crack. The study focuses on the physical features that can be used to recognize interface defects. The far scattering field for arbitrary plane wave illumination angle, including sliding and critical ones, the effects of the mutual edge waves diffraction, and the lateral wave excitation are analyzed. The approximate expressions of the stress intensity factors are obtained and establish their dependencies on the problem parameters. The elastic SH-mode diffraction from the finite interface crack in the joint of the half-space and the layer is considered. The mode transformation properties, as well as the spectral and displacement field characteristics, are analyzed. The problem of the interface crack recognition is discussed.

7.1 Introduction

The effects of interaction of the elastic waves with the defects of materials are one of the main sources of information about their damages. The necessary information is obtained from the analysis of the influence of the defect parameters on the displacement field. Since the modern optical methods make it possible to visualize the displacement fields on the surfaces of the testing sample with high accuracy, the problem of establishing a relationship between defect parameters and displacement fields arises in order to develop new methods of recognizing the defects. Therefore, to develop new methods of material diagnostics, it is important to study the physical properties of scattering of the elastic waves from the defects under different conditions of their sounding, visualizing the field displacement and analyzing its features. The novel theoretical methods of the field scattering analysis from the defects are based on the use of the most general numerical methods with minima restrictions on their shapes and properties. However, to use these methods in the form of commercial software packages does not always meet practical needs. It is known that finding the diffraction fields characteristics with the help of such tools requires significant

Z. Nazarchuk et al., *Optical Metrology and Optoacoustics in Nondestructive Evaluation of Materials*, Springer Series in Optical Sciences 242,
https://doi.org/10.1007/978-981-99-1226-1_7

computer time, especially when investigation is new and it is necessary to analyze dozens or even hundreds of different options. In addition, if we allow the possibility of obtaining all the necessary characteristics by the numerical methods, there is always the problem of their interpretation and formation of the physical imagination to understand the specifics of their interaction with the defects. Obviously, to study the physical properties of the elastic fields interaction with the defects, it is important to apply the simple models that allow for mathematically correct analysis of the problem for any parameters and which can be used to develop the approximate techniques for the solution of more complicated problems.

In this regard, we consider the diffraction of elastic waves on the interface crack-like defects that often appear in junctions of the construction elements, and these defects are ones of the most dangerous for the technical functioning of various types of structures. In order to form the theoretical bases of the interface defects detection, we suppose that such defects are thin, and as their models we use the flat interface strips or half-planes, on which we set the boundary conditions that ensure the absence of stress fields. Such defects are irradiated by a field of elastic SH-wave i.e. the transverse wave of horizontal polarization which can propagate in an elastic medium in the direction that is perpendicular to the direction of the vector of the particle displacement. Or, in other words, we consider the interface cracks in the structures with the plane-parallel boundaries and orient the sounding SH-wave so that the displacement vector of the particles of the layered media is parallel to the surfaces of the cracks faces and the boundaries of the media. Under such conditions, all the diffraction processes can be described by a single scalar function which must be found from the solution of the mixed boundary value problem for the Helmholtz equation. The main advantage of such models is that one of the most powerful mathematical technique of theoretical analysis, namely, the Wiener–Hopf method, can be applied for this purpose. This method involves reducing the boundary diffraction problem to a functional equation with respect to the Fourier transform of the displacements and stresses fields and solve this equation using the method of factorization and decomposition. The obtained solutions allow to find the necessary characteristics of the field with a given accuracy and provide the simple physical interpretation of their properties. For this purpose, we consider two key problems: the SH-wave diffraction from the finite crack at the plane interface of two homogeneous and isotropic semi-spaces, and the interface crack at the plane interface of the semi-space and layer. We regard this as an important prototype problem for the development of the solution methods for more complex problems. The chapter is laid out as follows. In §7.2, the general schemes of reduction of the wave diffraction problems to equations of the Wiener–Hopf are briefly considered based on the well-known monographs and numerous papers [1–7]. The problem of SH-wave diffraction from the finite interface crack at the plane junction of two semi-spaces is analyzed in §§ 7.3, 7.4, and the diffraction at the plane junction of semi-plane and layer is considered in § 7.5.

7.2 Wiener–Hopf Equation for Solution of the Wave Diffraction Problems

7.2.1 The Convolution Integral Equation and Wiener–Hopf Equation

The Wiener–Hopf technique is focused on solving the mixed boundary value problems for the Helmholtz equation, where the sought functions satisfy different types of boundary conditions on the coordinate surfaces. It is supposed that the integral transformations of the unknown functions exist and the boundary value problems can be reformulated for the transfomants. Using their analytical properties, these problems are reduced to the functional equations which are solved analytically or reduced to the Fredholm type integral equations of the second kind. Such a chain of transformations, including the solution of a functional equation, is called the Wiener–Hopf method. The necessary boundary value problems are formulated in the Cartesian coordinate system using the integral Fourier transform.

There are different ways to reduce the boundary value problems to the functional equations of the Wiener–Hopf type. As a rule, the main ideas for the solution of this equation are displayed using the convolution integral equation of the first type [2, 4, 6]:

$$\frac{1}{\sqrt{2\pi}}\int_{0}^{\infty} f(\xi)k(x-\xi)\mathrm{d}\xi = p(x), \quad 0 < x < \infty, \tag{7.1}$$

where $f(\xi)$ is the unknown function to be determined, $k(x-y) = k(|x-y|)$ is the kernel, and $p(x)$ is the known function; the variable x runs on a semi-infinite interval.

Let us consider the convolution integral equation that is given on an infinite interval $x \in (-\infty, \infty)$ as

$$\frac{1}{\sqrt{2\pi}}\int_{-\infty}^{\infty} f(\xi)k(x-\xi)\mathrm{d}\xi = p(x), \quad -\infty < x < \infty. \tag{7.2}$$

Next, we introduce the direct and inverse Fourier integral transform of the unknown function $f(x)$ as

$$F(\alpha) = \frac{1}{\sqrt{2\pi}}\int_{-\infty}^{\infty} f(x)e^{i\alpha x}\mathrm{d}x,$$

$$f(x) = \frac{1}{\sqrt{2\pi}} \int_{-\infty}^{\infty} F(\alpha) e^{-i\alpha x} d\alpha, \tag{7.3}$$

where $i = \sqrt{-1}$, $\alpha = \mathrm{Re}\,(\alpha) + i\mathrm{Im}(\alpha) \equiv \sigma + i\tau$. We apply the first transform (7.3) to the left part of the integral equation (7.2). Then, the obtained expression takes the form

$$\begin{aligned}
&\frac{1}{\sqrt{2\pi}} \int_{-\infty}^{\infty} e^{i\alpha x} \frac{1}{\sqrt{2\pi}} \int_{-\infty}^{\infty} f(\xi) k(x-\xi) d\xi dx \\
&= \frac{1}{2\pi} \int_{-\infty}^{\infty} f(\xi) \int_{-\infty}^{\infty} e^{i\alpha x} k(x-\xi) dx d\xi \\
&= \frac{1}{\sqrt{2\pi}} \int_{-\infty}^{\infty} f(\xi) e^{i\alpha\xi} d\xi \frac{1}{\sqrt{2\pi}} \int_{-\infty}^{\infty} e^{i\alpha y} k(y) dy = K(\alpha) F(\alpha),
\end{aligned} \tag{7.4}$$

where $K(\alpha)$ is the Fourier transform of the even kernel function $k(x)$. Here we suppose the possibility of changing the order of integration.

Using the expression (7.4), we reduce the integral equation (7.2) to the algebraic one as

$$K(\alpha) F(\alpha) = P(\alpha), \tag{7.5}$$

where $P(\alpha)$ is the Fourier transform of the known function $p(x)$.

Let us suppose that functions in Eq. (7.5) are free of zeros and singularities in the strip of the complex plane α covers the real axis. Then, using the inverse transform formula (7.3), we can represent the solution of the integral equation (7.2) in the form

$$f(x) = \frac{1}{\sqrt{2\pi}} \int_{-\infty}^{\infty} \frac{P(\alpha)}{K(\alpha)} e^{-i\alpha x} d\alpha. \tag{7.6}$$

The main idea of solving the integral equation (7.1) is to represent it in the form (7.2). For this purpose, it is necessary to make two steps: to supplement its unknown function and right part with zero, having set them on a full interval $-\infty < \xi, x < \infty$, and also to enter the additional unknown set on a semi-infinite interval $-\infty < x < 0$. This procedure allows us to rewrite Eq. (7.1) in the form

$$\frac{1}{\sqrt{2\pi}} \int_{-\infty}^{\infty} f_+(\xi) k(x-\xi) d\xi = p_+(x) + \varphi_-(x), \quad -\infty < x < \infty. \tag{7.7}$$

Here, we introduce new notations:

$$f_+(x) = \begin{cases} f(x) & \text{if } x > 0, \\ 0 & \text{if } x < 0, \end{cases} \quad \varphi_-(x) = \begin{cases} 0 & \text{if } x > 0, \\ \varphi(x) & \text{if } x < 0; \end{cases} \tag{7.8}$$

$$p_+(x) = \begin{cases} p(x) & \text{if } x > 0, \\ 0 & \text{if } x < 0, \end{cases} \tag{7.9}$$

where $\varphi_-(x)$ is the new unknown function.

Let us apply the Fourier transform to Eq. (7.7). This leads to the functional equation as

$$F^+(\alpha)K(\alpha) = \Phi^-(\alpha) + P^+(\alpha). \tag{7.10}$$

Here,

$$F^+(\alpha) = \frac{1}{\sqrt{2\pi}} \int_0^\infty f(x)e^{i\alpha x}\mathrm{d}x,\ \Phi^-(\alpha) = \frac{1}{\sqrt{2\pi}} \int_{-\infty}^0 \varphi(x)e^{i\alpha x}\mathrm{d}x, \tag{7.11}$$

$$P^+(\alpha) = \frac{1}{\sqrt{2\pi}} \int_0^\infty p(x)e^{i\alpha x}\mathrm{d}x. \tag{7.12}$$

Let us suppose that functions in Eq. (7.10) are free of zeros and singularities in the strip of the complex plane α covers the real axis $\tau_- < \tau < \tau_+$. In order to obtain the solution of Eq. (7.10), we suppose the exponential asymptotic behavior of functions (7.8) as $f(x) \sim e^{\tau_- x}$ if $x \to \infty$ and $\varphi(x) \sim e^{\tau_+ x}$ if $x \to -\infty$. Then, the Fourier transforms $F^+(\alpha)$, $\Phi^-(\alpha)$ are the regular functions in the overlapping semi-planes $\tau > \tau_-$, $\tau < \tau_+$. Their asymptotic behavior for $\alpha \to \infty$ are determined by the behavior of the originals at the edge of the semi-plane. Here, $f(x) \sim x^q$, $\varphi(x) \sim x^\delta$ if $x \to \pm 0$, where $-1 < q < 0, \delta \geq 0$. Then, $F^+(\alpha) \sim |\alpha|^{-(q+1)}$, $\Phi^-(\alpha) \sim |\alpha|^{-(\delta+1)}$ if $|\alpha| \to \infty$ for $\tau > \tau_-$, $\tau < \tau_+$ respectively. Function $K(\alpha)$ must be regular in the strip $\tau_- < \tau < \tau_+$; it allows for the canonical factorization as

$$K(\alpha) = K^+(\alpha)K^-(\alpha), \tag{7.13}$$

where $K^+(\alpha)$, $K^-(\alpha)$ are regular functions in the overlapping half-planes $\tau > \tau_-$, $\tau < \tau_+$ respectively; they have no zeros in the above-mentioned regions and decay algebraically at the infinity.

Taking this into account, let us rewrite Eq. (7.10) in the form as

$$F^+(\alpha)K^+(\alpha) = \frac{\Phi^-(\alpha)}{K^-(\alpha)} + \frac{P^+(\alpha)}{K^-(\alpha)}. \tag{7.14}$$

The term in the left side and the first term in the right side of this equation are the regular functions in the upper $\tau > \tau_-$ and in the lower $\tau < \tau_+$ half-planes respectively. The second term in the right-side of Eq. (7.14) is a regular function in the strip $\tau_- < \tau < \tau_+$, and out of this strip, the above-mentioned term allows for the singularities and zeros.

The next step for the solution of Eq. (7.1) is representing the second term in the right part of Eq. (7.14) in the form as

$$\frac{P^+(\alpha)}{K^-(\alpha)} = \left[\frac{P^+(\alpha)}{K^-(\alpha)}\right]^+ + \left[\frac{P^+(\alpha)}{K^-(\alpha)}\right]^-, \tag{7.15}$$

where $[\cdot]^+$, $[\cdot]^-$ are the functions regular in the half-planes $\tau > \tau_-$, $\tau < \tau_+$. Then, Eq. (7.14) can be represented in the form of

$$F^+(\alpha)K^+(\alpha) - \left[\frac{P^+(\alpha)}{K^-(\alpha)}\right]^+ = \frac{\Phi^-(\alpha)}{K^-(\alpha)} + \left[\frac{P^+(\alpha)}{K^-(\alpha)}\right]^-. \tag{7.16}$$

The obtained equation is the desirable Wiener–Hopf equation which is valid in the strip $\tau_- < \tau < \tau_+$. The terms in the left side of this equation are the regular functions in the upper half-plane $\tau > \tau_-$, and the terms in the right side are the regular functions in the lower half-plane $\tau < \tau_+$, and they are equal in the joint strip of regularity $\tau_- < \tau < \tau_+$. Let the terms of this equation tends to zero in the regularity regions as $|\sigma|^{-\widehat{p}}$ if $|\sigma| \to \infty$ with $\widehat{p} > 0$. Then the right and the left sides of the functional equation (7.16) determine the entire function which, according to the Liouville's theorem, is equal to zero in the whole complex plane α. Then, we can equate to zero the right and the left sides of this equations and represent its solution in form as

$$F^+(\alpha) = \frac{1}{K^+(\alpha)}\left[\frac{P^+(\alpha)}{K^-(\alpha)}\right]^+, \tag{7.17}$$

$$\Phi^-(\alpha) = -K^-(\alpha)\left[\frac{P^+(\alpha)}{K^-(\alpha)}\right]^-, \tag{7.18}$$

where $F^+(\alpha)$, $\Phi^-(\alpha)$ are the regular functions in the half-planes $\tau > \tau_-$, $\tau < \tau_+$, respectively.

Then, making use of the inverse Fourier transform of the expressions (7.17), (7.18), the solution of the integral equation (7.1) can be represented as

$$f(x) = \frac{1}{\sqrt{2\pi}} \int\limits_{-\infty+ic_1}^{\infty+ic_1} \frac{1}{K^+(\alpha)}\left[\frac{P^+(\alpha)}{K^-(\alpha)}\right]^+ e^{-i\alpha x} d\alpha, \tag{7.19}$$

$$\varphi(x) = -\frac{1}{\sqrt{2\pi}} \int\limits_{-\infty+ic_2}^{\infty+ic_2} K^-(\alpha) \left[\frac{P^+(\alpha)}{K^-(\alpha)} \right]^- e^{-i\alpha x} d\alpha. \tag{7.20}$$

Here, $\tau_- < c_1, c_2 < \tau_+$.

As follows from the above relations, the factorization and the decomposition of Eq. (7.10) is the key steps to obtaining its solution. Let us represent Eq. (7.15) as

$$S(\alpha) = S^+(\alpha) + S^-(\alpha), \tag{7.21}$$

where $S^+(\alpha) = [P^+(\alpha)/K^-(\alpha)]^+$ and $S^-(\alpha) = [P^+(\alpha)/K^-(\alpha)]^-$ are regular functions in the half-planes $\tau > \tau_-$ and $\tau < \tau_+$ respectively.

The admissibility of representation (7.21) is determined by a well-known theorem [2]: let $S(\alpha)$ be an analytic function that is regular in the strip $\tau_- < \tau < \tau_+$, and for all τ of this strip it is valid that $|S(\alpha)| < C_0|\sigma|^{-\widehat{p}}$ if $|\sigma| \to \infty$ and $\widehat{p} > 0$. Then, the functions $S^+(\alpha)$, $S^-(\alpha)$ are determined by the formulas

$$S^+(\alpha) = \frac{1}{2\pi i} \int\limits_{-\infty+i\varepsilon_1}^{\infty+i\varepsilon_1} \frac{S(t)}{t-\alpha} dt,$$

$$S^-(\alpha) = -\frac{1}{2\pi i} \int\limits_{-\infty+i\varepsilon_2}^{\infty+i\varepsilon_2} \frac{S(t)}{t-\alpha} dt. \tag{7.22}$$

Here, $\tau_- < \varepsilon_1 < \tau < \varepsilon_2 < \tau_+$.

Let us turn to factorization of the function $K(\alpha)$ in the form (7.13). The simplest representation of this kind can be made for even entire functions with the simple zeros $\alpha = \pm z_n$ in the complex plane. They can be written in the form of infinite Weierstrass products:

$$K(\alpha) = K(0) \prod_{n=1}^{\infty} \left(1 - \frac{\alpha^2}{z_n^2} \right). \tag{7.23}$$

The factorization of function (7.23) is performed elementarily and is written up to the exponential factor in the form as

$$K^{\pm}(\alpha) = \sqrt{K(0)} e^{\mp\chi(\alpha)} \prod_{n=1}^{\infty} \left(1 \pm \frac{\alpha}{z_n} \right) e^{\mp\frac{\alpha}{z_n}}. \tag{7.24}$$

Here, the functions $K^{\pm}(\alpha)$ are regular and have no zeros in the upper and lower half-planes $\tau \gtrless \mp z_1$. These functions are determined up to the factors $e^{\pm\chi(\alpha)}$, which are entire functions that have no zeros in any finite region of the complex plane α. The coefficient $\chi(\alpha)$ is taken from the condition that the split functions $K^{\pm}(\alpha)$ have an algebraic nature of behavior at $|\alpha| \to \infty$. If the main term of the asymptotic of zeros z_n is linearly dependent on the index, for example, $z_n = an + b + O(n^{-1})$ if $n \to \infty$, where $a,\ b$ are complex quantities, then, using the known formula for an infinite product:

$$\prod_{n=1}^{\infty}\left(1 \pm \frac{\alpha}{an+b}\right)e^{\mp\frac{\alpha}{an}} = \frac{e^{\mp C(\alpha/a)}\Gamma(b/a+1)}{\Gamma(\pm\alpha/a+b/a+1)}, \tag{7.25}$$

where $\Gamma(\cdot)$ is the gamma function, $C = 0.57721\ldots$ is the Euler's constant, the expression (7.24) can be rewritten as

$$\begin{aligned} K^{\pm}(\alpha) = {} & \frac{e^{\mp\chi(\alpha)\mp\frac{\alpha}{a}C}\sqrt{K(0)}}{\Gamma(\pm\alpha/a+b/a+1)} \\ & \times\left\{\Gamma(b/a+1)\prod_{n=1}^{\infty}\frac{1\pm\alpha/z_n}{1\pm\alpha/(an+b)}e^{\mp\alpha\left(\frac{1}{z_n}-\frac{1}{an}\right)}\right\}. \end{aligned} \tag{7.26}$$

Here, the infinite product is limited, and the asymptotic behavior of functions $K^{\pm}(\alpha)$ is determined using the well-known asymptotic Stirling formula for gamma function

$$\Gamma(\alpha) \sim \sqrt{2\pi}e^{-\alpha}\alpha^{\alpha-1/2},\ \alpha \to \infty,\ -\pi < \mathrm{Arg}\alpha < \pi \tag{7.27}$$

with result that [4]

$$\begin{aligned} & \frac{e^{-\chi(\alpha)-\frac{\alpha}{a}C}}{\Gamma(\alpha/a+b/a+1)} \\ & \sim \begin{cases} (2\pi)^{-1/2}(\alpha/a)^{-b/a-1/2}\exp\{(\alpha/a)[1-C-\ln(\alpha/a)]-\chi(\alpha)\}, \\ \qquad |\alpha| \to \infty, \quad -\pi < \mathrm{Arg}\,\alpha < \pi, \\ \\ (2/\pi)^{1/2}\sin(-\alpha\pi/2+b\pi/2)(-\alpha/a)^{-b/a-1/2}\times \\ \times\exp\{(\alpha/a)[1-C-\ln(-\alpha/a)]-\chi(-\alpha)\},\ |\alpha| \to \infty,\ \alpha = -|\alpha|. \end{cases} \end{aligned} \tag{7.28}$$

Let us consider function $K(\alpha)$ the logarithm of which allows for the decomposition as

$$\ln K(\alpha) = [\ln K(\alpha)]^{+} + [\ln K(\alpha)]^{-}. \tag{7.29}$$

As follows from [2, 4], if $K(\alpha)$ is an analytic function of the complex variable and regular in the strip $\tau_{-} < \tau < \tau_{+}$, has no zeros in this strip and $K(\alpha) \to 1$ if $|\alpha| \to \infty$, then the factorization of this function exists in the form (7.13) with the split functions as

$$K^{+}(\alpha) = \exp\left[\frac{1}{2\pi i}\int_{-\infty+ic}^{\infty+ic}\frac{\ln K(t)}{t-\alpha}dt\right], \quad \tau_{-} < c < \tau < \tau_{+},$$

$$K^{-}(\alpha) = \exp\left[\frac{-1}{2\pi i}\int_{-\infty+id}^{\infty+id}\frac{\ln K(t)}{t-\alpha}dt\right], \quad \tau_{-} < \tau < d < \tau_{+}. \tag{7.30}$$

Here, $K^{+}(\alpha)$, $K^{-}(\alpha)$ are regular and have no zeros in the upper and lower overlapping complex semi-planes $\tau > \tau_{-}$ and $\tau < \tau_{+}$ respectively.

For further convenience, let us consider the main techniques that are usually used for reducing the wave diffraction problems to the Wiener–Hopf functional equation.

7.2.2 *Reducing the Wave Diffraction Problem to the Wiener–Hopf Equation*

Three methods are most often used to reduce the boundary value problems of the diffraction theory to the Wiener–Hopf equation: Green's function method, the method of dual integral equations, and the Jones method [2, 4]. Here, we consider briefly the use of these methods to reduce the wave diffraction from the semi-infinite screen in homogeneous medium to the Wiener–Hopf equation. From a mathematical point of view, the diffraction problems are formulated as the mixed boundary value problems for the Helmholtz partial differential equation. Here, we consider the Neumann problems in the Cartesian coordinate system due to the fact that these problems include the diffraction of SH-waves from the cracks, which are the main subject of our study.

Let us consider the infinitely thin semi-plane in the Cartesian coordinate system (x, y, z):

$$S : \{-\infty < x < 0;\ y = 0\}. \tag{7.31}$$

Here, we omit the variable z $(-\infty < z < \infty)$ because all the wanted values do not depend on it.

Let the semi-plane S illuminated by the plane wave

$$u^{i}(x, y) = e^{-ik(x\cos\theta_0 + y\sin\theta_0)}, \tag{7.32}$$

where θ_0 is the angle of illumination; to simplify the analysis we assume that $0 < \theta_0 < \pi/2$.

The time dependence of $e^{-i\omega t}$ is assumed and has been omitted through all the chapter. The boundary value problem of the plane wave diffraction from the semi-plane S is formulated as follows: let us find in the plane (x, y) the scalar function/potential $u = u(x, y)$ which satisfies the Helmholtz equation

$$\Delta u + k^2 u = \frac{\partial^2 u}{\partial x^2} + \frac{\partial^2 u}{\partial y^2} + k^2 u = 0, \tag{7.33}$$

and the boundary condition at the semi-plane (7.31):

$$\left.\frac{\partial u^t}{\partial y}\right|_{y=\pm 0} = 0 \text{ for } -\infty < x < 0. \tag{7.34}$$

Here, $u^t = u^t(x, y)$ is the total field,

$$u^t(x, y) = u(x, y) + u^{\text{inc}}(x, y), \tag{7.35}$$

$u(x, y)$ is the diffracted field, $u^{\text{inc}}(x, y) = u^i(x, y)$; $k = \omega/c$ is the wave number, where ω is the angular frequency, $k = k' + ik''$ with $k',\ k'' > 0$; c is the wave velocity.

We seek the solution of the boundary value problem (7.33), (7.34) in the class of functions that satisfy the condition of energy limitation for any bounded volume V:

$$W = \frac{1}{2} \int\limits_V \{|\nabla u|^2 + k^2 |u|^2\} \mathrm{d}x\mathrm{d}y < \infty. \tag{7.36}$$

Here, $|\nabla u|^2 = u'_x u'^*_x + u'_y u'^*_y$, $|u|^2 = uu^*$, where the upper sign "$*$" denotes the complex conjugate value; $\mathrm{d}V = \mathrm{d}x\mathrm{d}y$.

Note, if the radiated source is located out of the edge of semi-plane S, then by contracting to zero the radius of the virtual cylindrical volume covering it, we obtain from (7.36) that $W \to 0$. This leads to the well-known Meixner condition [4] which determines the behavior for the diffracted field at the semi-plane edge as

$$\partial u/\partial r \sim r^{-1/2} \text{ if } r \to 0, \text{ where } r = \sqrt{x^2 + y^2}. \tag{7.37}$$

In addition, the diffracted field satisfies the radiation condition at infinity which forbids the incoming waves from infinity. When the scattering surfaces extend to infinity, the radiation condition is formulated as the limiting absorption condition [8]

$$u \sim e^{-\chi r}/\sqrt{r}, \tag{7.38}$$

where $\chi = \chi(k'') > 0$.

The solution of the problem for the case of real k we obtain in limiting case if $k'' \to 0$. These conditions ensure the unique solution of the boundary value problem (7.33), (7.34) and fully describe the diffraction phenomena that occur when a plane wave interacts with the perfect and infinitely thin half-plane. Next, we represent briefly the main ways for reducing the boundary value problem (7.33), (7.34) to the Wiener–Hopf equation [2, 4, 9, 10].

(a) Method of the Green functions

Let us consider the Green function which satisfies the equation as

$$\frac{\partial^2 G}{\partial x^2} + \frac{\partial^2 G}{\partial y^2} + k^2 G = -4\pi\delta(x - x')\delta(y - y'), \tag{7.39}$$

where $G = G(x, y; x', y')$; $\delta(\cdot)$ is the Dirac delta function.

Applying the double integral Fourier transform (7.3) to Eq. (7.39) for the variables x, y, we arrive at

$$\tilde{\tilde{G}}(\alpha, \beta; x', y') = \frac{2e^{i\alpha x' + i\beta y'}}{\alpha^2 + \beta^2 - k^2}. \tag{7.40}$$

Then, the double inverse Fourier transform of expression (7.40) leads to an integral representation of the Green's function in the form

$$G(x, y; x', y') = \int\limits_{-\infty+ia}^{\infty+ia} \frac{e^{-i\alpha(x-x') - \gamma|y-y'|}}{\gamma} d\alpha. \tag{7.41}$$

Here, $\gamma = \gamma(\alpha) = \sqrt{\alpha^2 - k^2}$, $-k'' < a < k''$.

Let us consider a two-valued function $\gamma(\alpha)$ with branch points $\alpha = \pm k$ on a two-sheeted Riemann surface with cuts; the cuts and the required sheet of the Riemann surface are chosen from the conditions: $\mathrm{Re}\gamma \geq 0$, $\sqrt{\alpha^2 - k^2} = -i\sqrt{k^2 - \alpha^2}$ ($\sqrt{\alpha - k} = -i\sqrt{k - \alpha}$, $\sqrt{k - \alpha} = i\sqrt{\alpha - k}$), which ensure the fulfilment of the radiation condition. Classical are the cuts in the complex plane that start from the branch points $\alpha = \pm k$; they path through the first and third quadrants to $\pm i\infty$ along the curves determined by the equation $\mathrm{Re}\gamma = 0$ [4]. However, a simpler system of the cuts is often used, where they path along the lines parallel to an imaginary axis; this ensures the fulfillment of the condition $\mathrm{Re}\gamma \geq 0$ in the strip $-k'' < \tau < k''$.

Next, changing of variables in the integral (7.41) as $x - x' = R\cos\theta$, $|y - y'| = R\sin\theta$, $\alpha = -k\cos(\theta + it)$, where $R = \sqrt{(x - x')^2 + (y - y')^2}$, $0 \leq \theta \leq \pi$ and $-\infty < t < \infty$, we express the integral representation of the Green function through the Hankel function of the first kind [11] as

$$G(x, y; x', y') = \int\limits_{-\infty}^{\infty} e^{ikR\,\mathrm{ch}\,t} dt = \pi i H_0^{(1)}(kR). \tag{7.42}$$

Let us apply the second Green's theorem to the Eqs. (7.33), (7.39) in a region bounded by the real axis and a semicircle of infinitely large radius at $y > 0$. Taking into account the fact that the functions $u(x, y)$ and $G(x, y; x', y')$ satisfy the radiation condition, we arrive at

$$\int_{0}^{\infty}\int_{-\infty}^{\infty}\{G(x,y;x',y')\Delta u(x',y')-u(x',y')\Delta G(x,y;x',y')\}\mathrm{d}x'\mathrm{d}y'$$

$$=\int_{-\infty}^{\infty}\left\{G(x,y;x',+0)\frac{\partial u(x',+0)}{\partial n'}-u(x',+0)\frac{\partial G(x,y;x',+0)}{\partial n'}\right\}\mathrm{d}x'. \quad (7.43)$$

Here, $\partial/\partial n' = -\partial/\partial y'$, $G(x, y; x', +0) = G(R_0)$, $R_0 = \sqrt{(x-x')^2 + y^2}$.

The equality (7.43) is preserved when the Green's function is replaced by the sum of the Hankel functions as

$$G_1(x,y;x',y') = G(x,y;x',y') + G(x,-y;x',y') = \pi i\,[H_0^{(1)}(kR) + H_0^{(1)}(kR_1)],$$

where $R_1 = \sqrt{(x-x')^2 + (y+y')^2}$ for which $\partial G_1/\partial y'|_{y'=0} = 0$ is correct.

Let us introduce the incident field as

$$\overset{\text{inc}}{\widehat{u}}(x,y) = \begin{cases} u^i(x,y) + u^i(x,-y), & y>0, \\ 0, & y<0, \end{cases} \quad -\infty < x < \infty. \quad (7.44)$$

Then, we reformulate the initial boundary value problem (7.33)–(7.35) as

$$\Delta\widehat{u} + k^2\widehat{u} = 0, \quad (7.45)$$

$$\left.\frac{\partial u^t}{\partial y}\right|_{y=\pm 0} = \left.\frac{\partial \widehat{u}}{\partial y}\right|_{y=\pm 0} = 0 \ \text{ for } \ -\infty < x < 0, \quad (7.46)$$

where $\widehat{u} = \widehat{u}(x, y)$,

$$u^t(x,y) = \widehat{u}(x,y) + \overset{\text{inc}}{\widehat{u}}(x,y), \quad (7.47)$$

$$\partial\overset{\text{inc}}{\widehat{u}}/\partial y|_{y=0} = 0. \quad (7.48)$$

Since Eq. (7.43) holds for arbitrary functions satisfying the Helmholtz equation, it also holds for the function $\widehat{u}(x, y)$. Therefore, taking this into account and the fact that $\partial G_1/\partial y'|_{y'=0} = 0$, we find that

$$\widehat{u}(x,y) = -\frac{i}{2}\int_{-\infty}^{\infty} H_0^{(1)}(kR_0)\frac{\partial\widehat{u}(x',+0)}{\partial y'}\mathrm{d}x', \quad -\infty < x < \infty, y > 0. \quad (7.49)$$

Using the relation (7.47), we get from (7.49) that

$$u^t(x, y) = \widehat{u}^{\text{inc}}(x, y) - \frac{i}{2}\int_0^\infty H_0^{(1)}(kR_0)\frac{\partial u^t(x', +0)}{\partial y'}\mathrm{d}x',$$

$$-\infty < x < \infty,\ y > 0. \tag{7.50}$$

Let us apply the second Green‘s theorem to the Eqs. (7.33), (7.39) in a region bounded by the real axis and a semicircle of infinitely large radius at $y < 0$. Using the boundary condition (7.46), the definition (7.47), and taking into account that $\partial/\partial n' = \partial/\partial y'$, we find that

$$u^t(x, y) = \frac{i}{2}\int_0^\infty H_0^{(1)}(kR_0)\frac{\partial u^t(x', -0)}{\partial y'}\mathrm{d}x', \quad -\infty < x < \infty,\ y < 0. \tag{7.51}$$

Here, $u^t|_{y=+0} = u^t|_{y=-0}$ and $\partial\, u^t/\partial y|_{y=+0} = \partial\, u^t/\partial y|_{y=-0} = \partial\, u^t/\partial y|_{y=0}$ if $0 < x < \infty$.

Therefore, if $y = 0$, we equate the right-hand part of the Eqs. (7.50), (7.51) and reduce the problem to the convolution integral equation as [2, 7, 10]

$$\frac{1}{\sqrt{2\pi}}\int_0^\infty f(\xi)H_0^{(1)}(k|x-\xi|)\mathrm{d}\xi = p(x), \quad 0 < x < \infty, \tag{7.52}$$

where

$$f(\xi) = \partial\, u^t(\xi, y)/\partial y|_{y=0}, \tag{7.53}$$

$$p(x) = -i\sqrt{2/\pi}\,e^{-ikx\cos\theta_0}. \tag{7.54}$$

Further, we represent this equation in the form (7.7) and reduce it to the functional Wiener–Hopf equation (7.10) with the kernel function as

$$K(\alpha) = \frac{1}{\sqrt{2\pi}}\int_{-\infty}^{\infty} H_0^{(1)}(k|x|)e^{i\alpha x}\mathrm{d}x = \frac{\sqrt{2/\pi}}{i\sqrt{\alpha^2-k^2}} = \frac{\sqrt{2/\pi}}{\sqrt{k^2-\alpha^2}}. \tag{7.55}$$

Here, the expression (7.55) directly follows from the Fourier transform of the integral representation of the Hankel function [11] as

$$H_0^{(1)}(k\sqrt{x^2+y^2}) = \frac{1}{\pi i}\int_{-\infty+ia}^{\infty+ia}\frac{e^{-itx-\sqrt{t^2-k^2}|y|}}{\sqrt{t^2-k^2}}\mathrm{d}t, \tag{7.56}$$

where $-k'' < a < k''$.

Using the notations of the Sect. 7.2, we factorize the kernel function (7.55) in form (7.13):

$$K^+(\alpha) = K^-(-\alpha) = \frac{\sqrt[4]{2/\pi}}{\sqrt{k+\alpha}}, \tag{7.57}$$

where $K^+(\alpha)$, $K^-(\alpha)$ are the split functions that are regular in the overlapping semi-planes $\tau > -k''$, $\tau < k''$ with the join strip of regularity $-k'' < \tau < k''$ and these functions tend to zero in the regularity regions as $\alpha^{-1/2}$.

Then, using the relations (7.9), (7.54), we represent the right-hand part of the obtained functional equation in the form

$$P^+(\alpha) = \frac{1}{\sqrt{2\pi}} \int\limits_0^{\infty} p(x) e^{i\alpha x} \mathrm{d}x = \frac{1}{\pi(\alpha - k\cos\theta_0)}, \tag{7.58}$$

where $P^+(\alpha)$ is a regular function for $\tau > \tau_- = k''\cos\theta_0$ and tends to zero as α^{-1} if $|\alpha| \to \infty$.

Next, we finally arrive at the following Wiener–Hopf equation

$$F^+(\alpha)\frac{\sqrt{2/\pi}}{\sqrt{k^2-\alpha^2}} = \Phi^-(\alpha) + \frac{1}{\pi(\alpha - k\cos\theta_0)}. \tag{7.59}$$

Taking into account the expression (7.57), we rewrite this equation in the equivalent form as

$$\frac{F^+(\alpha)}{\sqrt{k+\alpha}} - \frac{\sqrt{k-\alpha}}{\sqrt{2\pi}(\alpha - k\cos\theta_0)} = \sqrt{\pi/2}\,\Phi^-(\alpha)\sqrt{k-\alpha}. \tag{7.60}$$

Here, $F^+(\alpha)$, $\Phi^-(\alpha)$ are the Fourier transforms of the unknown functions:

$$f(x) = \begin{cases} \partial u^t(x,0)/\partial y & \text{if } x > 0, \\ 0 & \text{if } x > 0, \end{cases}$$

$$\varphi(x) = \begin{cases} 0 & \text{if } x > 0, \\ \varphi(x) & \text{if } x > 0. \end{cases} \tag{7.61}$$

Taking into account the representations (7.32), (7.44) and (7.47), we find that $u^t \sim e^{k''x\cos\theta_0}$ if $x \to \infty$, and, from the integral (7.52), we get the asymptotic estimation of the function $\varphi(x) \sim e^{k''x}$ if $x \to -\infty$. Therefore, the functions $F^+(\alpha)$ and $\Phi^-(\alpha)$ are regular in the semi-planes $\tau > k''\cos\theta_0$ and $\tau < k''$ of the complex plane α respectively. If $x \to \pm 0$, the behavior of the functions (7.61) looks as $f(x) \sim x^{-1/2}$ and $|\varphi(x)| \to \text{const} \geq 0$. Then, $F^+(\alpha) \sim |\alpha|^{-1/2}$ and $\Phi^-(\alpha) \sim |\alpha|^{-1}$ if $|\alpha| \to \infty$ in the regularity regions.

Let us turn to Eq. (7.60). The first term in the left-hand side of this equation is the regular function in the upper half-plane $\tau > k'' \cos\theta_0$ and tends to zero as α^{-1} if $|\alpha| \to \infty$. The term in the right-hand side is regular in the lower half-plane $\tau < k''$ and if $|\alpha| \to \infty$ tends to zero as $\alpha^{-1/2}$ in the regularity region. The second term in the left-hand side is the known function that is regular in the strip $k'' \cos\theta_0 < \tau < k''$, where it tends to zero as $\alpha^{-1/2}$ if $|\alpha| \to \infty$. Out of this strip, the known term has a simple pole singularity and the branch point. Therefore, Eq. (7.60) is valid at least in the strip $k'' \cos\theta_0 < \tau < k''$.

Let us represent the known term for Eq. (7.60) as a sum of two ones in form (7.15) that are regular in the semi-planes $\tau > k'' \cos\theta_0$ and $\tau < k''$ as in (7.21), (7.22). Next, the terms that are regular in the upper and lower half-planes we group in the left- and right-hand parts of the equation respectively. If $|\alpha| \to \infty$, all the terms of the obtained equation tend to zero in their regularity regions including the overlapping strip $k'' \cos\theta_0 < \tau < k''$. Then, according to the Liouville theorem, the terms in the right- and left-hand sides of this equation form the entire function that is identically equal to zero in the whole complex plane. Therefore, the solution of Eq. (7.60) looks like

$$F^+(\alpha) = \frac{\sqrt{k+\alpha}}{2\pi i\sqrt{2\pi}} \int\limits_{-\infty+i\varepsilon_1}^{\infty+i\varepsilon_1} \frac{\sqrt{k-t}\,\mathrm{d}t}{(t-k\cos\theta_0)(t-\alpha)},$$

$$\Phi^-(\alpha) = \frac{1}{2\pi^2 i\sqrt{k-\alpha}} \int\limits_{-\infty+i\varepsilon_2}^{\infty+i\varepsilon_2} \frac{\sqrt{k-t}\,\mathrm{d}t}{(t-k\cos\theta_0)(t-\alpha)}, \tag{7.62}$$

where $k'' \cos\theta_0 < \varepsilon_1 < \tau < \varepsilon_2 < k''$.

Complementing the integration contour with a semicircle of infinite radius into the lower half-plane and using the residues theorem for evaluation of the integrals (7.62), we find that

$$F^+(\alpha) = \frac{i\sqrt{k\cos\theta_0 - k}\sqrt{\alpha+k}}{\sqrt{2\pi}(\alpha - k\cos\theta_0)},$$

$$\Phi^-(\alpha) = \frac{1}{\pi(\alpha - k\cos\theta_0)}\left[1 - \frac{\sqrt{k\cos\theta_0 - k}}{\sqrt{\alpha - k}}\right]. \tag{7.63}$$

These solutions are the regular functions of the complex variable α in the semi-planes $\tau > k'' \cos\theta_0$, $\tau < k''$, have the needed behavior at infinity $|\alpha| \to \infty$ of the regularity regions. Using the expressions (7.44), (7.50), (7.51) and (7.63), we represent the total field as

$$u^t(x, y) = \widehat{u}^{\mathrm{inc}}(x, y) \mp \frac{i}{2} \int\limits_{-\infty}^{\infty} H_0^{(1)}(kR_0) \frac{\partial u^t(x', 0)}{\partial y'} \mathrm{d}x', \quad -\infty < x < \infty, \tag{7.64}$$

where

$$\frac{\partial u^t(x',0)}{\partial y'} = -\frac{\sqrt{k\cos\theta_0 - k}}{2\pi i}\int\limits_{-\infty+ib}^{\infty+ib}\frac{\sqrt{t+k}}{(t-k\cos\theta_0)}e^{-itx'}\mathrm{d}t \tag{7.65}$$

and $k''\cos\theta_0 < b < k''$. Then, for evaluation of the integral (7.64), we use the expression of the Fourier transform for the Hankel function as

$$\frac{1}{\sqrt{2\pi}}\int\limits_{-\infty}^{\infty}H_0^{(1)}(k\sqrt{x^2+y^2})e^{i\alpha x}\mathrm{d}x = \frac{\sqrt{2/\pi}\,e^{-\sqrt{\alpha^2-k^2}|y|}}{i\sqrt{\alpha^2-k^2}},$$

where $-k'' < \alpha < k''$, and finally, represent the total field as follows

$$u^t(x,y) = \widehat{u}^{\text{inc}}(x,y) \mp \frac{\sqrt{k-k\cos\theta_0}}{2\pi}\int\limits_{-\infty+ib}^{\infty+ib}\frac{e^{-i\alpha x \mp \sqrt{\alpha^2-k^2}y}}{(\alpha-k\cos\theta_0)\sqrt{\alpha-k}}d\alpha. \tag{7.66}$$

Here, the upper sign corresponds to $y > 0$ and the lower sign is for $y < 0$.

(b) Dual integral equations method

This method is often used for reducing the wave diffraction problem to the functional equations of the Wiener–Hopf type. Here, we briefly discuss the use of this method to reduce the diffraction problem to the Wiener–Hopf equation.

Let us turn to the wave diffraction problem (7.33), (7.34). Here, the total and incident fields are given by the expressions (7.35) and (7.32) respectively. The unknown diffracted field we represent through the Fourier integral as

$$u(x,y) = \pm\frac{1}{\sqrt{2\pi}}\int\limits_{-\infty}^{\infty}A(\alpha)\begin{Bmatrix} e^{-\gamma y}, & y>0 \\ e^{\gamma y}, & y<0 \end{Bmatrix}e^{-i\alpha x}\mathrm{d}x, \tag{7.67}$$

where the upper sign corresponds to $y > 0$ and the lower one is for $y < 0$; $A(\alpha)$ is the unknown function which is regular in the strip $\tau_- < \tau < \tau_+$; τ_-, τ_+ are to be determined.

Using diffracted field representation (7.67) as well as the boundary (7.34) and continuity conditions, we derive the dual integral equations that are given in the partial semi-infinite intervals as

$$\frac{1}{\sqrt{2\pi}}\int\limits_{-\infty}^{\infty}A(\alpha)e^{-i\alpha x}\mathrm{d}x = 0, \quad 0 < x < \infty, \tag{7.68}$$

$$\frac{1}{\sqrt{2\pi}}\int\limits_{-\infty}^{\infty} \gamma A(\alpha)e^{-i\alpha x}\mathrm{d}x = p(x), \quad -\infty < x < 0, \tag{7.69}$$

where $p(x) = \partial u^i(x,0)/\partial y = -ik \sin\theta_0 e^{-ikx\cos\theta_0}$.

Let us supplement these intervals with the infinite ones and rewrite the dual integral equations in the form

$$\frac{1}{\sqrt{2\pi}}\int\limits_{-\infty}^{\infty} A(\alpha)e^{-i\alpha x}\mathrm{d}x = q_-(x), \quad -\infty < x < \infty, \tag{7.70}$$

$$\frac{1}{\sqrt{2\pi}}\int\limits_{-\infty}^{\infty} \gamma A(\alpha)e^{-i\alpha x}\mathrm{d}x = h_+(x) + p_-(x), \quad -\infty < x < \infty. \tag{7.71}$$

Here, $q_-(x)$, $h_+(x)$ are the new unknown functions,

$$q_-(x) = \begin{cases} 0, & \text{for } x > 0, \\ q(x), & \text{for } x < 0, \end{cases}$$

$$h_+(x) = \begin{cases} h(x), & \text{for } x > 0, \\ 0, & \text{for } x < 0, \end{cases}$$

$$p_-(x) = \begin{cases} 0, & \text{for } x > 0, \\ p(x), & \text{for } x < 0. \end{cases} \tag{7.72}$$

At the edge $x \to 0$ of the semi-plane, the behavior of functions (7.72) looks like: $q(x) \sim x^{1/2}$, $h(x) \sim x^{-1/2}$, and $q(x) \sim e^{-ikx\cos\theta_0}$ if $x \to -\infty$ and $h(x) \sim e^{ikx}$ if $x \to +\infty$ are their asymptotic at the infinity.

Let us apply the Fourier integral transform to the Eqs. (7.70), (7.71) and reduce them to the Wiener–Hopf equation as

$$\sqrt{\alpha^2 - k^2}Q^-(\alpha) = H^+(\alpha) - \frac{1}{\sqrt{2\pi}}\frac{k\sin\theta_0}{\alpha - k\cos\theta_0}. \tag{7.73}$$

Here,

$$Q^-(\alpha) = \frac{1}{\sqrt{2\pi}}\int\limits_{-\infty}^{0} q(x)e^{i\alpha x}\mathrm{d}x, \tag{7.74}$$

$$H^+(\alpha) = \frac{1}{\sqrt{2\pi}}\int\limits_{0}^{\infty} h(x)e^{i\alpha x}\mathrm{d}x. \tag{7.75}$$

From the upper estimations, we find that $Q^-(\alpha)$ and $H^+(\alpha)$ are the regular functions in the overlapping semi-planes $\tau < \tau_+ = k'' \cos\theta_0$ and $\tau > \tau_- = -k''$, where $-k'' < \tau < k'' \cos\theta_0$ is the joint strip of regularity, and $Q^-(\alpha) \sim \alpha^{-3/2}$, $H^+(\alpha) \sim \alpha^{-1/2}$, if $|\alpha| \to \infty$ in the regularity regions. Then, applying the factorization procedure of the kernel function, we rewrite Eq. (7.73) as

$$\sqrt{\alpha - k}Q^-(\alpha) = \frac{H^+(\alpha)}{\sqrt{\alpha + k}} - \frac{1}{\sqrt{2\pi}} \frac{k \sin\theta_0}{(\alpha - k\cos\theta_0)\sqrt{\alpha + k}}. \tag{7.76}$$

The left part of this equation is the regular function in the lower half-plane $\tau < k'' \cos\theta_0$, and the first term in the right part is the regular function in the upper semi-plane $\tau > -k''$; they tend to zero as α^{-1} if $|\alpha| \to \infty$ in the regularity regions. The second term in the right part of Eq. (7.76) is the regular function in the strip $-k'' < \tau < k'' \cos\theta_0$ and tends to zero in this strip as $\alpha^{-3/2}$. Out of this strip of regularity, the known term of this equation involves the simple pole singularity and the branch point. Applying the decomposition procedure (7.21), (7.22) and using the Liouville theorem, we find the solution of the Wiener–Hopf equation (7.76) in the form

$$H^+(\alpha) = \frac{k \sin\theta_0 \sqrt{\alpha + k}}{\sqrt{2\pi}} \frac{1}{2\pi i} \int_{-\infty + i\varepsilon_1}^{\infty + i\varepsilon_1} \frac{dt}{(t - k\cos\theta_0)\sqrt{t + k}(t - \alpha)},$$

$$Q^-(\alpha) = \frac{k \sin\theta_0}{\sqrt{\alpha - k}\sqrt{2\pi}} \frac{1}{2\pi i} \int_{-\infty + i\varepsilon_2}^{\infty + i\varepsilon_2} \frac{dt}{(t - k\cos\theta_0)\sqrt{t + k}(t - \alpha)}. \tag{7.77}$$

Here, $-k'' < \varepsilon_1 < \tau < \varepsilon_2 < k'' \cos\theta_0$.

Complementing the integration contour with a semicircle of infinite radius into the upper half-plane and using the residues theorem for evaluation of the integrals (7.77), we find that

$$H^+(\alpha) = \frac{k \sin\theta_0}{\sqrt{2\pi}(\alpha - k\cos\theta_0)} \left[1 - \frac{\sqrt{\alpha + k}}{\sqrt{k\cos\theta_0 + k}} \right],$$

$$Q^-(\alpha) = -\frac{k \sin\theta_0}{\sqrt{2\pi}\sqrt{k\cos\theta_0 + k}(\alpha - k\cos\theta_0)\sqrt{\alpha - k}}. \tag{7.78}$$

These solutions are the regular functions in the semi-planes $\tau > -k''$, $\tau < k'' \cos\theta_0$ respectively and have the required asymptotic if $|\alpha| \to \infty$. Taking into account that $A(\alpha) = Q^-(\alpha)$, we find that

$$u^t = u^{\text{inc}} \mp \frac{k \sin\theta_0}{2\pi\sqrt{k\cos\theta_0 + k}} \int_{-\infty}^{\infty} \frac{e^{-i\alpha x \mp \sqrt{\alpha^2 - k^2} y}}{(\alpha - k\cos\theta_0)\sqrt{\alpha - k}} dx. \tag{7.79}$$

Here, the upper sign corresponds to $y > 0$ and the lower sign is for $y < 0$.

(c) Jones technique

Let us turn to the boundary value problem (7.33), (7.34). For further convenience, we represent the total and incident fields as in (7.35) and (7.32) respectively. Applying the Fourier transform to the Helmholtz equation (7.33) with respect to the variable x, we derive that

$$\frac{\mathrm{d}^2 U(\alpha, y)}{\partial y^2} + \gamma^2 U(\alpha, y) = 0. \tag{7.80}$$

Here, $U(\alpha, y)$ is the Fourier transform of the diffracted field,

$$U(\alpha, y) = \frac{1}{\sqrt{2\pi}} \int_{-\infty}^{\infty} u(x, y) e^{i\alpha x} \mathrm{d}x, \tag{7.81}$$

where $\tau_- < \tau < \tau_+$.

Let us introduce the Fourier integrals as

$$U^-(\alpha, y) = \frac{1}{\sqrt{2\pi}} \int_{-\infty}^{0} u(x, y) e^{i\alpha x} \mathrm{d}x,$$

$$U^+(\alpha, y) = \frac{1}{\sqrt{2\pi}} \int_{0}^{\infty} u(x, y) e^{i\alpha x} \mathrm{d}x. \tag{7.82}$$

These integrals are the regular functions in the semi-planes $\tau < \tau_+ = k'' \cos\theta_0$, $\tau > \tau_- = -k''$. Using this notation, we represent (7.81) in the form

$$U(\alpha, y) = U^-(\alpha, y) + U^+(\alpha, y). \tag{7.83}$$

Let us represent the solution of Eq. (7.80) as

$$U(\alpha, y) = \begin{cases} A(\alpha) e^{-\gamma y}, & y > 0, \\ B(\alpha) e^{\gamma y}, & y < 0. \end{cases} \tag{7.84}$$

Here, $A(\alpha)$, $B(\alpha)$ are the unknown functions of the complex variable α, regular in the strip $-k'' < \tau < k'' \cos\theta_0$.

Using the expressions (7.82)–(7.84), we get that

$$U'^-(\alpha, +0) + U'^+(\alpha, +0) = -\gamma A(\alpha),$$
$$U'^-(\alpha, -0) + U'^+(\alpha, -0) = \gamma B(\alpha), \tag{7.85}$$

where

$$U'^{-}(\alpha, y) = \frac{1}{\sqrt{2\pi}} \int_{-\infty}^{0} u'_y(x, y) e^{i\alpha x} dx, \tag{7.86}$$

$$U'^{+}(\alpha, y) = \frac{1}{\sqrt{2\pi}} \int_{0}^{\infty} u'_y(x, y) e^{i\alpha x} dx. \tag{7.87}$$

The functions $U'^{-}(\alpha, y)$, $U'^{+}(\alpha, y)$ are regular in the semi-planes $\tau < k'' \cos\theta_0$, $\tau > -k''$ respectively. Using the boundary condition (7.34), we arrive at

$$U'^{-}(\alpha, \pm 0) = U'^{-}(\alpha, 0) = \frac{k \sin\theta_0}{\sqrt{2\pi}(\alpha - k\cos\theta_0)}. \tag{7.88}$$

Taking into account that $U'^{+}(\alpha, +0) = U'^{+}(\alpha, -0) = U'^{+}(\alpha, 0)$, we represent the expression (7.84) in the form as

$$U(\alpha, y) = \mp \frac{e^{\mp \gamma y}}{\gamma} \left[U'^{+}(\alpha, 0) + \frac{k \sin\theta_0}{\sqrt{2\pi}(\alpha - k\cos\theta_0)} \right], \tag{7.89}$$

where the upper sign corresponds to $y > 0$ and the lower sign is for $y < 0$. We obtain the desirable Wiener–Hopf equation using the continuity condition of the total field and formulas (7.89):

$$\frac{U'^{+}(\alpha)}{\sqrt{\alpha + k}} + \frac{k \sin\theta_0}{\sqrt{2\pi}\sqrt{\alpha + k}(\alpha - k\cos\theta_0)} = -\sqrt{\alpha - k} J^{-}(\alpha)/2. \tag{7.90}$$

Here, $J^{-}(\alpha)$ is the unknown Fourier transform of the difference for the field potential at the opposite sides of the half-plane:

$$J^{-}(\alpha) = \frac{1}{\sqrt{2\pi}} \int_{-\infty}^{0} [u(x, +0) - u(x, -0)] e^{i\alpha x} dx; \tag{7.91}$$

$U'^{+}(\alpha)$, $J^{-}(\alpha)$ are the regular functions in the semi-planes $\tau > -k''$, $\tau < k'' \cos\theta_0$ respectively and tend to zero in the regularity regions as: $U'^{+}(\alpha) \sim \alpha^{-1/2}$, $J^{-}(\alpha) \sim \alpha^{-3/2}$ if $|\alpha| \to \infty$.

Equation (7.90) is valid for $-k'' < \tau < k'' \cos\theta_0$. Further, we apply the decomposition procedure (7.21), (7.22) for the solution of Eq. (7.90) and represent it in form

$$U'^{+}(\alpha) = \frac{k \sin\theta_0}{\sqrt{2\pi}(\alpha - k\cos\theta_0)} \left[\frac{\sqrt{\alpha + k}}{\sqrt{k\cos\theta_0 + k}} - 1 \right],$$

$$J^{-}(\alpha) = -\frac{\sqrt{2/\pi}k \sin\theta_0}{\sqrt{k\cos\theta_0) + k}(\alpha - k\cos\theta_0)\sqrt{\alpha - k}}. \tag{7.92}$$

Using the solution of the Wiener–Hopf equation (7.92), we find the total field representation as

$$u^t(x, y) = u^{\text{inc}} \mp \frac{k \sin\theta_0}{2\pi\sqrt{k\cos\theta_0 + k}} \int_{-\infty}^{\infty} \frac{e^{-i\alpha x \mp \sqrt{\alpha^2 - k^2} y}}{(\alpha - k\cos\theta_0)\sqrt{\alpha - k}} \mathrm{d}x. \tag{7.93}$$

The expression (7.93) coincides with the obtained in (7.79).

In this section, we have briefly considered the main methods used to reduce the boundary value problems of diffraction theory to the Wiener–Hopf equations. This consideration is absolutely necessary because it allows to see the general picture of finding the solutions based on the analysis of the analytical properties of their Fourier transforms. The choice of any of these methods to solve the physical problems, in our opinion, is the decision of the author. In our research, we use the Jones method which involves reformulating the original boundary value problem directly to find the necessary transformants and make deeper use of their analytical properties.

7.3 Plane SH-Wave Diffraction from the Finite Crack on the Interface of Materials

7.3.1 Short Overview

Since the defects often appear on the interfaces of different materials, the problem of the elastic waves diffraction from a finite crack formed on the plane interface of two elastic half-spaces is an important subject of nondestructive examination. The aim of this analysis is to study the perturbation introduced by a crack that is located on the interface of materials in the wave diffraction pattern. Indeed, the understanding of the effects induced by the crack in a broad frequency bands could help to optimize the inspection techniques and justify the procedures used to detect the crack. We consider a two-dimensional model and assume that it is illuminated by a plane elastic transverse wave of horizontal polarization or, in other words, by SH-wave. This is the simplest type of the excitation phenomenon that is fully described using a scalar function. This greatly simplifies the problem, as it allows us to obtain its mathematically correct solution in a wide frequency range, to conduct deeper analytical researches, and to obtain asymptotic expressions for the estimation of the fields of displacements and stresses.

The solutions of the problem of diffraction of shear SH-waves by a crack located on the plane interface of two isotropic homogeneous materials are analyzed in [12–17]. In these works, the authors study the phenomenon of scattering of waves by short cracks and the cracks whose width is comparable with wavelength by the method of integral equations. The problems of diffraction of SH-waves by systems of cracks located on the plane interfaces of layered materials are analyzed in [18–36].

At present, the problem of diffraction of waves by wide cracks from several to several dozens of wavelengths is investigated quite poorly. This is explained by the fact that, in the indicated frequency band, the method of integral equations and numerical methods lose their efficiency due to the problems encountered in computations. At the same time, the asymptotic methods traditionally used for the analysis of the fields scattered by large cracks fail to give adequate description of the fine effects of multiple reflections from the crack edges making significant contribution to the diffraction processes. The Wiener–Hopf technique proves to be one of the most adequate methods for the investigation of the diffraction of waves by cracks in the high-frequency region. It is frequently used for the investigation of diffraction of elastic waves by the cracks in the infinite media. Thus, the exact solutions of the plane problem of diffraction of elastic waves by semi-infinite cracks were obtained in [37–39]. In [40, 41], this method was also applied for the solution of wave diffraction of plane waves by finite cracks. Here, we study the SH-wave diffraction by a finite crack formed on the plane boundary of two isotropic elastic half-spaces and determine the scattered field in a broad frequency band. For the solution of this problem and in order to identify the cracks located on the boundary, we apply the modified Wiener–Hopf approach in terms of the Jones formulation. The similar approach was proposed early in the case of diffraction of electromagnetic waves [42]. SH-wave diffraction from the interface crack were analyzed in [43–47]. The authors would like to express their gratitude to M. Voytko, Ph.D. for agreeing to include in the monograph a number of jointly obtained results, as well as to Ya Kulynych, Ph.D. for his assistance in performing calculations.

7.3.2 Formulation of the Problem

Let us consider the interface crack in the Cartesian coordinates occupying the region $\{(x, y, z)| - L < x < 0,\ y = 0,\ z \in (-\infty, \infty)\}$ between the plane joint of two homogeneous and isotropic semi-spaces. Let this crack be illuminated by the transvers elastic SH-wave that is propagated in the plane xy, and its displacement vector $\mathbf{u}^{inc}(\equiv \vec{e}_z u^{inc}(x, y))$ is oriented along the axis Oz. The incident wave is assumed to have time harmonic variation $e^{-i\omega t}$; in the rest of the chapter, we shall drop the dependence on time. According to the given orientation of the sounding wave, the diffracted wave is also SH-wave. The geometrical scheme of the interface crack is shown in Fig. 7.1. Let the upper $(y > 0)$ and lower $(y < 0)$half-spaces be characterized by the densities ρ_1 and ρ_2, and the Lamé parameters μ_1 and μ_2, respectively. The wave numbers for the regions $y > 0$ and $y < 0$ are given as $k_l(\equiv \omega\sqrt{\rho_l/\mu_l})$, where $k_l = k_l' + ik_l''$, k_l', $k_l'' > 0$, $k_l' >> k_l''$; $l = 1, 2$.

Let us represent the total field $u^t(x, y)$ as

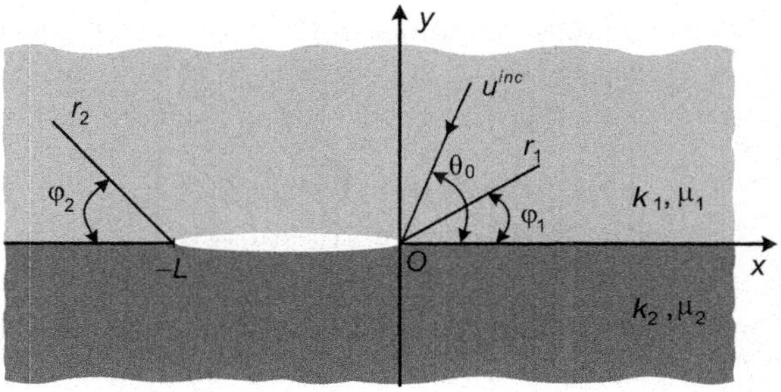

Fig. 7.1 Geometry of the problem

$$u^t(x, y) = u(x, y) + \begin{cases} u^{\text{inc}}(x, y) + R\, e^{-ik_1(x\cos\theta_0 - y\sin\theta_0)}, & y > 0, \\ T e^{-ik_2(x\cos\theta_1 + y\sin\theta_1)}, & y < 0. \end{cases} \tag{7.94}$$

Here, $u(x, y)$ is the unknown diffracted field $\mathbf{u}(\equiv \vec{e}_z u(x, y))$; the terms in parenthesis represent the total displacement field in the perfect join of semi-spaces illuminated by the plane wave

$$u^{\text{inc}}(x, y) = e^{-ik_1(x\cos\theta_0 + y\sin\theta_0)}; \tag{7.95}$$

R and T are reflection and transition coefficients of the plane wave (7.95):

$$R(\theta_0) = \frac{\mu_1 k_1 \sin\theta_0 - \mu_2 k_2 \sin\theta_1}{\mu_1 k_1 \sin\theta_0 + \mu_2 k_2 \sin\theta_1}, \tag{7.96}$$

$$T(\theta_0) = \frac{2\mu_1 k_1 \sin\theta_0}{\mu_1 k_1 \sin\theta_0 + \mu_2 k_2 \sin\theta_1}, \tag{7.97}$$

where θ_0 is the incident angle of the plane wave ($0 < \theta_0 < \pi/2$), θ_1 is the refraction angle; $k_2 \cos\theta_1 = k_1 \cos\theta_0$ is the Snell law.

The unknown function $u(x, y)$ is the solution of the boundary value problem for Helmholtz equation

$$\frac{\partial^2 u}{\partial x^2} + \frac{\partial^2 u}{\partial y^2} + k_l^2 u = 0, \tag{7.98}$$

where $l = 1$ if $y > 0$ and $l = 2$ if $y < 0$.

The function $u(x, y)$ should satisfy:

- the boundary conditions which ensure the absence of stresses on the crack faces $\tau_{zy}(x, \pm 0) = 0$:

$$\mu_1 \frac{\partial}{\partial y} u^t(x, +0) = \mu_2 \frac{\partial}{\partial y} u^t(x, -0) = 0 \text{ for } x \in (-L, 0); \tag{7.99}$$

– the continuity conditions of the displacement and stress fields on the join plane:

$$\left.\begin{array}{l} u^t(x,+0) - u^t(x,-0) = 0 \\ \mu_1 \frac{\partial}{\partial y} u^t(x,+0) - \mu_2 \frac{\partial}{\partial y} u^t(x,-0) = 0 \end{array}\right\} \text{ for } x \in (-\infty,-L) \cup (0,\infty). \quad (7.100)$$

In order to ensure the unicity of the solution of the boundary value problem (7.98)–(7.100), the unknown diffracted field should satisfy the edge condition (7.36), (7.37) as well as the condition at the infinity (7.38). For any $y = \text{const}$ the condition at the infinity can be reformulated as:

$$u(x,y),\ u'_y(x,y) = O(e^{-\chi|x|}) \text{ if } x \to \pm\infty. \quad (7.101)$$

Here, $\chi = \min(k''_1, k''_2)$; without the loss of generality, we suppose that $k''_1 < k''_2$. Further, we reduce the boundary value problem (7.98)–(7.100) to the Wiener–Hopf equation. For this purpose, let us introduce the Fourier transform of the diffracted field $u(x, y)$ as:

$$U(\alpha,y) = \frac{1}{\sqrt{2\pi}} \int_{-\infty}^{\infty} u(x,y) e^{i\alpha x} \mathrm{d}x, \quad (7.102)$$

where $U(\alpha, y)$ is the unknown function of the complex variable $\alpha = \operatorname{Re}\alpha + i \operatorname{Im}\alpha$ $(\equiv \sigma + i\tau)$, regular in the strip of the complex plane α that encompasses the real axis.

Let us represent this function as

$$U(\alpha,y) = e^{-i\alpha L} U^-(\alpha,y) + \Phi(\alpha,y) + U^+(\alpha,y). \quad (7.103)$$

Here,

$$U^-(\alpha,y) = \frac{1}{\sqrt{2\pi}} \int_{-\infty}^{-L} u(x,y) e^{i\alpha(x+L)} \mathrm{d}x,$$

$$U^+(\alpha,y) = \frac{1}{\sqrt{2\pi}} \int_{0}^{\infty} u(x,y) e^{i\alpha x} \mathrm{d}x,$$

$$\Phi(\alpha,y) = \frac{1}{\sqrt{2\pi}} \int_{-L}^{0} u(x,y) e^{i\alpha x} \mathrm{d}x. \quad (7.104)$$

Taking into account the asymptotic estimation (7.101), it is found that $U^+(\alpha, y)$, $U^-(\alpha, y)$ are regular functions of the complex variable α in the overlapping semi-planes $\tau > -k_1''$, $\tau < k_1'' \cos\theta_0$; $\Phi(\alpha, y)$ is the entire function. For further convenience, let us introduce the strip of regularity $\prod : (\tau_- < \tau < \tau_+)$, where $\tau_- = -k_1''$, $\tau_+ = k_1'' \cos\theta_0$. In the strip $\prod$, the function (7.102) satisfies the equation

$$\frac{d^2U(\alpha, y)}{dy^2} - \gamma_l^2 U(\alpha, y) = 0, \tag{7.105}$$

the solution of which we represent as

$$U(\alpha, y) = \begin{cases} A(\alpha)e^{-\gamma_1 y} \text{ if } y > 0, \\ B(\alpha)e^{\gamma_2 y} \text{ if } y < 0. \end{cases} \tag{7.106}$$

Here, $A(\alpha)$, $B(\alpha)$ are the unknown functions regular in the strip $\prod$; $\gamma_l = \sqrt{\alpha^2 - k_l^2} = -i\sqrt{k_l^2 - \alpha^2}$, $\mathrm{Re}\gamma_l \geq 0$, $l = 1, 2$.

Let us consider Fourier integrals as follows:

$$U'^-(\alpha, \pm 0) = \frac{1}{\sqrt{2\pi}} \int_{-\infty}^{-L} u_y'(x, \pm 0)e^{i\alpha(x+L)}\mathrm{d}x, \tag{7.107}$$

$$U'^+(\alpha, \pm 0) = \frac{1}{\sqrt{2\pi}} \int_{0}^{\infty} u_y'(x, \pm 0)e^{i\alpha x}\mathrm{d}x\ , \tag{7.108}$$

$$\Phi'(\alpha, \pm 0) = \frac{1}{\sqrt{2\pi}} \int_{-L}^{0} u_y'(x, \pm 0)e^{i\alpha x}\mathrm{d}x. \tag{7.109}$$

Here, $U'^-(\alpha, \pm 0)$ and $U'^+(\alpha, \pm 0)$ are regular functions of the complex variable α in the semi-planes $\tau < \tau_+$ and $\tau > \tau_-$ respectively, and $\Phi'(\alpha, \pm 0)$ is the entire function.

Let us differentiate the expression (7.106) with respect to the variable y, and taking into account notations (7.103), (7.107), and (7.108), (7.109) we find if $y \to \pm 0$, that

$$\begin{aligned} &e^{-i\alpha L}U'^-(\alpha, +0) + \Phi'(\alpha, +0) + U'^+(\alpha, +0) = -\gamma_1 A(\alpha), \\ &e^{-i\alpha L}U'^-(\alpha, -0) + \Phi'(\alpha, -0) + U'^+(\alpha, -0) = \gamma_2 B(\alpha). \end{aligned} \tag{7.110}$$

Further, we apply the Fourier transform to the boundary condition (7.99) with the result

$$\mu_1 \Phi'(\alpha, +0)$$

$$+ \mu_1 \frac{1}{\sqrt{2\pi}} \frac{\partial}{\partial y} \int_{-L}^{0} [u^{\mathrm{inc}}(x, y) + R\, e^{-ik_1(x\cos\theta_0 - y\sin\theta_0)}]\, e^{i\alpha x}\, \mathrm{d}x|_{y=+0} = 0,$$

$$\mu_2 \Phi'(\alpha, -0)$$

$$+ \mu_2 \frac{1}{\sqrt{2\pi}} \frac{\partial}{\partial y} \int_{-L}^{0} [T e^{-ik_2(x\cos\theta_1 + y\sin\theta_1)}]\, e^{i\alpha x}\, \mathrm{d}x|_{y=-0} = 0. \qquad (7.111)$$

The simple evaluation of these integrals leads to the explicit representation of the wanted functions as

$$\Phi'(\alpha, +0) = \frac{(1-R)k_1 \sin\theta_0}{\sqrt{2\pi}\,(\alpha - k_1\cos\theta_0)} \left[1 - e^{ik_1 L\cos\theta_0 - i\alpha L}\right],$$

$$\Phi'(\alpha, -0) = \frac{T k_1 \sqrt{(k_2/k_1)^2 - \cos^2\theta_0}}{\sqrt{2\pi}\,(\alpha - k_1\cos\theta_0)} \left[1 - e^{ik_1 L\cos\theta_0 - i\alpha L}\right]. \qquad (7.112)$$

Let us substitute the expressions (7.112) into the Eqs. (7.110) with the result:

$$e^{-i\alpha L} U'^{-}(\alpha, +0) + \frac{(1-R)k_1 \sin\theta_0}{\sqrt{2\pi}\,(\alpha - k_1\cos\theta_0)} \left[1 - e^{ik_1 L\cos\theta_0 - i\alpha L}\right]$$
$$+ U'^{-}(\alpha, +0) = -\gamma_1 A(\alpha),$$
$$e^{-i\alpha L} U'^{-}(\alpha, -0) + \frac{T k_1 \sqrt{(k_2/k_1)^2 - \cos^2\theta_0}}{\sqrt{2\pi}\,(\alpha - k_1\cos\theta_0)} \left[1 - e^{ik_1 L\cos\theta_0 - i\alpha L}\right]$$
$$+ U'^{-}(\alpha, -0) = \gamma_2 B(\alpha). \qquad (7.113)$$

Then, we rewrite these equations in form as

$$\Psi^{(+)}(\alpha, +0) + e^{-i\alpha L} \Psi^{-}(\alpha, +0) = -\gamma_1 A(\alpha),$$
$$\Psi^{(+)}(\alpha, -0) + e^{-i\alpha L} \Psi^{-}(\alpha, -0) = \gamma_2 B(\alpha). \qquad (7.114)$$

Here, we introduce the new notations:

$$\Psi^{(+)}(\alpha, +0) = U'^{+}(\alpha, +0) + \frac{N}{\alpha - k_1\cos\theta_0},$$
$$\Psi^{-}(\alpha, +0) = U'^{-}(\alpha, +0) - \frac{N e^{ik_1 L\cos\theta_0}}{\alpha - k_1\cos\theta_0},$$
$$\Psi^{(+)}(\alpha, -0) = U'^{+}(\alpha, -0) + \frac{N_1}{\alpha - k_2\cos\theta_1},$$
$$\Psi^{-}(\alpha, -0) = U'^{-}(\alpha, -0) - \frac{N_1 e^{ik_2 L\cos\theta_1}}{\alpha - k_2\cos\theta_1}, \qquad (7.115)$$

where $N = (2\pi)^{-1/2}k_1(1 - R)\sin\theta_0$, $N_1 = (2\pi)^{-1/2}k_2 T\sin\theta_1$; functions $\Psi^{(+)}(\alpha, \pm 0)$ are regular in the upper half-plane $\tau > \tau_-$, except the point $\alpha = k_1\cos\theta_0 = k_2\cos\theta_1$, where they have the simple pole; functions $\Psi^-(\alpha, \pm 0)$ are regular in the lower half-plane $\tau < \tau_+$.

Using the boundary conditions (7.99), the second continuity condition (7.100) and Eqs. (7.114) we derive

$$\begin{aligned} B(\alpha) &= -\frac{\mu_1\gamma_1}{\mu_2\gamma_2}A(\alpha),\\ \Psi^{(+)}(\alpha, +0) &= \frac{\mu_2}{\mu_1}\Psi^{(+)}(\alpha, -0) = \Psi^{(+)}(\alpha),\\ \Psi^-(\alpha, +0) &= \frac{\mu_2}{\mu_1}\Psi^-(\alpha, -0) = \Psi^-(\alpha). \end{aligned} \tag{7.116}$$

Taking into account Eqs. (7.114) and (7.116), we finally rewrite the field representation (7.106) as follows

$$U(\alpha, y) = \begin{cases} -\frac{1}{\gamma_1}\left[\Psi^{(+)}(\alpha) + e^{-i\alpha L}\Psi^-(\alpha)\right]e^{-\gamma_1 y}, & y > 0,\\ \frac{\mu_1}{\mu_2}\frac{1}{\gamma_2}\left[\Psi^{(+)}(\alpha) + e^{-i\alpha L}\Psi^-(\alpha)\right]e^{\gamma_2 y}, & y < 0. \end{cases} \tag{7.117}$$

Here, $U(\alpha, y)$ is the regular function in the strip $\prod$.

Then, using the first continuity condition (7.100) and field representation (7.117), leads us to the functional equation as

$$J_1(\alpha) + \frac{M(\alpha)}{\gamma_1}\left[\Psi^{(+)}(\alpha) + e^{-i\alpha L}\Psi^-(\alpha)\right] = 0, \tag{7.118}$$

where

$$J_1(\alpha) = \frac{1}{\sqrt{2\pi}}\int_{-L}^{0}[u(x, +0) - u(x, -0)]e^{i\alpha x}dx, \tag{7.119}$$

$$M(\alpha) = \frac{\mu_1\sqrt{\alpha^2 - k_1^2} + \mu_2\sqrt{\alpha^2 - k_2^2}}{\mu_2\sqrt{\alpha^2 - k_2^2}}. \tag{7.120}$$

Equation (7.118) is valid in the strip $\prod$; $\Psi^-(\alpha)$, $\Psi^{(+)}(\alpha)$, $J_1(\alpha)$ are unknown functions. In the regularity regions $\Psi^{(+)}(\alpha) = O(\alpha^{-1/2})$, $\Psi^-(\alpha) = O(\alpha^{-1/2})$ if $|\alpha| \to \infty$; $J_1(\alpha)$ is the entire function, $J_1(\alpha) = O(\alpha^{-3/2})$ if $\tau < \tau_+$, and $|\alpha| \to \infty$. Each term of Eq. (7.118) tends to zero in $\prod$ as $\alpha^{-3/2}$ if $|\alpha| \to \infty$. Therefore, the functional equation (7.118) is the Wiener–Hopf equation.

7.3.3 Factorization of the Kernel Function

To solve the Wiener–Hopf equation (7.118), we must address one of the key problems–the factorization of the kernel function (7.120):

$$M(\alpha) = M_+(\alpha)\ M_-(\alpha), \tag{7.121}$$

where the split functions $M_+(\alpha)$ and $M_-(\alpha)$ are regular in the overlapping semi-planes $\tau > -k_1''$ and $\tau < k_1''$ respectively and they have no zeroes in the regularity regions, and in these regions $M_\pm(\alpha) = O(1)$ if $|\alpha| \to \infty$.

Note that the function (7.120) is regular in $\prod$; outside of this, strip the kernel function has the branch points at $\alpha = \pm k_{1(2)}$. Let us represent this function as

$$M(\alpha) = \frac{\mu_1 + \mu_2}{\mu_2} \tilde{M}(\alpha), \tag{7.122}$$

where

$$\tilde{M}(\alpha) = \frac{\mu_2}{\mu_1 + \mu_2} + \frac{\mu_1}{\mu_1 + \mu_2} \frac{\sqrt{k_1^2 - \alpha^2}}{\sqrt{k_2^2 - \alpha^2}}. \tag{7.123}$$

Further, using the relations (7.122), (7.123), we represent the split functions as

$$M_\pm(\alpha) = \sqrt{\frac{\mu_1 + \mu_2}{\mu_2}} \exp\left[I^\pm(\alpha)\right]. \tag{7.124}$$

where

$$I^\pm(\alpha) = \pm \frac{1}{2\pi i} \int\limits_{L_\pm} \ln \tilde{M}(\nu) \frac{d\nu}{\nu - \alpha}, \tag{7.125}$$

L_+, L_- are the integration contours in the strip $\prod$ that are located along the real axis:

$$L_+ : \{\nu \in (-\infty + i\varepsilon_1, \infty + i\varepsilon_1)\}, L_- : \{\nu \in (-\infty + i\varepsilon_2, \infty + i\varepsilon_2)\},$$

$$\tau_- < \varepsilon_1 < \tau < \tau_+, \tau_- < \tau < \varepsilon_2 < \tau_+.$$

In order to simplify the expression (7.125), let us consider the logarithm of the kernel function $\ln \tilde{M}(\alpha)$ in the Riemann surface [48], formed by cuts of the complex plane α between the points $\pm k_1$ and $\pm k_2$ as in Fig. 7.2. Let us consider the integral $I^+(\alpha)$. For its evaluation, let us deform the contour L_+, enclosing it in the semicircle L_+^∞ of the infinity radius in the lower half-plane, and embrace the branch cut by the

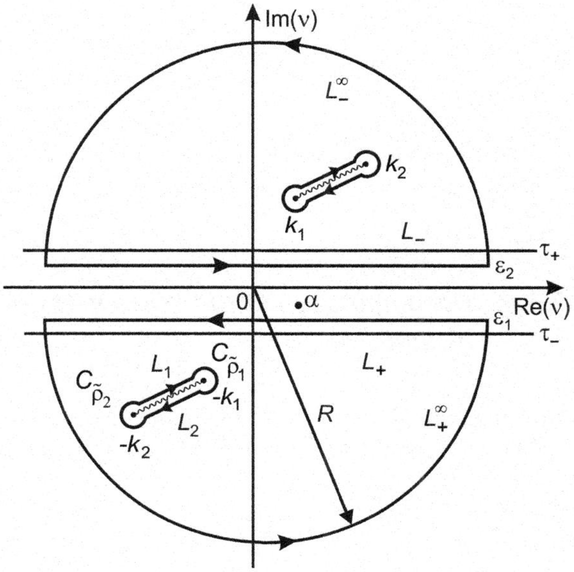

Fig. 7.2 Contours of integration

closed contour as is shown in Fig. 7.2. Then, we consider the region Ω_1 of complex plane bounded by contours $L_+, L_+^\infty, L_1, L_2, C_{\tilde\rho_1}, C_{\tilde\rho_2}$. In Ω_1, the integrand (7.125) is the single-valued and regular function. Taking into account that $\int_{L_+^\infty} \to 0$ if $R \to \infty$ and $\int_{C_{\tilde\rho_{1(2)}}} \to 0$ if $\tilde\rho_{1(2)} \to 0$, and using the properties of the integrals along the closed contour from the regular functions, we arrive at

$$I^+(\alpha) = \frac{1}{2\pi i}\int_{-k_1}^{-k_2} \ln\left(\frac{\mu_2\sqrt{k_2^2-\nu^2}-\mu_1\sqrt{k_1^2-\nu^2}}{\mu_2\sqrt{k_2^2-\nu^2}+\mu_1\sqrt{k_1^2-\nu^2}}\right)\frac{d\nu}{\nu-\alpha}. \tag{7.126}$$

Making use the replacement of the variable $\nu = -\sqrt{(k_1^2-k_2^2)t+k_2^2}$, we get

$$I^+(\alpha) = \frac{(k_2^2-k_1^2)}{4\pi i}\int_0^1 \frac{\ln\left[\frac{i\mu_2\sqrt{t}+\mu_1\sqrt{1-t}}{i\mu_2\sqrt{t}-\mu_1\sqrt{1-t}}\right]}{\sqrt{(k_1^2-k_2^2)t+k_2^2}\left(\sqrt{(k_1^2-k_2^2)t+k_2^2}+\alpha\right)}dt. \tag{7.127}$$

Finally, we arrive at

$$M_+(\alpha) = \sqrt{\frac{\mu_1+\mu_2}{\mu_2}}\exp\left[\frac{(k_2^2-k_1^2)}{4\pi i}q_1(\alpha)\right], \tag{7.128}$$

where

$$q_1(\alpha) = \int\limits_0^1 \frac{\ln\left[\frac{i\mu_2\sqrt{t}+\mu_1\sqrt{1-t}}{i\mu_2\sqrt{t}-\mu_1\sqrt{1-t}}\right]}{\sqrt{\left(k_1^2-k_2^2\right)t+k_2^2}\left(\sqrt{\left(k_1^2-k_2^2\right)t+k_2^2}+\alpha\right)}\,dt. \tag{7.129}$$

The desirable representation of the function $M_-(\alpha)$ through the integral $I^-(\alpha)$ can be found by way of enclosing the integration contour L_- in the upper half-plane.

7.3.4 Decomposition of the Wiener–Hopf Equation. Integral Equations

Dividing both sides of Eq. (7.118) by the function $M_-(\alpha)/\sqrt{\alpha - k_1}$ which is regular and has no zeros in the lower half-plane $\tau < \tau_+$ we obtain

$$\frac{J_1(\alpha)\sqrt{\alpha-k_1}}{M_-(\alpha)} + \frac{M_+(\alpha)}{\sqrt{\alpha+k_1}}\Psi^{(+)}(\alpha) + \frac{M_+(\alpha)e^{-i\alpha L}}{\sqrt{\alpha+k_1}}\Psi^-(\alpha) = 0, \tag{7.130}$$

where all functions are regular and have no zeroes in the strip $\prod$.

The first term of Eq. (7.130) is a regular function in the lower half-plane with the asymptotic behavior as α^{-1} if $|\alpha| \to \infty$. The second term (7.130) is a regular function in the upper half-plane $\tau > \tau_-$ except the point $\alpha = k_1 \cos\theta_0$, where it has the simple pole; its asymptotic behavior is as α^{-1} if $|\alpha| \to \infty$. The third term allows for the singularities in the upper and lower half-planes. Taking this into account, let us apply the decomposition procedure (7.22) to the second and the third terms and rewrite the functional equation (7.130) as

$$\begin{aligned} &\frac{J_1(\alpha)\sqrt{\alpha-k_1}}{M_-(\alpha)} + \left[\frac{M_+(\alpha)}{\sqrt{\alpha+k_1}}\Psi^{(+)}(\alpha)\right]^- + \left[\frac{M_+(\alpha)}{\sqrt{\alpha+k_1}}e^{-i\alpha L}\Psi^-(\alpha)\right]^- \\ &= -\left[\frac{M_+(\alpha)}{\sqrt{\alpha+k_1}}\Psi^{(+)}(\alpha)\right]^+ - \left[\frac{M_+(\alpha)}{\sqrt{\alpha+k_1}}e^{-i\alpha L}\Psi^-(\alpha)\right]^+, \end{aligned} \tag{7.131}$$

where $[\ldots]^+$ and $[\ldots]^-$ are the functions regular in the upper $\tau > \tau_-$ and lower $\tau < \tau_+$ half-planes respectively.

Let us consider the functions in the right-hand part of Eq. (7.131). We represent the first term there as follows:

$$\left[\frac{M_+(\alpha)}{\sqrt{\alpha+k_1}}\Psi^{(+)}(\alpha)\right]^+ = \frac{1}{2\pi i}\int\limits_{-\infty+i\varepsilon_1}^{\infty+i\varepsilon_1} \frac{M_+(\nu)}{\sqrt{\nu+k_1}}\Psi^{(+)}(\nu)\frac{d\nu}{\nu-\alpha}. \tag{7.132}$$

Here, $\tau_- < \varepsilon_1 < \tau < \tau_+$. The integrand (7.132) is the regular function anywhere in the upper half-plane $\tau > \varepsilon_1$ except the points $\nu = \alpha$ and $\nu = k_1 \cos\theta_0$, where this function has simple poles. Taking into account that integrand (7.132) in the upper half-plane decays as ν^{-2} if $|\nu| \to \infty$, we deform the integration contour in the upper half-plane, enclosing it in the semicircle of the infinity radius and applying the residue theorem, we find that

$$\left[\frac{M_+(\alpha)}{\sqrt{\alpha+k_1}}\Psi^{(+)}(\alpha)\right]^+ = \frac{M_+(\alpha)}{\sqrt{\alpha+k_1}}\Psi^{(+)}(\alpha) - \frac{M_+(k_1\cos\theta_0)N}{\sqrt{k_1+k_1\cos\theta_0}(\alpha-k_1\cos\theta_0)}. \tag{7.133}$$

The function (7.133) is regular in the upper half-plane. Note, that the obvious singularity in the second term for the right-hand part in (7.133) if $\alpha = k_1\cos\theta_0$ is compensated by the singularity of the first term at this point.

Let us consider the second term in the right-hand part of Eq. (7.131):

$$\left[\frac{M_+(\alpha)}{\sqrt{\alpha+k_1}}e^{-i\alpha L}\Psi^-(\alpha)\right]^+ = \frac{1}{2\pi i}\int\limits_{-\infty+i\,\varepsilon_1}^{\infty+i\,\varepsilon_1}\frac{M_+(\nu)e^{-i\nu L}\Psi^-(\nu)}{\sqrt{\nu+k_1}}\frac{d\nu}{\nu-\alpha}. \tag{7.134}$$

The integral (7.134) is convergent and tends to zero as α^{-1} if $|\alpha| \to \infty$ in the upper half-plane. For the terms in the left-hand side of Eq. (7.131), we find the similar estimations. Then, according to the Liouville's theorem, Eq. (7.131) determines the regular functions in the whole complex plane that are identically equal to zero. Therefore, we get the integral equation from (7.131) as

$$\frac{M_+(\alpha)}{\sqrt{\alpha+k_1}}\Psi^{(+)}(\alpha) + \frac{1}{2\pi i}\int\limits_{-\infty+i\,\varepsilon_1}^{\infty+i\,\varepsilon_1}\frac{M_+(\nu)e^{-i\nu L}\Psi^-(\nu)}{\sqrt{\nu+k_1}}\frac{d\nu}{\nu-\alpha} = \frac{P}{(\alpha-k_1\cos\theta_0)}, \tag{7.135}$$

where

$$P = \frac{NM_+(k_1\cos\theta_0)}{\sqrt{k_1+k_1\cos\theta_0}}.$$

Let us rewrite Eq. (7.118) in the form

$$\frac{J_1(\alpha)\sqrt{\alpha+k_1}}{M_+(\alpha)}e^{i\alpha L} + \frac{M_-(\alpha)}{\sqrt{\alpha-k_1}}\left[e^{i\alpha L}\Psi^{(+)}(\alpha) + \Psi^-(\alpha)\right] = 0. \tag{7.136}$$

Then, applying to this equation the decomposition procedure (7.22) and using the terms that are regular in the lower half-plane, we derive

$$\frac{M_-(\alpha)}{\sqrt{\alpha-k_1}}\Psi^-(\alpha)-\frac{1}{2\pi i}\int\limits_{-\infty+i\,\varepsilon_2}^{\infty+i\,\varepsilon_2}\frac{M_-(\nu)e^{i\nu L}\Psi^{(+)}(\nu)}{\sqrt{\nu-k_1}}\frac{d\nu}{\nu-\alpha}=0. \qquad (7.137)$$

Here, $\tau_- < \tau < \varepsilon_2 < \tau_+$.

The integral equations (7.135), (7.137) are the key equations for further analysis of the wave fields scattered by the interface crack.

7.3.5 Analysis of the Integral Equations

Let us rewrite the Eqs. (7.135), (7.137) as

$$X^{(+)}(\alpha)+\frac{1}{2\pi i}\int\limits_{-\infty+i\,\varepsilon_1}^{\infty+i\,\varepsilon_1}\frac{M_+(\nu)\sqrt{\nu-k_1}e^{-i\nu L}}{M_-(\nu)\sqrt{\nu+k_1}(\nu-\alpha)}X^-(\nu)\,d\nu=\frac{P}{(\alpha-k_1\cos\theta_0)},$$

$$X^-(\alpha)+\frac{1}{2\pi i}\int\limits_{-\infty+i\,\varepsilon_2}^{\infty+i\,\varepsilon_2}\frac{M_-(\nu)\sqrt{\nu+k_1}e^{i\,\nu L}}{M_+(\nu)\sqrt{\nu-k_1}(\nu-\alpha)}X^{(+)}(\nu)\,d\nu=0, \qquad (7.138)$$

where

$$X^{(+)}(\alpha)=\frac{M_+(\alpha)}{\sqrt{\alpha+k_1}}\Psi^{(+)}(\alpha);\quad X^-(\alpha)=\frac{M_-(\alpha)}{\sqrt{\alpha-k_1}}\Psi^-(\alpha).$$

Then, we reduce them to the next integral equation:

$$X^{(+)}(\alpha)=\int\limits_{-\infty+i\,\varepsilon_2}^{\infty+i\,\varepsilon_2}K(\alpha,\,t)X^{(+)}(t)\,dt+\frac{P}{(\alpha-k_1\cos\theta_0)}. \qquad (7.139)$$

Here,

$$K(\alpha,\,t)=\frac{M_-(t)\sqrt{t+k_1}e^{itL}}{4\pi^2M_+(t)\sqrt{t-k_1}}\int\limits_{-\infty+i\,\varepsilon_1}^{\infty+i\,\varepsilon_1}\frac{M_+(\nu)\sqrt{\nu-k_1}e^{-i\nu L}}{M_-(\nu)\sqrt{\nu+k_1}(t-\nu)(\nu-\alpha)}d\nu \qquad (7.140)$$

is the kernel of the equation; $\varepsilon_1 < \mathrm{Im}t$, $\mathrm{Im}\alpha=\tau$ $(t,\,\alpha\in\Pi)$.

In the lower half-plane $\mathrm{Im}\nu < \varepsilon_1$ the integrand (7.140) has the branch points at $\nu=-k_{1(2)}$. Let us consider this function in the Riemann surface formed by cuts of the complex plane ν, where the branch cut-lines run from the branch points to infinity parallel to the imaginary axis in the lower half-plane. Enclosing the integration contour in the lower half-plane, we reduce the initial integration path to the one along the contours that encompasses the branch cuts and branch points and find that

$$\int_{C_+}\int_{C_+} |K(\alpha,\, t)|^2\, |\mathrm{d}t|\, |\mathrm{d}\alpha| < \infty, \tag{7.141}$$

where C_+ integration contour in the upper half-plane.

The unknown function $X^{(+)}(\alpha) = O(\alpha^{-1})$ if $|\alpha| \to \infty$. Therefore, Eq. (7.139) allows for the solution in L_2 for any geometrical and frequency parameters except for the discrete points of its spectra.

7.3.6 Integration Along the Branch Cut in the Complex Plane

Let us turn to the integral equation (7.135) and consider the integral

$$I_1(\alpha) = \frac{1}{2\pi i}\int_{-\infty+i\,\varepsilon_1}^{\infty+i\,\varepsilon_1} \frac{M_+(\nu)e^{-i\nu L}\Psi^-(\nu)}{\sqrt{\nu+k_1}}\frac{\mathrm{d}\nu}{\nu-\alpha}. \tag{7.142}$$

Here, the integrand is the fast-oscillating function for large L and decays as ν^{-2} if $|\nu| \to \infty$.

Let us take into account that $M_+(\alpha) = M(\alpha)/M_-(\alpha)$ and $\tau = \mathrm{Im}\alpha > \mathrm{Im}\nu$. Therefore, in the lower half-plane, the integrand (7.142) has the branch points and the integrable singularities at $\nu = -k_{1(2)}$. Let us consider this integral in the Riemann surface of the complex plane ν, where the branch cuts run from the branch points to infinity along the lines parallel to the imaginary axis in the lower half-plane. Deforming the integration pass (7.142) in the lower half-plane, we reduce one to the contours that encompasses the branch cuts as is shown in Fig. 7.3. For further analysis, we use the sheet of the Riemann surface, where $\mathrm{Re}\gamma_{1(2)} \geq 0$ in $\prod$.

Taking into account the exponential decays of the integrand (7.142) in the lower half-plane $\mathrm{Im}(\nu) < 0$ as well as the rule for bypassing of the branch points, we reduce the integral (7.142) to the integration along the right-sides of the branch cuts. Then, the replacement of variables $\nu \to -\nu$ leads to the expression

$$I_1(\alpha) = \frac{1}{\pi}\int_{k_1}^{i\,\infty+k_1} \frac{e^{i\nu L}\Psi^-(-\nu)}{M_+(\nu)\sqrt{\nu-k_1}}\frac{\mathrm{d}\nu}{\nu+\alpha}$$

$$+\frac{\mu_1}{\pi\mu_2}\int_{k_2}^{i\,\infty+k_2} \frac{e^{i\nu L}\Psi^-(-\nu)\sqrt{\nu+k_1}}{M_+(\nu)\sqrt{\nu^2-k_2^2}}\frac{\mathrm{d}\nu}{\nu+\alpha}. \tag{7.143}$$

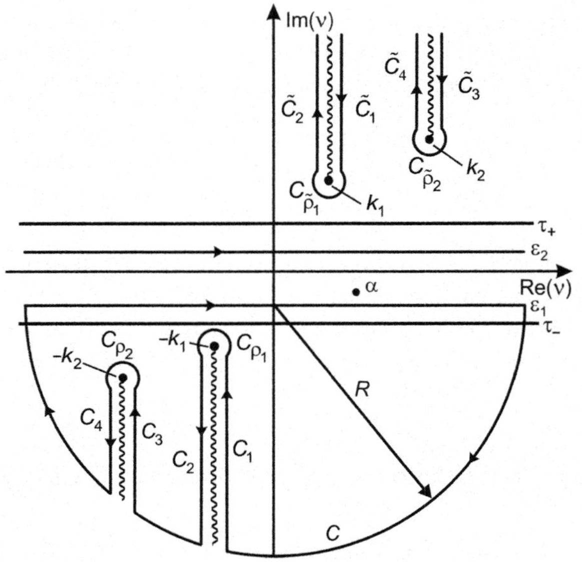

Fig. 7.3 Contour of integration

Here, we note that the contribution of the integrals along the circle $C_{\rho_{1(2)}}$ which encompass the branch points at $\nu = -k_{1(2)}$, where $\rho_{1(2)}$ are their radiuses, tend to zero if $\rho_{1(2)} \to 0$.

Next, we carry out similar transformations with the integral in (7.137), closing the integration contour into the upper half-plane and bypassing the branch points at $\nu = k_{1(2)}$, as is shown in Fig. 7.3. Note, the simple pole of the unknown function $\Psi^{(+)}(\nu)$ in the upper half-plane is taken into account for the evaluation. Therefore, taking into account the above analysis, we finally rewrite the integral equations (7.135), (7.137) as

$$\frac{M_+(\alpha)\Psi^{(+)}(\alpha)}{\sqrt{\alpha+k_1}} + \frac{1}{\pi}\int\limits_{k_1}^{i\infty+k_1} \frac{e^{i\nu L}\Psi^-(-\nu)}{M_+(\nu)\sqrt{\nu-k_1}}\frac{d\nu}{\nu+\alpha}$$

$$+\frac{\mu_1}{\pi\mu_2}\int\limits_{k_2}^{i\infty+k_2} \frac{\sqrt{\nu+k_1}e^{i\nu L}\Psi^-(-\nu)}{M_+(\nu)\sqrt{\nu^2-k_2^2}}\frac{d\nu}{\nu+\alpha} = \frac{P}{(\alpha-k_1\cos\theta_0)}, \tag{7.144}$$

$$\frac{M_-(\alpha)\Psi^-(\alpha)}{\sqrt{\alpha-k_1}} - \frac{1}{\pi i}\int\limits_{k_1}^{i\infty+k_1} \frac{e^{i\nu L}\Psi^{(+)}(\nu)}{M_+(\nu)\sqrt{\nu-k_1}}\frac{d\nu}{\nu-\alpha}$$

$$-\frac{\mu_1}{\pi i\mu_2}\int\limits_{k_2}^{i\infty+k_2} \frac{e^{i\nu L}\sqrt{\nu+k_1}\Psi^{(+)}(\nu)}{M_+(\nu)\sqrt{\nu^2-k_2^2}}\frac{d\nu}{\nu-\alpha} = \frac{P_1}{(\alpha-k_1\cos\theta_0)}. \tag{7.145}$$

Here

$$P_1 = \frac{NM_-(k_1 \cos\theta_0)e^{ik_1 L\cos\theta_0}}{i\sqrt{k_1 - k_1\cos\theta_0}}. \tag{7.146}$$

Consequently, we have reduced the problem to a system of integral equations of the second kind with the integration along the branch cuts in the complex plane.

7.3.7 Approximate Solutions for Wide Interface Crack

The integrands (7.144) and (7.145) decay exponentially along the integration pass and $|\nu| \to \infty$, and have integrable singularities at the lower integration boundaries. Further, we apply the asymptotic method for estimating these integrals for wide cracks ($k_{1(2)}L >> 1$). Let us consider the first integral (7.144). Then, after changing of the variable $\nu - k_1 = it/L$, we get

$$\int\limits_{k_1}^{i\infty + k_1} \frac{e^{i\nu L}\Psi^-(-\nu)}{M_+(\nu)\sqrt{\nu - k_1}}\frac{d\nu}{\nu + \alpha}$$

$$= \sqrt{L}e^{ik_1L - i\pi/4}\int\limits_0^\infty \frac{e^{-t}\,\Psi^-(-k_1 - it/L)}{M_+(k_1 + it/L)\sqrt{t}[t - iL(k_1 + \alpha)]}dt, \tag{7.147}$$

where $\sqrt{L} > 0$. Since the integrand (7.147) on the integration path decays exponentially with growing t, the main contribution to the integral is the value of the integrand in the vicinity of the point $t = 0$. Then, using the main terms of the Maclaurin series of functions $\Psi^-(-k_1 - it/L)$, $M_+(k_1 + it/L)$ for estimation (7.147), we find that

$$\int\limits_{k_1}^{i\infty + k_1} \frac{e^{i\nu L}\Psi^-(-\nu)}{M_+(\nu)\sqrt{\nu - k_1}}\frac{d\nu}{\nu + \alpha}$$

$$\approx \frac{e^{ik_1L}\sqrt{-iL}\Psi^-(-k_1)}{M_+(k_1)}\int\limits_0^\infty \frac{e^{-t}}{\sqrt{t}(t - iL(k_1 + \alpha))}dt = \Psi^-(-k_1)I_{11}(\alpha). \tag{7.148}$$

Here,

$$I_{11}(\alpha) = \frac{\sqrt{-iL}\,e^{ik_1L}}{M_+(k_1)}\Gamma_1[1/2, -iL\cdot(k_1 + \alpha)], \tag{7.149}$$

$$\Gamma_m(u, \upsilon) = \int_0^\infty \frac{t^{u-1}e^{-t}}{(t+\upsilon)^m} dt. \tag{7.150}$$

Note, the expression (7.150) defines a generalized gamma function introduced in [49] which has been used in [42, 50, 51].

Applying the similar transformations for estimation of the rest of the integrals in (7.144) and (7.145) we get

$$\begin{aligned}
&\int_{k_2}^{i\infty+k_2} \frac{\sqrt{\nu+k_1}e^{i\nu L}\Psi^-(-\nu)}{M_+(\nu)\sqrt{\nu^2-k_2^2}} \frac{d\nu}{\nu+\alpha} \sim \Psi^-(-k_2)I_{12}(\alpha),\\
&\int_{k_1}^{i\infty+k_1} \frac{e^{i\nu L}\Psi^{(+)}(\nu)}{M_+(\nu)\sqrt{\nu-k_1}} \frac{d\nu}{\nu-\alpha} \sim \Psi^{(+)}(k_1)I_{21}(\alpha),\\
&\int_{k_2}^{i\infty+k_2} \frac{e^{i\nu L}\sqrt{\nu+k_1}\Psi^{(+)}(\nu)}{M_+(\nu)\sqrt{\nu^2-k_2^2}} \frac{d\nu}{\nu-\alpha} \sim \Psi^{(+)}(k_2)I_{22}(\alpha).
\end{aligned} \tag{7.151}$$

Here,

$$\begin{aligned}
I_{12}(\alpha) &= \frac{\sqrt{-iL}\, e^{ik_2L}}{M_+(k_2)} \sqrt{\frac{k_1+k_2}{2k_2}} \Gamma_1[1/2, -iL\cdot(k_2+\alpha)],\\
I_{21}(\alpha) &= \frac{\sqrt{-iL}\, e^{ik_1L}}{M_+(k_1)} \Gamma_1[1/2, -iL\cdot(k_1-\alpha)],\\
I_{22}(\alpha) &= \frac{\sqrt{-iL}\, e^{ik_2L}}{M_+(k_2)} \sqrt{\frac{k_1+k_2}{2k_2}} \Gamma_1[1/2, -iL\cdot(k_2-\alpha)].
\end{aligned} \tag{7.152}$$

Note, the integrand (7.150) that determines the function $\Gamma_1[1/2, -iL\cdot(k_{1(2)}\pm\alpha)]$ has the integrable singularity at $t = 0$. For its accurate analysis, let us single out this singularity as

$$\begin{aligned}
f_1^*(\alpha) &= \int_0^\infty \frac{e^{-t}}{\sqrt{t}[t-iL\cdot(k_{1(2)}\pm\alpha)]} dt\\
&= \int_0^\infty \frac{e^{-t}-1}{\sqrt{t}[t-iL\cdot(k_{1(2)}\pm\alpha)]} dt + \int_0^\infty \frac{1}{\sqrt{t}[t-iL\cdot(k_{1(2)}\pm\alpha)]} dt\ .
\end{aligned} \tag{7.153}$$

Here, the second integral in the right-hand part, we can express in analytical form

$$\int\limits_0^\infty \frac{1}{\sqrt{t}[t - iL \cdot (k_{1(2)} \pm \alpha)]} \mathrm{d}t = \frac{\pi}{\sqrt{-iL(k_{1(2)} \pm \alpha)}}. \tag{7.154}$$

The integrand in the first integral of the right-hand part is free of the singularity and calculated numerically.

Let us turn to the Eqs. (7.144) and (7.145). Then, taking into account the approximate expressions (7.148), (7.151) we derive the approximate equations to determine the unknown functions $\Psi^{(+)}(\alpha)$ and $\Psi^{-}(\alpha)$ as

$$\Psi^{(+)}(\alpha) = \frac{\sqrt{\alpha + k_1}}{M_+(\alpha)} \times \left[\frac{P}{(\alpha - k_1 \cos\theta_0)} - \frac{I_{11}(\alpha)}{\pi} \Psi^{-}(-k_1) - \frac{\mu_1 I_{12}(\alpha)}{\pi \mu_2} \Psi^{-}(-k_2) \right], \tag{7.155}$$

$$\Psi^{-}(\alpha) = \frac{\sqrt{\alpha - k_1}}{M_-(\alpha)} \times \left[\frac{P_1}{(\alpha - k_1 \cos\theta_0)} + \frac{I_{21}(\alpha)}{\pi i} \Psi^{(+)}(k_1) + \frac{\mu_1 I_{22}(\alpha)}{\pi i \mu_2} \Psi^{(+)}(k_2) \right]. \tag{7.156}$$

Here, $\Psi^{(+)}(k_1)$, $\Psi^{(+)}(k_2)$, $\Psi^{-}(-k_1)$, $\Psi^{-}(k_2)$ are unknown.

Let us set $\alpha = k_1, \alpha = k_2$ and $\alpha = -k_1, \alpha = -k_2$ for the Eqs. (7.155) and (7.156) respectively. This leads to the fourth order system of linear algebraic equations with respect of the unknown values of the functions $\Psi^{(+)}(\alpha)$, $\Psi^{-}(\alpha)$ at the discrete points which we represent in the following matrix form

$$\begin{pmatrix} 1 & 0 & a_{13} & a_{14} \\ 0 & 1 & a_{23} & a_{24} \\ a_{13} & a_{14} & 1 & 0 \\ a_{23} & a_{24} & 0 & 1 \end{pmatrix} \begin{pmatrix} \Psi^{(+)}(k_1) \\ \Psi^{(+)}(k_2) \\ \Psi^{-}(-k_1) \\ \Psi^{-}(-k_2) \end{pmatrix} = \begin{pmatrix} f_1 \\ f_2 \\ f_3 \\ f_4 \end{pmatrix}. \tag{7.157}$$

Here,

$$a_{13} = \frac{\sqrt{-2ik_1 L}\, e^{ik_1 L}}{\pi\, M_+^2(k_1)} \Gamma_1[1/2,\ -2iLk_1],$$

$$a_{14} = \frac{\mu_1 \sqrt{-2ik_1 L}\, e^{ik_2 L}}{\mu_2 \pi\, M_+(k_1)\, M_+(k_2)} \sqrt{\frac{k_1 + k_2}{2\,k_2}}\, \Gamma_1[1/2,\ -iL(k_1 + k_2)], \tag{7.158}$$

$$a_{23} = \frac{\sqrt{-i\,(k_1 + k_2)\,L}\, e^{ik_1 L}}{\pi\, M_+(k_1)\, M_+(k_2)} \Gamma_1[1/2,\ -iL(k_1 + k_2)],$$

$$a_{24} = \frac{\mu_1 \sqrt{-i(k_1 + k_2)L}\, e^{ik_2 L}}{\mu_2 \pi\, M_+^2(k_2)} \sqrt{\frac{k_1 + k_2}{2\,k_2}}\, \Gamma_1[1/2,\ -iLk_2], \tag{7.159}$$

$$f_1 = \frac{\sqrt{2} N M_+(k_1 \cos\theta_0)}{k_1 M_+(k_1)(1-\cos\theta_0)\sqrt{1+\cos\theta_0}},$$
$$f_2 = \frac{N M_+(k_1 \cos\theta_0)\sqrt{k_1+k_2}}{M_+(k_2)(k_2 - k_1\cos\theta_0)\sqrt{k_1+k_1\cos\theta_0}}, \tag{7.160}$$

$$f_3 = \frac{\sqrt{2} N M_-(k_1 \cos\theta_0) e^{ik_1 L\cos\theta_0}}{k_1 M_+(k_1)(1+\cos\theta_0)\sqrt{1-\cos\theta_0}},$$
$$f_4 = -\frac{N M_-(k_1 \cos\theta_0) e^{ik_1 L\cos\theta_0}\sqrt{k_1+k_2}}{M_+(k_2)(k_2 + k_1\cos\theta_0)\sqrt{k_1-k_1\cos\theta_0}}. \tag{7.161}$$

Equation (7.157) splits into two independent second-order systems. Assuming that there is attenuation of the wave amplitude in the media ($\mathrm{Im}(k_{1(2)}) \neq 0$), we find that the determinants of both systems tend to unity with increasing crack width and frequency. Taking this into account, we get the approximate analytical solution of the system (7.157) in the form

$$\Psi^{(+)}(k_1) \approx f_1 - a_{13} f_3 - a_{14} f_4,$$
$$\Psi^{(+)}(k_2) \approx f_2 - a_{23} f_3 - a_{24} f_4, \tag{7.162}$$

$$\Psi^{-}(-k_1) \approx f_3 - a_{13} f_2 - a_{14} f_1,$$
$$\Psi^{-}(-k_2) \approx f_4 - a_{23} f_1 - a_{24} f_2. \tag{7.163}$$

Figure 7.4 shows the dependences of the maximum relative error of the approximate solutions (7.162), (7.163) on the wavelength of the crack at different angles of probing the iron-zinc joint with the parameters: *iron* $k_1 = 310\,\mathrm{m}^{-1}$, $\mu_1 = 8.2 \times 10^{10}$ Pa; *zinc* $k_2 = 415\,\mathrm{m}^{-1}$, $\mu_2 = 4.02 \times 10^{10}$ Pa. The exact solution is obtained numerically from (7.157). Note, for calculation we assume the attenuation in the media as $\mathrm{Im}(k_1)/\mathrm{Re}(k_1) = 0.0161$, $\mathrm{Im}(k_2)/\mathrm{Re}(k_2) = 0.0241$. As it can be seen from this figure if $L > 4\lambda$ ($\lambda = 2\pi/\mathrm{Re}(k_1)$ is the sounding wavelength), the maximum relative error for small angles of sounding does not exceed 1 %. However, with increasing angle θ_0, this error does not exceed 1 % already if $L \approx 2\lambda$. The error of approximate expressions (7.162), (7.163) increases significantly for the short cracks. Substituting formulas (7.162), (7.163) into the relation (7.155), (7.156), we obtain an approximate solution of the Wiener–Hopf equation (7.118).

7.3.8 *Structure of the Diffracted Fields. Far Field Approximation*

Let us represent the diffracted field through the solutions of the Eqs. (7.144), (7.145) as

Fig. 7.4 Dependence of the maximum relative error of approximate solutions (7.162), (7.163) on the dimensionless crack width for iron-zinc joint: 1—$\theta_0 = 45°$; 2—$\theta_0 = 20°$; 3—$\theta_0 = 10°$

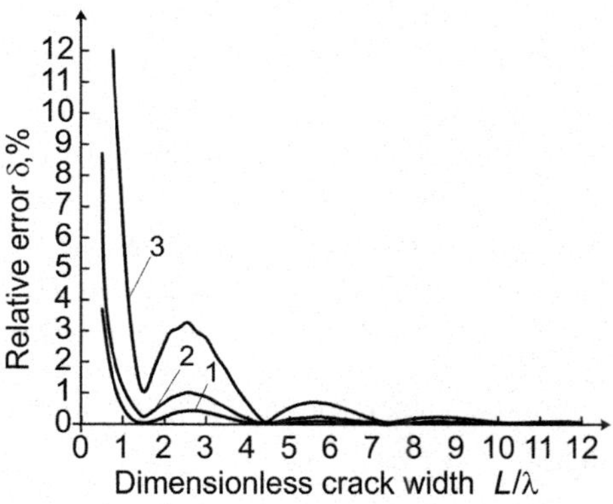

$$u(x, y) = \frac{1}{\sqrt{2\pi}} \int_{-\infty}^{\infty} \left[\Psi^{(+)}(\alpha) + \Psi^{-}(\alpha)e^{-i\alpha L}\right] \times \begin{Bmatrix} -\frac{1}{\sqrt{\alpha^2-k_1^2}} e^{-\sqrt{\alpha^2-k_1^2}y-i\alpha x}, & y > 0 \\ \frac{\mu_1}{\mu_2\sqrt{\alpha^2-k_2^2}} e^{\sqrt{\alpha^2-k_2^2}\,y-i\alpha x}, & y < 0 \end{Bmatrix} d\alpha. \tag{7.164}$$

Here, the features of the integrand terms are the branch points at $\alpha = k_{1(2)}$ and the simple poles at $\alpha = k_1 \cos\theta_0$ in the upper half-plane $\tau \geq 0$ and the branch points at $\alpha = -k_{1(2)}$ in the lower half-plane $\tau \leq 0$. Further, we apply the approximate solutions of the Eqs. (7.155) and (7.156) to analyze the diffracted field.

In order to analyse (7.164), let us introduce the polar coordinate system $x = r\cos\theta$, $y = r\sin\theta$, where $0 < \theta < \pi$, $0 \leq r < \infty$ and apply replacement of variable $\alpha = -k_1\cos\beta$, where $\beta = \mu + i\nu$ and $0 < \mu < \pi$, $-\infty < \nu < \infty$. Then, we find for $y > 0$ that

$$u(r,\theta) = -\frac{i}{\sqrt{2\pi}} \int_C \left[\Psi^{(+)}(-k_1\cos\beta)e^{ik_1 r\cos(\beta-\theta)} + \Psi^{-}(-k_1\cos\beta)e^{ik_1 r\cos(\beta-\theta)+ik_1 L\cos\beta}\right] d\beta, \tag{7.165}$$

where $C : \{(i\infty, 0) \cup (0, \pi) \cup (\pi, -i\infty)\}$ is the integration contour.

Let us assume that $|k_1| < |k_2|$. The integration contour in (7.165) bypasses the branch points at $\beta = 0\,(\pi)$; the simple pole singularities of functions $\mathbf{\Psi}^{(+)}$ and $\mathbf{\Psi}^{-}$ at $\beta = \pi - \theta_0$ are compensated and their contribution to the integral is equal to zero. Taking this into account, let us apply the saddle point method for an asymptotic estimate of the integral (7.165). For the far zone $r \to \infty$, $0 < \theta < \pi$ $(L/r << 1)$, the saddle point of the integrand is located at the point $\beta = \beta_s = \theta$. Then, deforming the integration path C through the saddle point, we arrive at the expression for the far diffracted field in the upper half-plane $0 < \theta < \pi$ as

$$u(r,\theta) \approx -\frac{e^{i(k_1 r+\pi/4)}}{\sqrt{k_1 r}}\left[\Psi^{(+)}(-k_1\cos\theta) + \Psi^{-}(-k_1\cos\theta)e^{ik_1 L\cos\theta}\right]. \tag{7.166}$$

To determine the diffracted field for $y < 0$ let us replace the variable $\alpha = -k_2\cos\beta$ in (7.164). Now the features of the integrands are the branch points $\beta = 0\,(\pi)$ and $\beta = \beta_{1(2)} = \arccos(\pm k_1/k_2)$, where the upper and lower signs correspond to β_1 and β_2 respectively. Then, for an asymptotic estimate of the integral (7.164), we deform the integration contour so that it passes through the saddle point $\beta = \beta_s = -\theta$, where $-\pi < \theta < 0$ and encloses the cuts of the complex plane emanating from the branch points at $\beta = \beta_{1(2)}$. If $r \to \infty$ and $-\pi < \theta < 0$, we find that

$$u(r,\theta) \approx \frac{\mu_1}{\mu_2}\frac{e^{i(k_2 r+\pi/4)}}{\sqrt{k_2 r}}\left[\Psi^{(+)}(-k_2\cos\theta) + \Psi^{-}(-k_2\cos\theta)e^{ik_2 L\cos\theta}\right] + u_{\text{lat}}(r,\theta), \tag{7.167}$$

where $u_{\text{lat}}(r,\theta)$ is the integral along the contour which encompasses the brunch cut; this integral describes the contribution of the lateral wave to the displacement field [52].

Let us consider the integral representation of the lateral wave $u_{\text{lat}}(r,\theta)$ excited by the edge of the semi-infinite interface crack occupying the region ($y = 0,\ -\infty < x < 0$)

$$u_{\text{lat}}(r,\theta) = \frac{i}{\sqrt{2\pi}}\frac{\mu_1}{\mu_2}\int_{C_1}\Psi^{(+)}(-k_2\cos\beta)e^{ik_2 r\cos(\beta+\theta)}d\beta. \tag{7.168}$$

Here C_1 is the fragment of the deformed integration contour $C = C_1 \cup \Gamma_d$ which encompasses the branch point β_1; contour Γ_d crosses the saddle point $\beta = \beta_s = -\theta$. The integration path is shown in Fig. 7.5 if $-\theta_c < \theta < 0$, where $\theta_c = \operatorname{Re}\beta_1$ is the critical angle. To obtain the expression for the displacement field of the lateral wave (7.168), we apply the exact solution of the Wiener–Hopf equation for the wave diffraction problem from the semi-infinite interface crack as

$$\Psi^{(+)}(\alpha) = \frac{P\sqrt{\alpha+k_1}}{M_+(\alpha)(\alpha-k_1\cos\theta_0)}. \tag{7.169}$$

Let us continue the integration contour C_1 along the branch cut to $\beta = \beta(\mu,\nu)$: $\{\mu \to 0,\ \nu \to -\infty\}$ as is shown in Fig. 7.5 and replace the integration contour in (7.168) by the continued one. Taking into account that the contribution of integration along the ring enclosing the branch point goes to zero during its contraction to the point and using the rule for bypassing the branch point β_1, we write the following formula

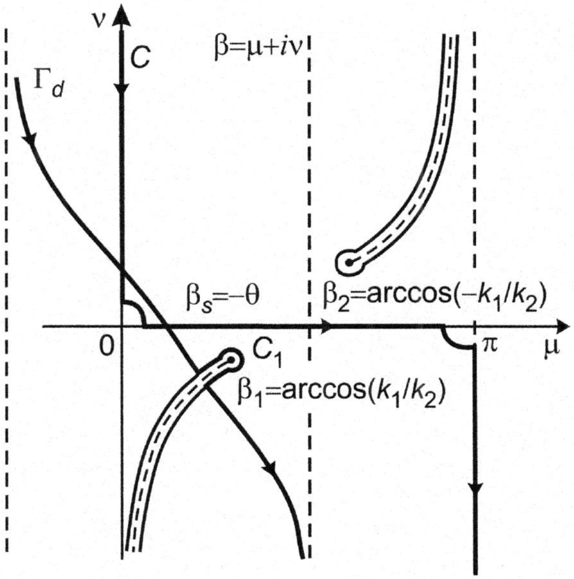

Fig. 7.5 Contour of integration in (7.168), $-\pi < \theta < 0$

$$u_{lat}(r,\theta) \approx \frac{P\mu_1/\mu_2}{\sqrt{2\pi}}$$

$$\times \int\limits_{-i\infty}^{\beta_1} \frac{\sqrt{k_2\cos\beta - k_1}\,2\mu_2^2 k_2^2 \sin^2\beta}{(k_2\cos\beta\ + k_1\cos\theta_0)} \frac{M_+(k_2\cos\beta)e^{ik_2 r\cos(\beta+\theta)}}{\mu_2^2 k_2^2\sin^2\beta + \mu_1^2(k_2^2\cos^2\beta - k_1^2)} \mathrm{d}\beta. \tag{7.170}$$

Let us apply the saddle point technique to evaluate this integral. Then, the integration contour should pass from the branch point $\beta = \beta_1$ to $\beta \to -i\infty$ along the curve on which the function $\exp[ik_2 r\cos(\beta+\theta)]$ decays fastest. This condition will be met when Re $(\cos(\beta - |\theta|)) = \text{const}$, Im $(\cos(\beta - |\theta|)) > 0$. The equation of this curve for small Im β_1 has the form $\cos(\beta_1 - |\theta|) = \cos(\mu - |\theta|)\mathrm{ch}\nu$. Let us introduce the approximate relations [52]:

$$\mu \approx \beta_1, d\beta \approx i\,d\nu, \beta \approx \beta_1,$$
$$\sqrt{k_2\cos\beta - k_1} = -\,i\,e^{-i\pi/4}\sqrt{k_2|\nu|\sin\beta_1}.$$

Taking this into account, we find that

$$u_{lat}(\rho, \theta) \approx \frac{P}{\sqrt{2\pi}} \frac{\mu_1}{\mu_2} \times \frac{e^{-i\pi/4}\sqrt{k_2 \sin\beta_1} M_+(k_1) e^{ik_2 r \cos(\beta_1 - |\theta|)}}{k_1(1+\cos\theta_0)} \int\limits_0^\infty \sqrt{\nu} e^{-\nu k_2 r \sin(\beta_1 - |\theta|)} d\nu. \tag{7.171}$$

Since

$$\int\limits_0^\infty \sqrt{\nu} e^{-\nu k_2 r \sin(\beta_1 - |\theta|)} d\nu = \frac{1}{2} \frac{\sqrt{\pi}}{[k_2 r \sin(\beta_1 - |\theta|)]^{3/2}}, \tag{7.172}$$

then the final expression for the displacement field caused by the lateral wave excited by the edge of the semi-infinite interface crack looks like

$$u_{\text{lat}}(r, \theta) = \frac{P}{2\sqrt{2}} \frac{\mu_1}{\mu_2} \frac{e^{-i\pi/4}\sqrt{k_2 \sin\beta_1} M_+(k_1)}{k_1(1+\cos\theta_0)} \frac{e^{ik_2 r \cos(\beta_1 - |\theta|)}}{[k_2 r \sin(\beta_1 - |\theta|)]^{3/2}}. \tag{7.173}$$

Here, $-\pi/2 < -\theta_c < \theta < 0$. It follows from (7.173) that the modulus of the displacement caused by the lateral wave increases when the observation angle approaches the critical angle $\theta \to -\theta_c$.

Therefore, the expressions (7.166) and (7.167) are the key formulas for far displacement field analysis. The plane wave sounding of the interface crack excites the lateral wave caused by the interface crack edges which can be considered as a mark of the interface defect. Formula (7.173) determines the lateral wave displacement for semi-infinite interface crack.

Next, we consider some examples of the far field diffracted from an interface crack formed at the junction of two different materials. The choice of the materials for research in general can be arbitrary, but we focused on joins that would the best illustrate the diffraction effects for their possible use in diagnosis.

7.3.9 Displacement Field Analysis

We now present the numerical examples of the far-field patterns to discuss the scattering characteristics of the interface crack located on the plane interface of two joined half-spaces made of iron and zinc with densities and the Lamé parameters $\rho_1 = 7.87$ Mg/m^3, $\rho_2 = 6.92$ Mg/m^3 and $\mu_1 = 82.0$ GPa, $\mu_2 = 40.2$ GPa respectively. Here, and through this chapter, the crack is illuminated with plane SH-waves (7.95) whose frequency $\omega = 1.0$ MHz and the incident waves propagate from the upper medium "1" (iron) into the lower medium "2" (zinc). The wave velocities in these media c_1 and c_2, and $c_1 > c_2$. For the sake of convenience, it is assumed that the

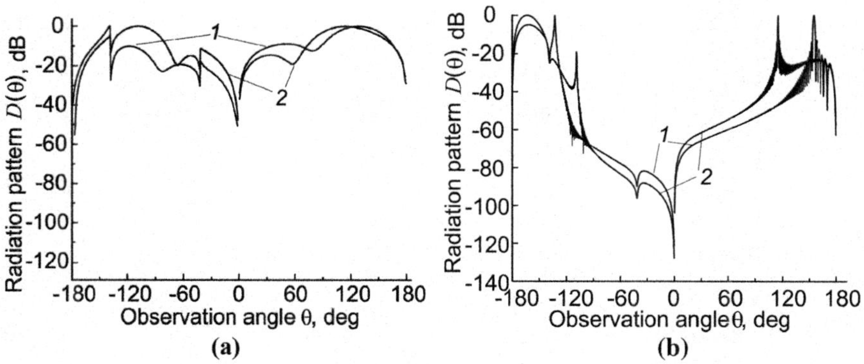

Fig. 7.6 Normalized far-field diffracted from interface crack in joint "iron-zinc": **a**—$L = \lambda_1$, **b**—$L = 64\lambda$; 1—$\theta_0 = 25°$, 2—$\theta_0 = 65°$

wave numbers have a small imaginary parts. In Fig. 7.6, we present the normalized far diffracted field depending on the observation angle for different width and angles of incidence determined as

$$D(\theta) = 20\log_{10}\frac{\lim_{r\to\infty}|e^{-ik_{1(2)}r}u(r,\theta)\sqrt{r}|}{\max_{-\pi<\theta<\pi}[\lim_{r\to\infty}|e^{-ik_{1(2)}r}u(r,\theta)\sqrt{r}|]}. \tag{7.174}$$

It follows from Fig. 7.6 that the number of oscillations of the diffracted field rapidly increases with the crack width. The presence of rapid oscillations for larger cracks is explained by the effect of multiple scattering of the elastic edge waves from the crack vertices. We also observe the effect of mirror reflection in medium "1" and the effect of refraction shifting the sharp main peak of the far-field pattern in medium "2" according to the Snell's law. In Fig. 7.6, we observe the presence of jumps in the far-field patterns for the lower half space and the observation angles $\theta \approx -42°$, and $\theta \approx -138°$ which coincide with the critical ones. This effect is observed under the sounding of the interface crack at different angles and crack width and can be explained by the fact that each edge of the crack is the effective local source which excites the lateral wave.

To elucidate the behavior of the displacement field scattered by a finite crack in the high-frequency region, methods of geometric optics or the physical theory of diffraction are usually used. However, these methods, as a rule, do not take into account the effect of mutual diffracted wave between the vertices of the interface crack [53]. In order to estimate this effect, the expressions (7.166), (7.167) as well as (7.155), (7.156) are used with the result that

$$u(r,\theta)=\begin{cases}\dfrac{e^{i(k_1 r+\pi/4)}}{\sqrt{k_1 r}}\Big[S_{11}^{(1)}(-k_1\cos\theta)+S_{22}^{(1)}(-k_1\cos\theta)e^{ik_1L\cos\theta}+S_1^{int}(-k_1\cos\theta)\Big], \\ \qquad 0<\theta<\pi; \\ \dfrac{e^{i(k_2 r+\pi/4)}}{\sqrt{k_2 r}}\Big[S_{11}^{(2)}(-k_2\cos\theta)+S_{22}^{(2)}(-k_2\cos\theta)e^{ik_2L\cos\theta}+S_2^{int}(-k_2\cos\theta)\Big], \\ \qquad -\pi<\theta<0. \end{cases} \tag{7.175}$$

Here, the term $u_{lat}(r,\theta)$ is omitted in view of its fast decay if $r\to\infty$;

$$S_{11}^{(1)}(-k_1\cos\theta)=\frac{\sqrt{2}(1-R)M_+(k_1\cos\theta_0)\sin(\theta/2)\sin(\theta_0/2)}{\sqrt{\pi}M_-(k_1\cos\theta)(\cos\theta+\cos\theta_0)}, \tag{7.176}$$

$$S_{22}^{(1)}(-k_1\cos\theta)=-\frac{\sqrt{2}(1-R)M_-(k_1\cos\theta_0)\cos(\theta/2)\cos(\theta_0/2)e^{ik_1L\cos\theta_0}}{\sqrt{\pi}M_+(k_1\cos\theta)(\cos\theta+\cos\theta_0)}; \tag{7.177}$$

$$S_{11}^{(2)}(-k_2\cos\theta)=-\frac{\mu_1}{\mu_2}\sqrt{\frac{k_1(k_1-k_2\cos\theta)}{\pi}}\,\frac{(1-R)M_+(k_1\cos\theta_0)\sin(\theta_0/2)}{M_-(k_2\cos\theta)(k_1\cos\theta_0+k_2\cos\theta)}, \tag{7.178}$$

$$S_{22}^{(2)}(-k_2\cos\theta)=\frac{\mu_1}{\mu_2}\sqrt{\frac{k_1(k_1+k_2\cos\theta)}{\pi}}\,\frac{(1-R)M_-(k_1\cos\theta_0)\cos(\theta_0/2)e^{ik_1L\cos\theta_0}}{M_+(k_2\cos\theta)(k_1\cos\theta_0+k_2\cos\theta)}; \tag{7.179}$$

$$\begin{aligned}S_1^{int}(-k_1\cos\theta)&=\frac{\sqrt{2k_1}\sin(\theta/2)}{M_-(k_1\cos\theta)}\\&\times\left[\frac{I_{11}(-k_1\cos\theta)\Psi^-(-k_1)}{\pi}+\frac{\mu_1 I_{12}(-k_1\cos\theta)\Psi^-(-k_2)}{\pi\mu_2}\right]\\&+\frac{\sqrt{2k_1}\cos(\theta/2)}{M_+(k_1\cos\theta)}\\&\times\left[\frac{I_{21}(-k_1\cos\theta)\Psi^{(+)}(k_1)}{\pi}+\frac{\mu_1 I_{21}(-k_1\cos\theta)\Psi^{(+)}(k_2)}{\pi\mu_2}\right]e^{ik_1L\cos\theta},\end{aligned} \tag{7.180}$$

$$\begin{aligned}S_2^{int}(-k_2\cos\theta)&=-\frac{\mu_1}{\mu_2}\frac{\sqrt{k_1-k_2\cos\theta}}{M_-(k_2\cos\theta)}\\&\times\left[\frac{I_{11}(-k_2\cos\theta)\Psi^-(-k_1)}{\pi}+\frac{\mu_1 I_{12}(-k_2\cos\theta)\Psi^-(-k_2)}{\pi\mu_2}\right]\\&-\frac{\mu_1}{\mu_2}\frac{\sqrt{k_1+k_2\cos\theta}}{M_+(k_2\cos\theta)}\end{aligned}$$

$$\times \left[\frac{I_{21}(-k_2 \cos\theta)\Psi^{(+)}(k_1)}{\pi} + \frac{\mu_1 I_{21}(-k_2 \cos\theta)\Psi^{(+)}(k_2)}{\pi \mu_2} \right] e^{ik_2 L \cos\theta}. \tag{7.181}$$

Here, expressions (7.176) and (7.177) determine the field in the upper half-space scattered from the vertices of semi-infinite cracks occupying the regions $-\infty < x < 0$ and $-L < x < \infty$. The field scattered by these cracks in the lower half-space is determined by formulas (7.178) and (7.179). The contribution to the diffracted field of multiple reflection of edge waves by the crack tips is described by formulas (7.180) and (7.181).

Let the crack be located on a flat interface between two metals: medium "1" (titanium) and medium "2" (lead), which are characterized by the densities and Lame parameters $\rho_1 = 4.4$ Mg/m^3, $\rho_2 = 11.34$ Mg/m^3 and $\mu_1 = 47.0$ GPa, $\mu_2 = 62.0$ GPa respectively. The velocities of elastic transverse waves in these materials differ significantly: $c_1 = 3270$ m/s, $c_2 = 740$ m/s. At a given frequency $\omega = 1.0$ MHz, the wavelength in titanium $\lambda_1 \approx 20$ mm and in lead $\lambda_2 \approx 5$ mm.

The far field diffracted from the interface crack are shown in Fig. 7.7. Curves 1 in this figure correspond to the case when the diffracted field is formed by two separate vertices, and in formula (7.175) only the terms $S_{11}^{(1)}$, $S_{11}^{(2)}$, $S_{22}^{(1)}$, $S_{22}^{(2)}$ are taken into account. We keep all the terms for the calculation of curves 2 in this formula, and the calculation of curves 3 we keep only the terms $S_1^{\rm int}$ and $S_2^{\rm int}$. From the behavior of the curves in this figure, we see that at small sounding angles $\theta_0 \to 0$ and for sounding angle approach to the critical one, the effects of mutual edge waves diffraction from the crack edges dominate over the whole range of observation angles. If we increase the crack width and the incidence angle, these lead to decrease of the terms $S_1^{\rm int}$, $S_2^{\rm int}$, which is observed in Fig. 7.8.

In Fig. 7.9, we observe the distribution of the diffracted field when the sounding angle is close to the normal one (curve 1) and is in the region, where the conditions for total internal reflection are satisfied (curve 2). If $\theta_0 = 85°$ (curve 1), the scattered field

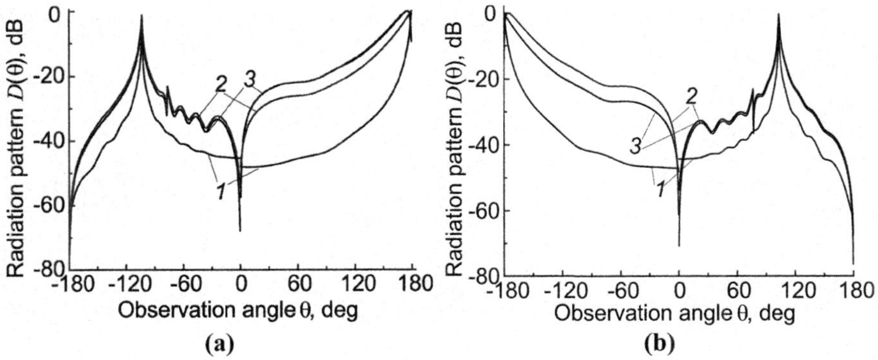

Fig. 7.7 Influence of the mutual edge waves interaction on the far field; 1—without interaction; 2—includes interaction; 3—pure interaction effects: **a** joint "titanium–lead", $L = \lambda_1$, $\theta_0 = 1.5°$; **b** joint "lead-titanium", $L = 5\lambda_1$ $\theta_0 = \theta_c \approx 77°$

Fig. 7.8 Normalized far field diffracted from the interface crack in joint "titanium-lead": $L = 10\,\lambda_1$, $\theta_0 = 40°$; 1—without edge-waves interaction; 2—includes edge-waves interaction

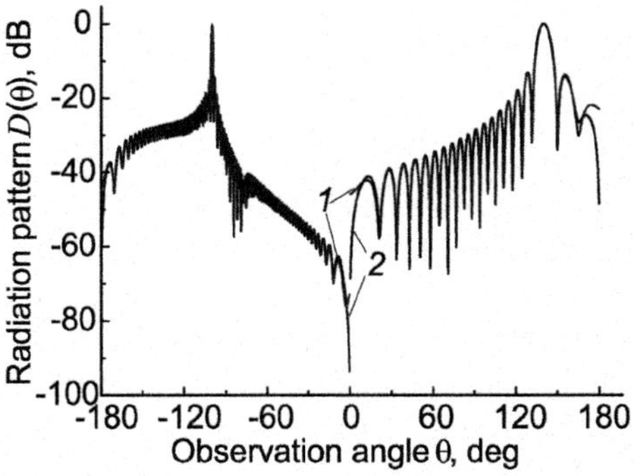

reaches maximum at $\theta = 95°$. The maximum of the refraction field is if $\theta \approx -113°$. This agrees with the laws of geometric optics. If the illumination angle $\theta_0 = 25°$, that is $0 < \theta_0 < \theta_c$ (curve *2*), then the total internal reflection occurs, and the refraction of waves at the join boundary can no longer be interpreted in terms of geometric optics.

In Fig. 7.10, we observe the dependences of the normalized amplitudes of the diffracted field on the dimensionless crack width in the "lead-titanium" join for irradiation at small and critical angles, respectively, and small angles of registration of the diffracted field. As can be seen from this figure, at the given angles of irradiation and registration of the elastic waves, the amplitude of the scattered displacement field monotonically depends on the crack width. This property can be used to find the width of the crack. As the irradiation angle and the angular coordinates of the registration points increase, the oscillations of the scattered field amplitude begin to appear as a function of the crack width.

In this section, we have considered the simple model of the elastic SH-plane wave scattering from finite interface crack and analyzed this problem on the rigorous level using the Wiener–Hopf technique. The approximate solution (7.155), (7.156) of the Wiener–Hopf equation for the wide interface crack is obtained and applied for the analysis of the displacement field. We focused our study on the physical features of

Fig. 7.9 Distribution of the far-field diffracted from interface crack in joint "lead-titanium": $L = 5\,\lambda_1$; 1—$\theta_0 = 85°$, 2—$\theta_0 = 25°$

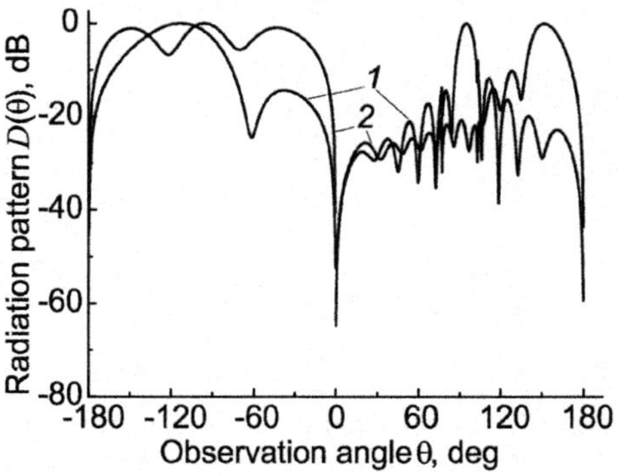

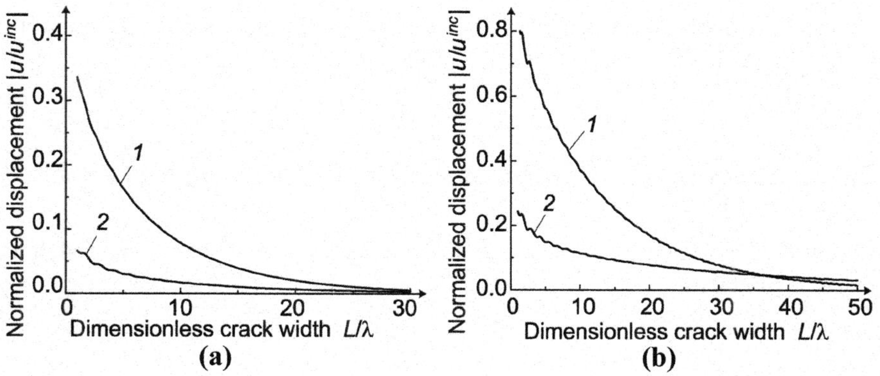

Fig. 7.10 Dependence of the magnitude of the far diffracted field in the registration angles $\theta_{r1} = -\theta_{r2} = 1.5°$ on the crack width for join "lead-titanium" in the upper "1" and in the lower "2" half-spaces; **a** sliding angle of illumination $\theta_0 = 1.5°$; **b** illumination under the critical angle

the diffracted fields which can be used for recognition of the interface cracks. The approximate expressions (7.173), (7.175) and (7.166) are obtained and applied for the analysis of the SH-lateral wave excitation, the effects of the mutual edge waves diffraction, and far field distribution for arbitrary angles of illumination, including sliding and critical angles.

Further, we have analyzed the dynamical stress intensity factors at the vertices of the interface crack to estimate their dependence on the problem parameters.

7.4 Stress Field at the Tips of the Interface Crack

7.4.1 Introduction

The stress intensity factors (SIF) are defined as coefficients for the singularity of the stress field components in the vicinity of the crack tips. In the case of harmonic dynamic loading of structures with the interface cracks, the SIF depends on the physical and mechanical characteristics of the joint materials, defect parameters, as well as the frequency and angle of incidence of the sounding wave. Information about these dependencies is important for choosing the optimal load characteristics for diagnostics of materials. In this section, we show the influence of the diffraction effects on the dependence of the SIF on various parameters of a dynamic system when it is irradiated by a plane SH wave. The dynamic properties of the SIF were studied in [54–56] when a crack was irradiated by a plane SH wave in a homogeneous medium, and for interfacial cracks under normal irradiation in [27, 57–59]. Despite a significant number of publications on the scattering of elastic waves by cracks that are on the verge of separation of materials, the physical and mechanical regularities of the formation of the complex SIFs in the vicinity of the crack tips under harmonic loading have not been clarified. Here we will consider the patterns of behavior of the dynamic SIFs in the vicinity of the tips of interfacial cracks.

7.4.2 Stress Intensity Factor at the Tips of the Interface Crack

For further convenience, let us express the generalized gamma function through the Erfc(z) integral as

$$\Gamma_1(1/2, -iL(k_{1(2)} \pm \alpha)) = \int_0^\infty \frac{e^{-t}}{\sqrt{t}\big(t - iL(k_{1(2)} \pm \alpha)\big)}\,\mathrm{d}t$$
$$= \frac{\pi e^{-iL(k_{1(2)} \pm \alpha)}}{\sqrt{-iL(k_{1(2)} \pm \alpha)}}\mathrm{Erfc}\Big[\sqrt{-iL(k_{1(2)} \pm \alpha)}\Big]. \quad (7.182)$$

The stress field singularities at the edge tips are determined by the asymptotic behavior of functions (7.155), (7.156) if $|\alpha| \to \infty$. To find them, we use the asymptotic of the Erfc (z) function for a large argument and represent the asymptotic expressin of function (7.182) in form

$$\Gamma_1(1/2, z) = \sqrt{\pi}\, z^{-1} + O(z^{-5/2}), \quad (7.183)$$

where $z = -iL(k_{1(2)} \pm \alpha)$ and $|z| \to \infty \quad (|\arg z| < \pi/4)$.

Using the relation (7.183), we find the asymptotic expressions of functions (7.149), (7.152) for $|\alpha| \to \infty$ as

$$I_{11}(\alpha) \sim \frac{\sqrt{\pi}\, e^{i\pi/4} e^{ik_1 L}}{\sqrt{L} M_+(k_1)(k_1 + \alpha)},$$
$$I_{12}(\alpha) \sim \frac{\sqrt{\pi}\, e^{i\pi/4} e^{ik_2 L}}{\sqrt{L} M_+(k_2)(k_2 + \alpha)} \sqrt{\frac{k_1 + k_2}{2k_2}}, \quad (7.184)$$

$$I_{21}(\alpha) \sim -\frac{i\sqrt{\pi}\, e^{-i\pi/4} e^{ik_1 L}}{\sqrt{L} M_+(k_1)(\alpha - k_1)},$$
$$I_{22}(\alpha) \sim -\frac{i\sqrt{\pi}\, e^{-i\pi/4} e^{ik_2 L}}{\sqrt{L} M_+(k_2)(\alpha - k_2)} \sqrt{\frac{k_1 + k_2}{2k_2}}. \quad (7.185)$$

Further, it is found that $M_+^{-1}(\alpha) \to \sqrt{(\mu_1 + \mu_2)/\mu_2}$ if $|\alpha| \to \infty$ and $(k_{1(2)}/|\alpha|) \to 0$. Then taking into account the asymptotic expressions (7.184), (7.185), we find the asymptotic representation of the relations (7.155), (7.156):

$$\Psi^{(+)}(\alpha) \approx Q_1(k_1, k_2; \theta_0) \begin{cases} \alpha^{-1/2} \\ -i\,(-\alpha)^{-1/2} \end{cases};$$

$$\Psi^{-}(\alpha) \approx Q_2(k_1, k_2; \theta_0)\begin{cases} \alpha^{-1/2}, & \alpha \to \infty, \\ i\,(-\alpha)^{-1/2}, & \alpha \to -\infty. \end{cases} \tag{7.186}$$

Here,

$$Q_1(k_1, k_2; \theta_0) = \sqrt{\frac{\mu_2}{\mu_1 + \mu_2}}\Bigg[\sqrt{\frac{k_1}{\pi}}(1 - R)M_{+}(k_1 \cos\theta_0)\sin\left(\frac{\theta_0}{2}\right) - A_1\Psi^{-}(-k_1) - A_2\Psi^{-}(-k_2)\Bigg], \tag{7.187}$$

$$Q_2(k_1, k_2; \theta_0) = \sqrt{\frac{\mu_2}{\mu_1 + \mu_2}}\,i\Bigg[\sqrt{\frac{k_1}{\pi}}(R - 1)M_{-}(k_1 \cos\theta_0)e^{ik_1 L\cos\theta_0}\cos\left(\frac{\theta_0}{2}\right) + A_1\Psi^{(+)}(k_1) + A_2\Psi^{(+)}(k_2)\Bigg], \tag{7.188}$$

$$A_1 = \frac{e^{i\pi/4}e^{ik_1 L}}{\sqrt{\pi L}M_{+}(k_1)}, \quad A_2 = \frac{\mu_1 e^{i\pi/4}e^{ik_2 L}}{\mu_2\sqrt{\pi L}M_{+}(k_2)}\sqrt{\frac{k_1 + k_2}{2k_2}}. \tag{7.189}$$

Let us substitute expressions (7.186) into (7.164) and, taking into account that $\tau_{zy} = \mu_{1(2)}\partial\, u(x,\ y)/\partial y$ and if $|\alpha| \to \infty$, $\sqrt{\alpha^2 - k_{1(2)}^2} \sim \alpha + O(k_{1(2)}/\alpha)$, we find the asymptotic representation τ_{zy} [58] for $y>0$ as

$$\tau_{zy} \approx \frac{\mu_1}{\sqrt{2\pi}}\int_0^{+\infty}\Big\{-i\big[Q_1 + Q_2 e^{i\alpha L}\big]e^{-\alpha|y| - i\alpha x} + \big[Q_1 + Q_2 e^{-i\alpha L}\big]e^{-\alpha|y| - i\alpha x}\Big\}\frac{d\alpha}{\sqrt{\alpha}}. \tag{7.190}$$

Let us rewrite (7.190) in the following form:

$$\tau_{zy} \approx \frac{\mu_1(1 - i)Q_1}{\sqrt{2\pi}}\int_0^{\infty}\frac{e^{-\alpha|y|}}{\sqrt{\alpha}}(\cos\alpha x + \sin\alpha x)d\alpha + \frac{\mu_1(1 + i)Q_2}{\sqrt{2\pi}}\int_0^{\infty}\frac{e^{-\alpha|y|}}{\sqrt{\alpha}}(\cos\alpha(x + L) + \sin\alpha(x + L))d\alpha. \tag{7.191}$$

Next, taking into account that [60]

$$\int_0^{\infty}\frac{e^{-\alpha|y|}}{\sqrt{\alpha}}\begin{pmatrix}\cos\alpha x \\ \sin\alpha x\end{pmatrix}d\alpha$$

$$= \begin{pmatrix} \sqrt{\pi}\cos(0.5\,\mathrm{arctg}(|x|/|y|))/(x^2+|y|^2)^{1/4} \\ \frac{|x|}{x}\sqrt{\pi}\sin(0.5\,\mathrm{arctg}(|x|/|y|))/(x^2+|y|^2)^{1/4} \end{pmatrix}, \tag{7.192}$$

finally, we get

$$\begin{aligned} \tau_{zy} &\approx \frac{\mu_1(1-i)Q_1}{\sqrt{2\pi}} \\ &\times \left[\frac{\sqrt{\pi}\cos(0.5\,\mathrm{arctg}(|x|/|y|))}{(x^2+|y|^2)^{1/4}} + \frac{|x|}{x}\frac{\sqrt{\pi}\sin(0.5\,\mathrm{arctg}(|x|/|y|))}{(x^2+|y|^2)^{1/4}} \right] \\ &+ \frac{\mu_1(1+i)Q_2}{\sqrt{2\pi}} \left[\frac{\sqrt{\pi}\cos(0.5\,\mathrm{arctg}(|x+L|/|y|))}{((x+L)^2+|y|^2)^{1/4}} \right. \\ &\left. - \frac{|x+L|}{x+L}\frac{\sqrt{\pi}\sin(0.5\,\mathrm{arctg}(|x+L|/|y|))}{((x+L)^2+|y|^2)^{1/4}} \right]. \end{aligned} \tag{7.193}$$

The expression (7.193) gives the behavior of the stress field in the vicinity of the interphase crack vertices. In each of the vertices, we introduce a local polar coordinate system (Fig. 7.1): for the right vertex ($x \to +0$) $x = r_1\cos\varphi_1$, $y = r_1\sin\varphi_1$, and for the left one ($x \to -L-0$) $x + L = -r_2\cos\varphi_2$, $y = r_2\sin\varphi_2$. In these local coordinate systems, the expression (7.193) takes the form

$$\tau_{zy} \approx \sqrt{2\pi}\mu_1 \left[\frac{(1-i)Q_1\cos(\varphi_1/2)}{\sqrt{2\pi r_1}} + \frac{(1+i)Q_2\cos(\varphi_2/2)}{\sqrt{2\pi r_2}} \right]. \tag{7.194}$$

Normalizing expression (7.194) to the maximum modulus of the stress field of a plane wave in a homogeneous medium ($\max|\tau_{zy}^{(inc)}| = \mu_1|k_1|$), we write

$$\tau_{zy}^* \approx \frac{Q_1^*\cos(\varphi_1/2)}{\sqrt{2\pi r_1}} + \frac{Q_2^*\cos(\varphi_2/2)}{\sqrt{2\pi r_2}}, \tag{7.195}$$

where

$$Q_1^* = \sqrt{2\pi}\,(1-i)Q_1/|k_1|, \tag{7.196}$$

$$Q_2^* = \sqrt{2\pi}\,(1+i)Q_2/|k_1| \tag{7.197}$$

are the normalized SIF for near and far (relative to the source) tips of the interface crack.

Substituting expressions (7.162), (7.163) into the relation (7.187), (7.188), we obtain from (7.196) and (7.197) the approximate analytical expressions for the SIF, which are valid for arbitrary sounding angles. Taking into account the first terms in formulas (7.162), (7.163), we obtain the simple expressions for the SIF. At some characteristic values of the incidence angle we get

$$Q_1(k_1,k_2;\theta_0=\pi/2)=Q_2(k_1,k_2;\theta_0=\pi/2)$$
$$=\sqrt{\frac{\mu_1k_1+\mu_2k_2}{(\mu_1+\mu_2)k_1k_2}}\left[\frac{k_1}{\sqrt{2\pi}}-\frac{k_1e^{i\pi/4}e^{ik_1L}}{\pi\sqrt{k_1L}[M_+(k_1)]^2}\right.$$
$$\left.+\frac{\mu_1^{i\pi/4}e^{ik_2L}(k_1+k_2)k_1}{2\pi k_2\mu_2\sqrt{k_2L}\,[M_+(k_2)]^2}+O((k_{1(2)}L)^{-1})\right] \tag{7.198}$$

for normal incidence of the sounding wave ($\theta_0=\pi/2,\ R=(\mu_1k_1-\mu_2k_2)/(\mu_1k_1+\mu_2k_2)$);

$$Q_1(k_1,k_2;\theta_0\to+0)=\sqrt{\frac{\mu_2k_1}{\mu_1+\mu_2}}\left[-\frac{e^{i\pi/4}e^{2ik_1L}}{\pi\sqrt{2k_1L}[M_+(k_1)]^3}\right.$$
$$\left.+\frac{\mu_1 2e^{i\pi/4}e^{i(k_1+k_2)L}}{\pi\mu_2\sqrt{2k_2L}\,[M_+(k_2)]^2M_+(k_1)}+O((k_{1(2)}L)^{-1})\right], \tag{7.199}$$

$$Q_2(k_1,k_2;\theta_0\to+0)=\sqrt{\frac{\mu_2k_1}{\mu_1+\mu_2}}\left[-\frac{2ie^{ik_1L}}{\sqrt{\pi}\,M_+(k_1)}\right.$$
$$+\frac{1}{\sin(\theta_0/2)}\frac{i\sqrt{2}e^{i\pi/4}e^{ik_1L}}{\pi\sqrt{k_1L}\,M_+(k_1)}$$
$$\left.+\frac{i\sqrt{2}\mu_1e^{i\pi/4}e^{ik_2L}M_+(k_1)}{\mu_2\pi\sqrt{k_2L}\,[M_+(k_2)]^2}+O((k_{1(2)}L)^{-1})\right] \tag{7.200}$$

for sliding incidence of the sounding wave ($\theta_0\to+0,\ R\to-1$);

$$Q_1(k_1,k_2;\theta_0\to\theta_c)$$
$$=\sqrt{\frac{\mu_2}{\mu_1+\mu_2}}\left[-\frac{2k_1^2e^{i\pi/4}e^{i(k_1+k_2)L}}{\pi\sqrt{k_1L}[M_+(k_1)]^2M_+(k_2)(k_1+k_2)\sqrt{k_1-k_2}}\right.$$
$$\left.+\frac{\mu_1e^{i\pi/4}e^{2ik_2L}(k_1+k_2)^{3/2}}{2k_2\mu_2\pi\sqrt{k_2L}[M_+(k_2)]^3}+O((k_{1(2)}L)^{-1})\right], \tag{7.201}$$

$$Q_2(k_1,k_2;\theta_0\to\theta_c)=\sqrt{\frac{\mu_2}{\mu_1+\mu_2}}\left[-\frac{i\sqrt{2(k_1^2+k_2^2)}\,e^{ik_2L}}{\sqrt{\pi k_1}M_+(k_2)}\right.$$
$$\left.+\frac{i\sqrt{2(k_1+k_2)}}{\sqrt{\cos\theta_c-\cos\theta_0}}\frac{e^{i\pi/4}e^{ik_1L}}{\pi\sqrt{k_1L}M_+(k_2)}+O((k_{1(2)}L)^{-1})\right] \tag{7.202}$$

for sounding at a critical angle ($|k_2|<|k_1|,\theta_0\to\text{arc}\cos(k_2/k_1),\ R\to1$).

For the sliding and critical incidence angles, the asymptotic expressions (7.200) and (7.202) have singularities that indicate a tendency to increase the modules of the SIF if $\theta_0 \to 0°$ or $\theta_0 \to \theta_c$. From (7.200), it follows that the module of the SIF increases as $1/\theta_0$ if $\theta_0 \to 0$ for the crack vertex distant from the radiation source. This feature is introduced by the second term of formula (7.200), which is responsible for the interaction of the SH-diffracted wave between the crack vertices. For the vertex closest to the radiation source, the behavior of the SIF module under the sliding irradiation is determined by (7.199), which indicates its slow attenuation with increasing crack width. As follows from the (7.202) if the interface crack is irradiated at a critical angle $\theta_c = \arccos(k_2/k_1)$, $|k_2| < |k_1|$, then the SIF in the vicinity of the remote vertex admits a singularity as $1/\sqrt{\theta - \theta_c}$. This feature is manifested when the materials do not lose of the strain energy $\left(\mathrm{Im} k_{1(2)} = 0\right)$, and it is associated with the mutual scattering of SH-edge waves between the crack vertices. Since there is always a loss of strain energy in real materials, the critical angle will be complex. Consequently, when the real structures are irradiated at a critical angle, only a limited increase of the SIF modulus is observed. For a crack tip closer to the source, the behavior of the SIF modulus in the case of a critical irradiation angle slowly decays with increasing crack width.

7.4.3 Numerical Examples

Let us consider a finite crack at a flat interface between two half-spaces filling by the different metals. Let the upper half-space be occupied by soft iron (medium "1"), and the lower one filled with copper (medium "2") with the densities and Lamé parameters as $\rho_1 = 7.8\ \mathrm{Mg/m^3}$, $\mu_1 = 77.0\,\mathrm{GPa}$ and $\rho_2 = 8.96\ \mathrm{Mg/m^3}$, $\mu_2 = 45.0\,\mathrm{GPa}$. Let this structure be irradiated by an SH-wave propagating normally up to the connection limit; under these conditions the SIF $Q_1^* = Q_2^* = Q^*$. In Fig. 7.11, the dependences $|Q^*|$ on the dimensionless crack width obtained in [27, 57] and our results are compared. For comparison, scale factors $\chi = 5.104$ and $\chi = 15.034$ were used. Curves 2 and 3 practically coincide in all diapason of analysis, and curve 1, which is obtained by the integral equations method, qualitatively coincides with them for $k_1 L/2 > 4$ $(L > 4\lambda/3)$.

Further, we study the dependencies SIF on the different parameters under high frequency irradiation of the interface crack. Let us consider the finite interface crack in joint of "iron-zinc" half-spaces; the iron fills the upper half-space (medium "1") and zinc fills the lower one (medium "2") with the densities and Lamé parameters as ρ_1=7.87 $\mathrm{Mg/m^3}$, μ_1=82.0 GPa and ρ_2=6.92 $\mathrm{Mg/m^3}$, μ_2=40.2 GPa. This structure is illuminated by the plane SH-wave which propagates from medium "1" to medium "2". Figure 7.12 presents the dependencies of the normalized modules of SIF on the dimensionless crack width L/λ for the different illumination angles. The curves in this figure oscillate with respect to the monotonous lines that correspond to the normalized module of SIF for the vertexes of the semi-infinite cracks that occupy the semi-planes $\{-\infty < x < 0,\ y = 0,\ -\infty < z < \infty\}$ and

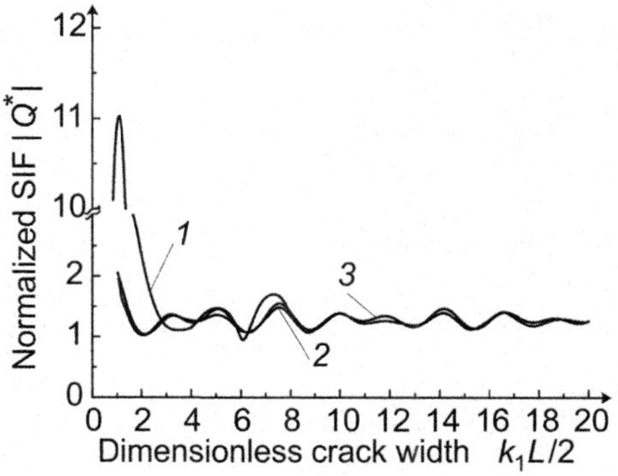

Fig. 7.11 Comparison of the normalized dependencies SIF on the dimensionless width for the vertices of the interface crack in an "soft iron-copper" joint under the normal ($\theta_0 = 90°$) illumination of a plane SH-wave: 1—obtained in [27]; 2—obtained in [57]; 3—our results

$\{-L < x < \infty, y = 0, -\infty < z < \infty\}$. The oscillations of the normalized modules of SIF attenuate if the wavelength of the crack increases; these oscillations tend to the values $|Q_1^*|$, $|Q_2^*|$ for the semi-infinite cracks. This behavior of the SIF is observed for different sounding angles. As follows from this, the oscillations of the SIF are caused by the multiple scattering of the edge waves from the crack vertexes [54].

Let us now consider the joint of "zinc-iron" half-spaces with a crack at the interface. Unlike in the case of "iron-zinc" joint, here the velocity of the SH-wave in the upper half-space (zinc) is less than in the lower one (iron). We also consider the "aluminum-steel" joint with the interface crack, where $\rho_1 = 2.7$ Mg/m^3, $\mu_1 = 26.0$ GPa and $\rho_2 = 7.86$ Mg/m^3, $\mu_2 = 80.0$ GPa. Note that for these two joints, the relation $|k_2| < |k_1|$ is valid, and their illumination at a critical angle leads to the total internal reflection phenomenon.

For these joints, the dependencies of the modules SIF on the illumination angle are presented in Fig. 7.13. It is seen from this figure that the above-mentioned dependencies of $|Q_1^*|$ and $|Q_2^*|$ for each couple of joints are mirror symmetrical with respect to $\theta_0 = \pi/2$. From the behavior of the curves, we observe a significant increase of $|Q_2^*|$ and decrease of $|Q_1^*|$ if the "zinc–iron" and "aluminum–steel" joints are irradiated at critical angles (for these couples $\theta_c \approx 42°$ and $\theta_c \approx 14°$, respectively). For irradiation at an angle $\theta_0 = \pi - \theta_c$, we observe the opposite behavior of the SIF: $|Q_2^*|$ decreases and $|Q_1^*|$ increases.

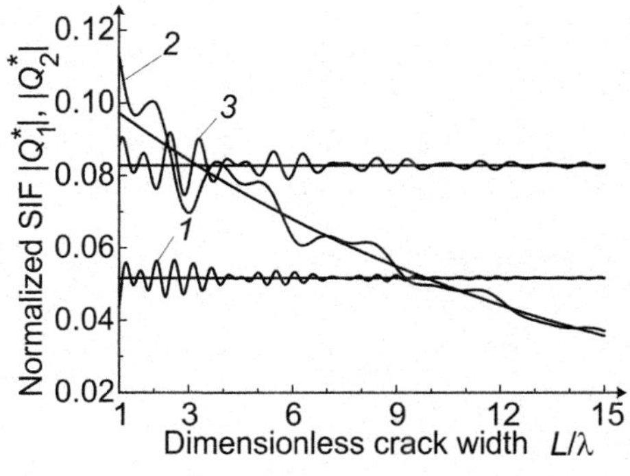

Fig. 7.12 Dependence of the modules of the normalized SIF on the parameter L/λ in the "iron-zinc" joint ($\text{Im}k_{1(2)} \neq 0$): 1—$|Q_1^*|$, 2—$|Q_2^*|$ for $\theta_0 = 45°$; 3—$|Q_1^*| = |Q_2^*|$ for $\theta_0 = 90°$

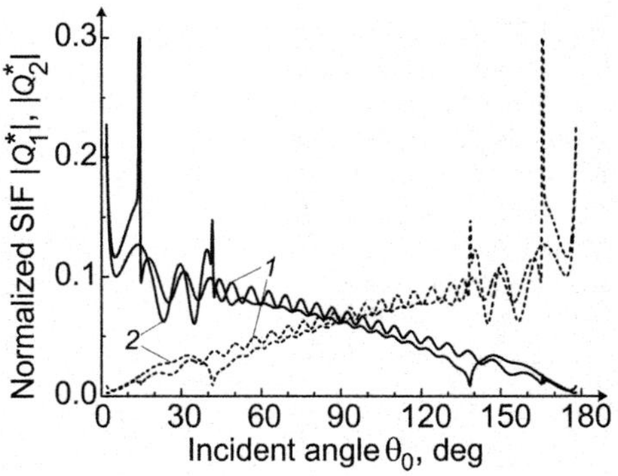

Fig. 7.13 Dependence of the modules of the normalized SIF on the sounding angle of joints with an interface crack: $L = 10\,\lambda$, $\mathrm{Im}k_{1(2)} = 0$: 1—"aluminum–steel"; 2—"zinc–iron; ----$|Q_1^*|$, —— $|Q_2^*|$

7.4.4 *Stress Intensity Factor of the Interface Semi-infinite Cracks*

Let us consider two semi-infinite interface cracks irradiated by the plane SH wave; the first one of them occupies the region $\{-\infty < x < 0,\ y = 0,\ -\infty < z < \infty\}$ and the second one $\{-L < x < \infty,\ y = 0, -\infty < z < \infty\}$. For the first and the second cracks, the SIF factors Q_1^* and Q_2^* are determined by the first terms of the expressions (7.187), (7.188) respectively. For these two cases we get that

$$Q_1(k_1, k_2; \theta_0 = \pi/2)$$
$$= Q_2(k_1, k_2; \theta_0 = \pi/2) = \frac{k_1}{\sqrt{2\pi}}\sqrt{\frac{\mu_1 k_1 + \mu_2 k_2}{(\mu_1 + \mu_2) k_1 k_2}} \tag{7.203}$$

for normal incidence of the sounding wave;

$$Q_1(k_1, k_2; \theta_0 \to +0) \to 0,$$
$$Q_2(k_1, k_2; \theta_0 \to +0) = -\frac{2i e^{ik_1 L}}{M_+(k_1)}\sqrt{\frac{\mu_2 k_1}{\pi(\mu_1 + \mu_2)}} \tag{7.204}$$

for sliding incidence of the sounding wave;

$$Q_1(k_1, k_2; \theta_0 \to \theta_c) \to 0,$$
$$Q_2(k_1, k_2; \theta_0 \to \theta_c) = -\frac{i e^{ik_2 L}}{M_+(k_2)}\sqrt{\frac{2\mu_2(k_1^2 + k_2^2)}{\pi k_1(\mu_1 + \mu_2)}} \tag{7.205}$$

for sounding at a critical angle.

Note the SIF Q_1^* and Q_2^* are expressed through Q_1 and Q_2 by the relations (7.196). As it follows from the expression (7.205) $Q_1(\theta_0 \to \theta_c) \to 0$. This is explained by the fact that the reflection coefficient $R \to 1$ if $\theta_0 \to \theta_c$, and Q_1 is proportional to

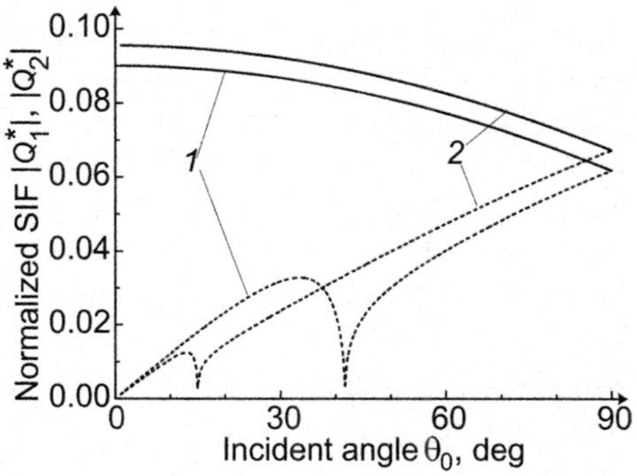

Fig. 7.14 Dependencies the modules of the normalized SIF on the incident angle for semi-infinite interface cracks: 1—"aluminum–steel" joint; 2—"zinc–iron" joint; ---- $|Q_1^*|$, — $|Q_2^*|$

$(R(\cos\theta_0) - 1)M_-(k_1 \cos\theta_0) \to 0$ (7.187). As follows from (7.188), the function $Q_2(\theta_0)$ is not equal to zero if $\theta_0 \to \theta_c$, because

$$\lim_{\theta_0 \to \theta_c} [(R(\cos\theta_0) - 1)M_-(k_1 \cos\theta_0)] = -\frac{2}{M_+(k_2)}. \tag{7.206}$$

Comparing the behavior of the curves in Figs. 7.14, 7.13 and 7.12, we conclude that the oscillatory dependence of the SIF moduli for the finite interface cracks is due to the mutual diffraction of the edge waves from the crack vertices.

As it follows from the above study, the approximate analytical expressions of the SIF for two vertices of the interface crack between the joint of two half-spaces illuminated by the plane SH-wave are obtained. The key expressions (7.187), (7.188) establish the dependencies of the dynamical SIF on the problem parameters. It is found that the SIF for the vertex that is closer to the radiated source is limited, but the SIF for distant vertex has two types of singularities; when the interface crack is illuminated under the sliding angle $\theta_0 \to 0°$ and if the crack is illuminated under the critical angle $\theta_0 \to \theta_c$. The cause of these singularities is the mutual edge wave diffraction from the crack vertexes. Note the SIF at the vertex of semi-infinite interface crack is limited.

7.5 SH-Waveguide Modes Diffraction from the Finite Interface Crack on the Rigid Junction of the Elastic Plate and the Semi-space

7.5.1 Introduction

In Sects. 7.3, 7.4, we considered a model of the elastic wave diffraction from the crack of finite width that is located at the interface between two homogeneous elastic half-spaces. The solution of this problem made it possible to elucidate the features of the formation of diffracted fields scattered from the interface defects in a wide frequency range. In order to elucidate the manifestation of the diffraction effects caused by the scattering of elastic waves from the defects in the limited structures, we consider the problem of SH-wave diffraction from the interface crack in the rigid joint of the elastic layer with the half-space. The propagation of the elastic waves in the layers with the defects on the interface is simulated by joints of the waveguides, on the walls of which different types of boundary conditions are specified [61–64]. The general approaches to studying the propagation of elastic waves in waveguide structures are given in [33, 65–69], and the features of their interaction with defects were considered in [70–73].

7.5.2 Formulation of the Problem

Let us consider the interface crack occupying in the Cartesian coordinates the region $\Gamma : \{x \in (-L, 0),\ y = 0,\ z \in (-\infty, \infty)\}$ between the rigid joint of the homogeneous elastic layer $\overline{P} : \{x \in (-\infty, \infty),\ y \in (-d, 0),\ z \in (-\infty, \infty)\}$ and the upper semi-spaces (Fig. 7.15). Let this crack be illuminated by the propagating SH-mode in the elastic layer without defects. Therefore, the diffraction problem is reduced to the boundary value problem for Helmholtz equation as

$$\Delta u + k^2 u = 0, \tag{7.207}$$

$$\tau_{zy} = \mu \frac{\partial (u + u^i)}{\partial y} = 0,\ y = -d;\ x \in (-\infty, \infty), \tag{7.208}$$

$$\tau_{zy} = \mu \frac{\partial (u + u^i)}{\partial y} = 0,\ y = 0;\ x \in (-L, 0), \tag{7.209}$$

$$u^t = u + u^i = 0,\ y = 0;\ x \in (-\infty, -L) \cup (0, \infty). \tag{7.210}$$

Here, $u = u(x, y)$ is the unknown diffracted field, $\mathbf{u}(\equiv \vec{e}_z u(x, y))$ is the displacement vector, $u^t = u^t(x, y)$ is the total field, $u^i(x, y) = e^{\gamma_j x} \sin(\beta_j y)$ is the incident

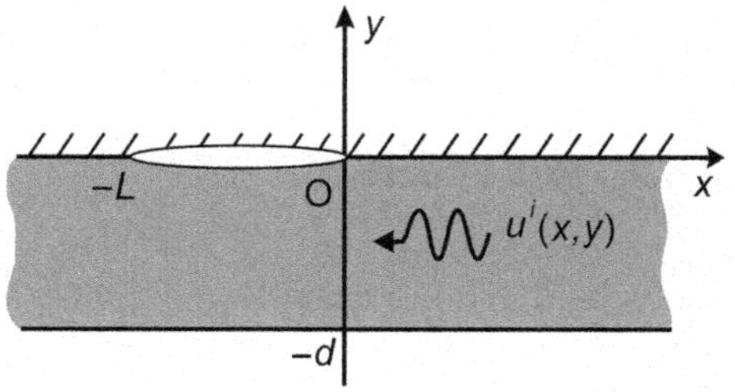

Fig. 7.15 Joint of the elastic layer and half-space with the interface crack

mode which is propagate in the negative direction of axis x; $\beta_j = \pi(2j-1)/2d$, $j = 1, 2, 3\ldots$; $\gamma_j = \sqrt{\beta_j^2 - k^2}$, $\mathrm{Re}\gamma_j > 0$; $k = k' + ik''$ ($k', k'' > 0$, $k' >> k''$).

In order to ensure the uniqueness of the solution of the boundary value problem (7.207)–(7.210), the unknown diffracted field should satisfy the edge condition

$$u \sim \rho^{1/2},\ \partial u/\partial y \sim \rho^{-1/2} \text{ if } \rho = [x^2 + y^2]^{1/2} \to 0$$

and

$$\rho = [(x+L)^2 + y^2]^{1/2} \to 0, \tag{7.211}$$

as well as the radiation condition in the limiting absorption condition form (7.38).

7.5.3 Solution for Semi-infinite Interface Crack

Let us introduce the Fourier transform of the diffracted field $u(x, y)$:

$$U(\alpha, y) = \frac{1}{\sqrt{2\pi}} \int_{-\infty}^{\infty} u(x, y) e^{i\alpha x} \mathrm{d}x. \tag{7.212}$$

Appling the Fourier transformation to Eq. (7.207), we arrive at the transformed wave equation, the solution of which we represent as

$$U(\alpha, y) = B(\alpha) e^{\gamma y} + C(\alpha) e^{-\gamma y}. \tag{7.213}$$

Here, $B(\alpha)$, $C(\alpha)$ are the unknown functions; $\gamma = \sqrt{\alpha^2 - k^2}$, $\mathrm{Re}\gamma > 0$; $U(\alpha, y)$ is the regular function in the strip $\alpha \in \Pi : \{-\tau_0 < \tau < \tau_0\}$ of a complex plane α, where $\tau_0 < k''$, $k'' < \mathrm{Re}\gamma_1$ if $j > 1$.

Let us suppose that $L \to \infty$ and introduce the Fourier integrals as

$$U'^{+}(\alpha, 0) = \frac{1}{\sqrt{2\pi}} \int_{0}^{\infty} u'_y(x, 0) e^{i\alpha x} \mathrm{d}x, \tag{7.214}$$

$$U'^{-}(\alpha, 0) = \frac{1}{\sqrt{2\pi}} \int_{-\infty}^{0} u'_y(x, 0) e^{i\alpha x} \mathrm{d}x, \tag{7.215}$$

where $U'^{+}(\alpha, 0)$ is the regular function of the parameter α in the upper half-plane $\tau > -\tau_0$; the function $U'^{-}(\alpha, 0)$ is regular in the lower half-plane $\tau < \tau_0$.

Differentiating the expression (7.213) according to the variable y and applying the integral Fourier transformation to the boundary condition (7.208), we find that

$$C(\alpha) = B(\alpha)e^{-2\gamma d}. \tag{7.216}$$

Taking into account the expressions (7.212)–(7.216) and using the boundary condition (7.209), we get the equation as

$$U'^{+}(\alpha, 0) + \frac{i\beta_j}{\sqrt{2\pi}(\alpha - i\gamma_j)} = \gamma B(\alpha)\left[1 - e^{-2\gamma d}\right]. \tag{7.217}$$

Next, using (7.217), we rewrite Eq. (7.213) in the form

$$U(\alpha, y) = \left[\frac{i\beta_j}{\sqrt{2\pi}(\alpha - i\gamma_j)} + U'^{+}(\alpha, 0)\right]\frac{\text{ch}\gamma(y + d)}{\gamma\,\text{sh}\,(\gamma d)}. \tag{7.218}$$

Using (7.218) and applying the Fourier transformation to the boundary condition (7.210), we arrive at the Wiener–Hopf equation as

$$M(\alpha)U'^{+}(\alpha, 0) + \frac{i\beta_j M(\alpha)}{\sqrt{2\pi}(\alpha - i\gamma_j)} + J_1(\alpha) = 0, \alpha \in \Pi. \tag{7.219}$$

Here, $U'^{+}(\alpha, 0)$, $J_1(\alpha)$ are the unknown functions; $J_1(\alpha)$ is regular in the lower half-plane $\tau < \tau_0$; $M(\alpha)$ is the kernel function, regular in the strip $\alpha \in \Pi$ and has no zeros there; outside of this strip, the function $M(\alpha)$ allows for the simple poles and zeros;

$$J_1(\alpha) = -\frac{1}{\sqrt{2\pi}}\int_{-\infty}^{0} u(x, 0)e^{i\alpha x}\text{d}x, \tag{7.220}$$

$$M(\alpha) = \frac{\text{ch}(\gamma d)}{\gamma\text{sh}(\gamma d)}. \tag{7.221}$$

Factorization of the kernel function (7.221) using the expressions (7.13), (7.23), (7.24) leads to its representation in the form

$$M(\alpha) = M_{+}(\alpha)M_{-}(\alpha). \tag{7.222}$$

Here,

$$M_{-}(\alpha) = M_{+}(-\alpha),$$

$$M_+(\alpha) = \frac{\sqrt{\cos(kd)} \prod_{n=1}^{\infty} \left[1 + \frac{\alpha}{i\gamma_{nc}}\right] e^{i\alpha\frac{d}{\pi n}}}{i\sqrt{k\sin(kd)}\left(1 + \frac{\alpha}{i\gamma_{0s}}\right) \prod_{n=1}^{\infty} \left[1 + \frac{\alpha}{i\gamma_{ns}}\right] e^{i\alpha\frac{d}{\pi n}}}, \tag{7.223}$$

$$\gamma_{nc} = \sqrt{\pi^2(2n-1)^2/(4d^2) - k^2},\ \gamma_{ns} = \sqrt{(\pi n/d)^2 - k^2}(\gamma_{0s} = -ik);$$

$M_+(\alpha)$, $M_-(\alpha)$ are the split functions, regular and have no zeros in the half-planes $\tau > -\tau_0$, $\tau < \tau_0$ respectively; in the regularity regions $M_\pm(\alpha) = O(\alpha^{-1/2})$ if $|\alpha| \to \infty$. Outside of the regularity regions, the functions $M_\pm(\alpha)$ have the simple zeros and poles at $\alpha = \mp i\gamma_{nc}$ and $\alpha = \mp i\gamma_{ns}$ $(n = 1, 2, 3, \ldots)$ respectively, and they have the simple pole at $\alpha = \mp k$.

Let us rewrite Eq. (7.219) as follows

$$M_+(\alpha)U'^+(\alpha, 0) + S^+(\alpha) = -\frac{J_1(\alpha)}{M_-(\alpha)} - S^-(\alpha),\ \alpha \in \Pi, \tag{7.224}$$

where

$$S^+(\alpha) = \frac{i\beta_j M_+(\alpha)}{\sqrt{2\pi}(\alpha - i\gamma_j)} - \frac{i\beta_j M_+(i\gamma_j)}{\sqrt{2\pi}(\alpha - i\gamma_j)}, \tag{7.225}$$

$$S^-(\alpha) = \frac{i\beta_j M_+(i\gamma_j)}{\sqrt{2\pi}(\alpha - i\gamma_j)}. \tag{7.226}$$

The asymptotic behavior of the function (7.214) in the regularity regions if $|\alpha| \to \infty$ is as: $U'^+(\alpha, 0) = O(\alpha^{-1/2})$. The function (7.220) tends to zero in the lower half-plane $\tau < \tau_0$ as $J_1(\alpha) = O(\alpha^{-3/2})$ if $|\alpha| \to \infty$. The terms in the left-side of Eq. (7.224) are regular functions of the complex variable α in the upper half-plane $\tau > -\tau_0$. The terms in the right side of this equation are regular for $\tau < \tau_0$. Each term in the functional equation (7.224) tends to zero in the regularity regions as α^{-1} if $|\alpha| \to \infty$. The terms in the right and in the left sides of Eq. (7.224) have the common strip of regularity Π. Therefore, applying the Liouville's theorem, we find the solution of the Wiener–Hopf equation (7.224) as

$$U'^+(\alpha, 0) = -\frac{i\beta_j}{\sqrt{2\pi}(\alpha - i\gamma_j)}\left[1 - \frac{M_+(i\gamma_j)}{M_+(\alpha)}\right], \tag{7.227}$$

$$J_1(\alpha) = -\frac{i\beta_j M_+(i\gamma_j) M_-(\alpha)}{\sqrt{2\pi}(\alpha - i\gamma_j)}. \tag{7.228}$$

Further, using the relations (7.218) and (7.227) and applying the inverse Fourier transformation (7.3), we arrive at the expression to determine the diffracted field in the layer $\overline{P}$ that is rigidly joint with the half-space; the interface crack in the junction occupy the half-plane surface $x < 0,\ y = 0$:

$$u(x, y) = \begin{cases} \sum_{q=1}^{\infty} R_{jq} e^{-\gamma_{qc} x} \sin\left(\frac{\pi(2q-1)}{2d} y\right), & x > 0; \\ -\sin(\beta_j y) e^{\gamma_j x} + \sum_{q=0}^{\infty} T_{jq} \cos\left(\frac{\pi q}{d} y\right) e^{\gamma_{qs} x}, & x < 0. \end{cases} \quad (7.229)$$

Here, we express the wave modes transformation coefficients in the form

$$R_{jq} = \frac{\pi \beta_j M_+(i\gamma_j)}{2d} \frac{(2q-1) M_+(i\gamma_{qc})}{(\gamma_{qc} + \gamma_j)\gamma_{qc}}, \quad (7.230)$$

$$T_{jq} = \frac{\varepsilon_q \beta_j M_+(i\gamma_j)}{d M_+(i\gamma_{qs})(\gamma_{qs} - \gamma_j)\gamma_{qs}}, \quad (7.231)$$

where $\varepsilon_q = 1/2$ if $q = 0$ and $\varepsilon_q = 1$, if $q > 0$.

From the expressions (7.230), (7.231) directly follows that $R_{jq},\ T_{jq} = O(q^{-3/2})$. Therefore, $u = O(x^{1/2})$ and $du/dy = O(x^{-1/2})$ if $x,\ y \to 0$.

7.5.4 Solution for the Finite Interface Crack

In this case the boundary value problem (7.207)–(7.210) is reduced to the following Wiener–Hopf equation

$$\left[\Psi^{(+)}(\alpha) + e^{-i\alpha L}\Psi^{-}(\alpha)\right] M(\alpha) + J_1(\alpha) = 0, \ \alpha \in \Pi. \quad (7.232)$$

Here, $\Psi^{-}(\alpha)$, $\Psi^{(+)}(\alpha)$, $J_1(\alpha)$ are unknown functions: $\Psi^{-}(\alpha)$ is regular in the complex half-plane $\tau < \tau_0$; $\Psi^{(+)}(\alpha)$ is regular in the region $\tau > -\tau_0$, except the point $\alpha = i\gamma_j$ $\left(\mathrm{Re}\gamma_j > \tau_0\right)$, where it has the simple pole; $J_1(\alpha)$ is the entire function; $M(\alpha)$ is given by the formula (7.221);

$$\Psi^{-}(\alpha) = U'^{-}(\alpha, 0) - \frac{i\beta_j e^{-\gamma_j L}}{\sqrt{2\pi}(\alpha - i\gamma_j)}; \quad (7.233)$$

$$\Psi^{(+)}(\alpha) = U'^{+}(\alpha, 0) + \frac{i\beta_j}{\sqrt{2\pi}(\alpha - i\gamma_j)}; \quad (7.234)$$

$$U'^{-}(\alpha) = \frac{1}{\sqrt{2\pi}} \int_{-\infty}^{-L} u'_y(x, 0) e^{i\alpha(x+L)} \mathrm{d}x; \quad (7.235)$$

$$J_1(\alpha) = -\frac{1}{\sqrt{2\pi}} \int_{-L}^{0} u(x, 0) e^{i\alpha x} \mathrm{d}x. \quad (7.236)$$

The asymptotic behaviour of the functions (7.233), (7.234) if $|\alpha| \to \infty$ in the regularity regions is as: $\Psi^{(+)}(\alpha) = O(\alpha^{-1/2})$, $\Psi^{-}(\alpha) = O(\alpha^{-1/2})$. The entire function (7.236) in the region $\tau < \tau_0$ decays as $J_1(\alpha) = O(\alpha^{-3/2})$ if $|\alpha| \to \infty$, and $e^{i\alpha L} J_1(\alpha) = O(\alpha^{-3/2})$ if $\tau > -\tau_0$ and $|\alpha| \to \infty$. Taking this into account, we rewrite Eq. (7.232) as

$$M_{+}(\alpha)\Psi^{(+)}(\alpha) + e^{-i\alpha L} M_{+}(\alpha)\Psi^{-}(\alpha) + \frac{J_1(\alpha)}{M_{-}(\alpha)} = 0,\ \alpha \in \Pi, \tag{7.237}$$

$$M_{-}(\alpha)e^{i\alpha L}\Psi^{(+)}(\alpha) + M_{-}(\alpha)\Psi^{-}(\alpha) + \frac{e^{i\alpha L} J_1(\alpha)}{M_{+}(\alpha)} = 0, \alpha \in \Pi. \tag{7.238}$$

Using the decomposition procedure (7.21), (7.22) of the functional equations (7.237), (7.238), we arrive at the integral equations of the second kind, and finally, we reduce these equations to the functional relations as

$$M_{+}(\alpha)\Psi^{(+)}(\alpha) - \sum_{n=0}^{\infty} \frac{\varepsilon_n e^{-\gamma_{ns} L}\Psi^{-}(-i\gamma_{ns})}{id\gamma_{ns} M_{+}(i\gamma_{ns})(i\gamma_{ns} + \alpha)} = \frac{i\beta_j}{\sqrt{2\pi}} \frac{M_{+}(i\gamma_j)}{\alpha - i\gamma_j}, \tag{7.239}$$

$$M_{-}(\alpha)\Psi^{-}(\alpha) + \sum_{n=0}^{\infty} \frac{\varepsilon_n e^{-\gamma_{ns} L}\Psi^{(+)}(i\gamma_{ns})}{id\gamma_{ns} M_{+}(i\gamma_{ns})(\alpha - i\gamma_{ns})} = 0, \tag{7.240}$$

where $\varepsilon_n = 1/2$ if $n = 0$ and $\varepsilon_n = 1$ if $n > 0$.

Putting $\alpha = i\gamma_{ls}$ in the first equation and $\alpha = -i\gamma_{ls}$ in the second one, $l = 0, 1, 2, \ldots$, we get from the functional equations (7.239), (7.240) the infinite system of linear algebraic equations (ISLAE) of the second kind that we represent in the matrix form as

$$[I + A]X = F. \tag{7.241}$$

Here, $X = \{x_n\}_{n=0}^{\infty}$, $x_n = M_{+}(i\gamma_{ns})\Psi^{(+)}(i\gamma_{ns})$, $x_n = O(n^{-1})$ if $n \to \infty$; I is the identical matrix, $A : \{a_{ln}\}_{n,l=0}^{\infty}$, $F = \{f_l\}_{l=0}^{\infty}$;

$$\begin{aligned} a_{ln} = & -\frac{\varepsilon_n d^{-2} e^{-\gamma_{ns} L}}{\gamma_{ns}[M_{+}(i\gamma_{ns})]^2} \\ & \times \sum_{m=0}^{\infty} \frac{\varepsilon_m e^{-\gamma_{ms} L}}{[M_{+}(i\gamma_{ms})]^2 \gamma_{ms}(\gamma_{ls} + \gamma_{ms})(\gamma_{ms} + \gamma_{ns})}; \end{aligned} \tag{7.242}$$

$$f_l = \frac{\beta_j}{\sqrt{2\pi}} \frac{M_{+}(i\gamma_j)}{\gamma_{ls} - \gamma_j}. \tag{7.243}$$

For the matrix elements (7.242), the next asymptotic estimation is valid

$$|a_{ln}| \leq B\frac{e^{-\pi nL/d}}{ln} \tag{7.244}$$

if $n,\ p \to \infty$, where $B = \text{const}$. Therefore, $\sum_{l,n} |a_{ln}|^2 < \infty$and ISLAE (7.241) have an unique solution, except the discrete values of the wave number for which the corresponding homogeneous equation allows for the non-trivial solution.

Let us represent the diffracted displacement field through the integral

$$u(x, y) = \frac{1}{\sqrt{2\pi}} \times \int_{-\infty}^{\infty} [\Psi^{-}(\alpha)e^{-i\alpha L} + \Psi^{(+)}(\alpha)]\frac{\text{ch}\gamma(y+d)}{\gamma\ \text{sh}(\gamma d)}e^{-i\alpha x}d\alpha. \tag{7.245}$$

Here, $\Psi^{-}(\alpha)$, $\Psi^{(+)}(\alpha)$ are determined by the expressions (7.239), (7.240). The field representation (7.245) can be evaluated through the residue series. For the region $D_1 = \{x, y : -L < x < 0,\ -d < y < 0\}$ the series representation of the diffracted field looks like

$$u^{(t)}(x, y) = \frac{\sqrt{2\pi}}{d} \times \sum_{q=0}^{\infty} \frac{\varepsilon_q}{\gamma_{qs}} \cos\frac{\pi q y}{d}[\Psi^{(+)}(i\gamma_{qs})e^{\gamma_{qs}x} + \Psi^{-}(-i\gamma_{qs})e^{-\gamma_{qs}(x+L)}]. \tag{7.246}$$

The dispersion equation to determine the spectra of the initial diffraction problem looks as

$$\det[I + A(\Omega)] = 0. \tag{7.247}$$

Here, $\Omega = kd = \omega d/c$, c is the velocity of the SH-wave in the elastic layer, $\omega = \text{Re}(\omega) + i\,\text{Im}(\omega)$ is the complex frequency, $\text{Re}(\omega) > 0$, $\text{Im}(\omega) \leq 0$.

7.5.5 *One Mode Approximation*

Let the dimensionless width of the elastic waveguide layer $kd < \pi/2$. Then, in the waveguide region $D_1 = \{x, y : -L < x < 0,\ -d < y < 0\}$ only one of the undamped modes is excited. Therefore, if $kL >> 1$, we find from the Eqs. (7.239), (7.240) one mode approximate equations:

$$\Psi^{(+)}(k) - \frac{e^{ikL}}{4k^2 d M_{+}^2(k)}\Psi^{-}(-k) = \frac{\beta_j}{\sqrt{2\pi}}\frac{M_{+}(i\gamma_j)}{M_{+}(k)(k-\gamma_j)}, \tag{7.248}$$

$$\Psi^{-}(-k) - \frac{e^{ikL}}{4k^2 d\, M_+^2(k)} \Psi^{(+)}(k) = 0. \tag{7.249}$$

We represent the solution of these equations as

$$\Psi^{(+)}(k) = \frac{\beta_j}{\sqrt{2\pi}} \frac{M_+(i\gamma_j)}{M_+(k)(k-\gamma_j)D(k)}, \tag{7.250}$$

$$\Psi^{-}(-k) = \frac{\beta_j M_+(i\gamma_j)e^{ikL}}{4\sqrt{2\pi}k^2 d M_+^3(k)(k-\gamma_j)D(k)}. \tag{7.251}$$

Here,

$$D(k) = 1 - \left[\frac{e^{ikL}}{4k^2 d M_+^2(k)}\right]^2. \tag{7.252}$$

The approximate dispersive equation directly follows from (7.252):

$$4k^2 d M_+^2(k) \pm e^{ikL} = 0 \tag{7.253}$$

which allows for determination of the complex resonant frequency in one mode approximation.

7.5.6 Numerical Analysis

Let us consider SH-wave mode diffraction from the edge of the semi-infinite interface crack in joint of the elastic layer and half-space. For convenience of conducting the numerical analysis, we accept that wave number k is the complex value with a small imaginary part, $k''d \le 10^{-4}$. Let us consider the mode transformation coefficients at the crack edge R_{jq} (7.230) and T_{jq} (7.231). By virtue of the incident j-th waveguide mode propagates in the negative direction, the physical sense of these coefficients are the complex amplitudes of the q-th waveguide mode scattered from the edge into the regions $x > 0$ and penetrates in the delamination area $x < 0$, respectively. In Fig. 7.16, we observe the dependencies $|R_{1q}|$, $|T_{1q}|$ on the dimensionless width of the layer $h = \mathrm{Re}(kd)$ for several scattered and penetrating mode numbers q. As follows from the Fig. 7.16a, if the dimensionless width of the layer tends to the critical one, $h = \pi/2$ for the first mode of the perfectly join layer $x > 0$, the coefficient $|R_{11}| \to 1$.

We also observe in Fig. 7.16a and b that $|R_{1q>1}|$ and $|T_{1q}| \to 0$, if $h \to \pi/2$. This explains practically pure back scattering of the first SH-mode from the crack edge and the absence of any mode penetration into the delamination area. We can see from the behavior of the rest of the curves in Fig. 7.16a that the transformation

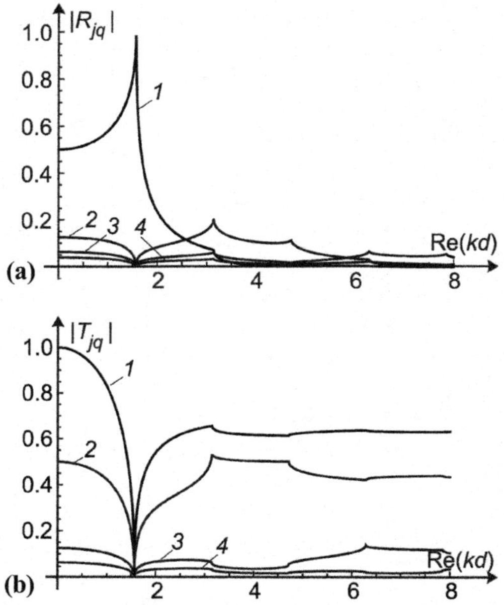

Fig. 7.16 Dependences $|R_{jq}|$ (**a**) and $|T_{jq}|$ (**b**) on the dimensionless width of the layer for $j = 1$; **a** 1—$q = 1$; 2—$q = 2$; 3—$q = 3$; 4—$q = 4$; **b** 1—$q = 0$; 2—$q = 1$; 3—$q = 2$; 4—$q = 3$

of the first mode into the higher ones is not effective for different h, except the slight growth $|R_{12}|$, if $h \to \pi$ (see curve 2) which is the critical dimensionless width of the layer for the crack area $x < 0$. From the behavior of the curves in Fig. 7.16b we can conclude that two lower modes dominantly penetrate into the delamination area. This follows from the dependence $|T_{10}|$, $|T_{11}|$ on the h (see curves 1 and 2). If $h < \pi$, only one mode with index $q = 0$ propagates without attenuation in the layer for $x < 0$. For $h > \pi$, the coefficient $|T_{10}|$ practically does not depends on h. Therefore, this wave propagates in the layer without the dispersion for $x < 0$. As it follows from the above analysis, the properties of wave modes transformation at the crack edge can be used to recognize the crack position using the displacement field analysis at the free layer surface.

In Fig. 7.17, we observe the essential difference between the behavior of the displacement field at the free face of the layer under the crack area $x < 0$ and under the area of the perfect joint $x > 0$ for different dimensionless frequency $\pi/2 < kd < 3\pi/2$. Under this condition in perfectly joint area $x > 0$ only the first mode propagates without attenuation. Then, if $kd = 2.5$, only the mode with index $q = 0$ is propagated under the crack area $x < 0$; the module of its complex amplitude practically does not depend on the normalized coordinate (see curve 1 in Fig. 7.17). Growing the dimensionless frequency kd leads to the excitation of the two modes in the region $x < 0$ that propagate without attenuation in the delaminated layer. This leads to the interference effects which we can observe in this figure (see curves 2) and explains the essential difference between the distribution of the displacement fields on the free layer face in perfectly joint area and in the damaged one. The above mentioned distribution essentially depends on the dimensionless frequency or on the

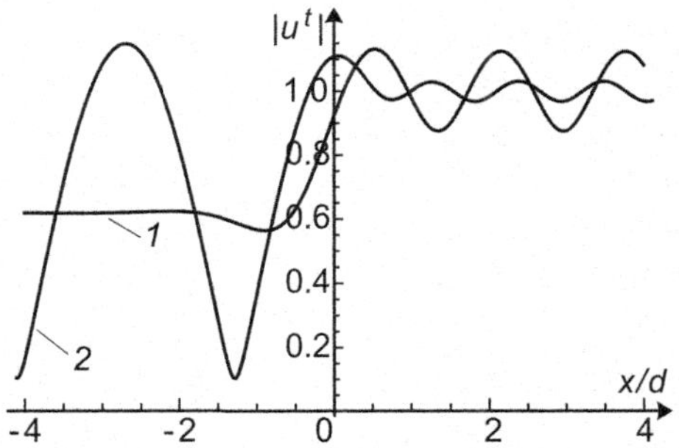

Fig. 7.17 Displacement field distribution on the free face of the layer for different dimensionless width/frequency; the structure is illuminated by the first waveguide mode for perfectly join layer: 1—$kd = 2.5$, 2—$kd = 3.3$

dimensionless layer width. Therefore, these effects can be applied for diagnostics of the interface damaged area in joint of the elastic layer and semi-space by using the SH-propagated modes.

Let us consider the finite interface crack at the boundary of the rigid joint between the half-space and the elastic layer. Let us consider this structure as chain of the rectangular waveguides $D_2 \cup D_1 \cup D_3$ with the different waveguide properties, where

$$\begin{aligned} &D_1 = \{x \in (-L, 0),\ y \in (-d, 0)\}, \\ &D_2 : \{x \in (-\infty, -L),\ y \in (-d, 0)\}, \\ &D_3 : \{x \in (0, \infty),\ y \in (-d, 0)\}. \end{aligned} \tag{7.254}$$

Taking into account the scattering properties of the open ends of waveguides (7.254), we consider the finite waveguide D_1 as an open resonator. It follows from the above analysis that the waveguide mode with a zero critical frequency can propagate in D_1 without the attenuation. The minimum critical frequency of the adjacent to D_1 the semi-infinite waveguides D_2, D_3 is $\min\{\omega_n\} = \pi c/(2d)$, where $\omega_n = \pi c(2n - 1)/(2d)$, $n = 1, 2, 3, \ldots$ and c is the SH-wave velocity in the layer. Therefore, the resonator D_1 can support the high quality vibration if $0 < \mathrm{Re}(\omega) < \min \omega_n$.

To determine the resonant regimes of excitation, let us find the complex roots of Eq. (7.247). For this purpose, let us consider Eq. (7.247) as a function of the complex spectral parameter Ω. This function is considered on the Riemann surface with the branch points on the real axis $\Omega = \pm\pi q$ and $\Omega = \pm\pi\,(2q - 1)/2$, $q = 1, 2, 3 \ldots$. The cuts of the complex plane ensure the suppression over time of the eigen oscillations of the resonator D_1 on the first sheet of the Riemann surface [74, 75]. The number of the complex roots Ω_n of Eq. (7.247) in the region $0 < \mathrm{Re}(\omega) < \mathrm{Re}(\omega_n)$, where $\omega_n = \Omega_n c/d$, for fixed parameter $p = L/d$ determines the brunches of the roots. Figure 7.18 a, b shows the dependences on the parameter p of the real $\left(\mathrm{Re}(\Omega_1^{(1,2,3)}) > 0\right)$ and imaginary $\left(\mathrm{Im}(\Omega_1^{(1,2,3)}) < 0\right)$ parts of the roots from the first three brunches in the region $\Omega < \pi/2$. It follows from the behaviour of the curves in Fig. 7.18b, if the parameter p growth, the module of the imaginary parts of the roots decrease, and consequently, the quality factor of oscillations in the resonant region D_1 increases.

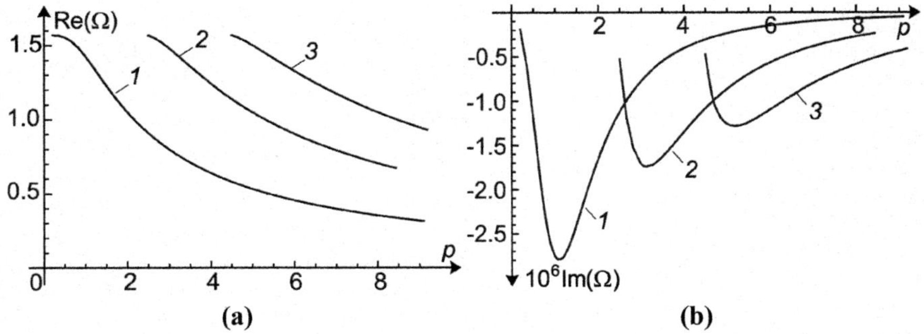

Fig. 7.18 Dependencies of the real (**a**) and the imaginary (**b**) parts of the complex roots of the dispersion equation (7.247) on the dimensionless parameter $p = L/d$; the curves 1, 2, 3 corresponds to the different branches of the complex roots

The data shown in Fig. 7.18 can be used to select the probing frequency for diagnosing the interface crack. The curve in Fig. 7.19 illustrates the typical distribution of the modulus of the displacement field on the free surface of the layer under the crack area when the structure is excited at resonant frequencies. To simplify the analysis, one mode approximate Eq. (7.253) can be used to determine the resonance frequencies; the curve in Fig. 7.20 shows the dependence of the relative error of calculation as a function of the parameter p:

$$\delta(p) = \frac{|\Omega_1^{(1)}(p) - \tilde{\Omega}_1^{(1)}(p)|}{|\Omega_1^{(1)}(p)|} \times 100\%.$$

Here, $\Omega_1^{(1)}(p)$ and $\tilde{\Omega}_1^{(1)}(p)$ are the complex roots from the first branch that are obtained of the Eqs. (7.247) and (7.253) respectively. As follows from this figure, the relative error does not exceed 0.04%.

Table 7.1 presents the complex resonance frequencies of the main mode that are excited in the open resonator formed by the interface finite cracks in the joint of the half-space with titanium ($c = 3270$ m/s) and nickel ($c = 2960$ m/s) layers; the layer thickness is $d = 10$ mm. Comparing the data given in Table 7.1, we see that the sensitivity of the resonant frequency to a change in the width of the crack is

Fig. 7.19 Displacement field distribution on the free face of the layer under the resonant excitation for $\mathrm{Re}(\Omega_1^{(1)}) = 0.453$, $p = L/d = 6$

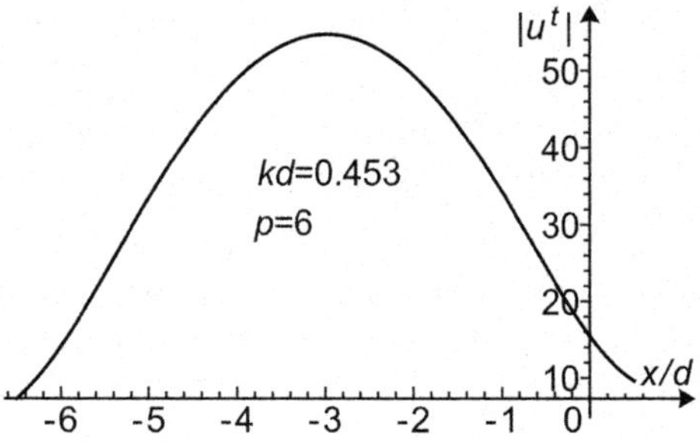

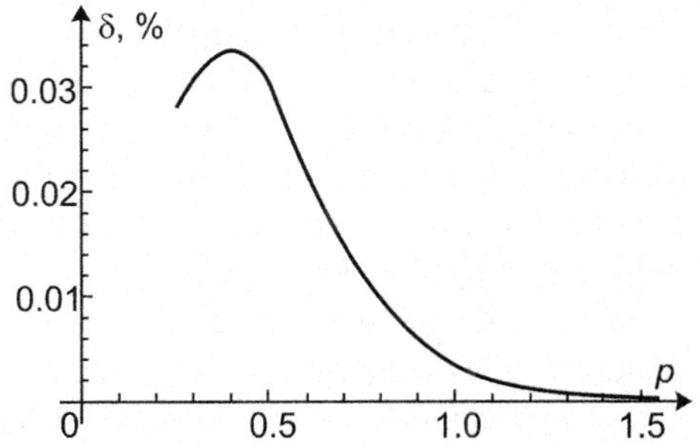

Fig. 7.20 Relative error in determining the roots from the first brunch by the approximate Eq. (7.253)

higher for the layers with a higher propagation velocity of SH-waves. Therefore, in such layers it is possible to more accurately detect changes in the width of the interfacial crack by the shift of the resonant frequency. These resonant frequencies can be applied for probing in the selective diagnosis of interfacial defects (tuning to detect cracks of a given width) or in determining the amount of crack growth from the shift of the resonant frequency. Note, for the short cracks, the resonant frequency is uniquely determined, and with an increase in the crack width ($p > 2.5$) the number of resonant frequencies corresponding to a fixed one increases (see Fig. 7.18).

Figure 7.21 shows the displacement field distribution on the free face of the layer joint with the semi-space if the interface crack of finite width exists. The layer width and the wave frequency are taken from the condition that only one lower SH-mode can propagate without attenuation in the perfectly joint waveguide areas D_2, D_3 (7.254) and at least two modes can propagate without attenuation in the waveguide

Table 7.1 Complex resonance frequencies

L, mm	Titanium $c = 3270$m/s ω, rad/sec	Nickel $c = 2960$m/s ω, rad/sec
1	$513645 - i\,0.0766074$	$464951 - i\,0.0693449$
2	$513520 - i\,0.376162$	$464838 - i\,0.34501$
3	$512967 - i\,0.886432$	$464337 - i\,0.802398$
5	$508704 - i\,2.44434$	$460478 - i\,2.21262$
10	$467195 - i\,5.68506$	$422905 - i\,5.14611$
15	$402326 - i\,4.97067$	$364185 - i\,4.49945$
20	$343917 - i\,3.44204$	$311313 - i\,3.11573$
30	$261295 - i\,1.60675$	$236524 - i\,1.45443$
	$484271 - i\,3.53072$	$438361 - i\,3.196$
40	$209259 - i\,0.839975$	$189421 - i\,0.760344$
	$406369 - i\,2.80089$	$367845 - i\,2.53536$
50	$174179 - i\,0.487761$	$157667 - i\,0.441521$
	$343239 - i\,1.79332$	$310700 - i\,1.62331$
	$491103 - i\,2.59858$	$444546 - i\,2.35223$

region under the crack area D_1 if $1.6 \leq kd \leq 3.3$. Note, the structure is illuminated by the first mode of the perfectly joint layer from the D_3 region. In Fig. 7.21, we observe the significant changes of the displacement field distribution under the crack area with a good contrast of the crack edges for wide band of the crack width. The fiches of the displacement field at the free face of the layer can be definitely used for detection of the damaged area.

Let us consider the total displacement fields on the free face of the plane layer in the waveguide region D_3. We observe the fields module oscillations along x-axis caused by the interference of the incident and reflected from the crack edge SH-modes of the first order. The amplitude and frequency of these oscillations are dependent on the parameter kd. For $kd = 1.6$ we observe practically total reflection of the incident mode from the crack edge; the interference of the reflected and incident modes leads to oscillations of the displacement field, the amplitude of which practically exceeds the incident wave twice. In this case we also observe the weak wave penetration into the damaged area D_1 and the formation there of the weak oscillations of the displacement field if the crack width is groused. These oscillations are caused by the interference of the zero-order modes scattered from the crack ends. In the region D_2 only the first mode is excited and we do not observe the interference pattern there. If the dimensionless layer thickness $kd = 2.5$, the structure of the field modes in

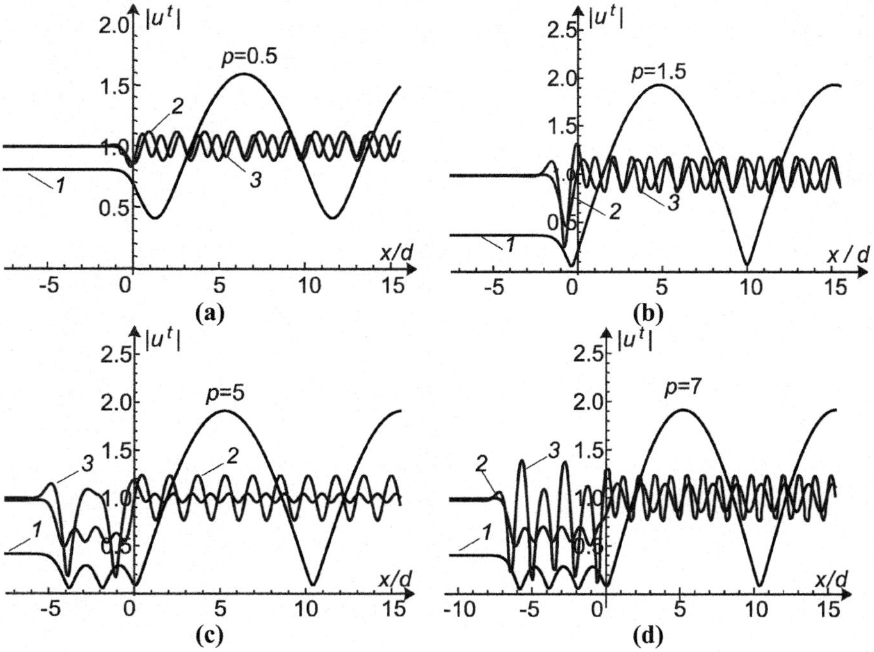

Fig. 7.21 Displacement field distribution on the free face of the layer for different dimensionless width/frequency and parameter $p = L/d$; the structure is illuminated by the first waveguide mode for perfectly joint layer: **a** $p = 0.5$; **b** $p = 1.5$; **c** $p = 5$; **d** $p = 7$; 1—$kd = 1.6$, 2—$kd = 2.5$, 3—$kd = 3.3$

the different areas of the layer is the same as in the previous case. The growth of the parameter kd leads to the increase of the frequency of the displacement field oscillations on the free layer surface and to the decrease of their amplitudes (see curves 2 in Fig. 7.21), which can be explain by the growth of the wave penetration into the damaged area D_1. The further growth of the parameter kd leads to the excitation of two modes (zero and first order) in D_1 and the qualitative change of the displacement field oscillation on the free layer surface under the crack in view of the more complicate interference processes (see curves 3 in Fig. 7.21c, d). All these transformations of the displacement field distribution on the layer free surface under the perfect junction as well as under the damaged one show the possibility to recognize the interface crack location. Note, here we used the one mode regime of excitation in the perfect junction area.

In this section, the simple model of the finite interface crack in joint of the half-space and the layer sounded by elastic SH-mode is considered. This problem is analyzed on the rigorous level using the Wiener–Hopf technique. The problem is reduced to the infinite systems of linear algebraic equations of the second kind (7.241), which is treated numerically. Based on the obtained solution, we found the dependencies of SH-wave mode transformation coefficients on the wave number, displacement field distribution in the free face on the problem parameters and spectral properties of the dynamical system: "simple joint with the finite interface crack". The obtained characteristics allow us to formulate the conditions for recognition of the interface crack.

References

1. Wiener N, Hopf E (1931) Uber eine klasse singularrer integralgelichungen. S.B. Preuss. Akad. Wiss. (in German)
2. Noble B (1958) Methods based on the Wiener–Hopf technique for the solution of partial differential equations. Pergamon, London
3. Fock VA (1965) Electromagnetic diffraction and propagation problems. International Series of monographs, vol 1, Oxford, Pergamon Press
4. Mittra R, Lee W (1971) Analytical techniques in the theory of guided Waves. Macmillan, New York
5. Analytical and numerical methods in electromagnetic wave theory (1993) Edited by Hashimoto M, Idemen M, Tretyakov O, Science House Co., Ltd., Tokyo
6. Direct and inverse electromagnetic scattering (1996) Edited by Serbest AH, Cloude SR, Pitman Research Notes in Mathematics Series 361, Addison Wesley Longman
7. Wainstein LA (1969) The theory of diffraction and factorization method. Golem Press, Boulder CA
8. Sveshnikov AG (1950) Radiation principle. Dokl Akad Nauk SSSR 75(5):917–920
9. Morse PM, Feshbach F (1953) Methods of theoretical physics, Part 1&2. McGraw Hill
10. Hönl H, Maue AW, Westpfahl K (1961) Theorie der beugung. In: Handbuch der Physik. Band XXV/1, S 218. Springer-Verlag, Berlin
11. Watson GN (1944) Treatise on the theory of Bessel functions. Cambridge University Press
12. Loeber JF, Sih GC (1968) Diffraction of antiplane shear waves by a finite crack. J Acoust Soc Am 44(1):90–98

13. Fil'shtinskii LA (1991) Diffraction of elastic waves by cracks, holes, and inclusions in isotropic media. Izv Ros Akad Nauk Mekh Tverd Tela (4):119–127
14. Popov VG, Ulanovskii AÉ (1995) Comparative analysis of the diffraction fields caused by elastic waves passing through defects of different nature. Izv Ros Akad Nauk Mekh Tverd Tela 4:99–109
15. Mal K (1970) Interaction of elastic waves with a Griffith crack. Int J Eng Sci 8:763–776
16. Van der Hijden JHMT, Neerhoff FL (1984) Scattering of elastic waves by a plane crack of finite width. J Appl Mech 51(3):646–651
17. Parton VZ, Borisovskii VG (1985) Dynamic fracture mechanics. Mashinostroenie, Moscow (in Russian)
18. Guz' AN, Zozulya VV (1993) Brittle fracture of materials under dynamic loads. Naukova Dumka, Kiev (in Russian)
19. Emets' VF (2001) Simulation of the process of scattering of ultrasound by local objects and interpretation of the data. Lviv (in Ukrainian)
20. Luong WC, Keer LM, Achenbach JD (1975) Elastodynamic stress intensity factors of a crack near an interface. Int J Solids Struct 11:919–925
21. Kirpichnikova NY, Svirkina LA, Filippov VB (2001) Scattering of plane elastic waves by a small inhomogeneity placed in an elastic layer. Zap Nauchn Sem POMI, 275:72–84
22. Kirpichnikova NY, Svirkina LA, Filippov VB (2002) Diffraction of plane elastic vertically polarized waves by a small inhomogeneity in a layer. Zap Nauchn Sem POMI, 285:88–108
23. Takei M, Shindo Y, Atsumi A (1982) Diffraction of transient horizontal shear waves by a finite crack at the interface of two bonded dissimilar elastic solids. Eng Fract Mech 16(6):799–807
24. Boström A (1987) Elastic wave scattering from an interface crack: antiplane strain. J Appl Mech 54(3):503–508
25. Loeber JF, Sih GC (1973) Transmission of antiplane shear waves past an interface crack in dissimilar media. Eng Fruct Mech 5:699–725
26. Palaiya RM, Majumdar P (1981) Interaction of antiplane shear waves by rigid strip lying at the interface of two dissimilar elastic half-spaces. Z Angew Math Mech 61(2):120–122
27. Srivastava KN, Palaiya RM, Karaulia RM (1980) Interaction of antiplane shear waves by a Griffith crack at the interface of two bonded dissimilar elastic half-spaces. Int J Fract 16(4):349–358
28. Singh M, Dhaliwal RS, Rokne JG (1980) Diffraction of antiplane shear waves by a Griffith crack or a rigid strip lying at the interface of homogeneous and nonhomogeneous elastic half-spaces. Acta Mech 37(1–2):137–134
29. Kundu T (1988) Dynamic interaction between two interface cracks in a three-layered plate. Int J Solids Struct 24(1):27–39
30. Kundu T (1987) The transient response of two cracks at the interface of a layered half space. Int J Eng Sci 25:1427–1439
31. Srivastava KN, Palaiya RM, Karaulia RM (1981) Diffraction of SH waves by two coplanar Griffith cracks at the interface of two bonded dissimilar elastic half-spaces. Indian J Pure Appl Math 12(2):242–252
32. Wang YS, Gross D (2001) Interaction of harmonic waves with a periodic array of interface cracks in a multi-layered medium: anti-plane case. Int J Solids Struct 38:4631–4655
33. Wang YS, Gross D (2001) Transfer matrix method of wave propagation in a layered medium with multiple interface cracks: anti-plane case. J Appl Mech 68:499–503
34. Zhang C (1991) Dynamic stress intensity factors for periodically spaced collinear antiplane shear cracks between dissimilar media. Theor Appl Fract Mech 15:219–227
35. Pryakhina OD, Smirnova AV (2005) Integral equations of the dynamic problems for multilayer media containing systems of cracks. Prikl Mekh Mat 69(2):345–351
36. Kuo AY (1984) Transient stress intensity factors of an interfacial crack between two dissimilar anisotropic half-spaces. Part 2. Fully Anisotropic Materials. J Appl Mech 51(4):780–786
37. Maue AW (1953) Die Beugung elastischer Wellen an der Halfbebene. Z Angew Math Mech 33(1/2):1–10

38. Cherepanov GP (1972) Diffraction of elastic waves by a cut. In: Continuum mechanics and related problems of analysis. Nauka, Moscow, pp 615–622 (in Russian)
39. Norris AN, Achenbach JD (1984) Elastic wave diffraction by a semiinfinite crack in a transversely isotropic material. Q J Mech Appl Math 37:566–580
40. Lewis PA, Temple JAG, Walker EJ, Wickham GR (1998) Calculation of the diffraction coefficients for a semi-infinite crack embedded in an infinite anisotropic linearly elastic body. Proc Roy Soc London, A 454:1781–1803
41. Thau SA, Lu TH (1970) Diffraction of the transient horizontal shear waves by a finite crack and a finite rigid ribbon. Int J Eng Sci 8:857–874
42. Sapmaz S, Kobayashi K, Büyükaksoy A, Uzgören G (1996) High–frequency diffraction by a strip located at the plane interface between two different media. IEICE Trans Electron E79–C(5):709–719
43. Kuryliak DB, Voytko MV (2005) Wiener–Hopf analysis of the elastic wave diffraction by the finite crack located at the plane interface between the elastic isotropic half-space medium. In: Direct and inverse problems of electromagnetic and acoustic wave theory (DIPED), Lviv–Tbilisi, pp 143–147
44. Kurylyak DB, Nazarchuk ZT, Voitko MV (2006) Analysis of the field of a plane SH-wave scattered by a finite crack on the interface of materials. Mater Sci 42(6):711–724
45. Kurylyak DB, Nazarchuk ZT, Voitko MV (2007) Stress field in a crack on the interface of materials irradiated with SH-waves. Mater Sci 43(4):464–478
46. Voitko MV, Nazarchuk ZT, Kurylyak DB (2008) Interaction of plane SH-waves with the tips of interface cracks. Mater Sci 44(2):260–267
47. Kurylyak DB, Nazarchuk ZT, Voitko MV (2009) On the Relationship between the stress intensity factors and the far field of SH-waves diffraction on interface cracks. Mater Sci 45(3):329–338
48. Lavrentev MA, Shabat BV (1987) Methods of the theory of function of complex variable. Nauka, Moscow
49. Kobayashi K (1991) On generalized Gamma functions occurring in diffraction theory. J Phys Soc Jpn 60:1501–1512
50. Virchenko NO, Rumyanceva OV (2008) On the generalized Gaussian hypergeometric function and its application. Rep NAS Ukraine 4:12–18
51. Prieto AI, Prieto AI, Matera J, Srivastava HM (2006) Use of the generalized Lommel-Wright function in a unified presentation of the Gamma-type functions occurring in diffraction theory and associated probability distributions. Integr Transform Spec Funct 17(5):365–378
52. Brekhovskikh LM (1973) Waves in Layered Media. Nauka, Moscow (in Russian)
53. Ufimtsev PY (2007) Fundamentals of the physical theory of diffraction. Wiley, USA
54. Chang SJ (1971) Diffraction of plane dilatational waves by a finite crack. Q J Mech Appl Math 24:423–443
55. Keogh PS (1985) High-frequency scattering by a Griffith crack I: a crack Green's function. Q J Mech Appl Math 38:185–204
56. Tan TH (1977) Scattering of plane, elastic waves by a plane crack of finite width. Appl Sci Rez 33(1):75–88
57. Pal SC, Ghosh ML (1990) High frequency scattering of antiplane shear waves by interface crack. Indian J Pure Appl Math 21(12):1107–1124
58. Pramanik R, Pal SC, Ghosh ML (1999) High frequency scattering due to a pair of time harmonic antiplane forces on the faces of a finite interface crack between dissimilar anisotropic materials. Eur J Mech A/Solids 18:1013–1026
59. Srivastava KN, Gupta OP, Palaiya RM (1978) Interaction of elastic wave in two dissimilar elastic half-planes having Griffith crack at interface. Int J Fract 14:145–154
60. Gradshteyn IS, Ryzhik IM (1980) Tables of integrals, series, and products, 4th edn. Academic, San Diego
61. Rokhlin SI (1980) Diffraction of Lamb waves by a finite crack in an elastic layer. J Acoust Soc Am 67(4):1157–1165

62. Kyrylova OI, Mykhas'kiv VV (2019) Harmonic vibration and resonance effects in the case of longitudinal shear of a hollow cylinder with crack. Mate Sci 55(1):114–123
63. Kuryliak DB, Nazarchuk ZT, Voytko MV, YaP K (2013) On the interaction of an elastic SH-wave with an interface crack in the perfectly rigid joint of a plate with a half-space. J Math Sci 192(6):609–622
64. Golub MV, Zhang C, Wang Y (2011) SH–wave propagation and resonance phenomena in a periodically layered composite structure with a crack. J Sound Vibr 330:3141–3154
65. Auld BA (1973) Acoustic fields and waves in solids, vol 2. Wiley Interscience Publication, New York, London, Sydney, Toronto
66. Miklowitz J (1978) The theory of elastic waves and wave guides. North-Holland, New York
67. Graff KF (1991) Wave motion in elastic solids. Dover Publications, New York
68. Zaman FD (2001) Diffraction of SH-waves across a mixed boundary in a plate. Mech Res Commun 28(2):171–178
69. Asghar S, Zaman FD (1987) Diffraction of SH-waves by a finite crack in a layer overlying a half space. Boll. Di Geofisica Teorica Ed Applicata 19(113):43–50
70. Glushkov E, Glushkova N, Golub M, Boström (2006) Natural resonance frequencies, wave blocking, and energy localization in an elastic half-space and waveguide with a crack. J Acoust Soc Am 119(6):3589–3598
71. Rokhlin SI (1981) Resonance phenomena of Lamb waves scattering by a finite crack in a solid layer. J Acoust Soc Am 69(4):922–928
72. Glushkov YV, Glushkova NV, Yeremin AA, Mykhas'kiv VV (2009) The layered element method in the dynamic theory of elasticity. J Appl Math Mech 73(4):449–456
73. Kuryliak DB, Voytko MV (2004) Wiener–Hopf analysis of the elastic wave diffraction by the finite crack located at the plane interface between the elastic isotropic slab and half-space medium. In: Direct and inverse problems of electromagnetic and acoustic wave theory (DIPED), Lviv–Tbilisi, pp 22–25
74. Shestopalov VP (1987) Spectral theory and excitation of open structures. Naukova Dumka, Kiev (in Russian)
75. Sirenko YK, Sukharevsky IV, Sukharevsky OI, Yashina NP (2000) Fundamental and applied problems for electromagnetic scattering waves theory. Krok, Kharkiv (in Russian)
76. Thau SA, Lu TH (1971) Transient stress intensity factors for a finite crack in an elastic solid caused by a dilatational wave. Int J Solids Struct 7:731–750

Index

Z. Nazarchuk et al., *Optical Metrology and Optoacoustics in Nondestructive Evaluation of Materials*, Springer Series in Optical Sciences 242,
https://doi.org/10.1007/978-981-99-1226-1

W

Y

The manufacturer's authorised representative in the EU is Springer Nature Customer Service Centre GmbH, Europaplatz 3, 69115 Heidelberg, Germany. If you have any concerns regarding our products, please contact ProductSafety@springernature.com

Printed and bound by CPI Group (UK) Ltd, Croydon, CR0 4YY
01/10/2026
02225870-0015